B

Monographs in Mathematics
Vol. 83

1988

Birkhäuser
Boston · Basel · Berlin

V. I. Arnold
S. M. Gusein-Zade
A. N. Varchenko

Singularities of Differentiable Maps

Volume II

Monodromy and Asymptotics of Integrals

Under the Editorship of V. I. Arnold

1988

Birkhäuser
Boston · Basel · Berlin

Originally published as
Osobennosti differentsiruemykh otobrazhenii
by Nauka, Moscow, 1984

Translated by Hugh Porteous

Translation revised by the authors
and James Montaldi

Library of Congress Cataloging in Publication Data
(Revised for vol. 2)

Arnol'd, V. I. (Vladimir Igorevich), 1937.
  Singularities of differentiable maps.

  (Monographs in matematics ; vol. 82 –)
  Translation of: Osobennosti differentsiruemykh
otobrazhenii.
  Includes bibliographies and indexes.
  Contents: v. 1. The classification of critical
points, caustics and wave fronts – – v. 2. Monodromy
and asymptotics of integrals.
    1. Differentiable mappings. 2. Singularities (Mathematics)
I. Gusein-Zade. S. M. (Sabir Medzhidovich)
II. Varchenko, A. N. (Aleksandr Nikolaevich) III. Title.

IV. Series.
QA614.58.A7513 1985   514'.72   84-12134
ISBN 0-8176-3187-9

CIP-Kurztitelaufnahme der Deutschen Bibliothek

Arnol'd, Vladimir I.:
Singularities of differentiable maps / V. I.
Arnold ; S. M. Gusein-Zade ; A. N. Varchenko. –
Boston ; Basel ; Stuttgart : Birkhäuser
  Einheitssacht.: Osobennosti differenciruemych
  otobraženij ⟨engl.⟩

NE: Gusejn-Zade, Sabir M.:; Varčenko, Aleksandr N.:

Vol. 2. Monodromy and asymptotics of integrals / under
the editorship of V. I. Arnold. Transl. by Hugh
Porteous. – 1987.

  (Monographs in mathematics ; Vol. 83)
  ISBN 3-7643-3185-2 (Basel...)
  ISBN 0-8176-3185-2 (Boston)

NE: GT

© 1988 Birkhäuser Boston, Inc.
Printed in Germany
ISBN 0-8176-3185-2
ISBN 3-7643-3185-2

Preface

The present volume is the second volume of the book "Singularities of Differentiable Maps" by V. I. Arnold, A. N. Varchenko and S. M. Gusein-Zade. The first volume, subtitled "Classification of critical points, caustics and wave fronts", was published by Moscow, "Nauka", in 1982. It will be referred to in this text simply as "Volume 1".

Whilst the first volume contained the zoology of differentiable maps, that is it was devoted to a description of what, where and how singularities could be encountered, this volume contains the elements of the anatomy and physiology of singularities of differentiable functions. This means that the questions considered in it are about the structure of singularities and how they function.

Another distinctive feature of the present volume is that we take a hard look at questions for which it is important to work in the complex domain, where the first volume was devoted to themes for which, on the whole, it was not important which field (real or complex) we were considering. Such topics as, for example, decomposition of singularities, the connection between singularities and Lie algebras and the asymptotic behaviour of different integrals depending on parameters become clearer in the complex domain.

The book consists of three parts. In the first part we consider the topological structure of isolated critical points of holomorphic functions. We describe the fundamental topological characteristics of such critical points: vanishing cycles, distinguished bases, intersection matrices, monodromy groups, the variation operator and their interconnections and method of calculation.

The second part is devoted to the study of the asymptotic behaviour of integrals of the method of stationary phase, which is widely met with in applications. We give an account of the methods of calculating asymptotics, we discuss the connection between asymptotics and various characteristics of critical points of the phases of integrals (resolution of singularities, Newton polyhedra), we give tables of the orders of asymptotics for critical points of the phase which were classified in Volume 1 of this book (in particular for simple, unimodal and bimodal singularities).

The third part is devoted to integrals evaluated over level manifolds in a neighbourhood of the critical point of a holomorphic function. In it we shall consider integrals of holomorphic forms, given in a neighbourhood of a critical point, over cycles, lying on level hypersurfaces of the function. Integral of a holomorphic form over a cycle changes holomorphically under continuous deformation of the cycle from one level hypersurface to another. In this way there arise many-valued holomorphic functions, given on the complex line in a neighbourhood of a critical value of the function. We show that the asymptotic behaviour of these functions (that is the asymptotic behaviour of the integrals) as the level tends to the critical one is connected with a variety of characteristics of the initial critical point of the holomorphic function.

The theory of singularities is a vast and rapidly developing area of mathematics, and we have not sought to touch on all aspects of it.

The bibliography contains works which are directly connected with the text (although not always cited in it) and also works connected with volume 1 but for some or other reason not contained in its bibliography.

References in the text to volume 1 refer to the above-mentioned book "Singularities of Differentiable Maps".

The authors offer their thanks to the participants in the seminar on singularity theory at Moscow State University, in particular A. M. Gabrielov, A. B. Givental, A. G. Kushnirenko, D. B. Fuks, A. G. Khovanski and S. V. Chmutov. The authors also wish to thank V. S. Varchenko and T. V. Ogorodnikova for rendering inestimable help in preparing the manuscript for publication.

The authors.

Contents

Part I

The topological structure of isolated critical points of functions

Introduction

In the topological investigation of isolated critical points of complex-analytic functions the problem arises of describing the topology of its level sets. The topology of the level sets or infra-level sets of smooth real-valued functions on manifolds may be investigated with the help of Morse theory (see [255]). The idea there is to study the change of structure of infra-level sets and level sets of functions upon passing critical values. In the complex case passing through a critical value does not give rise to an interesting structure, since all the non-singular level sets near one critical point are not only homeomorphic but even diffeomorphic. The complex analogue of Morse theory, describing the topology of level sets of complex analytic functions, is the theory of Picard-Lefschetz (which historically precedes Morse theory). In Picard-Lefschetz theory the fundamental principle is not passing through a critical point but going round it in the complex plane.

Let us fix a circle, going round the critical value. Each point of the circle is a value of the function. The level sets, corresponding to these values, give a fibration over the circle. Going round the circle defines a mapping of the level set above the initial point of the circle into itself. This mapping is called the (classical) *monodromy* of the critical point.

The simplest interesting example in which one can observe all this clearly and carry through the calculations to the end is the function of two variables given by

$$f(z, w) = z^2 + w^2, \quad (z, w) \in \mathbb{C}^2.$$

It has a unique critical point $z = w = 0$. The critical value is $f = 0$. The critical level set $V_0 = \{(z, w) : z^2 + w^2 = 0\}$ consists of two complex lines intersecting in the point 0. All the other level sets

$$V_\lambda = \{(z, w) : z^2 + w^2 = \lambda\} \qquad (\lambda \neq 0)$$

are topologically the same; they are diffeomorphic to a cylinder $S^1 \times \mathbb{R}^1$ (figure 1).

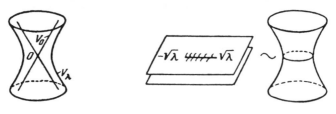

Fig. 1.   Fig. 2.

To show this, we consider the Riemann surface of the function $w = \sqrt{(\lambda - z^2)}$ (figure 2). This surface is glued together from two copies of the complex z-plane, joined along the cut $(-\sqrt{\lambda}, \sqrt{\lambda})$. Each copy of the cut plane is homeomorphic to a half cylinder; the line of the cut corresponds to a circumference of the cylinder. In this way, the whole (four-real-dimensional) space $\mathbb{C}^2$ decomposes into the singular fibre V_0 and the non-singular fibres V_λ, diffeomorphic to cylinders, mapping to the critical value 0 and the non-critical values $\lambda \neq 0$ by the mapping

$$f: (\mathbb{C}^2, 0) \to (\mathbb{C}, 0).$$

Let us proceed to the construction of the monodromy. We consider on the target plane a path going round the critical value 0 in the positive direction (anticlockwise):

$$\lambda(t) = \exp(2\pi i t)\alpha, \quad 0 \leqslant t \leqslant 1, \quad \alpha > 0 \text{ (figure 3)}$$

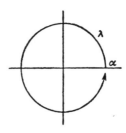

Fig. 3.

Let us observe how the fibre $V_{\lambda(t)}$ changes as t varies from 0 to 1. For this we consider the Riemann surfaces of the functions

$$w = \sqrt{(\lambda(t) - z^2)}.$$

As the parameter t increases, both the branch points $z = \pm\sqrt{\lambda(t)} = \exp(\pi i t)$ $\times(\pm\sqrt{\alpha})$ move around the point $z=0$ in the positive direction. As t varies from 0 to 1, each of these points performs a half turn and arrives at the other's place. In this way, as $\lambda(t)$ goes round the critical value 0, it corresponds to a sequence of Riemann surfaces, depicted in figure 4, beginning and ending with the same surface V_α.

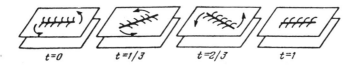

$$t=0 \qquad t=1/3 \qquad t=2/3 \qquad t=1$$

Fig. 4.

Now it is easy to construct a family continuous in t, of diffeomorphisms from the initial fibre $V_{\lambda(0)} = V_\alpha$ to the fibre $V_{\lambda(t)}$ over the point $\lambda(t)$

$$\Gamma_t : V_{\lambda(0)} \to V_{\lambda(t)}$$

beginning with the identity map, Γ_0, and ending with the *monodromy* $\Gamma_1 = h$. For example one may define Γ_t in the following fashion. Choose a smooth "bump function", $\chi(\tau)$, such that

$$\chi(\tau) = 1 \text{ for } 0 \leqslant \tau \leqslant 2\sqrt{\alpha},$$

$$\chi(\tau) = 0 \text{ for } \tau \geqslant 3\sqrt{\alpha}.$$

We let

$$g_t(z) = \exp\{\pi i t \cdot \chi(|z|)\} \cdot z.$$

The family of diffeomorphisms g_t from the complex z-plane into itself defines the desired family of diffeomorphisms Γ_t. The diffeomorphism $h = \Gamma_1 : V_\alpha \to V_\alpha$ of the cylinder is the identity outside a sufficiently large compact set (for $|z| > 3\sqrt{\alpha}$).

We consider now the action of the monodromy h on the homology of a non-singular fibre V_α. The first homology group $H_1(V_\alpha; \mathbb{Z}) \approx \mathbb{Z}$ of the cylinder V_α is generated by the homology class of the "gutteral" circle Δ (figure 5). As $\alpha \to 0$ the circle Δ tends to the point 0. Therefore it is called the *vanishing cycle of Picard-Lefschetz*.

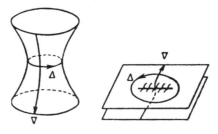

Fig. 5.

We consider further the first homology group $H_1^{cl}(V_\alpha; \mathbb{Z})$ of the fibre V_α with closed support. According to Poincaré duality, this group is also isomorphic to the group $\mathbb{Z}$ of integers. It is generated by the homology class of the "covanishing cycle" ∇ – a line on the cylinder going from infinity to infinity and intersecting the vanishing cycle, Δ, once transversely (see figure 5). We shall suppose that the cycle ∇ is oriented in such a way that its intersection number $(\nabla \circ \Delta)$ with the vanishing cycle Δ, determined by the complex orientation of the fibre V_α, is equal to $+1$.

Figure 4 allows us to observe the action of the diffeomorphisms Γ_t on the vanishing and covanishing cycles (figure 6).

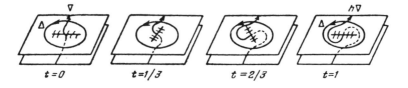

$t=0$ $t=1/3$ $t=2/3$ $t=1$

Fig. 6.

We notice that the diffeomorphism $h = \Gamma_1$ of the cylinder $V_\alpha \approx S^1 \times \mathbb{R}^1$ can be described as follows: it is fixed outside a certain annulus, the circles forming the annulus rotate through various angles varying from 0 at one edge to 2π at the other. In this way, under the action of the monodromy mapping h. the vanishing cycle Δ is mapped into itself, the covanishing cycle winds once around the cylinder (figure 7).

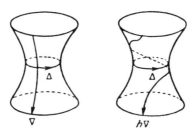

Fig. 7.

The diffeomorphism h is the identity outside some compact set. Outside this compact set the cycles V and hV coincide. Therefore the cycle $hV - V$ is concentrated in a compact part of the cylinder. From figure 7 (or from figure 6) it is clear that

$$hV - V = -\Delta.$$

In this way any cycle δ with closed support gives rise to a cycle $h\delta - \delta$ with compact support. This defines a mapping from the homology of the fibre V_α with closed support into its homology with compact support. It is called the *variation* and is denoted by

$$\text{Var}: H_1^{cl}(V_\alpha; \mathbb{Z}) \to H_1(V_\alpha; \mathbb{Z}).$$

From figure 7 or figure 6 it can be seen that we have

$$\text{Var } \delta = (\Delta \circ \delta)\, \Delta$$

for every cycle

$$\delta \in H_1^{cl}(V_\alpha; \mathbb{Z}).$$

Here $(\Delta \circ \delta)$ is the intersection number of the cylces Δ and δ, defined by the complex orientation of the fibre V_α. This relationship is called the *formula of Picard-Lefschetz*.

We notice that, generally speaking, the diffeomorphisms Γ_t are defined only up to homotopy and that there is no a priori reason why the mapping Γ_1 should be fixed outside a compact set. For example, the family of diffeomorphisms

$$\Gamma_t' : (z, w) \to (z \exp(\pi i t), w \exp(\pi i t))$$

defines the mapping

$$\Gamma_1' = h' : (z, w) \to (-z, -w),$$

which is not fixed outside a compact set and is not, therefore, suitable for defining the variation (though it is suitable for defining the action of the monodromy on the compact homology). Thus we have considered the fundamental concepts of the theory of Picard-Lefschetz: vanishing cycles, monodromy and variation for the simplest example of the function $f(z, w) = z^2 + w^2$.

In the general case of an arbitrary function of any number of variables, the topology of the fibre V_λ will not be as simple as in the example we analysed. The investigation of the topology of the fibre V_λ, the monodromy and the variation in the general case is a difficult problem, solved completely only for a few special cases. In this part we shall recount several methods and results which have been obtained along these lines.

The fundamental method which we shall make use of is the method of deformation (or perturbation). Under a small perturbation, a complicated critical point of a function of n variables breaks up into simple ones. These simple critical points look like the critical point 0 of the function

$$f(z_1, \ldots, z_n) = z_1^2 + \ldots + z_n^2$$

and can be investigated completely in the same way as we analysed the case $n = 2$ above. In place of the cylinders V_λ, which occured in the case $n = 2$, the non-singular fibre in the general case are smooth manifolds

$$V_\lambda = \{(z_1, \ldots, z_n) : z_1^2 + \ldots + z_n^2 = \lambda\}, \quad \lambda \neq 0,$$

which are diffeomorphic to the space TS^{n-1}, the tangent bundle of the $(n-1)$-dimensional sphere (giving a cylinder for $n = 2$). The vanishing cycle in V_1 is the real sphere

$$S^{n-1} = \{z \in \mathbb{R}^n \subset \mathbb{C}^n : z_1^2 + \ldots + z_n^2 = 1\}.$$

If complicated critical points break down under deformation into μ simple ones, then the perturbed function will have, in general, μ critical values (figure 8).

In this case it is possible in the target plane of the perturbed function to go round each of the μ critical values. In this way we get, not one monodromy diffeomorphism h, but a whole *monodromy group* $\{h_\gamma\}$, where γ runs through the fundamental group of the set of non-critical values.

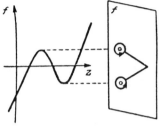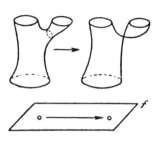

Fig. 8.   Fig. 9.

The non-singular fibre V_λ of the perturbed function have the same structure (inside some ball surrounding the critical point of the initial function) as the non-singular fibre of the initial function. When the value of λ tends to one of the critical values of the perturbed function, a certain cycle on the non-singular fibre vanishes. This cycle is a sphere whose dimension is a half of the (real) dimension of the fibre V_λ (figure 9). Tending in this manner to all μ critical values, we define in the non-singular fibre μ vanishing cycles, each a sphere of the middle dimension. It happens that the non-singular fibre is homotopy equivalent to a bouquet of these spheres.

In the case when the real dimension of the non-singular fibre is divisible by 4 (that is when the number n of variables is odd), the intersection number gives a symmetric bilinear form in the homology group $H_{n-1}(V_\lambda;\mathbb{Z})$ of the non-singular level manifold. The self-intersection number of each of the vanishing cycles is equal to 2 or -2, depending on the number of variables n. The action of going round the critical value corresponding to a vanishing cycle is equivalent to reflection in a mirror which is orthogonal to this cycle, where orthogonality is defined by the scalar product given by the intersection numbers.

For example, for the function of three variables

$$f(z_1,z_2,z_3)=z_1^{k+1}+z_2^2+z_3^2$$

a suitable perturbation is

$$\tilde{f}(z_1,z_2,z_3)=f(z_1,z_2,z_3)-\varepsilon z_1.$$

This function has $\mu = k$ critical points $(\sqrt[k]{\varepsilon/(k+1)}\,\xi_m, 0, 0)$, where ξ_m $(m = 1, \ldots, k)$ are the k-th roots of unity. The corresponding vanishing cycles $\Delta_1, \ldots, \Delta_k$ can be chosen so that they have the following intersection numbers

$$(\Delta_i \circ \Delta_i) = -2,$$

$$(\Delta_1 \circ \Delta_2) = (\Delta_2 \circ \Delta_3) = \ldots = (\Delta_{k-1} \circ \Delta_k) = 1,$$

and all other intersection numbers are zero. The monodromy group is generated by the resulting reflections in the orthogonal complements of the cycles Δ_m, and coincides with the Weyl group A_k (see [53]), that is with the group $S(k+1)$ of permutations of $(k+1)$ elements.

In this Part we shall generally (except in Chapter 5) be concerned with isolated singularities of functions. Therefore by the term "singularity" we shall understand the germ of a holomorphic function $f : (\mathbb{C}^n, 0) \to (\mathbb{C}, 0)$, having at the origin an isolated critical point (that is, a point at which all the partial derivatives of the function f are equal to zero).

Let $G : (\mathbb{C}^n, 0) \to (\mathbb{C}^p, 0)$ be the germ at the origin of an analytic function, let U be a neighbourhood of the origin in the space $\mathbb{C}^n$, in which a representative of the germ G is defined, and let G_λ be a family of functions from U to $\mathbb{C}^p$, analytic in λ in a neighbourhood of 0 in $\mathbb{C}$, such that $G_0 = G$. We shall refer to the function G_λ, for sufficiently small λ, as a *small perturbation* $\tilde{G}$ of the function G, without spelling out each time the dependence on the parameter λ.

Throughout, the absolute homology groups will be considered reduced modulo a point; the relative groups of a pair "manifold-boundary" will be modulo a fundamental cycle. (For this reason the tilde over the letter H, which usually indicates a reduced homology group, will be omitted.) All the homology will be considered with coefficients in the group $\mathbb{Z}$ of integers, unless we specifically indicate otherwise.

Let $\mathbb{C}^n$ be the n-dimensional complex vector space with coordinates

$$x_j = u_j + iv_j, \quad (j = 1, \ldots, n; \ u_j \text{ and } v_j \text{ real}).$$

The space $\mathbb{C}^n \cong \mathbb{R}^{2n}$, considered as a real $2n$-dimensional vector space, has a preferred orientation, which we shall call the complex one. This orientation is defined so that the system of coordinates in the space $\mathbb{R}^{2n}$ given by $u_1, v_1, u_2, v_2, \ldots, u_n, v_n$ has positive orientation. Complex manifolds will be considered to have this complex orientation unless we specifically indicate otherwise. With this choice of orientation the intersection numbers of complex submanifolds will always be non-negative.

Chapter 1

Elements of the theory of Picard-Lefschetz

In this chapter we shall define concepts of Picard-Lefschetz theory such as vanishing cycles, the monodromy and variation operators, the Picard-Lefschetz operators, etc. As we have already said, they are used to investigate the topology of critical points of holomorphic functions.

1.1 The monodromy and variation operators

Let $f: M^n \to \mathbb{C}$ be a holomorphic function on an n-dimensional complex manifold M^n, with a smooth boundary ∂M^n (in the real sense). Let U be a contractible compact region in the complex plane with smooth boundary ∂U. We shall suppose that the following conditions are satisfied:

(i) For some neighbourhood U' of the region U, the restriction of f to the preimage of U' is a proper mapping $f^{-1}(U') \to U'$, that is a mapping for which the preimage of any compact set is compact.

(ii) The restriction of f to $\partial M^n \cap f^{-1}(U')$ is a regular mapping into U', that is a mapping, the differential of which is an epimorphism.

(iii) The function f has in the preimage, $f^{-1}(U')$, of the region U' a finite number of critical points p_i $(i=1, \ldots, \mu)$ with critical values $z_i = f(p_i)$ lying inside the region U, that is in $U \setminus \partial U$.

From condition (ii) it follows that the restriction of the function f to $\partial M^n \cap f^{-1}(U)$ defines a locally trivial, and consequently (since the region U is assumed contractible) also a trivial fibration $\partial M^n \cap f^{-1}(U) \to U$. The direct product structure in the space of this fibration is unique up to homotopy. In addition, the restriction of the function f to the preimage $f^{-1}(U \setminus \{z_i\})$ of the set of non-critical values is a locally trivial fibration.

We will denote by F_z $(z \in U)$ the level set of the function f $(F_z = f^{-1}(z))$. If $z \in U$ is a non-critical value of the function f, then the corresponding level set F_z is a compact $(n-1)$-dimensional complex manifold with smooth boundary $\partial F_z = F_z \cap \partial M^n$. Let us fix a non-critical value z_0 lying on the boundary ∂U of the region U. Let γ be a loop in the complement of the set of critical values

$U \setminus \{z_i | i = 1, \ldots, \mu\}$ with initial and end points at z_0 ($\gamma : [0, 1] \to U \setminus \{z_i\}$, $\gamma(0)$ $= \gamma(1) = z_0$). (We can suppose, without loss of generality, that all the loops and paths we encounter are piecewise smooth.) Going round the loop γ generates a continuous family of mappings $\Gamma_t : F_{z_0} \to M^n$ (lifting homotopy), for which Γ_0 is the identity map from the level manifold F_{z_0} into itself, $f(\Gamma_t(x)) = \gamma(t)$, that is Γ_t maps the level manifold F_{z_0} into the level manifold $F_{\gamma(t)}$. The homotopy Γ_t can and will be chosen to be consistent with the direct product structure on $\partial M^n \cap f^{-1}(U)$. Indeed we can choose as $\{\Gamma_t\}$ a family of diffeomorphisms $F_{z_0} \to F_{\gamma(t)}$, but we shall not need this in the sequel. Thus the map

$$h_\gamma = \Gamma_1 : F_{z_0} \to F_{z_0}$$

is the identity map on the boundary ∂F_{z_0} of the level manifold F_{z_0}. It is defined uniquely up to homotopy (fixed on the boundary ∂F_{z_0}) by the class of the loop γ in the fundamental group $\pi_1(U \setminus \{z_i\}, z_0)$ of the complement of the set of critical values.

Definitions. The transformation h_γ of the non-singular level set F_{z_0} into itself is called the *monodromy* of the loop γ. The action $h_{\gamma *}$ of the transformation h_γ on the homology of the non-singular level set $H_*(F_{z_0})$ is called the *monodromy operator* of the loop γ.

The monodromy operator is uniquely defined by the class of the loop γ in the fundamental group of the complement of the set of critical values.

We shall discuss also the automorphism $h_\gamma^{(r)}$ induced by the transformation h_γ in the relative homology group $H_*(F_{z_0}, \partial F_{z_0})$ of the non-singular level set modulo its boundary. In the introduction to this Part we used, instead of the relative homology group $H_*(F_{z_0}, \partial F_{z_0})$, the homology group $H_1^{cl}(V_\alpha)$ with closed support (using the isomorphism

$$H_*(F_{z_0}, \partial F_{z_0}) \cong H_*^{cl}(F_{z_0} \setminus \partial F_{z_0})).$$

Let δ be a relative cycle in the pair $(F_{z_0}, \partial F_{z_0})$. Since the transformation h_γ is the identity on the boundary ∂F_{z_0} of the level manifold F_{z_0}, the boundary of the cycle $h_\gamma \delta$ coincides with the boundary of the cycle δ. Therefore the difference $h_\gamma \delta - \delta$ is an absolute cycle in the manifold F_{z_0}. It is not hard to see that the mapping $\delta \mapsto h_\gamma \delta - \delta$ gives the correct definition of the homomorphism

$$\mathrm{var}_\gamma : H_*(F_{z_0}, \partial F_{z_0}) \to H_*(F_{z_0}).$$

Definition. The homomorphism

$$\mathrm{var}_\gamma : H_*(F_{z_0}, \partial F_{z_0}) \to H_*(F_{z_0})$$

is called the *variation operator* of the loop γ.

It is not difficult to see that the automorphisms $h_{\gamma*}$ and $h^{(r)}_{\gamma*}$ are connected with the variation operator by the relations

$$h_{\gamma*} = id + \mathrm{var}_\gamma \cdot i_*,$$

$$h^{(r)}_{\gamma*} = id + i_* \cdot \mathrm{var}_\gamma,$$

where

$$i_* : H_*(F_{z_0}) \to H_*(F_{z_0}, \partial F_{z_0})$$

is the natural homomorphism induced by the inclusion

$$F_{z_0} \subset (F_{z_0}, \partial F_{z_0}).$$

If the class of the loop γ in the fundamental group $\pi_1(U \setminus \{z_i\}, z_0)$ of the complement of the set of critical values is equal to the product $\gamma_1 \cdot \gamma_2$ of the classes γ_1 and γ_2, then

$$\mathrm{var}_\gamma = \mathrm{var}_{\gamma_1} + \mathrm{var}_{\gamma_2} + \mathrm{var}_{\gamma_2} \cdot i_* \cdot \mathrm{var}_{\gamma_1},$$

$$h_{\gamma*} = h_{\gamma_2*} \cdot h_{\gamma_1*},$$

$$h^{(r)}_{\gamma*} = h^{(r)}_{\gamma_2*} \cdot h^{(r)}_{\gamma_1*}.$$

Therefore the mapping $\gamma \mapsto h_{\gamma*}$ is an (anti)homomorphism of the fundamental group $\pi_1(U \setminus \{z_i\}, z_0)$ of the complement of the set of critical values into the group $\mathrm{Aut}\, H_*(F_{z_0})$ of automorphisms of the homology group $H_*(F_{z_0})$ of the non-singular level set. We shall denote by $(a \circ b)$ the intersection number of the cycles (or homology classes) a and b. This notation will be used both in the case when both the cycles a and b are absolute and in the case when one of them is relative. Remember that the level manifold F_{z_0} is a complex manifold and therefore possesses the preferred orientation which defines the intersection number of the cycles on it.

Lemma 1.1.   Let

$$a, b \in H_*(F_{z_0}, \partial F_{z_0})$$

be relative homology classes,

$$\dim a + \dim b = 2n - 2,$$

$$\gamma \in \pi_1(U \setminus \{z_i\}, z_0).$$

Then

$$(\text{var}_\gamma a \circ \text{var}_\gamma b) + (a \circ \text{var}_\gamma b) + (\text{var}_\gamma a \circ b) = 0.$$

Proof.   Choose relative cycles which are representatives of the homology classes a and b so that their boundaries (lying in the boundary ∂F_{z_0} of the level manifold F_{z_0}) do not intersect. This can be done using the dimensional relationships

$$\begin{aligned} \dim \partial a + \dim \partial b &= 2n - 4 \\ &< 2n - 3 \\ &= \dim \partial F_{z_0}. \end{aligned}$$

The chosen cycles we shall also denote by a and b. For such cycles the intersection number makes sense, though, of course, it is not an invariant of the classes in the homology group $H_*(F_{z_0}, \partial F_{z_0})$. We have

$$\text{var}_\gamma a = h_\gamma a - a, \qquad \text{var}_\gamma b = h_\gamma b - b,$$

$$\begin{aligned} (\text{var}_\gamma a \circ \text{var}_\gamma b) + (\text{var}_\gamma a \circ b) + (a \circ \text{var}_\gamma b) &= \\ &= (h_\gamma a \circ h_\gamma b) - (a \circ h_\gamma b) - (h_\gamma a \circ b) + (a \circ b) + \\ &\quad + (h_\gamma a \circ b) - (a \circ b) + (a \circ h_\gamma b) - (a \circ b) = 0, \end{aligned}$$

since $(h_\gamma a \circ h_\gamma b) = (a \circ b)$.

Corollary.   For $a, b \in H_*(F_{z_0}, \partial F_{z_0})$

$$(h_{\gamma*}^{(r)} a \circ \text{var}_\gamma b) + (\text{var}_\gamma a \circ b) = 0.$$

Proof.

$$\begin{aligned} &(h_{\gamma*}^{(r)} a \circ \text{var}_\gamma b) + (\text{var}_\gamma a \circ b) \\ &= (i_* \cdot \text{var}_\gamma a \circ \text{var}_\gamma b) + (a \circ \text{var}_\gamma b) + (\text{var}_\gamma a \circ b) \\ &= 0 \end{aligned}$$

since

$$h_{\gamma *}^{(r)} = id + i_* \cdot \text{var}_\gamma,$$
$$(i_* \cdot \text{var}_\gamma a \circ \text{var}_\gamma b) = (\text{var}_\gamma a \circ \text{var}_\gamma b).$$

1.2 Vanishing cycles and the monodromy group

Let us suppose now that all critical points p_i of the function f are non-degenerate (that is that $\det(\partial^2 f/\partial x_j \partial x_k) \neq 0$), and all critical values $z_i = f(p_i)$ are different $(i = 1, \ldots, \mu)$. Remember that in this case the function f is said to be Morse.

Definition. The *monodromy group* of the (Morse) function f is the image of the homomorphism of the fundamental group $\pi_1(U \setminus \{z_i\}, z_0)$ of the complement of the set of critical values in the group Aut $H_*(F_{z_0})$ of automorphisms of the homology group $H_*(F_{z_0})$ of the non-singular level set F_{z_0} which is obtained by mapping the loop γ into the monodromy operator

$$h_{\gamma *} : H_*(F_{z_0}) \to H_*(F_{z_0}).$$

Let us be given in the region U a path $u : [0, 1] \to U$, joining some critical value z_i with the non-critical value z_0 ($u(0) = z_i$, $u(1) = z_0$) and not passing through critical values of the function f for $t \neq 0$. By the Morse lemma, there exists a local coordinate system $x_1, \ldots, x_n$ in a neighbourhood of the non-degenerate critical point p_i on the manifold M^n, in which the function f can be written in the form $f(x_1, \ldots, x_n) = z_i + \Sigma_{j=1}^n x_j^2$. For values of the parameter t near zero, we fix in the level manifold $F_{u(t)}$ the sphere $S(t) = \sqrt{(u(t) - z_i)}\, S^{n-1}$, where

$$S^{n-1} = \{(x_1, \ldots, x_n) : \Sigma_j x_j^2 = 1, \text{Im } x_j = 0\}$$

is the standard unit $(n-1)$-dimensional sphere.

Lifting the homotopy t from zero to one defines a family of $(n-1)$-dimensional spheres $S(t) \subset F_{u(t)}$ in the level manifolds $F_{u(t)}$ for all $t \in (0, 1]$. Note that for $t = 0$ the sphere $S(t)$ reduces to the critical point p_i.

Definition. The homology class $\Delta \in H_{n-1}(F_{z_0})$, represented by the $(n-1)$-dimensional sphere $S(1)$ in the chosen non-singular level manifold F_{z_0} is called a *vanishing* (along the path u) *cycle of Picard-Lefschetz*.

It is easy to see that the homotopy class of the path u in the set of all paths in the region U joining the critical value z_i with the non-critical value z_0 and not passing through critical values of the function f for $t \neq 0$, defines the homology class of the vanishing cycle Δ modulo orientation.

Definition. The set of cycles $\Delta_1, \ldots, \Delta_\mu$ from the $(n-1)$st homology group $H_{n-1}(F_{z_0})$ of the non-singular level set F_{z_0} is called *distinguished* if:

(i) the cycles $\Delta_i (i = 1, \ldots, \mu)$ are vanishing along non-self-intersecting paths u_i, joining the critical value z_i with the non-critical value z_0;

(ii) the paths u_i and u_j have, for $i \neq j$, a unique common point $u_i(1) = u_j(1) = z_0$;

(iii) the paths $u_1, \ldots, u_\mu$ are numbered in the same order in which they enter the point z_0, counting clockwise, beginning at the boundary ∂U of the region U (see figure 10).

Remark. The need to choose a non-critical value z_0 on the boundary ∂U of the region U was dictated by the need to number the elements of the distinguished set of vanishing cycles according to condition (iii).

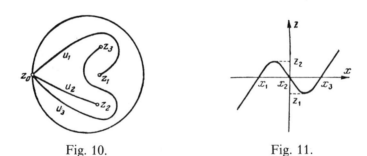

Fig. 10.   Fig. 11.

Examples. 1. Let us consider the Morse function $f(x) = x^3 - 3\lambda x$, where λ is a small positive number. This function is a perturbation of the function $f_0(x) = x^3$ (having the singularity type A_2 in the sense of volume 1), but we do not need that fact just now. The function f has two critical points ($x = \sqrt{\lambda}$ and $x = -\sqrt{\lambda}$) with critical values $z_1 = -2\lambda\sqrt{\lambda}$ and $z_2 = 2\lambda\sqrt{\lambda}$ respectively. As the non-critical value of the function f we take $z_0 = 0$. Let us join the critical values z_i ($i = 1, 2$) with the non-critical value z_0 by line segments u_1 and u_2. The level manifold $\{f = 0\}$ consists of three points $x_1 = -\sqrt{3\lambda}$, $x_2 = 0$ and $x_3 = \sqrt{3\lambda}$ (see figure 11). It is easy to see that the cycles, vanishing along the described paths u_1 and u_2 joining the critical values z_1 and z_2 with the non-critical value 0, are the differences

$\Delta_1 = \{x_3\} - \{x_2\}$ and $\Delta_2 = \{x_2\} - \{x_1\}$ of zeroth homology class represented by the points x_1, x_2 and x_3. Note that the orientation of the cycles was chosen by us arbitrarily: any of them can be multiplied by -1.

For greater clarity we chose the non-critical value $z_0 = 0$. It presupposes, certainly, a special choice for the region U. On this occasion it is not very important, but later, for the definition of a distinguished basis of vanishing cycles in the homology of a non-singular level manifold of a degenerate singularity, we shall consider the region U to be a disk of sufficiently large radius, in comparison with the critical values of the perturbed function. The need to choose a non-critical value on the boundary of the sufficiently large disk is dictated as, otherwise, firstly, the identification of the homology group of the non-singular level manifold of a singularity and its perturbation would not be unique and, secondly, the order in which the vanishing cycles must enter the distinguished basis would not be unique. In order to "correct" the example we considered above, we can choose a non-critical value z_0^* sufficiently large in absolute value $(|z_0^*| \gg 2\lambda\sqrt{\lambda})$, joining it with the non-critical value $z_0 = 0$ by a path which does not pass through the critical values of the function f, and observe the change of the non-singular level manifolds $f = z$ as z moves along this path from $z_0 = 0$ to z_0^*. We shall consider later an analogous construction in a more general case (§2.9). Here for simplicity we modify our example somewhat.

1*. We consider the Morse function $f(x) = x^3 + 3\lambda x$, where λ is a positive number. The critical points of the function f are $x = -\sqrt{\lambda}i$ and $x = \sqrt{\lambda}i$, the critical values are $z_1 = -2\lambda\sqrt{\lambda}i$ and $z_2 = 2\lambda\sqrt{\lambda}i$. We choose as the region U a disk of sufficiently large radius r with centre at zero $(r \gg 2\lambda\sqrt{\lambda})$. We consider two non-critical values of the function f: $z_0 = 0$ and $z_0^* = r$. The critical values $z_{1,2}$ are joined to $z_0 = 0$ by segments, going along the imaginary axis, $z_0 = 0$ is joined to $z_0^* = r$ by a segment of the positive real half-axis. In this way we get paths, u_1 and u_2, joining the critical values $z_{1,2}$ with the non-critical value z_0^*. As before the zero level manifold of the function f consists of three points

$$x_1 = -\sqrt{3\lambda}i, \; x_2 = 0 \text{ and } x_3 = +\sqrt{3\lambda}i.$$

The level manifold $\{f = z_0^*\}$ is near to the level manifold $\{f_0 = z_0^*\}$ of the function $f_0(x) = x^3$ (since $|z_0^*| = r \gg 2\lambda\sqrt{\lambda}$). Therefore it consists of three points

$$x_1^* \approx \exp(-2\pi i/3)\sqrt[3]{z_0^*},$$

$$x_2^* \approx \sqrt[3]{z_0^*},$$

$$x_3^* \approx \exp(2\pi i/3)\sqrt[3]{z_0^*}.$$

It is not difficult to see that, along the line segment joining the critical value $z_1 = -2\lambda\sqrt{\lambda}i$ (respectively $z_2 = 2\lambda\sqrt{\lambda}i$) with the non-critical value $z_0 = 0$, the cycle $\{x_2\} - \{x_1\}$ (respectively $\{x_3\} - \{x_2\}$) vanishes. Further, it is clear that as the non-critical value z moves along the segment of the positive half-axis from $z_0 = 0$ to $z_0^* = r$ the points of the manifold $\{f = z\}$ change in such a manner that the point x_2 remains on the real axis, the point x_1 is in the lower and the point x_3 is in the upper half-plane. Therefore as z moves from z_0 to z_0^*, the points x_1, x_2 and x_3 go to the points x_1^*, x_2^* and x_3^* respectively. Consequently, along the paths u_1 and u_2 which we described joining the critical values z_1 and z_2 of the function f with the non-critical value z_0^* the cycles

$$\Delta_1 = \{x_2^*\} - \{x_1^*\}$$

and

$$\Delta_2 = \{x_3^*\} - \{x_2^*\}$$

respectively vanish. It is easy to see that the vanishing cycles Δ_1 and Δ_2 form a distinguished set.

2. As another example we consider the function of two variables $f(x, y) = x^3 - 3\lambda x + y^2$ (λ is a small positive number). This function is a perturbation of the function $f_0(x, y) = x^3 + y^2$, which also has singularity type A_2 in the sense of volume 1. The function has the same critical values, $z_1 = -2\lambda\sqrt{\lambda}$ and $z_2 = 2\lambda\sqrt{\lambda}$, as the function in the first example. These values are taken at the points $(\sqrt{\lambda}, 0)$ and $(-\sqrt{\lambda}, 0)$ respectively. We join the critical values z_1 and z_2 with the non-critical value $z_0 = 0$ by segments u_1 and u_2 of the real axis. The zero level manifold of the function f (the complex curve $\{f = 0\}$) is the graph of the two-valued function $y = \pm\sqrt{(-x^3 + 3\lambda x)}$ and therefore is a double covering of the plane of the complex variable x, branching at the points $x_1 = -\sqrt{3\lambda}, x_2 = 0$ and $x_3 = \sqrt{3\lambda}$. It can be obtained from two copies of the plane of the complex variable x with cuts from the point x_1 to the point x_2 and from the point x_3 to infinity (see figure 12), glued together criss-cross along these cuts.

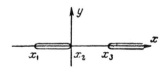

Fig. 12.

As z moves along the real axis from $z_1 = -2\lambda\sqrt{\lambda}$ to $z_2 = 2\lambda\sqrt{\lambda}$ the manifold $\{f = z\}$ is deformed. The movement of the branch points $\tilde{x}_1 = \tilde{x}_1(z)$, $\tilde{x}_2 = \tilde{x}_2(z)$, and $\tilde{x}_3 = \tilde{x}_3(z)$ as a double covering of the plane of the complex variable x is illustrated in figure 13.

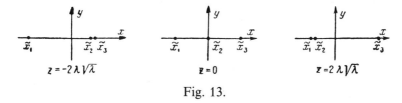

Fig. 13.

From this it is clear that the vanishing cycles corresponding to the critical values $z_1 = -2\lambda\sqrt{\lambda}$ and $z_2 = 2\lambda\sqrt{\lambda}$ and the paths u_1 and u_2 which we described joining them to the non-critical value 0 are the one-dimensional cycles Δ_1 and Δ_2 portrayed in figure 14 (we have indicated by dashes the part of the cycle lying on the second sheet of the surface; the orientation of the vanishing cycles again can be chosen arbitrarily).

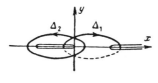

Fig. 14.

Once again let u be a path joining some critical value z_i with a non-critical value z_0.

Definition. A *simple loop* corresponding to the path u is an element of the fundamental group $\pi_1(U\setminus\{z_i\}, z_0)$ of the complement of the set of critical values represented by the loop going along the path u from the point z_0 to the point z_i, going round the point z_i in the positive direction (anticlockwise) and returning along the path u to the point z_0.

The region U, with the μ critical values $\{z_i | i = 1, \ldots, \mu\}$ of the function f removed from it is homotopically equivalent to a bouquet of μ circles. Therefore the fundamental group $\pi_1(U\setminus\{z_i\}, z_0)$ of the complement of the set of critical values of the function f is a free group on μ generators. If $\{u_i | i = 1, \ldots, \mu\}$ is a system of loops, defining a distinguished set of vanishing cycles $\{\Delta_i\}$, then the group $\pi_1(U\setminus\{z_i\}, z_0)$ is generated by the simple loops $\tau_1, \ldots, \tau_\mu$ corresponding to the paths $u_1, \ldots, u_\mu$.

Definition. The set of vanishing cycles $\Delta_1, \ldots, \Delta_\mu$, defined by the set of paths $\{u_i\}$, is called *weakly distinguished* if the fundamental group $\pi_1(U \setminus \{z_i\}, z_0)$ of the complement of the set of critical values is the free group on the generators $\tau_1, \ldots, \tau_\mu$, corresponding to the paths $u_1, \ldots, u_\mu$.

We note that permutation of the elements preserves weak distinguishment of a set, but does not preserve its distinguishment.

If the set of paths $\{u_i | i = 1, \ldots, \mu\}$ defines a weakly distinguished set of vanishing cycles $\{\Delta_i\}$ in the $(n-1)$st homology group of the non-singular level manifold, then the monodromy group of the function f is generated by the monodromy operators $h_{\tau_i *}$ of the simple loops τ_i $(i = 1, \ldots, \mu)$, corresponding to the paths u_i. Therefore the monodromy group of the (Morse) function f is always a group generated by μ generators.

Definition. The monodromy operator

$$h_i = h_{\tau_i *} : H_*(F_{z_0}) \rightarrow H_*(F_{z_0})$$

of the simple loop τ_i is called the *Picard-Lefschetz operator* corresponding to the path u_i (or the vanishing cycle Δ_i).

Examples. 1. We consider the Morse function $f(x) = x^3 + 3\lambda x$ of example 1* following the definition of distinguished sets of vanishing cycles. Let τ_i be the simple loop (with initial and final points at the point z_0^*) corresponding to the path u_i. As the non-critical value z moves along the loop τ_1, the level manifold $\{f = z\}$ changes in the following manner: The points x_1^* and x_2^* approach each other, make a half-turn about a common centre, changing places, then move apart to the other's former place; the point x_3^* returns to its own place. Therefore the monodromy h_{τ_1} of the loop τ_1 exchanges the points x_1^* and x_2^* and fixes the points x_3^*. From this it follows that

$$h_1 \Delta_1 = h_{\tau_1 *} \Delta_1 = h_{\tau_1 *}(\{x_2^*\} - \{x_1^*\}) = \{x_1^*\} - \{x_2^*\} = -\Delta_1,$$
$$h_1 \Delta_2 = h_{\tau_1 *} \Delta_2 = h_{\tau_1 *}(\{x_3^*\} - \{x_2^*\}) = \{x_3^*\} - \{x_1^*\} = \Delta_2 + \Delta_1.$$

Similarly

$$h_2 \Delta_2 = -\Delta_2,$$
$$h_2 \Delta_1 = \Delta_2 + \Delta_1.$$

The homology group $H_{n-1}(F_z, \partial F_z)$ of the non-singular level manifold modulo its boundary is the dual group to the group $H_{n-1}(F_z)$ (reduced modulo a point for $n=1$). In the given case its rôle is filled by the ordinary zeroth homology group of the level manifold $\{f = z_0^*\}$ (consisting of the three points x_1^*, x_2^* and x_3^*), factored by the subgroup generated by the "maximal cycle"

$$\{x_1^*\} + \{x_2^*\} + \{x_3^*\}.$$

It is generated by two cycles V_1 and V_2 such that

$$(V_i \circ \Delta_j) = \delta_{ij}.$$

We can take as these cycles

$$V_1 = -\{x_1^*\}, \quad V_2 = \{x_3^*\}.$$

From the description of the monodromy transformation h_{τ_1} it follows that

$$\mathrm{var}_{\tau_1} V_1 = -\{x_2^*\} + \{x_1^*\} = -\Delta_1,$$
$$\mathrm{var}_{\tau_1} V_2 = 0.$$

For the loop τ_2 we have

$$\mathrm{var}_{\tau_2} V_1 = 0,$$
$$\mathrm{var}_{\tau_2} V_2 = -\Delta_2.$$

We consider now the loop τ, defined by the formula $\tau(t) = z_0^* \exp(2\pi i t)$. The loop τ goes once round the critical values of the function f in the positive direction (anticlockwise) along a circle of large radius. From the fact that for large $|z|$ the level set $\{f = z\}$ is close to the level set $\{x^3 = z\}$, it follows that the monodromy transformation h_τ, of the loop τ cyclically permutes the points x_1^*, x_2^* and x_3^*

$$(x_1^* \to x_2^* \to x_3^* \to x_1^*).$$

From this it follows that

$$h_{\tau*} \Delta_1 = h_{\tau*}(\{x_2^*\} - \{x_1^*\}) = \{x_3^*\} - \{x_2^*\} = \Delta_2,$$
$$h_{\tau*} \Delta_2 = h_{\tau*}(\{x_3^*\} - \{x_2^*\}) = \{x_1^*\} - \{x_3^*\} = -\Delta_1 - \Delta_2.$$

These relationships can be deduced also from the fact that the loop τ is homotopic to the product $\tau_2\tau_1$ of the simple loops τ_2 and τ_1 and consequently $h_{\tau*}=h_1 \cdot h_2$. For the variation operator we have:

$$\mathrm{var}_\tau V_1 = -\{x_2^*\}+\{x_1^*\} = -\varDelta_1,$$

$$\mathrm{var}_\tau V_2 = \{x_1^*\}-\{x_3^*\} = -\varDelta_1 - \varDelta_2.$$

The monodromy group of the Morse function f is generated by the monodromy operators

$$h_1 = h_{\tau_1*} \text{ and } h_2 = h_{\tau_2*}$$

of the loops τ_1 and τ_2 (the operators of Picard-Lefschetz). All the elements of this group preserve the intersection form in the homology group $H_0(F_{z\delta})$ of the nonsingular level manifold (reduced modulo a point), generated by the vanishing cycles $\varDelta_1$ and $\varDelta_2$. In order to describe the monodromy group visually, we consider on the Euclidean plane six vectors of length $\sqrt{2}$ making with each other the angle $\pi/3$ (figure 15). It is not difficult to see that the group $H_0(F_{z\delta})$ (as a module over the ring of integers with an integral bilinear form on it) is isomorphic to the integer lattice on the plane, spanned by the vectors $\varDelta_1$ and $\varDelta_2$ (see figure 15; the six vectors, portrayed on the figure, are the elements of the lattice, the square of whose length equals 2). The operator h_1 is realised by reflection in the line L_1, orthogonal to the vector $\varDelta_1$, whilst the operator h_2 is realised by reflection in the line L_2, orthogonal to the vector $\varDelta_2$. From this it follows that the group of transformations of the lattice, generated by the operators h_1 and h_2 (the

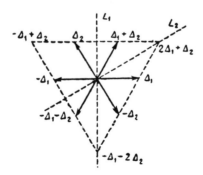

Fig. 15.

monodromy group of the Morse function f) is isomorphic to the group $S(3)$ of permutations of three elements (namely the three vectors

$$(2\Delta_1 + \Delta_2),\ (-\Delta_1 + \Delta_2)\ \text{and}\ (-\Delta_1 - 2\Delta_2)).$$

2. In example 2 the description of the level manifold $\{f = z\}$ as a double covering of the plane of the complex variable x, branching in three points, allows us to observe the operation of the monodromy transformations h_{τ_1} and h_{τ_2} and obtain the relationships

$$h_1 \Delta_1 = \Delta_1,$$

$$h_1 \Delta_2 = \Delta_2 + \Delta_1,$$

$$h_2 \Delta_1 = \Delta_1 - \Delta_2,$$

$$h_2 \Delta_2 = \Delta_2,$$

$$\text{var}_{\tau_1} V_1 = -\Delta_1,$$

$$\text{var}_{\tau_1} V_2 = 0,$$

$$\text{var}_{\tau_2} V_1 = 0,$$

$$\text{var}_{\tau_2} V_2 = -\Delta_2.$$

We will not go over here the corresponding geometrical considerations, leaving them to the reader.* We indicate only the relative cycles, which can be taken as V_1 and V_2 (generating the relative homology group $H_1(F_{z_\delta}, \partial F_{z_\delta})$ of the non-singular level manifold of the function f modulo the boundary so that $(V_i \circ \Delta_j) = \delta_{ij})$: see figure 16.

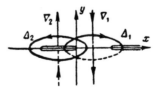

Fig. 16.

The monodromy group of the Morse function $f(x, y) = x^3 - 3\lambda x + y^2$ is isomorphic to the subgroup of the group of non-singular (2×2) matrices

---

* They are analogous to those which were discussed in the introduction for a simpler case.

generated by the matrices

$$\begin{bmatrix} 1 & 1 \\ 0 & 1 \end{bmatrix} \text{ and } \begin{bmatrix} 1 & 0 \\ -1 & 1 \end{bmatrix},$$

corresponding to the Picard-Lefschetz operators h_1 and h_2. This group coincides with the group of all integer matrices with determinant $+1$.

1.3 The Picard-Lefschetz Theorem

Let τ be the simple loop corresponding to the path u, which joins the critical value z_i with the non-critical value z_0, and let $\Delta \in H_{n-1}(F_{z_0})$ be a cycle, vanishing along the path u. We want to define the action of the operators var_τ and $h_{\tau*}$ in the respective homology groups.

Without loss of generality we can suppose that the critical value $z_i = u(0)$ is equal to zero and that in a neighbourhood of the critical point p_i the function f has the form

$$f(x_1, \ldots, x_n) = \Sigma_j x_j^2$$

(for sufficiently small $\|(x_1, \ldots, x_n)\|$, for example for

$$\Sigma_j |x_j|^2 \leqslant 4\varepsilon^2;$$

at the point p_i all local coordinates x_j equal zero), the non-critical value z_0 is sufficiently close to the critical value 0 (for example $|z_0| = \varepsilon^2$), and $u(t) = tz_0$. In addition, we shall suppose that all non-zero critical values of the function f are greater than $4\varepsilon^2$ in absolute value. A linear change of coordinates allows us to suppose that $\varepsilon = 1$, and $z_0 = 1$. The loop τ can be changed by a homotopy into the loop $\tau' : \tau'(t) = \exp(2\pi i t)$, $t \in [0, 1]$.

Let

$$r = r(x_1, \ldots, x_n) = \|x\| = (\Sigma_j |x_j|^2)^{1/2} = (\Sigma_j x_j \bar{x}_j)^{1/2}$$

be the norm of the vector $x = (x_1, \ldots, x_n)$. We denote by $\tilde{F}_z$ the intersection of the level set F_z with the closed ball $\bar{B}_2 = \{(x_1, \ldots, x_n) : r \leqslant 2\}$ of radius 2 in the space $\mathbb{C}^n$.

Lemma 1.2. For $|z| < 4$ the level set F_z is transverse to the $(2n-1)$-dimensional sphere $S_2 = \partial \bar{B}_2$ (the level set F_z is a manifold for $z \neq 0$, the zero level set F_0 is a manifold everywhere except zero).

Proof. Let $x \in F_z \cap S_2$ and suppose that the level set F_z is not transverse to the sphere S_2 at the point x. Then $dr^2(x)$ is linearly dependent on $df(x)$ and $d\bar{f}(x)$, that is $dr^2(x) = \alpha df(x) + \beta d\bar{f}(x)$, where $\alpha, \beta \in \mathbb{C}$. We have

$$df(x) = 2\Sigma x_j \, dx_j,$$

$$d\bar{f}(x) = 2\Sigma \bar{x}_j \, d\bar{x}_j,$$

$$dr^2(x) = \Sigma \bar{x}_j \, dx_j + \Sigma x_j \, d\bar{x}_j,$$

from which it follows that

$$\bar{x}_j = 2\alpha x_j, \quad (j = 1, \ldots, n).$$

But not all coordinates x_j equal zero. Therefore $|2\alpha| = 1$, and so

$$r^2(x) = 2\alpha f(x), \quad |f(x)| = r^2(x) = 4,$$

which is what we had to prove.

From this lemma it follows that, for $0 < |z| < 4$, the sets $\tilde{F}_z = F_z \cap \bar{B}_2$ are manifolds with boundary, which are diffeomorphic to each other. It is clear that the set $\tilde{F}_0$ is a cone with vertex at zero and is therefore contractible.

Lemma 1.3. For $0 < |z| < 4$ the manifold $\tilde{F}_z$ is diffeomorphic to the disk sub-bundle of the tangent bundle of the standard $(n-1)$-dimensional sphere S^{n-1}.

Proof. Without loss of generality it is possible to suppose that $z = 1$, that is to consider the manifold $\tilde{F}_1$. Let $x_j = u_j + iv_j$, where u_j and v_j are real. Using u_j and v_j as coordinates, the manifold $\tilde{F}_1$ is given by the equations

$$\Sigma u_j^2 - \Sigma v_j^2 = 1,$$

$$\Sigma u_j v_j = 0,$$

$$\Sigma u_j^2 + \Sigma v_j^2 \leq 4.$$

In the real vector space $\mathbb{R}^{2n}$ with coordinates $\tilde{u}_j, \tilde{v}_j \, (j = 1, \ldots, n)$ the space of the disc bundle of the tangent bundle of the standard $(n-1)$-dimensional sphere, lying in the space $\mathbb{R}^n$, can be given in the form

$$\Sigma \tilde{u}_j^2 = 1,$$

$$\Sigma \tilde{u}_j \tilde{v}_j = 0$$

$$\Sigma \tilde{v}_j^2 \leqslant \varrho^2.$$

($\varrho > 0$ is the radius of the discs in the bundle, on which the type of the space of the bundle as a differentiable manifold does not, of course, depend). It is not difficult to check that the transformation $\tilde{u}_j = u_j/\sqrt{\Sigma u_j^2}$, $\tilde{v}_j = v_j$ gives the required diffeomorphism (for $\varrho^2 = 3/2$).

From this it follows that $H_k(\tilde{F}_1) = 0$ for $k \neq n-1$, $H_{n-1}(\tilde{F}_1) = \mathbb{Z}$. Moreover the homology group $H_{n-1}(\tilde{F}_1)$ is generated by the Picard-Lefschetz vanishing cycle Δ, represented by the standard $(n-1)$-dimensional sphere

$$S^{n-1} = \{(x_1, \ldots, x_n) : \Sigma u_j^2 = 1, v_j = 0\}.$$

Let us temporarily consider the manifold $\tilde{F}_1$ with the orientation defined by the structure of the tangent bundle of the sphere. This means that at the point $(1, 0, \ldots, 0)$ of the manifold $\tilde{F}_1$ a positively oriented coordinate system is $u_2, u_3, \ldots, u_n, v_2, v_3, \ldots, v_n$. The orientation of $\tilde{F}_1$ as a complex manifold is defined by the following ordering of the coordinates: $u_2, v_2, u_3, v_3, \ldots, u_n, v_n$. It is easy to see that these two orientations differ by the sign $(-1)^{(n-1)(n-2)/2}$.

The self-intersection number of the zero section of the tangent bundle of a manifold coincides with the Euler characteristic χ of this manifold. This statement can be proved in the following manner. According to one of the definitions the Euler characteristic of a manifold N is the number of singular points of a general vector field v on the manifold N, counted with the multiplicity $+1$ or -1 according to their index ($v: N \to TN$, $v(x) \in T_x N$). In order to count the number of self-intersections ($N \circ N$) of the manifold N considered as the zero section of its tangent bundle TN, we can choose a perturbation $\tilde{N}$ of the manifold N in the space TN, which intersects N transversally in a finite number of points, and define the intersection number ($N \circ \tilde{N}$) of the cycles N and $\tilde{N}$ at these points. As such a perturbation we can take $\tilde{N} = \{(x, v(x))\}$, where $x \in N$, $v(x) \in T_x N$, v is a vector field in general position on the manifold N. The points of intersection of the cycles N and $\tilde{N}$ coincide with the singular points of the vector field v. Moreover, simple counting shows that the intersection number of the cycles N and $\tilde{N}$ at a point of intersection coincides with the index of this point as a singular point of the vector field v.

The Euler characteristic $\chi(S^{n-1})$ of the $(n-1)$-dimensional sphere S^{n-1} is equal to $1 + (-1)^{n-1}$, that is it is equal to 0 for even n and 2 for odd n. From the

fact that the manifold $\tilde{F}_1$ is diffeomorphic to the space of the disc subbundle of the tangent bundle of the $(n-1)$-dimensional sphere, it follows that the self-intersection number of the vanishing cycle $\varDelta$ in the manifold $\tilde{F}_1$ oriented according to the structure of the space of the tangent bundle of the sphere, is equal to $\chi(S^{n-1})=1+(-1)^{n-1}$. From this follows the following result:

Lemma 1.4. The self-intersection number of the vanishing cycle $\varDelta$ in the *complex* manifold $\tilde{F}_1$ is equal to

$$(\varDelta \circ \varDelta)=(-1)^{(n-1)(n-2)/2}(1+(-1)^{n-1})$$

$$=\begin{cases} 0 & \text{for} \quad n\equiv 0 \ \mathrm{mod}\, 2, \\ +2 & \text{for} \quad n\equiv 1 \ \mathrm{mod}\, 4, \\ -2 & \text{for} \quad n\equiv 3 \ \mathrm{mod}\, 4. \end{cases}$$

By the theorem on Poincaré duality the relative homology group $H_k(\tilde{F}_1, \partial\tilde{F}_1)$ is zero for $k\neq n-1$, and the group $H_{n-1}(\tilde{F}_1, \partial\tilde{F}_1)$ is isomorphic to the group $\mathbb{Z}$ of integers. Moreover the group $H_{n-1}(\tilde{F}_1, \partial\tilde{F}_1)$ is generated by the relative cycle V, dual to the vanishing cycle $\varDelta$, that is such that the intersection number $(V\circ \varDelta)$ is equal to one. As a representative of the cycle V we can choose the non-singular submanifold

$$T=\{(x_1,\ldots,x_n)\in\tilde{F}_1 : u_1>0, u_2=\ldots=u_n=0\}$$

of the manifold $\tilde{F}_1$, oriented in a suitable manner. By the diffeomorphism of the level manifold $\tilde{F}_1$ with the space of the tangent bundle of the sphere, constructed in Lemma 1.3, the submanifold T corresponds to a fibre of this bundle, that is the ball in the tangent space of the sphere S^{n-1} at the point $(1,0,\ldots,0)$.

We consider the restriction of the function f to $f^{-1}(\bar{D}_1)\backslash B_2$ where B_2 is the open ball of radius 2, $\bar{D}_1$ is the closed disk of radius 1 in the space $\mathbb{C}$. It defines the locally trivial and hence trivial fibration

$$f^{-1}(\bar{D}_1)\backslash B_2 \to \bar{D}_1$$

over the unit disk $\bar{D}_1$. A lifting Γ_t of the homotopy $t\mapsto \tau'(t)=\exp(2\pi it)$ to a homotopy of the fibre F_1 can be chosen consistent with the structure of a direct product on the space $f^{-1}(\bar{D}_1)\backslash B_2$ of this fibration. A relative cycle δ of dimension k from $(F_1, \partial F_1)$ can be represented in the form

$$\delta = \delta_1 + \delta_2,$$

where δ_1 is a relative cycle in $(\tilde{F}_1, \partial \tilde{F}_1)$ and δ_2 is a chain in $F_1 \setminus B_2$. The transformation $h_{\tau'} = \Gamma_1$ of going round is the identity on $F_1 \setminus B_2$. Therefore it preserves the chain δ_2 and acts non-trivially only on the cycle δ_1. In this way

$$\operatorname{var}_{\tau'}(\delta) = \operatorname{var}_{\tau'}(\delta_1)$$

(here we use one and the same notation for the variation operators $\operatorname{var}_{\tau'}$, corresponding to the pairs $(M^n, \partial M^n)$ and $(\bar{B}_2, \partial \bar{B}_2)$). If the dimension k of the cycle δ is different from $(n-1)$, then $\delta_1 = 0$ in the relative homology group $H_k(\tilde{F}_1, \partial \tilde{F}_1)$ (this group is itself zero). From this follows the following assertion:

Lemma 1.5. In all dimensions except the $(n-1)$st the variation operator $\operatorname{var}_{\tau'}$ is zero, and the operators $h_{\tau' *}$ and $h^{(r)}_{\tau' *}$ are identical.

If $k = \dim \delta = (n-1)$, then $\delta_1 = m \cdot V$ in the homology group $H_{n-1}(\tilde{F}_1, \partial \tilde{F}_1)$. Here $m = (\delta \circ \Delta)$. Therefore in order to determine the action of the variation operator $\operatorname{var}_{\tau'}$ it is sufficient to calculate the homology class $\operatorname{var}_{\tau'}(V)$.

Theorem of Picard-Lefschetz.

$$\operatorname{var}_{\tau'}(V) = (-1)^{n(n+1)/2} \Delta.$$

Corollaries. For $a \in H_{n-1}(F_{z_0}, \partial F_{z_0})$

$$\operatorname{var}_{\tau}(a) = (-1)^{n(n+1)/2}(a \circ \Delta)\Delta,$$

$$h^{(r)}_{\tau *}(a) = a + (-1)^{n(n+1)/2}(a \circ \Delta)i_*(\Delta);$$

for $a \in H_{n-1}(F_{z_0})$

$$h_{\tau *}(a) = a + (-1)^{n(n+1)/2}(a \circ \Delta)\Delta.$$

The last formula is usually called the *Picard-Lefschetz formula*. When the number of variables n is odd, it, together with Lemma 1.4 shows that the Picard-

Lefschetz operator $h_{\tau *}$ is the reflection of the space $H_{n-1}(F_{z_0})$ in the hyperplane orthogonal (with respect to the intersection form) to the corresponding vanishing cycle Δ.

A proof of the Picard-Lefschetz theorem can be obtained by elementary, though fairly heavy, straight calculation (see, for example, [150]). We give below a more invariant proof (see §2.4). Here we give a proof for the case when the number of variables n is odd.

There exists the natural lifting Ω_t of the homotopy $t \mapsto \tau'(t) = \exp(2\pi i t)$ $(0 \leqslant t \leqslant 1)$ to a homotopy of the fibre $\tilde{F}_1 \mapsto \tilde{F}_{\tau'(t)}$, which does not, however, agree with the structure of a direct product on the boundary. This lifting is given by the formula $\Omega_t(x) = \exp(\pi i t)x$. The homotopy Ω_t is not suitable for the determination of the variation operator $\mathrm{var}_{\tau'}$, but it is not hard to see that with its help we can determine the action of the monodromy operator $h_{\tau' *}$ on the homology group $H_{n-1}(\tilde{F}_1)$ of the fibre. It is clear that the transformation Ω_1 is multiplication by -1. In particular, on the vanishing sphere Δ it coincides with reflection in the centre. From this it follows that

$$\Omega_{1 *}(\Delta) = h_{\tau' *}(\Delta) = (-1)^n \Delta.$$

Let

$$i_* : H_{n-1}(\tilde{F}_1) \to H_{n-1}(\tilde{F}_1, \partial \tilde{F}_1)$$

be the natural homomorphism induced by the inclusion $\tilde{F}_1 \hookrightarrow (\tilde{F}_1, \partial \tilde{F}_1)$. Since

$$(\Delta \circ \Delta) = (i_*(\Delta) \circ \Delta), \quad (\nabla \circ \Delta) = 1,$$

then from Lemma 1.4 it follows that

$$i_*(\Delta) = (-1)^{(n-1)(n-2)/2}(1 + (-1)^{n-1})\nabla$$

$$= \begin{cases} 0 & \text{for } n \equiv 0 \bmod 2, \\ 2(-1)^{(n-1)/2}\nabla & \text{for } n \equiv 1 \bmod 2. \end{cases}$$

Since the homology group $H_{n-1}(\tilde{F}_1)$ of the fibre is isomorphic to the group of integers and is generated by the vanishing cycle Δ, then $\mathrm{var}_{\tau'}(\nabla) = m\Delta$ for some integer m. From the fact that

$$h_{\tau' *} = id + \mathrm{var}_{\tau'} \cdot i_*,$$

when the number of variables n is odd, we get

$$-\Delta = h_{\tau'*}(\Delta)$$
$$= \Delta + 2(-1)^{(n-1)/2}\,\mathrm{var}_{\tau'}(V)$$
$$= \Delta + 2(-1)^{(n-1)/2}\,m\Delta.$$

From this it follows that $m = (-1)^{(n+1)/2}$, which, for odd n, coincides with the statement of the theorem of Picard-Lefschetz.

Chapter 2

The topology of the non-singular level set and the variation operator of a singularity

2.1   The non-singular level set of a singularity

Let $f : (\mathbb{C}^n, 0) \to (\mathbb{C}, 0)$ be a singularity, that is the germ of a holomorphic function, with an isolated critical point at the origin. It follows from implicit function theorem that in a neighbourhood of the origin in the space $\mathbb{C}^n$ the level set $f^{-1}(\varepsilon)$ for $\varepsilon \neq 0$ is a non-singular analytic manifold and the level set $f^{-1}(0)$ is a non-singular manifold away from the origin. At the point $0 \in \mathbb{C}^n$ the level set has a singular point.

Lemma 2.1. There exists a $\varrho > 0$, such that the sphere $S_r \subset \mathbb{C}^n$ of radius $r \leqslant \varrho$ with centre at the origin intersects the level set $f^{-1}(0)$ transversely.

Indeed, the function $\|x\|^2$ on the set $f^{-1}(0)$ (in a neighbourhood of the point $0 \in \mathbb{C}^2$) can take only a finite number of critical values (for the case where f is a polynomial this assertion follows, for example, automatically from the 'curve selection lemma' of [256]; in the general case it can be derived from analogous reasoning). We choose as ϱ a number such that its square is less than all critical values of the function $\|x\|^2$ on the manifold $f^{-1}(0) \backslash 0$. The fact that all critical values of the function $\|x\|^2$ on $f^{-1}(0) \backslash 0$ are greater than ϱ^2 is equivalent to the fact that for $r \leqslant \varrho$ the sphere S_r of radius r with centre at the origin (which is a level manifold of the function $\|x\|^2$) intersects the manifold $f^{-1}(0) \backslash 0$ transversely.

From Lemma 2.1 it follows that for sufficiently small $\varepsilon_0 > 0$ the level manifold $f^{-1}(\varepsilon)$ is also transverse to the sphere S_ϱ for $|\varepsilon| \leqslant \varepsilon_0$. Thus the function $f : \bar{B}_\varrho \to \mathbb{C}$ satisfies conditions (i)–(iii) of § 1.1 (with the ball $\bar{B}_\varrho$ of radius ϱ with centre at the origin as M^n, the disk $\bar{D}_{\varepsilon_0}$ of radius ε_0 with centre at zero in the plane $\mathbb{C}$ as U and the unique critical point 0). We shall be interested in the topology of the level set $f^{-1}(\varepsilon)$ in a neighbourhood of the origin.

Definition. The *non-singular level set* of the singularity f near the critical point 0 is the set

$$V_\varepsilon = f^{-1}(\varepsilon) \cap \bar{B}_\varrho = \{x \in \mathbb{C}^n : f(x) = \varepsilon, \|x\| \leqslant \varrho\}$$

for $0 < |\varepsilon| \leqslant \varepsilon_0$, which is a complex manifold with boundary.

The manifold V_ε is defined uniquely up to diffeomorphism. It is known ([256]) that it has the homotopy type of a bouquet of spheres of dimension $(n-1)$. The number $\mu = \mu(f)$ of these spheres is called the *multiplicity* or *Milnor number* of the singularity f. The homology group $H_k(V_\varepsilon)$ of the non-singular level manifold is zero for $k \neq (n-1)$, $H_{n-1}(V_\varepsilon) \cong \mathbb{Z}^\mu$ is a free abelian group with μ generators. The assertions about the homology groups $H_k(V_\varepsilon)$ of the non-singular level set we prove below (see Theorem 2.1). With a small addition (the proof of the simple connectedness of the manifold V_ε, which arises from the same considerations as in Theorem 2.1), from this follows also the result on the homotopy type of the non-singular level set (for $n > 2$). The fundamental group $\pi_1(\bar{D}_{\varepsilon_0} \backslash 0)$ of the complement of the set of critical values is isomorphic to the group of integers and is generated by the class of the loop γ_0, which goes once round the critical value 0 in the positive direction (anticlockwise). We can, for example, set

$$\gamma_0(t) = \varepsilon \cdot \exp(2\pi i t) \qquad (|\varepsilon| \leqslant \varepsilon_0, t \in [0, 1]).$$

Definition. The *classical monodromy* $h : V_\varepsilon \to V_\varepsilon$ of the singularity f is the monodromy h_{γ_0} of the loop γ_0. The *classical monodromy operator* of the singularity f is the automorphism $h_* = h_{\gamma_0 *}$ of the homology group $H_{n-1}(V_\varepsilon)$ of the non-singular level set V_ε. The *variation operator* of the singularity f is the variation operator

$$\mathrm{Var}_f = \mathrm{var}_{\gamma_0} : H_{n-1}(V_\varepsilon, \partial V_\varepsilon) \to H_{n-1}(V_\varepsilon)$$

of the loop γ_0.

A basis of the homology group $H_{n-1}(V_\varepsilon) \cong \mathbb{Z}^{\mu(f)}$ of the non-singular level manifold V_ε of the singularity f can be constructed in the following manner. Let $\tilde{f} = f_\lambda$ be a perturbation of the function f, defined in a neighbourhood of the ball $\bar{B}_\varrho$ (we can, for example, take as $\tilde{f}$ the perturbation $f_\lambda = f + \lambda g$, where g is a linear function: $\mathbb{C}^n \to \mathbb{C}$). For sufficiently small λ ($|\lambda| \leqslant \lambda_0$) the level set $\tilde{f}^{-1}(\varepsilon)$ is transverse to the sphere S_ϱ for $|\varepsilon| \leqslant \varepsilon_0$ and the critical values of the function $\tilde{f}$ on the ball $\bar{B}_\varrho$ are less than ε_0 in modulus. It is easy to show that the non-singular level set $\tilde{f}^{-1}(\varepsilon) \cap \bar{B}_\varrho$ is diffeomorphic to the non-singular level set V_ε of the function f for $|\varepsilon| \leqslant \varepsilon_0$. From Sard's theorem it follows that almost all perturbations $\tilde{f}$ of the function f have in the ball $\bar{B}_\varrho$ only non-degenerate critical points with distinct critical values (in the example this will take place for almost all linear functions g).

Let us prove, for example, that the function $\tilde{f} = f + g$ is Morse for almost all linear functions $g : \mathbb{C}^n \to \mathbb{C}$. For this we consider the mapping $df : \mathbb{C}^n \to \mathbb{C}^n$, given by the formula

$$df(x) = (\partial f / \partial x_1(x), \ldots, \partial f / \partial x_n(x))$$

$(x = (x_1, \ldots, x_n) \in \mathbb{C}^n)$. Almost all the values $(l_1, \ldots, l_n) \in \mathbb{C}^n$ are non-critical for this mapping (Sard's theorem). If $(l_1, \ldots, l_n) \in \mathbb{C}^n$ is a non-critical value of the mapping df, then the function $f - \Sigma_j l_j x_j$ has only non-degenerate critical points. Indeed the critical points of the function $f - \Sigma_j l_j x_j$ are those points at which $\partial f / \partial x_j - l_j = 0$ $(j = 1, \ldots, n)$, that is these are the preimages of the point $(l_1, \ldots, l_n)$ for the mapping df. Since the value $(l_1, \ldots, l_n)$ is non-critical for the mapping df, then at these points the matrix $(\partial^2 f / \partial x_j \partial x_k)$ has non-zero determinant, which means that the corresponding critical points of the function $\tilde{f} = f - \Sigma_j l_j x_j$ are non-degenerate. The set of non-critical values of the mapping df is open. Therefore the addition to $\tilde{f}$ of a suitably small linear function does not remove it from the class of functions with non-degenerate critical points and allows us to obtain the fact that the critical values become pairwise distinct.

We again get the situation described in Chapter 1. As before, let

$$F_z = \tilde{f}^{-1}(z) \cap \bar{B}_\varrho \quad (|z| \leq \varepsilon_0),$$

the function $\tilde{f}$ has in the ball $\bar{B}_\varrho$ several critical points p_i with distinct critical values z_i $(|z_i| < \varepsilon_0, i = 1, \ldots, \mu)$, and $\{u_i\}$ is a system of paths joining the critical values z_i with the non-critical value z_0 $(|z_0| = \varepsilon_0)$ and defining in the homology group $H_{n-1}(F_{z_0})$ of the non-singular level set of the function $\tilde{f}$ a distinguished set of vanishing cycles $\{\Delta_i\}$. Remember that the last condition means that the paths u_i are not self-intersecting and pairwise do not have common points except the point z_0.

Theorem 2.1. The distinguished set of vanishing cycles $\{\Delta_i\}$ forms a basis of the (free abelian) homology group $H_{n-1}(F_{z_0}) \cong H_{n-1}(V_\varepsilon)$ of the non-singular level set of the singularity f. In particular the number of non-degenerate critical points of the function $\tilde{f}$ in $\bar{B}_\varrho \cap \tilde{f}^{-1}(\bar{D}_{\varepsilon_0})$ (into which the critical point of the function f decomposes) is equal to the multiplicity $\mu(f)$ of the singularity f. The group $H_k(F_{z_0})$ is zero for $k \neq (n-1)$.

Proof. Let

$$X = \bar{B}_\varrho \cap f^{-1}(\bar{D}_{\varepsilon_0})$$

$$\tilde{X} = \bar{B}_\varrho \cap \tilde{f}^{-1}(\bar{D}_{\varepsilon_0}),$$

where $\varrho > 0$ and $\varepsilon_0 > 0$ as described above. We shall show that the space X is contractible. From the fact that the zero level set $f^{-1}(0)$ of the function f is transverse to the spheres S_r of radius $r \leqslant \varrho$ with centre at the origin in the space $\mathbb{C}^n$, it immediately follows that the set $f^{-1}(0) \cap \bar{B}_\varrho$ is homeomorphic to the cone over the manifold $f^{-1}(0) \cap S_\varrho$ and consequently contractible. The contraction of the set $f^{-1}(0) \cap \bar{B}_\varrho$ to the point 0, belonging to it, can be realised with the help of a vector field on it, orthogonal to the submanifolds $f^{-1}(0) \cap S_r, r \leqslant \varrho$ (remember that the set $f^{-1}(0) \cap \bar{B}_\varrho$ is a manifold everywhere except zero).

In its turn the space $f^{-1}(0) \cap \bar{B}_\varrho$ is a deformation retract of the space $X = f^{-1}(\bar{D}_{\varepsilon_0}) \cap \bar{B}_\varrho$. We can construct the required deformation retraction of the space X, for example, in the following manner. Choose a sequence $\varrho = r_0 > r_1 > r_2 > \ldots > 0$, monotonically decreasing to zero. Let ε_i be numbers such that $\varepsilon_0 > \varepsilon_1 > \varepsilon_2 > \ldots > 0$ and the level set $f^{-1}(\varepsilon)$ is transverse to the sphere S_{r_i} of radius r_i with centre at zero for $|\varepsilon| \leqslant \varepsilon_i$. The function f determines locally trivial, and hence also trivial, fibrations

$$E_i = f^{-1}(\bar{D}_{\varepsilon_i}) \cap (\bar{B}_{r_0} \setminus B_{r_i}) \to \bar{D}_{\varepsilon_i}.$$

Moreover trivialisations of these fibrations can be chosen so that they will coincide on the intersections

$$E_i \cap E_{i-1} = f^{-1}(\bar{D}_{\varepsilon_i}) \cap (\bar{B}_{r_0} \setminus B_{r_{i-1}}).$$

We consider the deformation g_t of the disk $\bar{D}_{\varepsilon_0}$, defined for $0 \leqslant t \leqslant \varepsilon_0$ and given by the formula

$$g_t(x) = \begin{cases} t \cdot x / \|x\| & \text{for } \|x\| \geqslant t, \\ x & \text{for } \|x\| \leqslant t. \end{cases}$$

The mapping g_t maps the disk $\bar{D}_{\varepsilon_0}$ of radius ε_0 into the disk of radius t, keeping the latter fixed. The mapping g_0 is a deformation retraction of the disk $\bar{D}_{\varepsilon_0}$ into the point 0. Since the function f defines the locally trivial fibration

$$f^{-1}(\bar{D}_{\varepsilon_0} \setminus 0) \cap \bar{B}_{r_0} \to \bar{D}_{\varepsilon_0} \setminus 0,$$

there exists a family G_t $(0 < t \leqslant \varepsilon_0)$ of mappings of the set $X = f^{-1}(\bar{D}_{\varepsilon_0}) \cap \bar{B}_{r_0}$ into itself lifting the homotopy g_t. This family can be chosen in accordance with the structure of the direct product on the sets

$$E_i = f^{-1}(\bar{D}_{\varepsilon_i}) \cap (\bar{B}_{r_0} \setminus B_{r_i})$$

for $t \leqslant \varepsilon_i$. It is not difficult to see that the family G_t $(0 < t \leqslant \varepsilon_0)$ determines in a natural way the family G_t with $0 \leqslant t \leqslant \varepsilon_0$ in which the mapping G_0 is a deformation retraction of the set X into the zero level set $f^{-1}(0) \cap \bar{B}_{r_0}$.

If $\tilde{f}$ is a sufficiently small perturbation of the function f, then the space $\tilde{X}$ is diffeomorphic to the space X (as smooth manifolds with corners; indeed it is sufficient that $\tilde{X}$ is homotopy equivalent to X) and therefore is also contractible.

The function $\tilde{f}$ maps the space $\tilde{X}$ into the disk $\bar{D}_{\varepsilon_0}$ and away from the critical points $z_1, z_2, \ldots, z_\mu$ is a locally trivial fibration with fibre F_{z_0}. We consider the union $\cup_{i,t} u_i(t) = V$ of images of paths u_i. It is a deformation retract of the disk $\bar{D}_{\varepsilon_0}$. It is not difficult to see that a deformation retraction of the disk $\bar{D}_{\varepsilon_0}$ onto the space V can be lifted to a deformation retraction of the space $\tilde{X}$ onto the space $Y = \tilde{f}^{-1}(V)$ (analogous to the way that the deformation retraction of the disk $\bar{D}_{\varepsilon_0}$ to the point 0 is lifted to a deformation retraction of the space X to the zero level set $f^{-1}(0) \cap \bar{B}_{\varrho}$). If the singular fibres $\tilde{f}^{-1}(z_i)$ are cut out from the space Y, then the remaining space $Y \setminus \cup_{i=1}^\mu \tilde{f}^{-1}(z_i)$ will be a fibration over the contractible space $V \setminus \{z_i | i = 1, \ldots, \mu\}$. Consequently, it is homeomorphic to the direct product of the fibre F_{z_0} and the space $V \setminus \{z_i | i = 1, \ldots, \mu\}$ and therefore homotopy equivalent to the fibre F_{z_0}.

It is not difficult to show that up to homotopy type the space Y is obtained from the fibre F_{z_0} by gluing n-dimensional balls B_i to the vanishing spheres Δ_i. Here we define the mapping in one direction, determining the homotopy equivalence of the considered spaces. Let

$$s_i(t): S_i^{n-1} \to S_i(t) \subset F_{u_i(t)} \quad (0 \leqslant t \leqslant 1)$$

be a family of maps of the $(n-1)$-dimensional sphere (the index i simply fixes the number of the copy), defining the vanishing cycle $\Delta_i = s_i(1)$ $(s_i(0): S_i^{n-1} \to p_i)$. Let B_i be the n-dimensional ball, which is the cone over the sphere S_i^{n-1}

$$(B_i = [0, 1] \times S_i^{n-1} / 0 \times S_i^{n-1}).$$

The space

$$F_{z_0} \bigcup_{\{\Delta_i\}} \{B_i\},$$

obtained from the fibre F_{z_0} by gluing the n-dimensional balls B_i to the vanishing cycles Δ_i is the quotient space of the space

$$F_{z_0} \cup \bigcup_{i=1}^{\mu} B_i$$

by the equivalence relation

$$s_i(1)(a) \sim (1, a) \quad (a \in S_i^{n-1}, \ (1, a) \in B_i, \ i=1, \ldots, \mu).$$

The map ϕ of it into the space Y, which is a homotopy equivalence, can be given in the following fashion:

$$\phi(x) = x \quad \text{for} \quad x \in F_{z_0} \subset Y,$$

$$\phi(t, a) = s_i(t)(a) \quad \text{for} \quad (t, a) \in B_i, \ 0 \leqslant t \leqslant 1, \ a \in S_i^{n-1}.$$

There is the exact homology sequence of the pair $(Y, Y - \bigcup_{i=1}^{\mu} \tilde{f}^{-1}(z_i))$:

$$\ldots \to H_{k+1}(Y) \to H_{k+1}(Y, Y - \bigcup_{i=1}^{\mu} \tilde{f}^{-1}(z_i)) \to H_k(Y - \bigcup_{i=1}^{\mu} \tilde{f}^{-1}(z_i)) \to$$

$$\to H_k(Y) \to \ldots$$

Here $H_i(Y) = 0$ (since the space Y is homotopically equivalent to the contractible space X; remember that the homology is considered to be reduced modulo a point),

$$H_{k+1}(Y, Y \setminus \bigcup_{i=1}^{\mu} \tilde{f}^{-1}(z_i)) = \oplus_{i=1}^{\mu} H_{k+1}(B_i, \partial B_i)$$

$$= \begin{cases} 0 & \text{for} \quad k \neq n-1, \\ \mathbb{Z}^{\mu} & \text{for} \quad k = n-1, \end{cases}$$

$$H_k(Y \setminus \bigcup_{i=1}^{\mu} \tilde{f}^{-1}(z_i)) = H_k(F_{z_0}).$$

From the exactness of the sequence it follows that

$$H_k(F_{z_0}) \cong H_{k+1}(Y, Y \setminus \bigcup_{i=1}^{\mu} \tilde{f}^{-1}(z_i))$$

$$= \begin{cases} 0 & \text{for} \quad k \neq n-1, \\ \mathbb{Z}^{\mu} & \text{for} \quad k = n-1, \end{cases}$$

the generators of the group $H_n(B_i, \partial B_i)$ mapping into the vanishing cycles Δ_i. From this the result of the theorem follows.

It is not difficult to see that by considering the exact homotopy sequence of the pair $(Y, Y \setminus \bigcup\limits_{i=1}^{\mu} \tilde{f}^{-1}(z_i))$ we can deduce the simple-connectedness of the space $Y \setminus \bigcup\limits_{i=1}^{\mu} \tilde{f}^{-1}(z_i)$ or, which is the same thing, the simple-connectednes of the non-singular level set F_{z_0} for $n > 2$.

From theorem 2.1 it follows that the multiplicity of the critical point of a singularity f is equal to the number of non-degenerate critical points into which it decomposes under a perturbation of general form. This number is equal to the number of preimages (near zero) of a general point under the map

$$df : \mathbb{C}^n \to \mathbb{C}^n$$

$(df(x_1, \ldots, x_n) = (\partial f/\partial x_1(x), \ldots, \partial f/\partial x_n(x))$. From this we can obtain the following formula for the multiplicity of an isolated critical point of a function f:

$$\mu(f) = \dim_{\mathbb{C}} \ {}_nO/(\partial f/\partial x_1, \ldots, \partial f/\partial x_n),$$

where ${}_nO$ is the ring of germs at zero of holomorphic functions in n variables, $(\partial f/\partial x_1, \ldots, \partial f/\partial x_n)$ is the ideal in the ring ${}_nO$, generated by the partial derivatives of the function f (the Jacobian ideal of the germ f). This result was proved in Chapter 5 of Volume 1.

2.2 Vanishing Cycles and the Monodromy Group of a singularity

It was shown in §2.1 that almost all perturbations $\tilde{f}$ of the singularity $f : (\mathbb{C}^n, 0) \to (\mathbb{C}, 0)$ are Morse, that is in a neighbourhood of zero in the space $\mathbb{C}^n$ they have only non-degenerate critical points, equal in number to the multiplicity of the singularity f, all critical values $z_1, \ldots, z_\mu$ of the function $\tilde{f}$ being different. The non-singular level set V_ε of the singularity f is diffeomorphic to the non-singular level manifold $F_{z_0} = \tilde{f}^{-1}(z_0) \cap B_\varrho$ of the function $\tilde{f}$. The presence of such a diffeomorphism allows us to introduce the following definition.

Definition. A *vanishing cycle* Δ in the homology group $H_{n-1}(V_\varepsilon)$ of the non-singular level set of the singularity f is an element of this group corresponding to a cycle in the homology group $H_{n-1}(F_{z_0})$ of the non-singular level set of the function $\tilde{f}$, vanishing along a path joining some critical value z_i of the function $\tilde{f}$ with the non-critical value z_0.

Definition. A basis of the homology group $H_{n-1}(V_\varepsilon)$ of the non-singular level manifold consisting of a distinguished set of vanishing cycles $\{\Delta_i\}$ is called a *distinguished basis*. A basis consisting of a weakly distinguished set of vanishing cycles is called *weakly distinguished*.

Theorem 2.1 asserts that any distinguished set of vanishing cycles forms a basis. It will be shown later that any weakly distinguished set also forms a basis (see 2.6).

Remark. The terms "distinguished" and "weakly distinguished" were introduced by A.M. Gabrielov. In the work [205] a distinguished basis was called "geometrical".

Definition. The *monodromy group* of the singularity f is the monodromy group of the (Morse) function $\tilde{f}$.

It is not difficult to show that the set of vanishing cycles and the monodromy group of a singularity f do not depend on the choice of the Morse perturbation $\tilde{f} = f_\lambda$ of the function f. To see this we consider another such perturbation $\tilde{\tilde{f}} = f'_v$. The perturbations f_λ and f'_v can be included in one two-parameter family of functions $f_{\lambda,v}$ ($f_{\lambda,0} = f_\lambda$, $f_{0,v} = f'_v$). We can, for example, take $f_{\lambda,v} = f_\lambda + f'_v - f$ as this family. In the space $\mathbb{C}^2$ with coordinates (λ, v) the values of the parameters (λ, v) which correspond to non-Morse functions $f_{\lambda,v}$ form (in a neighbourhood of the point $(0,0) \in \mathbb{C}^2$) a set which is the image of an analytic set of complex dimension one. It does not, therefore, disconnect the space $\mathbb{C}^2$ of values of the parameters (λ, v). From this it follows that the perturbations $\tilde{f} = f_\lambda$ and $\tilde{\tilde{f}} = f'_v$ can be joined by a continuous one-parameter family of Morse functions $f_{\lambda(t), v(t)}$ ($t \in [0, 1]$, $f_{\lambda(0), v(0)} = \tilde{f}$, $f_{\lambda(1), v(1)} = \tilde{\tilde{f}}$). It is easy to see that along such a family of Morse functions the set of vanishing cycles and the monodromy group do not change.

For the same reason the concepts of distinguished and weakly distinguished bases are independent of the choice of perturbation.

From the results of chapter 1 it follows that the monodromy group of a singularity f is generated by the Picard-Lefschetz operators h_i corresponding to the elements Δ_i of a weakly distinguished basis in the homology of a non-singular level set of the function f near the critical point. If the number of variables n is odd, this operator is the reflection in a hyperplane, orthogonal (in the sense of intersection forms) to the vanishing cycle Δ_i. When, therefore, the number of variables is odd, the monodromy group of a singularity is a group generated by reflections.

Examples. As examples we can consider the functions $f(x) = x^3$ and $f(x, y) = x^3 + y^2$, which have the singularity type A_2 in the sense of volume 1. Their Morse perturbations can be chosen in the form $\tilde{f}(x) = x^3 \pm 3\lambda x$ and $\tilde{f}(x, y) = x^3 - 3\lambda x + y^2$ respectively, where λ is a small positive number. The distinguished bases in the homology of the non-singular level manifolds and the monodromy groups of the Morse functions $\tilde{f}(x)$ and $\tilde{f}(x, y)$ (coinciding with the distinguished bases and monodromy groups of the singularities $f(x)$ and $f(x, y)$) were considered in the examples of §1.2.

2.3   The Variation Operator and Seifert Form of a Singularity

In §2.1 the concept of the variation operator of a singularity was introduced. In order to study the properties of this operator we give another intrepretation of it ([101], [217]).

As above let $f : (\mathbb{C}^n, 0) \to (\mathbb{C}, 0)$ be a singularity, that is the germ of a holomorphic function, with an isolated critical point at zero, let ϱ be a sufficiently small positive number, and let S_ϱ^{2n-1} be a sphere of radius ϱ with centre at the origin in the space $\mathbb{C}^n$. Put $K = f^{-1}(0) \cap S_\varrho^{2n-1}$. From the fact that the level set $f^{-1}(0)$ intersects the sphere S_ϱ^{2n-1} transversely, it follows that K is smooth submanifold of the sphere S_ϱ^{2n-1} of codimension two. We denote by T a sufficiently small open tubular neighbourhood of the manifold K in the sphere S_ϱ^{2n-1}. We define the mapping $\Phi : S_\varrho^{2n-1} \setminus T \to S^1 \subset \mathbb{C}$ from the complement of the tubular neighbourhood of the manifold K to the circle by the formula $\Phi(x) = f(x)/|f(x)| = \exp(i \arg f(x))$. In [256] (§4) it is shown that the mapping Φ is a smooth fibration. Moreover the restriction of the mapping Φ to the boundary $\partial(S_\varrho^{2n-1} \setminus T) = \partial T$ has a natural structure of a trivial fibration $K \times S^1 \to S^1$.

The restriction of the function f to $f^{-1}(S_{\varepsilon_0}^1) \cap \bar{B}_\varrho$ defines a fibration over the circle $S_{\varepsilon_0}^1$ of radius ε_0, lying in the complex line $\mathbb{C}$, the fibre of which is the non-singular level manifold $V_{\varepsilon_0} = f^{-1}(\varepsilon_0) \cap \bar{B}_\varrho$ of the singularity f. As we explained above, the restriction of the function f to the boundary $f^{-1}(S_{\varepsilon_0}^1) \cap S_\varrho$ of the manifold $f^{-1}(S_{\varepsilon_0}^1) \cap \bar{B}_\varrho$ also has the structure of a trivial fibration. The classical monodromy and variation operators of the singularity are defined by way of the fibration

$$f^{-1}(S_{\varepsilon_0}^1) \cap \bar{B}_\varrho \to S_{\varepsilon_0}^1.$$

Lemma 2.2 (see [256] § 5). The two fibrations over the circles S^1 and $S_{\varepsilon_0}^1$ described above are equivalent (relative to the isomorphism of the circles given by multiplication by ε_0). In particular, the fibre $\Phi^{-1}(z)$ of the fibration Φ is diffeomorphic to the non-singular level set of the singularity f near the critical point.

In this way we can use the fibration Φ to define the variation operator Var_f of the singularity f. As before, we shall denote by

$$\Gamma_t : \Phi^{-1}(1) \to \Phi^{-1}(\exp(2\pi i t))$$

the family of diffeomorphisms which lifts the homotopy

$$t \mapsto \exp(2\pi i t) \quad (\Gamma_0 = id, t \in [0, 1])$$

and agree with the structure of the direct product on the boundary.

We digress a little to recall some definitions.

Let M be a (real) oriented n-dimensional manifold with boundary ∂M, let $e_1, \ldots, e_{n-1}$ be a frame in the tangent space to the boundary ∂M at some point, and let e_0 be the outward normal to the boundary ∂M in the manifold M at the same point. We say that the frame $e_1, \ldots, e_{n-1}$ defines the orientation of the boundary ∂M, if the frame $e_0, e_1, \ldots, e_{n-1}$ is a positively oriented frame in the tangent space of the manifold M. There is an analogous convention for chains and their boundaries.

Let a and b be non-intersecting $(n-1)$-dimensional cycles in the $(2n-1)$-dimensional sphere S^{2n-1}. When $n=1$ we shall suppose additionally that the cycles a and b are homotopic to zero. When $n>1$ this condition is satisfied automatically. We choose in the sphere S^{2n-1} an n-dimensional chain A, the boundary of which coincides with the cycle a. It is easy to see that the intersection number $(A \circ b)$ of the chains A and b in the sphere S^{2n-1} (which is well-defined, since the boundary of the chain A, which is equal to a, does not intersect the cycle b) does not depend on the choice of the chain A. Indeed if A' is another such chain then the difference $(A - A')$ will be an absolute n-dimensional cycle in the sphere S^{2n-1}, from which it follows that $((A - A') \circ b) = 0$, that is that $(A \circ b) = (A' \circ b)$. The intersection number $(A \circ b)$ of the chains A and b is called the *linking number* of the cycles a and b and denoted $l(a, b)$.

Another method of calculating the linking number goes as follows: Let D^{2n} be the ball, the boundary of which is the sphere S^{2n-1}. We choose two n-dimensional chains $\tilde{A}$ and $\tilde{B}$ in the ball D^{2n}, the boundaries of which coincide with the cycles a and b respectively and which lie wholly inside the ball D^{2n}, with the exception of their boundaries. In this case we can make sense of the intersection number $(\tilde{A} \circ \tilde{B})_D$ of the chains $\tilde{A}$ and $\tilde{B}$ in the ball D^{2n} and

$$l(a, b) = (A \circ b)_S$$
$$= (-1)^n (\tilde{A} \circ \tilde{B})_D$$
$$= (\tilde{B} \circ \tilde{A})_D$$
$$= (-1)^n l(b, a).$$

In order to prove this result we must remark that the intersection number $(\tilde{A} \circ \tilde{B})_D$ is well-defined, that is it does not depend on the concrete choice of the chains $\tilde{A}$ and $\tilde{B}$ for which $\partial \tilde{A} = a$ and $\partial \tilde{B} = b$. We can consider the ball D^{2n} as a cone over the sphere S^{2n-1}, that is as the quotient space obtained from the product $[0, 1] \times S^{2n-1}$ of the interval $[0, 1]$ and the sphere S^{2n-1} by factoring out the subspace $\{0\} \times S^{2n-1}$ (the slices $\{t\} \times S^{2n-1}$ $(0 \leqslant t \leqslant 1)$ corresponding in the ball to concentric spheres of radius t). Then as the chain $\tilde{B}$ we can take the cone over the cycle b with vertex at the centre of the ball D^{2n} $(\tilde{B} = [0, 1] \times b / \{0\} \times b)$, and as the chain $\tilde{A}$ we can take the union of the cylinder $[1/2, 1] \times a$ over the cycle a and the chain $\{1/2\} \times A$, lying in the sphere $\{1/2\} \times S^{2n-1}$ of radius $1/2$ $(A \subset S^{2n-1}, \partial A = a$; for $n = 1$ see figure 17). In this case the chains $\tilde{A}$ and $\tilde{B}$ will intersect at points of the form $(1/2, x)$, where x is an intersection point of the chain A with the cycle b. The sign, which differs the corresponding intersection numbers can be calculated without difficulty.

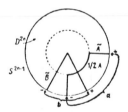

Fig. 17.

We return to our consideration of the singularity f. Let a and b be $(n-1)$-dimensional cycles in the fibre $\Phi^{-1}(1)$ of the fibration

$$\Phi : S_\varrho^{2n-1} \setminus T \to S^1.$$

The cycle $\Gamma_{1/2*} b$ lies in the fibre $\Phi^{-1}(-1)$ and therefore does not intersect the cycle a. Consequently, it makes sense to talk about the linking number of the cycles a and $\Gamma_{1/2*} b$ as cycles lying in the $(2n-1)$-dimensional sphere.

Definition. The *Seifert Form* of the singularity f is the bilinear form L on the homology group $H_{n-1}(\Phi^{-1}(1))$ ($\cong H_{n-1}(V_\varepsilon)$), defined by the formula

$$L(a,b) = l(a, \Gamma_{1/2*}b),$$

where $a, b \in H_{n-1}(\Phi^{-1}(1))$.

The theorem of Alexander duality asserts that the linking number defines a duality between the homology groups $H_{n-1}(\Phi^{-1}(1))$ and $H_{n-1}(S^{2n-1}\setminus\Phi^{-1}(1))$.

It is not difficult to see that the fibre $\Phi^{-1}(-1)$ is a deformation retract of the space $S^{2n-1}\setminus\Phi^{-1}(1)$. Consequently, the homology group $H_{n-1}(S^{2n-1}\setminus\Phi^{-1}(1))$ is isomorphic to the group $H_{n-1}(\Phi^{-1}(-1))$.

Since the transformation $\Gamma_{1/2*}$ is an isomorphism between the groups $H_{n-1}(\Phi^{-1}(1))$ and $H_{n-1}(\Phi^{-1}(-1))$, then the Seifert form defines a duality between the homology group $H_{n-1}(\Phi^{-1}(1))$ and itself, that is it is a nondegenerate integral bilinear form with determinant equal to (± 1). We remark that the Seifert form L, generally speaking, does not possess the property of symmetry.

Let $b \in H_{n-1}(\Phi^{-1}(1))$ be an absolute homology class and $a \in H_{n-1}(\Phi^{-1}(1), \partial\Phi^{-1}(1))$ be a relative homology class modulo the boundary.

Lemma 2.3. $L(\text{Var}_f a, b) = (a \circ b)$.

Proof. Let us choose a relative $(n-1)$-cycle in the pair $(\Phi^{-1}(1), \partial\Phi^{-1}(1))$ which is a representative of the homology class a (we shall denote it also by a). Let us consider the mapping $[0,1] \times a \to S^{2n-1}$ from the cylinder over the cycle a into the sphere, mapping $(t,c) \in [0,1] \times a$ to $\Gamma_t(c) \in S^{2n-1}$. Under this mapping the lower end $\{0\} \times a$ of the cylinder $[0,1] \times a$ maps to the chain a, the upper end $\{1\} \times a$ to the chain $\Gamma_1 a$, $[0,1] \times \partial a$ maps to the boundary ∂T of the tubular neighbourhood of the manifold K. Therefore this mapping defines an n-chain in the sphere S^{2n-1} (its image), the boundary of which consists of two parts: the variation $\text{Var}_f a = \Gamma_1 a - a$ of the cycle a (lying in the fibre $\Phi^{-1}(1)$) and a cycle lying on ∂T. Contracting the second part of its boundary inside the tubular neighbourhood T along radii, we obtain a chain A in the sphere S^{2n-1}, the boundary of which lies in the fibre $\Phi^{-1}(1) \subset S^{2n-1}$ and is equal to $\text{Var}_f(a)$. The intersection of the chain A with the cycle $\Gamma_{1/2*}b$ is the same as the intersection of the cycles $\Gamma_{1/2*}a$ and $\Gamma_{1/2*}b$

in the fibre $\Phi^{-1}(-1)$. Therefore

$$L(\mathrm{Var}_f\, a, b) = l(\mathrm{Var}_f\, a, \Gamma_{1/2*} b) =$$
$$= (A \circ \Gamma_{1/2*} b)_S = (\Gamma_{1/2*} a \circ \Gamma_{1/2*} b)_{\Phi^{-1}(-1)} = (a \circ b)_{\Phi^{-1}(1)},$$

which is what we were trying to prove.

Since the Seifert form L defines a duality between the homology group $H_{n-1}(\Phi^{-1}(1))$ and itself, and the intersection number defines a duality between the groups $H_{n-1}(\Phi^{-1}(1))$ and $H_{n-1}(\Phi^{-1}(1), \partial\Phi^{-1}(1))$, then we have

Theorem 2.2. The variation operator Var_f of the singularity f is an isomorphism of the homology groups

$$H_{n-1}(\Phi^{-1}(1), \partial\Phi^{-1}(1)) \xrightarrow{\sim} H_{n-1}(\Phi^{-1}(1))$$

or, which is the same thing, of the groups

$$H_{n-1}(V_\varepsilon, \partial V_\varepsilon) \xrightarrow{\sim} H_{n-1}(V_\varepsilon).$$

Remark. If we already had a proof of the Picard-Lefschetz theorem in the general case, then this result could be obtained by assigning the matrix of the operator Var_f in a distinguished basis of the homology group $H_{n-1}(V_\varepsilon)$ and the basis of the group $H_{n-1}(V_\varepsilon, \partial V_\varepsilon)$ dual to it (see §2.5).

From this theorem and Lemma 2.3 follows

Theorem 2.3. If $a, b \in H_{n-1}(V_\varepsilon)$, then

$$L(a, b) = (\mathrm{Var}_f^{-1} a \circ b).$$

Remark. The definition of the linking number and the Seifert form sometimes differs from that given here either in sign or by a permutation of the arguments (for example in [101]).

The Seifert form is very useful for studying the topological structure of singularities. In particular, it can be shown that the Seifert form (or the variation operator $(H_{n-1}(V_\varepsilon))^* \to H_{n-1}(V_\varepsilon)$) determines the intersection form on the homology group $H_{n-1}(V_\varepsilon)$ of the non-singular level manifold.

Theorem 2.4. For $a, b \in H_{n-1}(V_\varepsilon)$

$$(a \circ b) = -L(a, b) + (-1)^n L(b, a).$$

Proof. Since the variation operator of a singularity is an isomorphism, there exist relative cycles $a', b' \in H_{n-1}(V_\varepsilon, \partial V_\varepsilon)$ such that $a = \mathrm{Var}_f\, a'$ and $b = \mathrm{Var}_f\, b'$. It remains to apply the result of lemma 1.1 to the cycles a' and b'.

In addition to the intersection form the variation operator also determines the action of the classical monodromy operator of a singularity. The inverse of the variation operator, acts from the homology group $H_{n-1}(V_\varepsilon)$ of the non-singular level manifold to the group $H_{n-1}(V_\varepsilon, \partial V_\varepsilon)$ dual to it. To it corresponds the operator

$$(\mathrm{Var}_f^{-1})^T : H_{n-1}(V_\varepsilon) \to H_{n-1}(V_\varepsilon, \partial V_\varepsilon)$$

defined so that

$$((\mathrm{Var}_f^{-1})^T a \circ b) = L(b, a) = (\mathrm{Var}_f^{-1} b \circ a)$$

for $a, b \in H_{n-1}(V_\varepsilon)$. In matrix form it means that the matrix of the operator $(\mathrm{Var}_f^{-1})^T$ is obtained from the matrix of the operator Var_f^{-1} by transposition.

Theorem 2.5 ([199]). The classical monodromy operator h_* of a singularity can be expressed in terms of its variation operator Var_f by the formula $h_* = (-1)^n \mathrm{Var}_f (\mathrm{Var}_f^{-1})^T$.

Proof. We have the equality $(x \circ y) = (i_* x \circ y)$, where $x, y \in H_{n-1}(V_\varepsilon)$, i_* is the homomorphism $H_{n-1}(V_\varepsilon) \to H_{n-1}(V_\varepsilon, \partial V_\varepsilon)$, induced by the inclusion $V_\varepsilon \hookrightarrow (V_\varepsilon, \partial V_\varepsilon)$. Together with Theorem 2.4 it gives

$$i_* = -\mathrm{Var}_f^{-1} + (-1)^n (\mathrm{Var}_f^{-1})^T.$$

For the classical monodromy operator of the singularity we have

$$h_* = id + \mathrm{Var}_f\, i_*$$

$$= id - \mathrm{Var}_f\, \mathrm{Var}_f^{-1} + (-1)^n \mathrm{Var}_f (\mathrm{Var}_f^{-1})^T$$

$$= (-1)^n \mathrm{Var}_f (\mathrm{Var}_f^{-1})^T$$

which is what we had to prove.

There is an analogous result for the action of the classical monodromy in the relative homology group.

Theorem 2.6.

$$h^{(r)}_* = (-1)^n (\text{Var}^{-1})^T \text{Var}.$$

2.4   Proof of the Picard-Lefschetz theorem

We shall use here the notation of §1.3.

From the fact that the variation operator

$$\text{var}_{\tau'} : H_{n-1}(\tilde{F}_1, \partial \tilde{F}_1) \to H_{n-1}(\tilde{F}_1),$$

being the variation operator of the singularity

$$f(x_1, \ldots, x_n) = x_1^2 + \ldots + x_n^2,$$

is an isomorphism (Theorem 2.2), it follows that $\text{var}_{\tau'}(V) = \pm \Delta$. To determine the sign in this formula we use Theorem 2.3. In the definition of the fibration Φ (for the critical point 0 of the function $x_1^2 + \ldots + x_n^2$) we can suppose that $\varrho = 1$. The fibration $\Phi : S^{2n-1} \setminus T \to S^1$ is given by the formula

$$\Phi(x_1, \ldots, x_n) = (x_1^2 + \ldots + x_n^2)/|x_1^2 + \ldots + x_n^2|$$

$(|x_1|^2 + \ldots + |x_n|^2 = 1)$. The fibre $\Phi^{-1}(1)$ of this fibration is diffeomorphic to the level manifold $\tilde{F}_1$. The vanishing cycle Δ in the manifold $\tilde{F}_1$ corresponds in the fibre $\Phi^{-1}(1)$ to the cycle defined by the equations

$$x_1^2 + \ldots + x_n^2 = 1, \quad \text{Im } x_j = 0.$$

We shall denote this cycle by Δ also.

We have

$$(\text{Var}^{-1} \Delta \circ \Delta) = L(\Delta, \Delta)$$

$$= l(\Delta, \Gamma_{1/2} * \Delta)$$

$$= (-1)^n (\tilde{A} \circ \tilde{B})_D,$$

where $\tilde{A}$ and $\tilde{B}$ are n-dimensional chains in the ball $D = D^{2n}$, the boundaries of which lie on the sphere S^{2n-1} and are equal to Δ and $\Gamma_{1/2} * \Delta$ respectively. It is not difficult to see that in order to calculate the linking number $l(\Delta, \Gamma_{1/2} * \Delta)$ it is possible to use the family of diffeomorphisms

$$\Gamma_t : \Phi^{-1}(1) \to \Phi^{-1}(\exp(2\pi it)),$$

which do not necessarily agree with the structure of the direct product on the boundary. We can take for this the family defined by the formula

$$\Gamma_t(x_1, \ldots, x_n) = (\exp(\pi it)x_1, \ldots, \exp(\pi it)x_n).$$

Then the cycle $\Gamma_{1/2 *} \varDelta$ will be determined by the equations

$$x_1^2 + \ldots + x_n^2 = -1 \quad \text{and} \quad \operatorname{Re} x_j = 0.$$

We can take as the chains $\tilde{A}$ and $\tilde{B}$ the chains in the ball D^{2n}, given by the equations $\{\operatorname{Im} x_j = 0\}$ and $\{\operatorname{Re} x_j = 0\}$ respectively. The orientations of the chains $\tilde{A}$ and $\tilde{B}$ are in agreement with the help of a mapping from $\tilde{A}$ to $\tilde{B}$, which is multiplication by i. If a positively oriented system of coordinates on the disc $\tilde{A}$ is the set $u_1, \ldots, u_n$ $(x_j = u_j + iv_j)$ then a positively oriented system of coordinates on the disc $\tilde{B}$ will be $v_1, \ldots, v_n$. The chains $\tilde{A}$ and $\tilde{B}$ are smooth manifolds (n-dimensional discs) and intersect transversely at the point 0. From this it follows that

$$(\tilde{A} \circ \tilde{B})_D = (-1)^{n(n-1)/2}.$$

Therefore

$$(\operatorname{Var}^{-1} \varDelta \circ \varDelta) = (-1)^{n(n+1)/2},$$

that is

$$\operatorname{Var}^{-1} \varDelta = (-1)^{n(n+1)/2} \, V,$$

$$\operatorname{Var} V = (-1)^{n(n+1)/2} \, \varDelta,$$

which is what we had to prove.

2.5   The intersection matrix of a singularity

As we have already said, the monodromy group of a singularity is generated by the Picard-Lefschetz operators h_i, corresponding to the elements $\varDelta_i$ of a weakly distinguished basis in the homology of the non-singular level manifold of the singularity f near a critical point. By the Picard-Lefschetz theorem we have

$$h_i(a) = a + (-1)^{n(n+1)/2}(a \circ \Delta_i)\Delta_i.$$

Thus the matrix of pairwise intersections of elements of a weakly distinguished basis determines the monodromy group of the singularity.

Definition. The matrix $S = (\Delta_i \circ \Delta_j)$ is called the *intersection matrix* of the singularity f (with respect to the basis $\{\Delta_i\}$).

Remark. Here we use i for the number of the column, and j for the number of the row. This way of writing down the matrix of the bilinear form coincides with the way of writing it down as the matrix of an operator (in this case i_*) from the homology space $H_{n-1}(V_\varepsilon)$ to its dual space $H_{n-1}(V_\varepsilon, \partial V_\varepsilon)$ with bases $\{\Delta_i\}$ and its dual $((\Delta_i \circ \Delta_j) = (i_* \Delta_i \circ \Delta_j))$.

Definition. The *bilinear form associated with the singularity f* is an integral bilinear form defined on the homology group $H_{n-1}(V_\varepsilon)$ of the non-singular level manifold of the singularity f by the intersection number.

The bilinear form associated with the singularity is symmetric for an odd number of variables n and antisymmetric for an even number of variables. The intersection matrix of the singularity is the matrix of the form with respect to the basis $\{\Delta_j\}$. The diagonal elements of the intersection matrix are determined in Lemma 2.4 of §2.3 and are equal to 0 for even n and ± 2 for odd n.

If $\tilde{f}$ is a perturbation of the function f, and $\{\Delta_i\}$ is a distinguished basis of vanishing cycles, defined by a system of paths $u_1, \ldots, u_\mu$, then the loop τ', which goes in a positive direction round all the critical values into which the zero critical

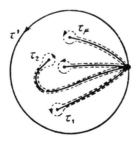

Fig. 18.

value of the function f decomposes, is homotopic to the product $\tau_\mu \ldots \tau_1$ of simple loops, corresponding to the paths $u_\mu, \ldots, u_1$ (figure 18).

From this follows

Lemma 2.4. The classical monodromy operator h_* of the singularity f is equal to the product $h_1 \ldots h_\mu$ of Picard-Lefschetz operators, corresponding to the elements $\{\Delta_i\}$ of a distinguished basis in the homology of the non-singular level manifold.

The action of the variation operator of the singularity f can be defined by the formulae

$$\mathrm{Var}_f = \mathrm{var}_{\tau_\mu \cdot \ldots \cdot \tau_1}$$

$$= \sum_{r=1}^{\mu} \; \sum_{i_1 < i_2 < \ldots < i_r} \mathrm{var}_{\tau_{i_1}} \cdot i_* \cdot \mathrm{var}_{\tau_{i_2}} \cdot i_* \cdot \ldots \cdot i_* \cdot \mathrm{var}_{\tau_{i_r}}, \qquad (*)$$

$$\mathrm{var}_{\tau_i}(a) = (-1)^{\frac{n(n+1)}{2}} (a \circ \Delta_i) \Delta_i \quad (a \in H_{n-1}(V_\varepsilon, \partial V_\varepsilon)).$$

Choose in the group $H_{n-1}(V_\varepsilon, \partial V_\varepsilon)$, which is the dual of the group $H_{n-1}(V_\varepsilon)$, the basis $\{V_i\}$, dual to the basis $\{\Delta_i\}$, that is such that

$$(V_i \circ \Delta_j) = \delta_{ij}.$$

From the formula $(*)$ it follows that

$$\mathrm{Var}_f(V_i) = (-1)^{n(n+1)/2} \Delta_i + \sum_{j<i} c_i^j \Delta_j,$$

where c_i^j are certain integers. So we have proved

Lemma 2.5. With respect to a distinguished basis the matrix of the variation operator Var_f of a singularity f is an upper triangular matrix with diagonal entries equal to $(-1)^{n(n+1)/2}$.

The same properties are possessed by the matrix of the operator Var_f^{-1}, which by theorem 2.3 coincides with the matrix of the Seifert form L of the singularity f (see the remark at the beginning of the section, defining the matrix entries of a bilinear form).

Let S be the intersection matrix of the singularity f with respect to any basis, L be the matrix of the Seifert form (or of the operator Var_f^{-1}) of this singularity with respect to the same basis, H be the matrix of the classical monodromy operator h_*, $H^{(r)}$ be the matrix of the operator $h_*^{(r)}$ with respect to the dual basis. Theorems 2.4, 2.5 and 2.6 show that

$$S = -L + (-1)^n L^T,$$

$$H = (-1)^n L^{-1} L^T,$$

$$H^{(r)} = (-1)^n L^T L^{-1}$$

(the symbol ^{T} means the transpose of the matrix). If $\{\varDelta_i\}$ is a distinguished basis of vanishing cycles, then the matrix L with respect to it is upper triangular and the matrix L^T is lower triangular. Thus the intersection matrix with respect to a distinguished basis has an invariant decomposition into the sum of an upper triangular and a lower triangular matrix.

It was stated above that the intersection matrix of a singularity with respect to a distinguished basis determines its classical monodromy operator (with respect to the same basis). The converse is also true. Before proving this we formulate one useful general result.

Lemma 2.6. Let A and B be upper triangular matrices with ones on the diagonal, and let $C = AB^T$. Then the matrices A and B can be reconstructed from the matrix C.

The following formulation of this result is equivalent to the previous one.

Lemma 2.7. Let A and B be upper triangular matrices with ones on the diagonal. If AB^T is the identity matrix then A and B are also identity matrices.

The proof of this lemma does not present any difficulty.

Theorem 2.7 ([205]). The matrix of the classical monodromy operator of a singularity with respect to a distinguished basis determines its variation operator and its intersection matrix.

The proof applies Lemma 2.6 to the identity

$$H = (-1)^n \tilde{L}^{-1} \tilde{L}^T,$$

where

$$\tilde{L} = (-1)^{n(n+1)/2} L,$$

in which $\tilde{L}$ and $\tilde{L}^{-1}$ are upper triangular matrices with ones on the diagonal.

2.6   Change of basis

The system of paths $\{u_i\}$, defining a distinguished or weakly distinguished basis, can be chosen in more than one way. If we change the initial system of paths, we can get different bases of vanishing cycles in the homology group $H_{n-1}(V_\varepsilon)$ of the non-singular level set of the singularity near the critical point. We describe several elementary operations of change of basis, preserving its distinguished or weakly distinguished character. Let $\{u_i\}$ be a system of paths, defining the distinguished basis $\{\varDelta_i\}$ in the homology group

$$H_{n-1}(F_{z_0}) \cong H_{n-1}(V_\varepsilon)$$

of the non-singular level manifold. This means that the u_i are non-self-intersecting paths, joining the critical values z_i of the perturbation $\tilde{f}$ of the function f with the non-critical value z_0 and intersecting each other only at the point z_0. Let τ_i be a simple loop corresponding to the path u_i.

Definition of the operation α_m $(1 \leqslant m < \mu)$. We define a new system of paths $\{\tilde{u}_i\}$ in the following manner:

$$\tilde{u}_i = u_i \quad \text{for} \quad i \neq m, m+1;$$

$$\tilde{u}_{m+1} = u_m;$$

$$\tilde{u}_m = u_{m+1}\tau_m.$$

Here by $u_{m+1}\tau_m$ we understand the path obtained by traversing the path u_{m+1} followed by the loop τ_m. It is clear (see below) that the system of paths $\{\tilde{u}_i\}$ defines a weakly distinguished set of vanishing cycles $\{\tilde{\varDelta}_i\}$. It is not difficult to see that the system of paths $\{\tilde{u}_i\}$ can be deformed a little so that it satisfies the conditions of the definition of a distinguished basis (figure 19). Therefore the basis $\{\tilde{\varDelta}_i\}$ is distinguished. The basis $\{\tilde{\varDelta}_i\}$ is related to the basis $\{\varDelta_i\}$ by the following formulae:

$$\tilde{\varDelta}_i = \varDelta_i \quad \text{for} \quad i \neq m, m+1;$$

$$\tilde{\varDelta}_{m+1} = \varDelta_m;$$

$$\tilde{\varDelta}_m = h_m(\varDelta_{m+1}) = \varDelta_{m+1} + (-1)^{n(n+1)/2} \cdot (\varDelta_{m+1} \circ \varDelta_m)\varDelta_m$$

(the Picard-Lefschetz transformation). The operation of transferring from the distinguished basis $\{\Delta_i\}$ to the distinguished basis $\{\tilde{\Delta}_i\}$, described by these formulae, is denoted by α_m.

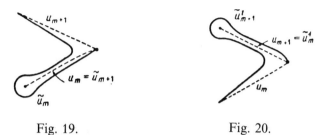

Fig. 19.   Fig. 20.

Definition of the operation β_{m+1} $(1 \leqslant m < \mu)$. Let the system of paths $\{\tilde{u}'_i\}$ be defined in the following manner:

$$\tilde{u}'_i = u_i \quad \text{for} \quad i \neq m, m+1;$$

$$\tilde{u}'_m = u_{m+1};$$

$$\tilde{u}'_{m+1} = u_m \tau_{m+1}^{-1}$$

(figure 20). This system of paths defines a distinguished basis $\{\tilde{\Delta}'_i\}$, related to the basis $\{\Delta_i\}$ by the fomulae:

$$\tilde{\Delta}'_i = \Delta_i \quad \text{for} \quad i \neq m, m+1;$$

$$\tilde{\Delta}'_m = \Delta_{m+1};$$

$$\tilde{\Delta}'_{m+1} = h_{m+1}^{-1}(\Delta_m) = \Delta_m + (-1)^{n(n+1)/2} \cdot (\Delta_{m+1} \circ \Delta_m) \Delta_{m+1}$$

(the inverse Picard-Lefschetz transformation). The operation of transferring from the distinguished basis $\{\Delta_i\}$ to the distinguished basis $\{\tilde{\Delta}'_i\}$, described by these formulae, is denoted by β_{m+1}.

It is not difficult to see that the operation β_{m+1} is the inverse of the operation α_m in the sense that the successive application of these in either order brings one back to the initial basis. We consider the free group generated by the elements α_m $(m = 1, \ldots, \mu - 1)$. To each element of this group (a word in the symbols α_m and α_m^{-1}) there corresponds an operation of change of distinguished basis (taking into consideration the fact that the action of α_m^{-1} on the basis coincides with the action of the operation β_{m+1}). It is clear that the actions of the operations $\alpha_m \alpha_{m'}$ and

$\alpha_{m'}\alpha_m$ are the same when $|m-m'| \geqslant 2$. In addition, the actions of the operations $\alpha_m\alpha_{m+1}\alpha_m$ and $\alpha_{m+1}\alpha_m\alpha_{m+1}$ are the same for any m from 1 to $(\mu-2)$. The "proof" of this fact is given in figure 21.

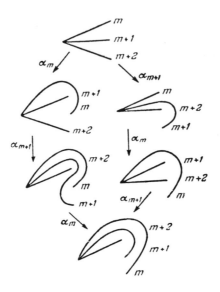

Fig. 21.

In this way we get an action, on the set of distinguished bases of the homology group of the non-singular level set near a critical point, of the quotient group of the free group on the $(\mu-1)$ generators α_m $(m=1,\ldots,\mu-1)$ by the relations

$$\alpha_{m+1}\alpha_m\alpha_{m+1} = \alpha_m\alpha_{m+1}\alpha_m \quad \text{for} \quad 1 \leqslant m < \mu-1,$$

$$\alpha_m\alpha_{m'} = \alpha_{m'}\alpha_m \quad \text{for} \quad |m-m'| \geqslant 2.$$

This group is the braid group with μ strands (see, for example [57]; see also Section 3.3).

We consider an operation which preserves the property of being weakly distinguished for a set of vanishing cycles. As a preliminary we show that any such set forms a basis in the homology of the non-singular level manifold.

Theorem 2.8. Any weakly distinguished set of vanishing cycles forms a basis of the homology group $H_{n-1}(V_\varepsilon)$ of the non-singular level manifold.

Let $\{\Delta_i\}$ be a weakly distinguished set of vanishing cycles defined by the system of paths $\{u_i\}$, and let τ_i be the corresponding simple loops. The set of loops $\{\tau_i\}$ is a system of free generators of the fundamental group $\pi_1(U\setminus\{z_i\};z_0)$ of the complement of the set of critical values. In order to prove that the set of vanishing cycles $\{\Delta_i\}$ forms a basis of the group $H_{n-1}(V_\varepsilon)$, it is sufficient to prove that any vanishing cycle Δ (defined with the help of the path v, joining a critical value z_j with the non-critical value z_0) is linearly dependent on the cycles $\Delta_1,\ldots,\Delta_\mu$ with integer coefficients. We can suppose that the paths u_j and v coincide near the critical value z_j. In this case the loop $\gamma=u_j v^{-1}$ can be considered as an element of the fundamental group $\pi_1(U\setminus\{z_i\};z_0)$ of the complement of the set of critical values. We have $\Delta=\pm h_{\gamma^*}\Delta_j$ (the sign depending on the orientation of the vanishing cycles Δ and Δ_j). In the group $\pi_1(U\setminus\{z_i\};z_0)$ the loop γ can be expressed in terms of the generators $\tau_1,\ldots,\tau_\mu$. Consequently the vanishing cycle Δ can be obtained from the cycle Δ_j by the successive application of some Picard-Lefschetz operators h_i and their inverses, and is therefore linearly dependent on the cycles $\Delta_1,\ldots,\Delta_\mu$ with integer coefficients.

Definition of the operations $\alpha_m(m')$ and $\beta_m(m')$ of change of weakly distinguished basis. Let $\{u_i\}$ be a system of paths, defining the weakly distinguished basis $\{\Delta_i\}$ of the homology group $H_{n-1}(V_\varepsilon)$ of the non-singular level manifold. For $m\neq m'$ we define the operation of change of basis $\alpha_m(m')$ [$\beta_m(m')$], corresponding to the change of the path $u_{m'}$ to the path $u_{m'}\tau_m$ [$u_{m'}\tau_m^{-1}$], that is transforming the weakly distinguished basis $\{\Delta_i\}$ into the basis $\{\tilde{\Delta}_i\}$ defined by the formulae

$$\tilde{\Delta}_i=\Delta_i \quad \text{for} \quad i\neq m',$$

$$\tilde{\Delta}_{m'}=h_m(\Delta_{m'})=\Delta_{m'}+(-1)^{n(n+1)/2}(\Delta_{m'}\circ\Delta_m)\Delta_m$$

$$[\tilde{\Delta}_{m'}=h_m^{-1}(\Delta_{m'})=\Delta_{m'}+(-1)^{n(n+1)/2}(\Delta_m\circ\Delta_{m'})\Delta_m].$$

The action of the operations $\alpha_m(m')$ and $\beta_m(m')$ on the system of simple loops $\{\tau_i\}$ consists of changing the loop $\tau_{m'}$ into a conjugate of it in the fundamental group $\pi_1(U\setminus\{z_i\};z_0)$ of the complement of the set of critical values (that is $\tau_m^{-1}\tau_{m'}\tau_m$ for $\alpha_m(m')$ and $\tau_m\tau_{m'}\tau_m^{-1}$ for $\beta_m(m')$). For this reason, if the initial system of simple loops is a system of free generators of the group $\pi_1(U\setminus\{z_i\};z_0)$, then the same property will be possessed also by the system of simple loops, obtained after application of the operation $\alpha_m(m')$ or $\beta_m(m')$. Therefore the operations $\alpha_m(m')$ and $\beta_m(m')$ preserve the property of a basis being weakly distinguished.

It is easy to see that the operations $\alpha_m(m')$ and $\beta_m(m')$ are inverses of each other. When the number of variables n is odd, these operations coincide. If a distinguished basis is considered as weakly distinguished, and in particular we forget the order of the vanishing cycles, then the action of the operation α_m coincides with the action of the operation $\alpha_m(m+1)$, and β_{m+1} with $\beta_{m+1}(m)$.

It can be shown (see [150]) that any two distinguished bases can be obtained one from the other by iterations of operations α_m and β_m and a change of orientations of some elements. It was proved that any two weakly distinguished bases can be obtained one from the other in an analogous way with the help of operations $\alpha_m(m')$ and $\beta_m(m')$ (Humphries, S. P. [170] and, apparently, already Whitehead J. H. C., 1936).

2.7 The Variation Operator and the Intersection Matrix of a "Direct Sum" of Singularities

Definition. The *direct sum* of the singularities $f: (\mathbb{C}^n, 0) \to (\mathbb{C}, 0)$ and $g: (\mathbb{C}^m, 0) \to (\mathbb{C}, 0)$ of functions of n and m variables respectively is the singularity of the function $f \oplus g: (\mathbb{C}^{n+m}, 0) \to (\mathbb{C}, 0)$ of $(n+m)$ variables defined by the formula

$$f \oplus g(x, y) = f(x) + g(y)$$

$(x \in \mathbb{C}^n, y \in \mathbb{C}^m, (x, y) \in \mathbb{C}^{n+m} \approx \mathbb{C}^n \oplus \mathbb{C}^m)$.

Lemma 2.8. The multiplicity $\mu(f \oplus g)$ of the direct sum of the singularities f and g is equal to the product $\mu(f) \mu(g)$ of their multiplicities.

Indeed if $\tilde{f}(x)$ is a perturbation of the singularity f, with $\mu(f)$ non-degenerate critical points p_i, and $\tilde{g}(y)$ is a perturbation of the singularity g, with $\mu(g)$ non-degenerate critical points q_j, then $\tilde{f}(x) + \tilde{g}(y)$ is a perturbation of the singularity $f \oplus g$ with $\mu(f) \mu(g)$ non-degenerate critical points (p_i, q_j) $(i = 1, \ldots, \mu(f);$ $j = 1, \ldots, \mu(g))$.

M. Sebastiani and R. Thom ([322]) proved that the classical monodromy operator of the singularity $f \oplus g$ is equal to the tensor product of the classical monodromy operators of the singularities f and g. A. M. Gabrielov ([116]) obtained a description of the intersection matrix of the singularity $f \oplus g$ under the condition that the intersection matrices of the singularities f and g, with respect to distinguished bases, are known. We give an account of these results in a form somewhat different from that found in [322] and [116].

We need one topological concept.

Definition. The *join* $X * Y$ of the topological spaces X and Y is the quotient space of the direct product $X \times I \times Y$ ($I = [0, 1]$) by the equivalence relation:

$$(x, 0, y_1) \sim (x, 0, y_2) \quad \text{for any} \quad y_1, y_2 \in Y, \ x \in X;$$

$$(x_1, 1, y) \sim (x_2, 1, y) \quad \text{for any} \quad x_1, x_2 \in X, \ y \in Y.$$

We can consider that the spaces X and Y lie in their join $X * Y$ as the lower and upper bases respectively ($\{(x, 0, y)\}$ and $\{(x, 1, y)\}$). Therefore the join $X * Y$ can be represented as the space swept out by non-intersecting segments joining every point of the space X to every point of the space Y. If we consider the projection $(x, t, y) \mapsto t$ of the join $X * Y$ to the line segment $I = [0, 1]$, then the preimage of the point 0 coincides with the space X, the preimage of the point 1 coincides with the space Y and that of a point $t \in (0, 1)$ with the product $X \times Y$.

If Y is the space consisting of one point, then the join $X * Y$ coincides with the cone over the space X. If Y is the space consisting of two points, then the join $X * Y$ is homeomorphic to the suspension of the space X (the quotient space of the cylinder $[-1, 1] \times X$ over the space X by the equivalence relations

$$(-1, x_1) \sim (-1, x_2), \ (1, x_1) \sim (1, x_2)$$

for all $x_1, x_2 \in X$). If the space X is homeomorphic to the k-dimensional sphere S^k, and Y to the l-dimensional sphere S^l, then the join $X * Y$ is homeomorphic to the $(k + l + 1)$-dimensional sphere S^{k+l+1}.

Lemma 2.9. Let the homology groups of the spaces X and Y either not have torsion or be considered with coefficients in a field. Then the homology group $H_n(X * Y)$ of the join of the spaces X and Y is isomorphic to

$$\oplus_{0 \leqslant k \leqslant n-1} H_k(X) \otimes H_{n-k-1}(Y).$$

In other words $H_*(X * Y) = H_*(X) \otimes H_*(Y)$ if we consider that $\dim(a \otimes b) = \dim a + \dim b + 1$ for $a \in H_*(X), b \in H_*(Y)$. If α is a cycle in the space X and β is a cycle in the space Y, then the cycle corresponding to $\alpha \otimes \beta$ in the space $X * Y$ is the join of the cycles α and β. Here it is essential that the homology groups are supposed reduced modulo a point.

The embedding

$$H_k(X) \otimes H_{n-k-1}(Y) \hookrightarrow H_n(X*Y),$$

generally speaking, is defined only up to multiplication by (± 1). It's choice is determined by a method of orientating the join of cycles. We can, for example, suppose that the orientation of the join $a * b$ is induced from the usual orientation of the direct product $a \times I \times b$. We note, however, that the results formulated below do not depend on this choice.

Let f be a singularity $(\mathbb{C}^n, 0) \to (\mathbb{C}, 0)$, let V_ε be a non-singular level manifold of the singularity f near the critical point $(V_\varepsilon = f^{-1}(\varepsilon) \cap \bar{B}_\varrho)$, and let u be a path joining the non-critical value ε with the critical value 0.

Lemma 2.10. There exists a continuous family of mappings
$$H_t: V_\varepsilon \to V_{u(t)} = f^{-1}(u(t)) \cap \bar{B}_\varrho (t \in [0, 1]),$$

such that
1) $H_0 = id: V_\varepsilon \to V_\varepsilon$;
2) H_t is the inclusion $V_\varepsilon \to V_{u(t)}$ for $0 \leqslant t < 1$;
3) H_1 maps V_ε into the point $0 \in \mathbb{C}^n$.

The proof of this result can be constructed in an analogous way to the way that, in Theorem 2.1, it was shown that the space $f^{-1}(0) \cap \bar{B}_\varrho$ is a deformation retract of the space $f^{-1}(\bar{D}_{\varepsilon_0}) \cap \bar{B}_\varrho$.

The family of mappings H_t is determined uniquely up to isotopy. It gives an embedding of the cone over the non-singular level manifold V_ε into the space $\mathbb{C}^n$ $((x, t) \to H_t(x) \text{ for } 0 \leqslant t \leqslant 1)$.

Now let f and g be two singularities in n and m variables respectively, let
$$V_\varepsilon(f) = f^{-1}(\varepsilon) \cap \bar{B}_{\varrho_1}$$

and
$$V_\varepsilon(g) = g^{-1}(\varepsilon) \cap \bar{B}_{\varrho_2}$$

be the non-singular level manifolds of the singularities f and g respectively and let u be a non-self-intersecting path in the target plane of the function f, joining ε with zero (without loss of generality, we can suppose that $u(t) = (1 - t)\varepsilon$). We define a path v, joining ε with zero in the target plane of the function g, by the formula $v(t) = \varepsilon - u(1 - t)$. Let $H_t(f): V_\varepsilon(f) \to V_{u(t)}(f)$ and $H_t(g): V_\varepsilon(g) \to V_{v(t)}(g)$ be the families of functions described in Lemma 2.10. We define the inclusion j of the join $V_\varepsilon(f) * V_\varepsilon(g)$ of the non-singular level manifolds $V_\varepsilon(f)$ and $V_\varepsilon(g)$ into the level set
$$(f \oplus g)^{-1}(\varepsilon) \subset \mathbb{C}^{n+m}$$

by the formula

$$j(x, t, y) = (H_t(f)x, H_{1-t}(g)\, y)$$

for $x \in V_\varepsilon(f)$, $y \in V_\varepsilon(g)$, $t \in [0, 1]$. If we impose natural limits on the radii ϱ_1, ϱ_2 and ϱ (for example,

$$\varrho_1 \leqslant \varrho/\sqrt{2},\ \varrho_2 \leqslant \varrho/\sqrt{2})$$

then j is an embedding of the join $V_\varepsilon(f) * V_\varepsilon(g)$ into the level manifold

$$V_\varepsilon(f \oplus g) = (f \oplus g)^{-1}(\varepsilon) \cap \bar{B}_\varrho$$

of the singularity of $f \oplus g$ near the critical point.

The mapping

$$j\colon V_\varepsilon(f) * V_\varepsilon(g) \to V_\varepsilon(f \oplus g)$$

together with the isomorphism

$$H_{n+m-1}(V_\varepsilon(f) * V_\varepsilon(g)) \cong H_{n-1}(V_\varepsilon(f)) \otimes H_{m-1}(V_\varepsilon(g))$$

defines the homomorphism

$$j_*\colon H_{n-1}(V_\varepsilon(f)) \otimes H_{m-1}(V_\varepsilon(g)) \to H_{n+m-1}(V_\varepsilon(f \oplus g)).$$

In the work [322] was proved

Theorem 2.9. The homomorphism j_* is an isomorphism and the inclusion

$$j\colon V_\varepsilon(f) * V_\varepsilon(g) \to V_\varepsilon(f \oplus g)$$

is a homotopy equivalence.

The fact that the non-singular level manifold $V_\varepsilon(f \oplus g)$ of the singularity $f \oplus g$ is homotopically equivalent to the join $V_\varepsilon(f) * V_\varepsilon(g)$ can be explained in the following manner. We consider the function f on the manifold $V_\varepsilon(f \oplus g)$ (more precisely we consider the function $f \circ \pi_1$, where

$$\pi_1\colon V_\varepsilon(f \oplus g) \subset \mathbb{C}^{n+m} \approx \mathbb{C}^n \oplus \mathbb{C}^m \to \mathbb{C}^n$$

is the projection on the first factor). The preimage $(f \circ \pi_1)^{-1}(z)$ of the point $z \in \mathbb{C}$ consists of points $(x, y) \in \mathbb{C}^n \oplus \mathbb{C}^m$, for which $f(x) = z$, $g(y) = \varepsilon - z$. Therefore (if we ignore the details connected with the radii of the balls in which we consider the non-singular level manifolds of the functions)

$$(f \circ \pi_1)^{-1}(z) = f^{-1}(z) \times g^{-1}(\varepsilon - z).$$

The mapping

$$(f \circ \pi_1): V_\varepsilon(f \oplus g) \to \mathbb{C}$$

is non-degenerate outside the preimages of the points 0 and ε. We consider in the plane $\mathbb{C}$ the segment $J = u([0, 1])$ which is the image of the path u. It joins the points 0 and ε ($u(0) = \varepsilon$, $u(1) = 0$). Over the complement of the segment J the mapping $f \circ \pi_1$ defines a locally trivial fibration. The segment J is a deformation retract of the plane $\mathbb{C}$. From this it follows that the space $(f \circ \pi_1)^{-1}(J)$ is a deformation retract of the space $V_\varepsilon(f \oplus g)$ and is therefore homotopy equivalent to it. The space $(f \circ \pi_1)^{-1}(u(t))$, for $t \in (0, 1)$, is diffeomorphic to the product $V_{\varepsilon_1}(f) \times V_{\varepsilon_2}(g)$ of the non-singular level manifolds of the singularities f and g. The space $(f \circ \pi_1)^{-1}(u(0))$ is diffeomorphic to $f^{-1}(\varepsilon) \times g^{-1}(0)$. The space $g^{-1}(0)$ is contractible to a point. Therefore $(f \circ \pi_1)^{-1}(u(0))$ is contractible to a space diffeomorphic to the non-singular level manifold $V_\varepsilon(f)$. Similarly the space

$$(f \circ \pi_1)^{-1}(u(1)) = f^{-1}(0) \times g^{-1}(\varepsilon)$$

is contractible to a space diffeomorphic to the non-singular level manifold $V_\varepsilon(g)$. This description of the fibres of the mapping

$$(f \circ \pi_1): (f \circ \pi_1)^{-1}(J) \to J$$

over the points of the segment J coincides with the description of the preimage of the points $t \in I = [0, 1]$ under the projection $V_\varepsilon(f) * V_\varepsilon(g) \to I$ (see the definition of the join). Therefore the space $(f \circ \pi_1)^{-1}(J)$ is homotopy equivalent to the join $V_\varepsilon(f) * V_\varepsilon(g)$. A little more accurate reasoning allows one to turn this explanation into a proof.

From now on we shall identify the homology group $H_{n+m-1}(V_\varepsilon(f \oplus g))$ of the non-singular level manifold of the singularity $f \oplus g$ with the tensor product of the groups $H_{n-1}(V_\varepsilon(f))$ and $H_{m-1}(V_\varepsilon(g))$. This identification also determines an identification of the relative homology group

$$H_{n+m-1}(V_\varepsilon(f \oplus g), \partial V_\varepsilon(f \oplus g))$$

(which is the dual of the group $H_{n+m-1}(V_\varepsilon(f \oplus g)))$ with the tensor product of the groups $H_{n-1}(V_\varepsilon(f), \partial V_\varepsilon(f))$ and $H_{m-1}(V_\varepsilon(g), \partial V_\varepsilon(g))$.

Theorem 2.10 (P. Deligne, see [94]).

$$\mathrm{Var}_{f \oplus g} = (-1)^{nm} \, \mathrm{Var}_f \otimes \mathrm{Var}_g.$$

For the proof it is sufficient to show that for any homology classes

$$a_1, a_2 \in H_{n-1}(V_\varepsilon(f)), b_1, b_2 \in H_{m-1}(V_\varepsilon(g))$$

we have the equality

$$([\mathrm{Var}_{f \oplus g}^{-1}(a_1 \otimes b_1)] \circ [a_2 \otimes b_2]) = (-1)^{nm}(\mathrm{Var}_f^{-1}a_1 \circ a_2) \cdot (\mathrm{Var}_g^{-1}b_1 \circ b_2).$$

We shall not carry out this proof in detail, but only outline its main steps (though it would not be difficult to reconstruct the whole proof).

Let

$$H_t(f): V_\varepsilon(f) \to V_{(1-t)\varepsilon}(f) = f^{-1}((1-t)\varepsilon) \cap \bar{B}_\varrho$$

be the family of mappings described in Lemma 2.10 (for the sake of definiteness we suppose that $u(t) = (1-t)\varepsilon$. As we have already said, the family $H_t(f)$ defines an embedding of the cone over the level manifold $V_\varepsilon(f)$ into the space $\mathbb{C}^n$. Let A_1 be the cone over the cycle a_1, determined by the family $H_t(f)$. Then A_1 is an n-dimensional chain, the boundary of which lies in the non-singular level manifold $V_\varepsilon(f)$ and coincides with the cycle a_1. Let

$$\Gamma_t(f): V_\varepsilon(f) \to V_{\exp(2\pi it)\varepsilon}(f)$$

be the family of mappings obtained by lifting the homotopy

$$\varepsilon \mapsto \exp(2\pi it)\varepsilon \quad (0 \leqslant t \leqslant 1),$$

and let

$$\tilde{a}_2 = \Gamma_{1/2}(f)(a_2)$$

be the $(n-1)$-dimensional cycle in the level manifold $V_{-\varepsilon}(f)$. Let $\tilde{A}_2$ be the cone over the cycle $\tilde{a}_2$, by an analogous construction. From the considerations of §2.3 it follows that

$$(\mathrm{Var}_f^{-1} a_1 \circ a_2) = (-1)^n (A_1 \circ \tilde{A}_2),$$

the chains A_1 and $\tilde{A}_2$ intersecting only at zero. The chains B_1 and $\tilde{B}_2$ are defined in the same way, with

$$(\mathrm{Var}_g^{-1} b_1 \circ b_2) = (-1)^m (B_1 \circ \tilde{B}_2).$$

In order to define

$$([\mathrm{Var}_{f \oplus g}^{-1}(a_1 \otimes b_1)] \circ [a_2 \otimes b_2])$$

by the same method it is necessary to construct cones C_1 and $\tilde{C}_2$ over the cycles $a_1 \otimes b_1$ and $\widetilde{a_2 \otimes b_2} = \Gamma_{1/2}(f \oplus g)(a_2 \otimes b_2)$. It is not difficult to see that we can take

$$(A_1 \times B_1) \cap \{(x, y) : (f(x) + g(y))/\varepsilon \leqslant 1\}$$

as C_1. We have an analogous result for $\tilde{C}_2$: we can take

$$C_2 = \tilde{A}_2 \times \tilde{B}_2 \cap \{(x, y) : (f(x) + g(y))/(-\varepsilon) \leqslant 1\}.$$

(Here we use the fact that $\Gamma_t(f \oplus g)(a_2 \otimes b_2) = \Gamma_t(f)(a_2) \otimes \Gamma_t(g)(b_2)$.) It follows that

$$
\begin{aligned}
([\mathrm{Var}_{f \oplus g}^{-1}(a_1 \otimes b_1)] \circ [a_2 \otimes b_2]) &= (-1)^{n+m}(C_1 \circ \tilde{C}_2) \\
&= (-1)^{n+m}([A_1 \times B_1] \circ [\tilde{A}_2 \times \tilde{B}_2]) \\
&= (-1)^{n+m+nm}(A_1 \circ \tilde{A}_2)(B_1 \circ \tilde{B}_2) \\
&= (-1)^{nm}(\mathrm{Var}_f^{-1} a_1 \circ a_2)(\mathrm{Var}_g^{-1} b_1 \circ b_2),
\end{aligned}
$$

which is what we were trying to prove.

Let $\{\Delta_i\}$ $(i=1,\ldots,\mu(f))$ be a distinguished basis in the homology group $H_{n-1}(V_\varepsilon(f))$ of the non-singular level manifold of the singularity f, let $\{\Delta'_j\}$ $(j=1,\ldots,\mu(g))$ be a distinguished basis in the homology group $H_{m-1}(V_\varepsilon(g))$. From Theorem 2.9 it follows that the elements

$$\tilde{\Delta}_{ij}=j_*(\Delta_i \otimes \Delta'_j)$$

form a basis of the homology group

$$H_{n+m-1}(V_\varepsilon(f \oplus g))$$

of the non-singular level manifold of the singularity $f \oplus g$. The intersection matrix S of the singularity $f \oplus g$ with respect to this basis can be obtained with the help of Theorem 2.10 from the formula $S= -L+(-1)^{n+m}L^T$ where L is the matrix of the operator $\mathrm{Var}_{f \oplus g}^{-1}$ (or the Seifert form). From this follows

Theorem 2.11. The intersection numbers of the cycles $\tilde{\Delta}_{ij}$ are given by the following formulae:

$$(\tilde{\Delta}_{if_1} \circ \tilde{\Delta}_{ij_2})=\mathrm{sgn}\,(j_2 -j_1)^n(-1)^{nm+\frac{n(n-1)}{2}}(\Delta'_{j_1} \circ \Delta'_{j_2}) \qquad \text{for} \quad j_1 \neq j_2,$$

$$(\tilde{\Delta}_{i_1 j} \circ \tilde{\Delta}_{i_2 j})=\mathrm{sgn}\,(i_2 -i_1)^m(-1)^{nm+\frac{m(m-1)}{2}}(\Delta_{i_1} \circ \Delta_{i_2}) \quad \text{for} \quad i_1 \neq i_2,$$

$$(\tilde{\Delta}_{i_1 j_3} \circ \tilde{\Delta}_{i_2 j_2})=0 \quad \text{for} \quad (i_2 -i_1)(j_2 -j_1)<0,$$

$$(\tilde{\Delta}_{i_1 j_1} \circ \tilde{\Delta}_{i_2 j_2})=\mathrm{sgn}\,(i_2 -i_1)(-1)^{nm}(\Delta_{i_1} \circ \Delta_{i_2})(\Delta'_{j_1} \circ \Delta'_{j_2})$$

$$\text{for} \quad (i_2 -i_1)(j_2 -j_1)>0.$$

This result was obtained by A. M. Gabrielov in [116]. In addition the following result was proved.

Theorem 2.12. The cycles $\tilde{\Delta}_{ij}$ are vanishing cycles and form a distinguished basis of the homology group $H_{n+m-1}(V_\varepsilon(f \oplus g))$ of the non-singular level manifold of the singularity $f \oplus g$. It is implied that they are ordered lexicographically, that is that the cycle $\tilde{\Delta}_{i_1 j_1}$ precedes the cycle $\tilde{\Delta}_{i_2 j_2}$ if

$$i_1 <i_2, \quad \text{or} \quad i_1 =i_2, \quad j_1 <j_2.$$

Theorem 2.10 is a generalisation of the above-mentioned results of M. Sebastiani and R. Thom ([322]), describing the classical monodromy operator $h_{*(f \oplus g)}$ of the singularity $f \oplus g$.

Theorem 2.13.

$$h_{*(f \oplus g)} = h_{*f} \otimes h_{*g}.$$

This theorem is an immediate corollary of Theorem 2.10 and the relation $h_* = (-1)^n \mathrm{Var}(\mathrm{Var}^{-1})^T$ (Theorem 2.5). Conversely, Theorem 2.10 follows from Theorems 2.13, 2.12, the relation $h_* = (-1)^n \mathrm{Var}(\mathrm{Var}^{-1})^T$ and Theorem 2.7, confirming that the matrix of the classical monodromy operator of a singularity with respect to a distinguished basis determines the matrix of its variation operator.

Theorems 2.11 and 2.12 give the following description of the Dynkin diagram of the singularity $f \oplus g$ (for the definition see the following section). Its vertex set coincides with the direct product of the vertex sets of the diagrams corresponding to the singularities f and g. Two vertices (i_1, j_1) and (i_2, j_2) are joined to each other

(i) by an edge of the same multiplicity as that joining the vertices j_1 and j_2 in the second diagram, if $i_1 = i_2$;

(ii) by an edge of the same multiplicity as that joining the vertices i_1 and i_2 in the first diagram, if $j_1 = j_2$;

(iii) by an edge of multiplicity equal to minus the product of the multiplicities of the edges, joining the vertices i_1 and i_2 in the first diagram and j_1 and j_2 in the second diagram, if $(i_2 - i_1)(j_2 - j_1) > 0$.

If $(i_2 - i_1)(j_2 - j_1) < 0$, then the vertices (i_1, j_1) and (i_2, j_2) are not joined to each other.

2.8   The stabilisation of a singularity

Let $f : (\mathbb{C}^n, 0) \to (\mathbb{C}, 0)$ be the germ of a holomorphic function with an isolated critical point at the origin.

Definition. The germ of the function

$$f(x) + \Sigma_{j=1}^m y_j^2 : (\mathbb{C}^{n+m}, 0) \to (\mathbb{C}, 0)$$

is called a *stabilisation* of the germ f.

The multiplicity of the singularity is equal to the multiplicity of its stabilisation. Indeed, if $\tilde{f}$ is a perturbation of the singularity f, decomposing the critical point at zero into μ non-degenerate ones, then $\tilde{f}(x) + \Sigma_{j=1}^m y_j^2$ is the perturbation of its stabilisation, possessing the same property. Moreover, the functions $\tilde{f}(x)$ and $\tilde{f}(x) + \Sigma_{j=1}^m y_j^2$ have the same set of critical values. The connection between the intersection matrix of a singularity and the intersection matrix of its stabilisation is given by the following theorem, which is a special case of Theorem 2.11.

Theorem 2.14. Let $\{\varDelta_i\}$ be a distinguished basis of vanishing cycles in the homology of the non-singular level manifolds of the singularity $f(x)$. Then there exists a distinguished basis $\{\tilde{\varDelta}_i\}$ of the singularity $f(x) + \Sigma_{j=1}^m y_j^2$, such that the intersection matrix of its elements is defined by the relation

$$(\tilde{\varDelta}_i \circ \tilde{\varDelta}_j) = [\operatorname{sgn}(j-i)]^m (-1)^{nm + m(m-1)/2} (\varDelta_i \circ \varDelta_j) \quad \text{for } i \neq j.$$

Moreover the distinguished bases $\{\varDelta_i\}$ and $\{\tilde{\varDelta}_i\}$ correspond to identical sets of paths joining the critical values of the perturbations $\tilde{f}(x)$ and $\tilde{f}(x) + \Sigma_{j=1}^m y_j^2$ with the non-critical value.

From Theorem 2.14 it follows that the intersection matrices of the stabilisations of the singularities determine each other. In addition for $m \equiv 0 \bmod 4$ the intersection numbers $(\tilde{\varDelta}_i \circ \tilde{\varDelta}_j)$ and $(\varDelta_i \circ \varDelta_j)$ are equal for all i and j, for $m \equiv 2 \bmod 4$ they differ by a sign. In this way we associate with each singularity two symmetric and two antisymmetric bilinear forms (the intersection forms of its stabilisation). Moreover the symmetric (and antisymmetric) forms differ only by a sign. With each singularity we associate also two groups of transformations of the integral lattice $\mathbb{Z}^\mu$ (the monodromy groups of its stabilisations). The classical monodromy operator of the singularity $f(x)$ coincides with the classical monodromy operator of its stabilisation $f(x) + \Sigma_{j=1}^m y_j^2$ for even m and differs from it by a sign for odd m.

Theorem 2.14 allows us to formulate results on intersection matrices of singularities restricted to cases the dimensions of which have fixed residue modulo four. In the majority of cases it will be convenient to suppose that the number of variables is conjugate to three modulo four.

Definition. The *quadratic form of a singularity* is the quadratic form, defined by the intersection numbers in the homology of the non-singular level manifolds of its stabilisation with number of variables $N \equiv 3 \bmod 4$.

For this stabilisation the self-intersection number of the vanishing cycles $(\varDelta_i \circ \varDelta_i)$ is equal to -2, the Picard-Lefschetz operator acts on the homology group of the non-singular level manifold according to the formula $h_i(a) = a + (a \circ \varDelta_i)\varDelta_i$. From this it is clear that $h_i(\varDelta_i) = -\varDelta_i$ and that the transformation h_i is reflection in the hyperplane orthogonal to the vector $\varDelta_i$. The orthogonality is in terms of the scalar product, defined by the quadratic form of the singularity. Thus we can see that the corresponding monodromy groups are groups generated by reflections. Such groups (or the corresponding quadratic forms which determine them) are most conveniently described with the help of certain graphs.

Definition. The *Dynkin diagram* (or *D-diagram*) of a singularity is a graph defined as follows:

(i) its vertices are in one-to-one correspondence with the elements $\varDelta_i$ of a weakly distinguished basis of the homology of the non-singular level manifold of the stabilisation of the singularity with number of variables $N \equiv 3 \bmod 4$;

(ii) the ith and the jth vertices of the graph are joined by an edge of multiplicity $(\varDelta_i \circ \varDelta_j)$ (The edges of negative multiplicity are depicted by dashed lines).

The D-diagram of a singularity determines its monodromy group (although an effective description of the latter is obviously quite a hard problem). If the D-diagram of a singularity (with a known number of variables) is given relative to a distinguished basis and with its vertices numbered in the same order, then from it we can determine the bilinear form of the singularity, and also its variation operator, its classical monodromy operator, etc.

2.9   Example

We consider the singularity $f(x) = x^{k+1}$ (a singularity of type A_k in the terminology of part II of volume 1). The level manifold V_ε consists of $k+1$ points, the $(k+1)$th roots of ε. The multiplicity of this singularity is equal to k, and the homology group $H_0(V_\varepsilon)$ (reduced modulo a point) is isomorphic to $\mathbb{Z}^k$.

The function $\tilde{f}(x) = x^{k+1} - \lambda x$ $(\lambda \neq 0)$ is a Morse perturbation of the singularity f. We shall suppose that λ is real and greater than zero. The zero level manifold $\tilde{f}^{-1}(0)$ of the function $\tilde{f}$ also consists of $k+1$ points: $x_0 = 0$, $x_m = \sqrt[k]{\lambda}\,\xi_m$ $(m = 1, \ldots, k)$. Here ξ_m are the kth roots of unity, enumerated clockwise: $\xi_m = \exp(-2\pi i m/k)$. The critical points of the function $\tilde{f}$ are determined by the equation $\tilde{f}'(x) = (k+1)x^k - \lambda = 0$. Therefore $\tilde{f}$ has k critical points

$$p_m = \sqrt[k]{(\lambda/(k+1))}\,\xi_m$$

with critical values

$$z_m = -\left(\frac{\lambda k}{(k+1)}\right)^k \sqrt[k]{(\lambda/(k+1))}\,\xi_m \qquad (m=1,\dots,k).$$

We choose as the non-critical value z_0 a negative number of large modulus

$$(|z_0| \gg \left(\frac{\lambda k}{(k+1)}\right)^k \sqrt[k]{(\lambda/(k+1))}).$$

Let u_m be the path joining the critical value z_m of the function $\tilde{f}$ with zero along the radius $(u_m(t)=(1-t)z_m, t\in[0,1])$, and let v be the path going from zero to z_0 along the negative real axis and going round the critical value

$$z_k = -\left(\frac{\lambda k}{(k+1)}\right)^k \sqrt[k]{(\lambda/(k+1))}\,\xi_k$$

in the positive direction (anticlockwise). See figure 22.

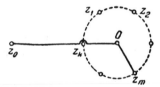

Fig. 22.

It is easy to see that the system of paths $\{u_m v\}$ define a distinguished basis of vanishing cycles $\{\Delta_m\}$ in the homology group $H_0(\tilde{f}^{-1}(z_0))$ (because by a small perturbation it can be reduced to a system of paths, satisfying the definition of a distinguished basis). In order to calculate the intersection numbers $(\Delta_m \circ \Delta_{m'})$ of the vanishing cycles in the homology group $H_0(\tilde{f}^{-1}(z_0))$ it is convenient to homotop the point z_0 along the path v to zero. In this way we reduce the problem to the calculation of the intersection numbers of the vanishing cycles, defined in the group $H_0(\tilde{f}^{-1}(0))$ by the system of paths $\{u_m\}$ (we shall denote these cycles by Δ_m too).

It is easy to show that the cycles $\Delta_m = x_m - x_0$ vanish along the paths u_m (that is as we move in the target plane of the function $\tilde{f}$ along the path u_m from zero to the critical value z_m the points x_m and x_0 merge). Consequently we have $(\Delta_m \circ \Delta_m)=2$, $(\Delta_m \circ \Delta_{m'})=1$ for any $m\neq m'$. For the stabilisation $f(x)+y_1^2+y_2^2$ the appropriate formula is $(\Delta_m \circ \Delta_m) = -2$, $(\Delta_m \circ \Delta_{m'}) = -1$ for any $m' \neq m$. We obtain the D-diagram of the singularity f in the following form: there are k vertices, every pair of which are joined by a dashed line (that is by an edge of multiplicity -1).

We simplify this diagram with the help of operations which change the distinguished basis. The operation α_{k-1} ($\Delta'_{k-1} = \Delta_k - \Delta_{k-1}$, $\Delta'_k = \Delta_{k-1}$) reduces the diagram to the following form: all the vertices except the $(k-1)$th are joined pairwise by dashed lines (edges of multiplicity -1), the $(k-1)$th vertex is joined only to the kth by a line of multiplicity $+1$. The operations α_{k-2} ($\Delta''_{k-2} = \Delta'_{k-1}$, $\Delta''_{k-1} = \Delta_{k-2}$), $\alpha_{k-3}, \ldots, \alpha_1$ do not change the form of the diagram, but lead only to the renumbering of the vertices. The application of the following sequence of operators

$$\alpha_{k-1}, \alpha_{k-2}, \ldots, \alpha_2, \alpha_{k-1}, \ldots, \alpha_3, \ldots, \alpha_{k-1}, \alpha_{k-2}, \alpha_{k-1}$$

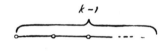

Fig. 23.

reduces the diagram to the classical Dynkin diagram A_k (figure 23). The basis of vanishing cycles we get can be described by the formulae

$$\Delta_1^0 = (x_k - x_{k-1}), \quad \Delta_2^0 = (x_{k-1} - x_{k-2}), \ldots,$$
$$\Delta_{k-1}^0 = (x_2 - x_1), \quad \Delta_k^0 = (x_1 - x_0).$$

As we move in the target plane of the function $\tilde{f}$ along the path v from zero to the non-critical value z_0, the points x_m ($m = 0, 1, \ldots, k$) move in the complex plane $\mathbb{C}$ tending to the rays

$$\arg x = \pi(2s+1)/(k+1)$$

(as $z_0 \to -\infty$). The points x_0 and x_k approach each other along the real axis, do a quarter rotation, anticlockwise, round the critical point p_k, and go apart again. It is easy to show that on the ray

$$\arg x = \pi(2s+1)/(k+1)$$

(that is $x = t \exp(\pi i(2s+1)/(k+1))$, $t > 0$) the function $\tilde{f}(x) = x^{k+1} - \lambda x$ does not take negative real values except in the case when k is even and $s=k/2$. Indeed,

$$\tilde{f}(x) = -t^{k+1} - \lambda t \exp(\pi i(2s+1)/(k+1)),$$

where the second term is real only when k is even and $s=k/2$. In this case the point $x_{k/2}$ moves along the negative real axis. From this it follows that as we move in the target plane of the function $\tilde{f}$ along the path v from zero to z_0 the points x_m approach the points

$$\tilde{x}_m = \sqrt[k+1]{(-z_0)} \exp(-\pi i(2m+1)/(k+1))$$

($m=0, 1, \ldots, k$). If in addition we travel in the target plane of the function in the negative direction (clockwise) from the point z_0 to the point $z_0' = -z_0$, then the points $\tilde{x}_m$ will cross over to the points

$$\tilde{\tilde{x}}_m = \sqrt[k+1]{(-z_0)} \exp(-2\pi i(m+1)/(k+1))$$

($m=0, 1, \ldots, k$).

We arrive at the following result:

Theorem 2.15. On the level manifold $V_1 = \{x : x^{k+1} = 1\}$ of the singularity $f(x) = x^{k+1}$ the distinguished basis is formed by the vanishing cycles

$$\Delta_1 = \zeta_1 - \zeta_2, \Delta_2 = \zeta_2 - \zeta_3, \ldots, \Delta_k = \zeta_k - \zeta_{k+1},$$

where

$$\zeta_j = \exp(2\pi i(j-1)/(k+1))$$

are the $(k+1)$th roots of unity ($j=1, \ldots, (k+1)$). The intersection numbers of these cycles are given by the formulae

$$(\Delta_j \circ \Delta_j) = 2,$$

$$(\Delta_j \circ \Delta_{j+1}) = -1,$$

$$(\Delta_j \circ \Delta_{j'}) = 0 \quad \text{for} \quad |j-j'| \geq 2.$$

The first calculation of the intersection forms and the classical monodromy operator for functions of several variables was given by F. Pham ([284]) for singularities of the form $f(x) = \sum_{k=1}^{n} x_k^{a_k}$ ($a_k \geq 2$). The multiplicity of this singularity is equal to $\Pi_{k=1}^{n}(a_k-1)$. F. Pham proved that in the homology group $H_{n-1}(V_e)$ of the non-singular level set of the function f there exists a basis $e_{i_1\ldots i_n}$ $(0 \leqslant i_k \leqslant a_k - 2)$ (in the notation of F. Pham $e_{i_1\ldots i_n}=(\Pi_{k=1}^{n}\omega_k^{i_k})e$), such that

$$(e_{i_1\ldots i_n} \circ e_{i_1\ldots i_n}) = (-1)^{\frac{n(n-1)}{2}}(1+(-1)^{n-1});$$

$$(e_{i_1\ldots i_n} \circ e_{j_1\ldots j_n}) = (-1)^{\frac{n(n-1)}{2}}(-1)^{\sum\limits_{k}(j_k-i_k)},$$

if $i_k \leqslant j_k \leqslant i_k + 1$ for all k. In the remaining cases (except those arising from the previous ones by a permutation of cycles)

$$(e_{i_1\ldots i_n} \cdot e_{j_1\ldots j_n}) = 0.$$

The result of F. Pham can be obtained from Theorem 2.11 (§2.7). Applying it to the singularity $f(x)=\Sigma_{k=1}^{n} x_k^{a_k}$ gives the same intersection matrix as that of F. Pham, if as a distinguished basis of the singularity $f_k(x_k)=x_k^{a_k}$ we use the basis described in theorem 2.15. For the singularity $f_k(x_k)=x_k^{a_k}$ we put

$$\varepsilon = 1, \ u(t)=(1-t), \ H(t)x_k = \sqrt[a_k]{(1-t)}\,x_k.$$

The application in sequence of the constructions described in §2.7 to the distinguished basis of the singularity $f_k(x_k)$ given by Theorem 2.15 reduces it, as it is not difficult to convince oneself, to the basis constructed by F. Pham in [284]. We obtain the following result.

Assertion. The basis of F. Pham is distinguished relative to the lexicographic ordering of its elements.

This means that the D-diagram of the singularity of F. Pham has the form depicted in figure 24 ($n=2$, $a_1 = 6$, $a_2 = 5$).

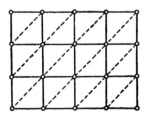

Fig. 24.

Chapter 3

The bifurcation sets and the monodromy group of a singularity

The characteristics of a singularity which were discussed in the second chapter (the multiplicity of a singularity, its intersection matrix, its monodromy group...) are closely linked to such objects as the level and function bifurcation sets of the singularity, its resolution and its polar curve. Several of these links will be discussed in this chapter.

3.1 The bifurcation sets of a singularity

In order to define the bifurcation sets of a singularity, we recall the definition of its versal deformation (for a more detailed exposition see Volume 1 Chapter 8).

Definition. A *deformation of* the singularity $f : (\mathbb{C}^n, 0) \to (\mathbb{C}, 0)$ is the germ of a holomorphic function $F(x, v)$ $(v \in \mathbb{C}^l)$,

$$F : (\mathbb{C}^n \oplus \mathbb{C}^l, 0) \to (\mathbb{C}, 0),$$

such that $F(x, 0) = f(x)$.

The space $\mathbb{C}^l$ is called the parameter space or base space of the deformation F.

Definition. The deformation $F(x, v)$ of the singularity f is *versal*, if any deformation $G(x, \eta)$ $(\eta \in \mathbb{C}^m)$ of the singularity f $(G(x, 0) = f(x))$ is "equivalent to the deformation induced by F", that is there exists an analytic map $\psi : (\mathbb{C}^m, 0) \to (\mathbb{C}^l, 0)$ of the parameter spaces and an analytic family $g(x, v)$

$$g : (\mathbb{C}^n \oplus \mathbb{C}^m, 0) \to (\mathbb{C}^n, 0),$$

$g(\cdot, 0) = id : \mathbb{C}^n \to \mathbb{C}^n$ of local changes of coordinates such that

$$G(x, v) = F(g(x, v), \psi(v)).$$

The dimension l of the base space $\mathbb{C}^l$ of the versal deformation $F(x, v)$ is not less than the multiplicity μ of the singularity f. There exists a (unique in a natural sense) versal deformation of the singularity f with a base, the dimension of which is exactly equal to μ. This deformation is said to be *miniversal*.

A miniversal deformation $F(x, v)$ of the singularity f can be constructed in the following fashion. In § 2.1 it was indicated that the quotient ring of the ring $_nO$ of germs of holomorphic functions on $(\mathbb{C}^n, 0)$ by the ideal generated by the partial derivatives of the function f (the Jacobian ideal) has dimension, as a complex vector space, equal to the multiplicity μ of the singularity f. Let the germs of the functions $\phi_i : (\mathbb{C}^n, 0) \to \mathbb{C}$ $(i = 0, 1, \ldots, \mu - 1)$ give a basis of this space. Then the deformation

$$F(x, v) = f(x) + \sum_{i=0}^{\mu-1} v_i \phi_i(x)$$

is miniversal $(v = (v_0, v_1, \ldots, v_{\mu-1}))$. We can take as ϕ_0 the germ of the function identically equal to unity.

Let $F(x, v)$ be a miniversal deformation of the singularity f $(v \in \mathbb{C}^\mu)$, let

$$W_v = \{x \in \mathbb{C}^n : F(x, v) = 0, \; \|x\| \leqslant \varrho\}$$

be the zero level set of the function $F(\cdot, v)$. Since $F(x, 0) = f(x)$, and the set $\{x \in \mathbb{C}^n : f(x) = 0\}$ is transverse to the sphere S_ϱ of sufficiently small radius ϱ, there exists an $\varepsilon > 0$ such that for $\|v\| \leqslant \varepsilon$ the set $\{x \in \mathbb{C}^n : F(x, v) = 0\}$ is transverse to the sphere S_ϱ. From this it follows that if the set W_v is non-singular, then it is diffeomorphic to the non-singular level set of the function f near the critical point. The set of those values of the parameter v for which W_v is singular forms a set of (complex) codimension one.

Definition. The *level bifurcation set* (or *set bifurcation diagram*) of the singularity f is the space $\Sigma_\varepsilon = \{v \in \mathbb{C}^\mu : \|v\| \leqslant \varepsilon, 0 \text{ is a critical value of the function } F(\cdot, v) \text{ in the ball } \|x\| \leqslant \varrho\}$.

Examples. (i) A miniversal deformation of the singularity A_2 $(f(x) = x^3)$ can be chosen in the form

$$F(x; \lambda_1, \lambda_2) = x^3 + \lambda_1 x + \lambda_2.$$

The local zero level set of the function $F(\cdot; \lambda_1, \lambda_2)$ does not have a singularity if the polynomial

$$x^3 + \lambda_1 x + \lambda_2$$

does not have multiple roots. Therefore the level bifurcation set Σ consists of those values $(\lambda_1, \lambda_2) \in \mathbb{C}^2$ for which the polynomial $x^3 + \lambda_1 x + \lambda_2$ has the form $(x-a)^2 (x-b)$, where $2a+b=0$. We have

$$\lambda_1 = 2ab + a^2 = -3a^2, \ \lambda_2 = -a^2 b = 2a^3.$$

Consequently, the bifurcation set Σ is described by the equation

$$\lambda_1^3 + \frac{27}{4} \lambda_2^2 = 0.$$

It (more precisely, of course, its real part) is depicted in figure 25.

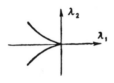

Fig. 25.

(ii) A miniversal deformation of the singularity A_3 $(f(x)=x^4)$ can be chosen in the form

$$F(x; \lambda_1, \lambda_2, \lambda_3) = x^4 + \lambda_1 x^2 + \lambda_2 x + \lambda_3 .$$

The local zero level set of the function $F(\cdot\,; \lambda_1, \lambda_2, \lambda_3)$ does not have a singularity if the polynomial

$$x^4 + \lambda_1 x^2 + \lambda_2 x + \lambda_3$$

does not have multiple roots. Therefore the level bifurcation set Σ consists of those values $(\lambda_1, \lambda_2, \lambda_3) \in \mathbb{C}^3$ for which the polynomial $x^4 + \lambda_1 x^2 + \lambda_2 x + \lambda_3$ has the form $(x-a)^2 (x-b)(x-c)$, where $2a+b+c=0$. The bifurcation set Σ has the name "swallow tail". It is depicted in figure 26.

The topological type of the pair $(D_\varepsilon, \Sigma_\varepsilon)$ $(D_\varepsilon = \{v \in \mathbb{C}^\mu : \|v\| \leq \varepsilon\}$ is a ball of radius ε in the base of the miniversal deformation F) does not depend on ε for

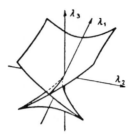

Fig. 26.

sufficiently small ε and does not depend on the choice of miniversal deformation of the singularity f. The space $(D_\varepsilon \backslash \Sigma_\varepsilon)$ (an open subspace of the ball D_ε) is the base of a locally trivial fibration

$$\{(x, v) \in \mathbb{C}^n \oplus \mathbb{C}^\mu : \|x\| \leqslant \varrho, \|v\| \leqslant \varepsilon, v \notin \Sigma_\varepsilon, F(x, v) = 0\} \rightarrow D_\varepsilon \backslash \Sigma_\varepsilon$$

with projection $(x, v) \mapsto v$. The fibre

$$W_v = \{x \in \mathbb{C}^n : F(x, v) = 0, \|x\| \leqslant \varrho\}$$

of this fibration is diffeomorphic to the non-singular level set of the singularity f.

As in any fibration, the fundamental group of its base acts on the homology of the fibre. In this way we get a natural representation

$$\pi_1 (D_\varepsilon \backslash \Sigma_\varepsilon) = \pi_1 (D_\varepsilon \backslash \Sigma_\varepsilon, v)$$

$$\rightarrow \operatorname{Aut} H_{n-1} (W_v) = \operatorname{Aut} H_{n-1} (V_\lambda).$$

Theorem 3.1. The image of the representation

$$\pi_1 (D_\varepsilon \backslash \Sigma_\varepsilon) \rightarrow \operatorname{Aut} H_{n-1} (V_\lambda)$$

of the fundamental group of the complement of the level bifurcation set of the singularity f in the homology of the non-singular level manifold coincides with the monodromy group of the singularity f.

For the proof we must choose a miniversal deformation $F(x, v)$ of the singularity f in the form

$$F_0 (x, v') - v_0,$$

where $v' \in \mathbb{C}^{\mu-1}$, $v_0 \in \mathbb{C}$, $v = (v_0, v')$. As the perturbation $f_\lambda(x)$ of the singularity f we can take a perturbation of the form $F_0(x, v'(\lambda))$. Let p be the natural projection of the base $\mathbb{C}^\mu$ of the miniversal deformation onto the space $\mathbb{C}^{\mu-1}$, mapping $v = (v_0, v')$ into v'. If $f_\lambda(x)$ is a Morse function (λ is sufficiently small), then the line $L = p^{-1}(v'(\lambda))$ is in general position with the manifold Σ_ε. The line L intersects the bifurcation set Σ_ε in those points $(v_0, v'(\lambda)) \in \mathbb{C}^\mu$ for which v_0 is a critical value of the function $f_\lambda(x)$. The number of such points is equal to the multiplicity $\mu(f)$ of the singularity f. The space $L \setminus \Sigma_\varepsilon$ coincides with the complement of the set of critical values of the function f_λ. The restriction of the fibration

$$\{(x, v) : v \notin \Sigma_\varepsilon, F(x, v) = 0\} \to D_\varepsilon \setminus \Sigma_\varepsilon$$

described above, to $L \setminus \Sigma_\varepsilon$ coincides with the fibration of the non-singular level manifolds of the function f_λ over the complement of the set of its critical values. From this it follows that the natural representation

$$\pi_1(L \setminus \Sigma_\varepsilon) \to \mathrm{Aut}\, H_{n-1}(V_\lambda),$$

the image of which is the monodromy group of the singularity f, factors through the fundamental group of the complement of the bifurcation set:

$$\pi_1(L \setminus \Sigma_\varepsilon) \xrightarrow{i_*} \pi_1(D_\varepsilon \setminus \Sigma_\varepsilon) \to \mathrm{Aut}\, H_{n-1}(V_\lambda)$$

where i_* is the homomorphism of fundamental groups induced by the inclusion

$$L \setminus \Sigma_\varepsilon \hookrightarrow D_\varepsilon \setminus \Sigma_\varepsilon .$$

From the fact that the line L is in general position with the manifold Σ_ε it turns out that the homomorphism

$$i_* : \pi_1(L \setminus \Sigma_\varepsilon) \to \pi_1(D_\varepsilon \setminus \Sigma_\varepsilon)$$

is an epimorphism. From this it follows that the image of the representation

$$\pi_1(D_\varepsilon \setminus \Sigma_\varepsilon) \to \mathrm{Aut}\, H_{n-1}(V_\lambda)$$

coincides with the monodromy of the singularity f.

The fact that the homomorphism

$$i_* : \pi_1(L \setminus \Sigma_\varepsilon) \to \pi_1(D_\varepsilon \setminus \Sigma_\varepsilon)$$

is an epimorphism is a variant of a theorem of Zariski ([413]), which goes as follows. Let M be a non-singular affine algebraic hypersurface in the space $\mathbb{C}^n$, let L be a (complex) line in general position in the space $\mathbb{C}^n$. Such a line intersects transversely the hypersurface M in m points $p_1, \ldots, p_m$. The theorem of Zariski asserts, in particular, that for the line L in general position the homomorphism of fundamental groups

$$i_* : \pi_1(L \backslash M) \to \pi_1(\mathbb{C}^n \backslash M),$$

induced by the inclusion $i : L \backslash M \to \mathbb{C}^n \backslash M$, is an epimorphism. The fundamental group $\pi_1(L \backslash M) = \pi_1(L \backslash \{p_i\})$ of the line with m points removed is a free group on m generators. We can take as these generators simple loops, corresponding to a system of non-intersecting paths in the complex line L, joining the points p_i with the base point $p_0 \in L \backslash M$. Thus the fundamental group of the complement of the hypersurface M is a group generated by the m generators described above.

The theorem of Zariski also describes all the relations between the generators. In order to give this description, we consider the projection $\pi : \mathbb{C}^n \to \mathbb{C}^{n-1}$ of the space $\mathbb{C}^n$ along the line L and its restriction $\pi|_M : M \to \mathbb{C}^{n-1}$ to the hypersurface M. The fact that the line L is in general position allows us, in particular, to suppose that the discriminant set of the map $\pi|_M$ (the image of the set of its critical points) is a reduced hypersurface in the space $\mathbb{C}^{n-1}$. In more detail this means the following. The closure of the set of critical values of the map $\pi|_M$ is a complex hypersurface N in the space $\mathbb{C}^{n-1}$. If $q \in \mathbb{C}^{n-1} \backslash N$, then the preimage $\pi|_M^{-1}(q)$ consists of m points, at each of which the differential of the map $\pi|_M$ is non-degenerate. Over each regular point of the hypersurface N apart from a set of codimension 1, a pair of points will merge. Near such points the hypersurface M can be given locally by the equation $x_0 + x_1^2 = 0$, where the projection π maps the point $(x_0, x_1, \ldots, x_{n-1}) \in \mathbb{C}^n$ to $(x_1, \ldots, x_{n-1}) \in \mathbb{C}^{n-1}$, and N is given locally in the space $\mathbb{C}^{n-1}$ by the equation $x_1 = 0$. Let $q_0 \in \mathbb{C}^{n-1} \backslash N$ be the image of the line L under the projection π, let L_1 be a line in general position in the space $\mathbb{C}^{n-1}$ passing through the point q_0. We can suppose that the line L_1 intersects the hypersurface N only in regular points, to which are mapped merged pairs of points from the preimage $\pi|_M^{-1}(q)$, all these intersections being transversal. In this case $\pi^{-1}(L_1) \cap M$ is a non-singular curve in the two-dimensional complex space $\pi^{-1}(L_1)$,

$$\pi|_{\pi^{-1}(L_1) \cap M} : \pi^{-1}(L_1) \cap M \to L_1$$

is an m-fold branched cover over the line L_1. Let $q_1, \ldots, q_k$ be the points of intersection of the line L_1 with the discriminant set N, and let τ be an arbitrary loop in the space $L_1 \setminus \{q_1, \ldots, q_k\}$ with beginning and end at the point q_0. Going round the loop τ corresponds to a homeomorphism T_τ, of the pair of spaces $(\pi^{-1}(q_0), \pi^{-1}(q_0) \cap M) = (L, L \cap M)$ into itself, defined, of course, only up to isotopy. This homeomorphism induces a transformation

$$T_{\tau *} : \pi_1(L \setminus M) \to \pi_1(L \setminus M)$$

of the fundamental group of the line with m deleted points, $L \setminus M$, into itself. It is clear that if $a \in \pi_1(L \setminus M)$, then $i_* a = i_* T_{\tau *} a$ where

$$i_* : \pi_1(L \setminus M) \to \pi_1(\mathbb{C}^n \setminus M).$$

The theorem of Zariski asserts that the homomorphism

$$\pi_1(\pi^{-1}(L_1) \setminus M) \to \pi_1(\mathbb{C}^n \setminus M)$$

induced by inclusion is an isomorphism, and the relations of the form we described generate all the relations between generators of the fundamental group $\pi_1(\mathbb{C}^n \setminus M)$. Naturally we can take, as generators of the system of relations, the relations $i_* a = i_* T_{\tau_i *} a$, where $\{\tau_i\}$ is a system of simple loops, corresponding to a system of non-intersecting paths joining the points $q_1, \ldots, q_k$ with the point q_0. From this it follows that the group $\pi_1(\mathbb{C}^n \setminus M)$ is a group with m generators and mk relations.

A local variant of this theorem, part of which we use here, is formulated in an analogous manner. The proof of this is given in [157].

If γ is a loop in the complement $D_\varepsilon \setminus \Sigma_\varepsilon$ of the level bifurcation set of the singularity f, then by analogy with §2.1 we shall denote by $h_{\gamma *}$ the corresponding automorphism of the homology group of the non-singular level manifold of the singularity f ($h_{\gamma *}$ belongs to the monodromy group of the singularity f).

There is, corresponding to the singularity, one more bifurcation set – the function bifurcation set. For its definition we consider a miniversal deformation $F_0(x, v)$ of the singularity f in the class of functions equal to zero at the point 0. Such a deformation has $\mu - 1$ parameters. We shall call it a *restricted miniversal deformation*. We can, for example, take for it the deformation

$$F_0(x, v) = f(x) + \Sigma_{i=1}^{\mu-1} v_i \phi_i(x),$$

where $v = (v_1, \ldots, v_{\mu-1})$, and $\phi_0, \phi_1, \ldots, \phi_{\mu-1}$ are germs generating a basis of the quotient ring of the ring of germs at zero of holomorphic functions by the Jacobian ideal $(\partial f / \partial x_1, \ldots, \partial f / \partial x_n)$ of the singularity f, $\phi_0 \equiv 1$, $\phi_i(0) = 0$ for $i \geqslant 1$ (the deformation $F_0(x, v)$ differs from the miniversal deformation $F(x, v)$ by the absence of the term $v_0 \cdot 1$, which does not influence the type of the function).

In a small ball-like neighbourhood D_ε of the origin in the base $\mathbb{C}^{\mu-1}$ of the restricted miniversal deformation we consisder the set of those values of the parameter v for which the function $F_0(\cdot, v)$ in a neighbourhood B_ϱ of the origin in the space $\mathbb{C}^n$ is Morse, that is has only non-degenerate critical points (μ in number) with distinct critical values. Its complement $\tilde{\Sigma}_\varepsilon$ is called the *function bifurcation set* of the singularity f. The topological type of the pair $(D_\varepsilon, \tilde{\Sigma}_\varepsilon)$ does not, of course, depend on ε for sufficiently small ε. The set $\tilde{\Sigma}_\varepsilon$ is a hypersurface in the space $\mathbb{C}^{\mu-1}$. It is, clearly, reducible, as it is the union of two hypersurfaces. One of these is the set of values of the parameter v for which the function $F_0(\cdot, v)$ has degenerate critical points, and the other is the set of those values v for which it has critical points with coincident critical values.

Examples.

(i) The function bifurcation set of the singularity A_2 consists, clearly, of one point $\lambda = 0$ in the base $\mathbb{C}^1$ of the restricted miniversal deformation.

(ii) The function bifurcation set of the singularity A_3 consists of those values $(\lambda_1, \lambda_2) \in \mathbb{C}^2$ for which the polynomial $x^4 + \lambda_1 x^2 + \lambda_2 x$ either has a degenerate critical point or has two non-degenerate critical points with the same critical value. The second of these sets is $\{\lambda_2 = 0\} \subset \mathbb{C}^2$. The first set is described by the condition that the polynomial $4x^3 + 2\lambda_1 x + \lambda_2$ (the derivative of the polynomial $x^4 + \lambda_1 x^2 + \lambda_2 x$) has multiple roots. This will be true for $\lambda_1^3 + \frac{27}{8}\lambda_2^2 = 0$. The function bifurcation set of the singularity is depicted in figure 27.

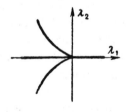

Fig. 27.

There is a natural map (projection) p from the base, $\mathbb{C}^\mu$, of the miniversal deformation to the base, $\mathbb{C}^{\mu-1}$, of the restricted miniversal deformation. It can be shown that the germ of the space $\tilde{\Sigma}_\varepsilon$ coincides with the set of non-regular values of the map $\Sigma_\varepsilon \to \mathbb{C}^{\mu-1}$, obtained by composing the inclusion $\Sigma_\varepsilon \hookrightarrow \mathbb{C}^\mu$ and the projection p. Over the complement of the function bifurcation set $\tilde{\Sigma}_\varepsilon$ of the singularity f this map defines an μ-fold cover.

3.2 The connectedness of the D-diagram and the 'irreducibility' of the classical monodromy operator of a singularity

In §3.1 we indicated that the function bifurcation set of a singularity is always reducible (with the exception of trivial cases of singularities of multiplicity one or two). In contrast to this the level bifurcation set of a singularity is irreducible.

Theorem 3.2 (see, for example, [117]). The level bifurcation set Σ_ε of a singularity f is an irreducible analytic set. Furthermore, there exists a germ of a proper map $(\mathbb{C}^{\mu-1}, 0) \to (\mathbb{C}^\mu, 0)$, the image of which coincides with the space Σ_ε and which is an isomorphism outside the set of singular points of Σ_ε.

The map of the space $\mathbb{C}^{\mu-1}$ into the base of the miniversal deformation $\mathbb{C}^\mu$ of the singularity f, mentioned in the theorem, can be constructed in the following manner. We consider the set of germs of functions $g \colon (\mathbb{C}^n, 0) \to (\mathbb{C}, 0)$, satisfying the conditions $g(0) = 0$, $dg(0) = 0$. On it acts the group of germs of analytic diffeomorphisms of the space $\mathbb{C}^n$, fixing the point 0. The orbit of the singularity f under the action of this group is a non-singular complex manifold of codimension $\mu - 1$ (a rigorous approach would consider everything in the space of jets of sufficiently high order). A transversal to the orbits at the point f has dimension $(\mu - 1)$ and defines a $(\mu - 1)$-parameter deformation of the singularity f. Like any deformation of the singularity f, it is equivalent to the deformation induced from a miniversal one by a map of its base $\mathbb{C}^{\mu-1}$ into the base $\mathbb{C}^\mu$ of a miniversal deformation. Since all the functions of the deformation we are considering have 0 as a critical value, the whole space $\mathbb{C}^{\mu-1}$ is carried by this map into the level bifurcation set Σ_ε. This is the map described in Theorem 3.2.

From this assertion A. M. Gabrielov ([117]) and F. Lazzeri ([205]) derive the following result.

Theorem 3.3. The D-diagram of any singularity relative to a distinguished basis is connected.

This assertion also follows from Theorem 3.4 (see below). The same result is true also for a weakly distinguished basis.

Corollary. Let $f_t(x)$ $(t \in [0, t_0])$ be a deformation of the singularity f and suppose that for small t the function $f_t(x)$ has, in a neighbourhood of zero in the space $\mathbb{C}^n$, k distinct critical points $p_1(t), \ldots, p_k(t)$. We suppose that all the critical values

$f_t(p_i(t))$ $(i=1,\ldots,k)$ of the function f_t coincide. Then $k=1$, that is the function f_t has only one critical point (of multiplicity $\mu(f)$).

Indeed, it is easy to see that if $k>1$ then the vanishing cycles, corresponding to distinct critical points of the deformation $f_t(x)$ of the singularity f, will have intersection number zero. Therefore the D-diagram of the singularity f decomposes into k disconnected components, which contradicts theorem 3.3. We shall prove here several results which are stronger than theorem 3.3.

Theorem 3.4. The monodromy group of a singularity acts transitively on the set of vanishing cycles in the homology of the non-singular level set near the critical point, that is for any vanishing cycles Δ_1 and Δ_2 there exists an element of the monodromy group of the singularity mapping Δ_1 to $\pm\Delta_2$.

Proof. As for the proof of Theorem 3.1, we consider a miniversal deformation of the singularity f of the form

$$F(x,v)=F_0(x,v')-v_0,$$

where

$$v'\in\mathbb{C}^{\mu-1}, \ v_0\in\mathbb{C}, \ v=(v_0,v').$$

As a perturbation $\tilde{f}$ of the singularity f we take the function $\tilde{f}=F_0(x,v')$ with fixed value of the parameter v'. We choose v' so that the function $F_0(x,v')$ will be Morse. This will be satisfied for almost all values of the parameter v' (except those that lie in the function bifurcation set of the singularity). If L is the (complex) line $p^{-1}(v')$ ($p:\mathbb{C}^{\mu}\to\mathbb{C}^{\mu-1}$ is the projection of the base of the miniversal deformation), then the intersection $L\cap\Sigma_{\varepsilon}$ consists of the points (z_i,v') $(i=1,\ldots,\mu)$, where z_i are the critical values of the function $\tilde{f}$.

The vanishing cycles Δ_k ($k=1,2$) in the homology of the non-singular level set $\{\tilde{f}=z_0\}$ are defined by paths u_k, joining the critical values z_{i_k} with the non-critical value z_0 and not passing through the critical values of the function $\tilde{f}$. We suppose, for simplicity, that for very small t we have $u_k(t)=z_{i_k}+t$. The paths u_1 and u_2 can be considered as paths in the complex line $L\subset\mathbb{C}^{\mu}$. From the irreducible of the bifurcation set Σ_{ε} it follows that the set of non-singular points of the space Σ_{ε} is connected. Those (non-singular) points of the space Σ_{ε}, at which the projection $p:\Sigma_{\varepsilon}\to\mathbb{C}^{\mu-1}$ is degenerate, form a subset of (complex) codimension one. Therefore their removal from the set of non-singular points of the space does not destroy its connectedness. From this it follows that the points (z_{i_1},v') and (z_{i_2},v')

can be joined by a path v $(v(0)=(z_{i_1}, v'), v(1)=(z_{i_2}, v'))$, which will lie in its entirety in the set of non-singular points of the space Σ_ε, at which its projection into the space $\mathbb{C}^{\mu-1}$ is non-degenerate. We consider a loop w in the complement of the level bifurcation set Σ_ε in the space $\mathbb{C}^\mu$ (beginning and ending at the point (z_0, v')), which is defined in the following manner. It goes from the point (z_0, v') to the point $(z_{i_1}+t_0, v')=(u_1(t_0), v')$ with sufficiently small t_0 along the path u_1, then it goes from the point $(z_{i_1}+t_0, v')$ to the point $(z_{i_2}+t_0, v')=(u_2(t_0), v')$ along the path $v+(t_0, 0)$, going parallel to the path v, and finally returns to the point (z_0, v') along the path u_2. It is not hard to see that the monodromy operator h_{w*}, corresponding to the loop w, maps the vanishing cycle Δ_1 into the vanishing cycle Δ_2 (maybe with changed orientation), which is what we were trying to prove.

Theorem 3.3 is an immediate corollary of Theorem 3.4. Indeed, let the D-diagram of the singularity f with respect to the (weakly distinguished) basis $\{\Delta_i\}$ be disconnected. From the theorem of Picard-Lefschetz it follows that the Picard-Lefschetz operators (and composites of them), acting on a basic vanishing cycle, map it into a cycle which is a linear combination of basic vanishing cycles from the same connected component of the diagram. Therefore in this case there will not exist an operator from the monodromy group of the singularity f which maps a basic vanishing cycle into a basic vanishing cycle from another connected component of the diagram, which contradicts Theorem 3.4. From this reasoning it follows that the D-diagram of the singularity is connected also in the case when the multiplicity of its edges are considered modulo $m>1$.

From Theorem 3.3 we can deduce several properties of the classical monodromy operator of the singularity. We formulate one proposition about diagonal matrices which we need for this.

Lemma 3.1. Let A and B be upper triangular $\mu \times \mu$ matrices with ones on the diagonal. We suppose that $A \cdot B^T$ is a matrix such that the intersection of its first k columns with its last $\mu-k$ rows contains only zeros. Then the same property is true also for the matrix B^T, that is the matrix B is the direct sum of upper triangular matrices of dimension $k \times k$ and $(\mu-k) \times (\mu-k)$.

The proof does not present any difficulty.

Theorem 3.5. Let $\Delta_1, \ldots, \Delta_\mu$ be a distinguished basis of the homology group $H_{n-1}(V_\varepsilon)$ of the non-singular level set of a singularity, and let I be a subset of the set of indices $\{1, \ldots, \mu\}$ such that linear span of the basis elements Δ_i with $i \in I$ is invariant relative to the classical monodromy operator h_*. Then either $I=\emptyset$ or $I=\{1, \ldots, \mu\}$.

We prove what appears at first sight to be a slightly stronger assertion, but which in fact is exactly equivalent to Theorem 3.5.

Let $\{\varDelta_1, \ldots, \varDelta_k\}$ be a set of vanishing cycles in the homology of the non-singular level manifold of the singularity f, defined by a system of paths $\{u_i\}$ $(i=1, \ldots, k)$, joining some of the critical values of the perturbation $\tilde{f}$ of the singularity f with its non-critical value z_0 and not passing (for $t \neq 0$) through the critical values of the function $\tilde{f}$. We suppose that the paths u_i have no self-intersections and do not intersect each other at points distinct from their ends which coincide with z_0.

Theorem 3.6. If the linear span of the vanishing cycles $\varDelta_1, \ldots, \varDelta_k$ in the homology group $H_{n-1}(F_{z_0})$ is invariant relative to the classical monodromy operator of the singularity, then either $k=0$ or $k=\mu(f)$.

Proof. We shall suppose that the cycles $\varDelta_1, \ldots, \varDelta_k$ (and the paths $(u_1, \ldots, u_k)$) are numbered in the order which is fixed by condition (iii) of the definition of a distinguished basis (see § 1.2). It is easy to see that the system of paths $\{u_i, i=1, \ldots, k\}$ can be increased to the system of paths $\{u_i; i=1, \ldots, k, \ldots, \mu\}$, defining the distinguished basis $\varDelta_1, \ldots, \varDelta_k, \ldots, \varDelta_\mu$ in the homology group $H_{n-1}(F_{z_0})$. The condition of invariance of the linear span of the elements $\varDelta_1, \ldots, \varDelta_k$ relative to the classical monodromy operator h_* means that in the matrix H of the operator h_* relative to the basis $\varDelta_1, \ldots, \varDelta_k, \ldots, \varDelta_\mu$ there are zeros in the intersection of the first k columns with the last $\mu - k$ rows. Applying Lemma 3.1 to the equality $H=(-1)^n L^{-1} L^T$ (where L is the matrix of the Seifert form of the singularity), we obtain the result that the matrix L is the direct sum of matrices of dimensions $k \times k$ and $(\mu-k) \times (\mu-k)$. Consequently the same property is possessed also by the intersection matrix of the singularity f with respect to the distinguished basis $\varDelta_1, \ldots, \varDelta_\mu$, which is equal to $-L+(-1)^n L^T$. For $k \neq 0, \mu$ this means that the D-diagram of the singularity f decomposes into a disjoint union of two diagrams (with k and $\mu - k$ vertices respectively), which contradicts Theorem 3.3.

Corollary. If the classical monodromy operator of a singularity is multiplication by one or minus one, then the singularity is non-degenerate, that is its multiplicity μ is equal to one.

This result (as the Sebastiani conjecture) was proved by N. A'Campo in [4]. There it was deduced from the following result.

Theorem 3.7. The trace tr h_* of the classical monodromy operator of the singularity $f : (\mathbb{C}^n, 0) \to (\mathbb{C}, 0)$ of a function of n variables is equal to $(-1)^{n-1}$.

3.3   The bifurcation sets of simple singularities

It is well known (see § 3.6) that for the simple singularities A_k, D_k, E_6, E_7, E_8 with an odd number of variables, the monodromy group is the same as the corresponding classical Weyl group (of the same name) (see [53]).

This group is the image of the fundamental group of the complement of the level bifurcation set Σ_ε of the singularity. For simple singularities the space Σ_ε can be obtained as follows.

Let $\mathbb{R}^k$ be a vector space on which the Weyl group W (A_k, D_k or E_k respectively) acts canonically and let $\mathbb{C}^k = \mathbb{R}^k \otimes_{\mathbb{R}} \mathbb{C}$ be its complexification. The action of the group W on the space $\mathbb{R}^k$ extends in a natural way to an action of W on the complexification $\mathbb{C}^k$. Let S be the union of the non-regular orbits of the action of the group W, that is the set of points on which the action of the group W is not free (has a non-trivial stabiliser subgroup). It is the same as the union of (complex) mirrors, reflection in which belongs to the group W. We consider the quotient space $\mathbb{C}^k / W$. It is known ([53]) that it is isomorphic as an analytic space to k-dimensional complex linear space.

Theorem 3.8 ([21]). For the simple singularities A_k, D_k, E_k the pair $(\mathbb{C}^k / W, S/W)$ is isomorphic (in a neighbourhood of zero) to the pair $(D_\varepsilon, \Sigma_\varepsilon)$, where Σ_ε is the level bifurcation set of the singularity.

Example. Let $f(x) = x^{k+1}$ (the singularity A_k). In this case the Weyl group W is the group of permutations of $k + 1$ elements. Its action on the space $\mathbb{C}^k$ is defined in the following manner. The space $\mathbb{C}^k$ is embedded in the space $\mathbb{C}^{k+1}$ of one larger dimension in the form of the hyperplane $\Sigma_{i=1}^{k+1} x_i = 0$, and the action of the group W on it is obtained by restricting its action on the space $\mathbb{C}^{k+1}$ as the group of permutations of coordinates. The quotient space $\mathbb{C}^{k+1} / W$ is mapped isomorphically onto the space $\mathbb{C}^{k+1}$ by mapping the class of the point $(x_1, \ldots, x_{k+1})$ to the point $(\sigma_1, \ldots, \sigma_{k+1})$, where $\sigma_i = \sigma_i(x_1, \ldots, x_{k+1})$ is the ith elementary symmetric function of the variables $x_1, \ldots, x_{k+1}$

$$(\sigma_1 = x_1 + \ldots + x_{k+1}, \ldots, \sigma_{k+1} = x_1 \cdot \ldots \cdot x_{k+1}).$$

The fact that this map is an isomorphism of complex manifolds follows from the fundamental theorem on symmetric functions (any analytic symmetric function of the variables $x_1, \ldots, x_{k+1}$ can be uniquely represented as an analytic function of the symmetric polynomials $\sigma_1, \ldots, \sigma_{k+1}$.) Under the isomorphism the space $\mathbb{C}^k/W$ maps isomorphically onto the coordinate hyperplane $\sigma_1 = 0$. The mirrors (non-regular orbits) are defined by the condition $x_i = x_j$. A miniversal deformation of the singularity $f(x)$ has the form

$$F(x, t_0, \ldots, t_{k-1}) = x^{k+1} + t_{k-1} x^{k-1} + \ldots + t_1 x + t_0.$$

Here

$$t_i = (-1)^{k+1-i} \sigma_{k+1-i}(x_1, \ldots, x_{k+1}),$$

where $(x_1, \ldots, x_{k+1})$ are the roots of the equation

$$F(x, t_0, \ldots, t_{k-1}) = 0, \quad t_k = x_1 + \ldots + x_{k+1} = 0.$$

The level bifurcation set consists of those values of the parameters $t = (t_0, t_1, \ldots, t_{k-1})$ for which the function $F(\cdot, t)$ has a critical point with critical value equal to zero, that is has a multiple root $x_i = x_j$. Hence in this case it is clear how to define the isomorphism mentioned in Theorem 3.8.

Approximately the same arguments are used to prove Theorem 3.8 for the other simple singularities.

Remember that a space of type $K(\pi, 1)$ is a space with fundamental group π and all of whose subsequent homotopy groups $(\pi_2, \pi_3, \ldots)$ are trivial. A space of type $K(\pi, 1)$ is the base of a principal fibre bundle with group π and with homotopically trivial total space.

In [57] it was proved that the space $\mathbb{C}^k/W \setminus S/W$ of regular orbits of the action of the group W is a space of type $K(\pi, 1)$, where π is the generalised Brieskorn braid group of the Weyl group W. If W is a group of Weyl type A_k, then π is the ordinary Artin braid group with $k+1$ strands.

Short digression. Braid groups.

To make our account more independent, we quote several definitions and results from the theory of braids. A more detailed exposition can be found in [57].

The graphico-geometric definition of a braid is that a braid is an object as depicted in figure 28. The braid consists of n non-intersecting strands in the space

$\mathbb{R}^3$, joining n fixed points on the lower base (line segment) with the same set of points on the upper base and going monotonically upwards from the lower base to the upper one. Two braids are considered equivalent if one can be deformed into the other without losing monotonicity and without allowing the strands to intersect each other. The braids can be multiplied, by placing one of them on top of the other (figure 29). With this multiplication, the braids of n strands (more precisely, their equivalence classes) form a group $B(n)$. The identity of this group is the "untangled" braid, consisting of vertical line segments joining points in the upper and lower bases. The braid inverse to a given one is obtained from it by reflection in a horizontal plane.

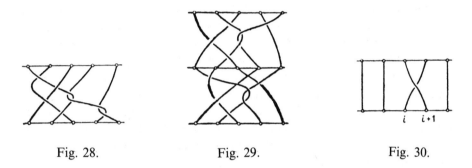

Fig. 28.   Fig. 29.   Fig. 30.

It is not hard to see that the braid group $B(n)$ of n strands is generated by $n-1$ generators $g_1,\ldots,g_{n-1}$, where g_i is the braid which 'crosses over" the ith and the $(i+1)$th strands (figure 30). These generators are connected by the relations

$$g_i g_j = g_j g_i \quad \text{for} \quad |i-j|>1, \quad \text{and}$$

$$g_i g_{i+1} g_i = g_{i+1} g_i g_{i+1} \quad (i=1,\ldots,(n-2)).$$

It can be shown that the indicated generators and relations define the group $B(n)$. To each braid there corresponds in an obvious way a permutation of n elements. Therefore there is a natural epimorphism of the braid group $B(n)$ onto the group $S(n)$ of permutations of n elements. The kernel $\hat{B}(n)$ of this homomorphism is called the coloured braid group on n strands. A coloured braid is a braid such that each of its strands returns to the same point as it started from.

A more formal definition of the braid group, which makes clear why it is valuable for problems in analysis, can be obtained in the following way. That part of the space $\mathbb{R}^3$ enclosed between the horizontal planes containing the lower and upper bases can be identified with the product $I \times \mathbb{C}$ of the segment $I=[0,1]$ and the plane of complex numbers $\mathbb{C}$. Under this identification, for each number

$t \in [0, 1]$, the braid gives rise in a continuous way to an unordered set of n distinct complex numbers. To $t=0$ and $t=1$ there correspond the same fixed set of numbers. In this way the braid group $B(n)$ is identified with the fundamental group of the space of all unordered sets of n distinct complex numbers. In exactly the same way, the coloured braid group $\hat{B}(n)$ is identified with the fundamental group of the space of all ordered sets of n distinct complex numbers.

Let $\mathbb{C}^n = \{(x_1, \ldots, x_n) : x_i \in \mathbb{C}\}$ be the space of ordered sets of n complex numbers, let S be the union of all hyperplanes, defined by the equations $x_i = x_j$, and let $\mathbb{C}_n \setminus S$ be the space of ordered sets of n distinct complex numbers. We have $\hat{B}(n) = \pi_1(\mathbb{C}^n \setminus S)$.

On the space $\mathbb{C}^n$, the group $S(n)$ of permutations of n elements acts by permuting the coordinates. The space $\mathbb{C}^n/S(n)$ is the space of unordered sets of n complex numbers. It is isomorphic to an n-dimensional complex vector space. The isomorphism is established by associating with an unordered set of complex numbers $(x_1, \ldots, x_n) \in \mathbb{C}^n/S(n)$ the polynomial $p(x) = \Pi_{i=1}^n (x - x_i)$ with roots $x_1, \ldots, x_n$ (or its coefficients, which up to sign are the elementary symmetric functions $\sigma_1, \ldots, \sigma_n$ of the variables $x_1, \ldots, x_n$: $\sigma_1 = x_1 + \ldots + x_n, \ldots,$ $\sigma_n = x_1 \cdot \ldots \cdot x_n$). The space $\Sigma = S/S(n)$ corresponds to the set of polynomials with multiple roots. In this way $(\mathbb{C}^n \setminus S)/S(n) = \mathbb{C}^n/S(n) \setminus \Sigma$ is the space of unordered sets of n distinct complex numbers; the braid group $B(n)$ coincides with its fundamental group $\pi_1(\mathbb{C}^n/S(n) \setminus \Sigma)$.

There is a stronger result which asserts that the space $(\mathbb{C}^n/S(n) \setminus \Sigma)$ is a space of type $K(\pi, 1)$ for the braid group $B(n)$ on n strands. This means that

$$\pi_1(\mathbb{C}^n/S(n) \setminus \Sigma) = B(n) \quad \text{and} \quad \pi_k(\mathbb{C}^n/S(n) \setminus \Sigma) = 0$$

for $k > 1$. Since the space $\mathbb{C}^n \setminus S$ is an ($n!$-fold) covering space of $(\mathbb{C}^n/S(n) \setminus \Sigma)$, the assertion about the space $\mathbb{C}^n/S(n) \setminus \Sigma$ is equivalent to the assertion that the space $\mathbb{C}^n \setminus S$ is a space of type $K(\pi, 1)$ (for the coloured braid group $\hat{B}(n)$ on n strands).

For the proof of the last assertion, we consider the map of the space $\mathbb{C}^n \setminus S = \{(x_1, \ldots, x_n) \in \mathbb{C}^n : x_i \neq x_j\}$ into the space $\{(x_1, \ldots, x_{n-1}) \in \mathbb{C}^{n-1} : x_i \neq x_j\}$, mapping the point $(x_1, \ldots, x_{n-1}, x_n)$ into $(x_1, \ldots, x_{n-1})$. It is easy to see that this map is a fibration with fibre $\mathbb{C} \setminus \{x_1, \ldots, x_{n-1}\}$. Since the fibre of this fibration has trivial homotopy groups from the second one upwards, the required assertion is proved by induction on the dimension n.

The group $S(n)$ of permutations of n elements is one of the finite groups generated by reflections. The action of the group $S(n)$ on the space $\mathbb{C}^n$ by permuting the coordinates is reducible. It decomposes into the direct sum of two actions: the action on the subspace $\mathbb{C}^{n-1}$, given by the equation $x_1 + \ldots + x_n = 0$, and the trivial action on the one-dimensional space $x_1 = \ldots = x_n$. On the subspace $\mathbb{C}^{n-1}$ its action is just the usual action of the Weyl group of type A_{n-1}, which is isomorphic to the group $S(n)$ of permutations of n elements. The union of all the mirrors, corresponding to the reflections in the group $S(n)$, in the space $\mathbb{C}^{n-1}$ (which are hyperplanes given by the equations $x_i = x_j$) and its quotient space under the action of this group we shall denote, as before, by S and Σ (this need not cause confusion). We have

$$\mathbb{C}^n \backslash S = (\mathbb{C}^{n-1} \backslash S) \times \mathbb{C}^1,$$
$$\mathbb{C}^n / S(n) \backslash \Sigma = (\mathbb{C}^{n-1} / S(n) \backslash \Sigma) \times \mathbb{C}^1.$$

Therefore

$$\hat{B}(n) = \pi_1(\mathbb{C}^n \backslash S) = \pi_1(\mathbb{C}^{n-1} \backslash S),$$

and

$$B(n) = \pi_1(\mathbb{C}^n / S(n) \backslash \Sigma) = \pi_1(\mathbb{C}^{n-1} / S(n) \backslash \Sigma).$$

The description of the coloured Braid group $\hat{B}(n)$ and the braid group $B(n)$ as fundamental groups of the the spaces $\mathbb{C}^{n-1} \backslash S$ and $\mathbb{C}^{n-1} / S(n) \backslash \Sigma$ respectively prompts a generalisation of this definition.

Let W be a finite irreducible group, generated by reflections, acting on the real vector space $\mathbb{R}^n$ of dimension n. The group W acts also on its complexification $\mathbb{C}^n$. We can show that the quotient space $\mathbb{C}^n / W$ is isomorphic to a complex vector space of dimension n (see [53]). Let $\{V_i\}$ be the set of all hyperplanes in the space $\mathbb{R}^n$, the reflections in which belong to the group W, and let $V_{i\mathbb{C}} \subset \mathbb{C}^n$ be their complexifications. Outside the subspace $S = \cup_i V_{i\mathbb{C}}$ the group W acts freely. Let $\Sigma = S/W$. The fundamental group

$$B_W = \pi_1(\mathbb{C}^n / W \backslash \Sigma)$$

of the space $\mathbb{C}^n / W \backslash \Sigma$ is called the (generalised) Brieskorn braid group of the group W; the fundamental group $\hat{B}_W = \pi_1(\mathbb{C}^n \backslash S)$ is called the (generalised) coloured Brieskorn braid group of the group W. There exists an exact sequence

$$1 \to \hat{B}_W \to B_W \to W \to 1.$$

Lemma 3.2. The spaces $\mathbb{C}^n / W \backslash \Sigma$ and $\mathbb{C}^n \backslash S$ are spaces of type $K(\pi, 1)$ (for the groups B_W and $\hat{B}_W$ respectively).

When the group W is of type A_{n-1} (the group of permutations of n elements) this lemma is already proved. We show how it can be proved when W is of type B_n (isomorphic to the groups of type C_n) and D_n. Naturally it is sufficient to prove the assertion of the lemma only for the space $\mathbb{C}^n \backslash S$.

When the group W is of type B_n, the mirrors of the reflections which belong to the group W are the hyperplanes $\{x_i \pm x_j = 0\}$ and $\{x_i = 0\}$ in the space $\mathbb{C}^n$. Using induction, we can suppose that the space

$$\{(x_1, \ldots, x_{n-1}) \in \mathbb{C}^{n-1} : x_i \pm x_j \neq 0, \ x_i \neq 0\}$$

is a space of type $K(\pi, 1)$. The natural projection

$$\mathbb{C}^n \backslash S = \{(x_1, \ldots, x_n) \in \mathbb{C}^n : x_i \pm x_j \neq 0, \ x_i \neq 0\}$$
$$\rightarrow \{(x_1, \ldots, x_{n-1}) \in \mathbb{C}^{n-1} : x_i \pm x_j \neq 0, \ x_i \neq 0\}$$

is a locally trivial fibration with fibre

$$\mathbb{C} \backslash \{0, \pm x_1, \pm x_2, \ldots, \pm x_{n-1}\}.$$

Since the fibre of this fibration has trivial homotopy groups from the second upwards, we can deduce the required asssertion about the space $\mathbb{C}^n \backslash S$.

When the group W is of type D_n the mirrors of the reflections belonging to the group W are the hyperplanes $\{x_i \pm x_j = 0\}$ in the space $\mathbb{C}^n$. We consider the map

$$\mathbb{C}^n \backslash S = \{(x_1, \ldots, x_n) \in \mathbb{C}^n : x_i \pm x_j \neq 0\}$$
$$\rightarrow \{(y_1, \ldots, y_{n-1}) \in \mathbb{C}^{n-1} : y_i \neq y_j, \ y_i \neq 0\},$$

given by the formula $y_i = x_n^2 - x_i^2$. This map is a locally trivial fibration. Its fibre is an affine complex curve and therefore has trivial homotopy groups from the second up. Just as for the space $\mathbb{C}^n \backslash S$, corresponding to the group of type B_n, (by considering the projection $\mathbb{C}^n \rightarrow \mathbb{C}^{n-1}$) it can be proved that the base

$$\{(y_1, \ldots, y_{n-1}) \in \mathbb{C}^{n-1} : y_i \neq y_j, \ y_i \neq 0\}$$

of this fibration is a space of type $K(\pi, 1)$. Therefore it follows that the space $\mathbb{C}^n \backslash S$ of this fibration is also a space of type $K(\pi, 1)$.

In the general case the lemma follows from the following general results of Deligne ([91]). Let us be given in the space $\mathbb{R}^n$ a finite number of hyperplanes V_i. Let $V_{i\mathbb{C}} \supset \mathbb{C}^n$ be their complexifications. Let us suppose that all the components of the complement of the union $\cup_i V_i$ in the space $\mathbb{R}^n$ are open simplicial cones (that is they have exactly n faces). Then the space $\mathbb{C}^n\backslash(\cup_i V_{i\mathbb{C}})$ is a space of type $K(\pi, 1)$. From Theorem 3.8 and Lemma 3.2 follows.

Theorem 3.9. For simple singularities the complement $D_\varepsilon\backslash\Sigma_\varepsilon$ of the level bifurcation set is a space of type $K(\pi, 1)$.

O. Lyashko and E. Looijenga (see [234], [231]) proved that for simple singularities the complement of the function bifurcation set is also a space of type $K(\pi, 1)$, where π is a subgroup of index $\mu! N^\mu |W|^{-1}$ in the Artin braid group on μ strands (here $|W|$ is the order of the corresponding Weyl group and N is the Coxeter number, that is, in the language of singularity theory, the order of the classical monodromy operator).

The inclusion of the fundamental group $\pi_1(\mathbb{C}^{\mu-1}\backslash\tilde{\Sigma}_\varepsilon)$ of the complement of the function bifurcation set of a simple singularity into the braid group on μ strands is constructed in the following manner. The braid group on μ strands can be considered as the fundamental group of the space of polynomials of the form

$$x^\mu + a_{\mu-2}x^{\mu-2} + \ldots + a_1 x + a_0,$$

which do not have multiple roots. In the complex vector space $\mathbb{C}^{\mu-1}_{(a)}$, with coordinates $(a_0, a_1, \ldots, a_{\mu-2})$, the points corresponding to polynomials with multiple roots form an algebraic variety Ξ. Its complement is a space of type $K(\pi, 1)$, where π is the braid group on μ strands. To each point v of the base $\mathbb{C}^{\mu-1}$ of the restricted miniversal deformation of the singularity f corresponds a function $F(\cdot, v)$, a perturbation of the singularity f. If each critical value is counted as many times as its multiplicity, then in a neighbourhood of zero in the space $\mathbb{C}^n$ this function has exactly μ critical values. The function bifurcation set $\tilde{\Sigma}_\varepsilon$ of the singularity f is distinguished in the base $\mathbb{C}^{\mu-1}$ of the restricted miniversal deformation by the condition that its points correspond to the functions $F(\cdot, v)$ which have less than μ distinct critical values. In this way when $v \in \tilde{\Sigma}_\varepsilon$ some critical values of the function $F(\cdot, v)$ coincide.

Let v be a point of the base $\mathbb{C}^{\mu-1}$ of the restricted miniversal deformation, let $F(\cdot, v)$ be the function corresponding to it, let $z_1, \ldots, z_\mu$ be its critical values in a neighbourhood of zero in the space $\mathbb{C}^n$ (the values z_i are not necessarily all distinct), let

$$\bar{z} = \sum_{i=1}^\mu z_i/\mu$$

be their arithmetic mean and let $\tilde{z}_i = z_i - \bar{z}$ $(i = 1, \ldots, \mu)$. Let

$$p_v(x) = \Pi_{i=1}^{\mu}(x - \tilde{z}_i)$$

be the polynomial of degree μ with roots

$$\tilde{z}_1, \ldots, \tilde{z}_\mu.$$

Since

$$\sum_{i=1}^{\mu} \tilde{z}_i = 0,$$

the coefficient of the monomial $x^{\mu-1}$ in the polynomial $p_v(x)$ equals zero. Therefore the polynomial $p_v(x)$ belongs to the space $\mathbb{C}_{(a)}^{\mu-1}$ of polynomials of type

$$x^{\mu} + a_{\mu-2} x^{\mu-2} + \ldots + a_1 x + a_0.$$

Setting in correspondence with the point $v \in \mathbb{C}^{\mu-1}$ the polynomial $p_v(x) \in \mathbb{C}_{(a)}^{\mu-1}$ we obtain a map

$$\psi : \mathbb{C}^{\mu-1} \to \mathbb{C}_{(a)}^{\mu-1}$$

from the base of the restricted miniversal deformation to the space $\mathbb{C}_{(a)}^{\mu-1}$. The map ψ maps the function bifurcation set $\tilde{\Sigma}_\varepsilon$ into the space $\varXi$ of polynomials with multiple roots, and the complement $\mathbb{C}^{\mu-1} \setminus \tilde{\Sigma}_\varepsilon$ of the bifurcation set into the space $\mathbb{C}_{(a)}^{\mu-1} \setminus \varXi$ of polynomials which do not have multiple roots. Direct calculation shows that in the complement of the bifurcation set $\tilde{\Sigma}_\varepsilon$ the mapping ψ is non-degenerate, that is it has rank equal to $\mu - 1$. The preimage of zero under the map ψ is the set of values of the parameter $v \in \mathbb{C}^{\mu-1}$ for which the function $F(\cdot, v)$ has a unique critical value. It follows from the corollary to Theorem 3.3 that it has a unique critical point. In this way the preimage of zero under the map ψ coincides with the stratum $\mu = \text{const}$ in the base of the restricted miniversal deformation. For simple singularities (and only for them!) this stratum consists of one point $v = 0$. Therefore it follows that the map

$$\psi : \mathbb{C}^{\mu-1} \to \mathbb{C}_{(a)}^{\mu-1}$$

is proper in a neighbourhood of zero and its restriction to the complement $\mathbb{C}^{\mu-1} \setminus \tilde{\Sigma}_\varepsilon$ of the function bifurcation set defines a cover of the space $\mathbb{C}_{(a)}^{\mu-1} \setminus \varXi$ of polynomials without multiple roots. So the complement of the function bifurcation set $\tilde{\Sigma}_\varepsilon$ of a simple singularity is a cover of a space of type $K(\pi, 1)$, from which it follows that it itself is a space of the same type. Moreover the map ψ induces an inclusion of the fundamental group of the complement of the bifurcation set $\tilde{\Sigma}_\varepsilon$ into the fundamental group of the space $\mathbb{C}_{(a)}^{\mu-1} \setminus \Xi$ of polynomials without multiple roots, which is the braid group on μ strands.

If $p : E \to B$ is a cover then its group of covering transformations is the group $\text{Aut}\,(p) = \{h : E \to E | h$ is a homeomorphism, $ph(x) = p(x)$ for $x \in E\}$. It is not difficult to see that the group of covering transformations $\text{Aut}\,(p)$ of the cover p is isomorphic to the quotient group

$$N(\pi_1(E))/\pi_1(E),$$

where $N(\pi_1(E))$ is the normaliser of the subgroup $\pi_1(E)$ in the group $\pi_1(B)$, that is

$$\{g \in \pi_1(B) : g\pi_1(E)g^{-1} = \pi_1(E)\}.$$

For simple singularities the group of covering transformations $\text{Aut}\,(\psi)$ of the cover

$$\psi : \mathbb{C}^{\mu-1} \setminus \tilde{\Sigma}_\varepsilon \to \mathbb{C}_{(a)}^{\mu-1} \setminus \Xi$$

of the complement of the function bifurcation set over the space of polynomials without multiple roots, which we constructed above, is described in [220]. It is cyclic for all simple singularities except A_1 and D_4. Its order is equal to the Coxeter number of the corresponding Weyl group (or, which amounts to the same thing, the order of the classical monodromy operator) for singularities of type $A_\mu(\mu \neq 1)$, $D_\mu(\mu \neq 4)$ and E_6, whilst for singularities of type E_7 and E_8 it is half the Coxeter number. For singularities of type D_4 the group $\text{Aut}\,(\psi)$ is isomorphic to $\mathbb{Z}_3 \oplus S(3)$, where $S(3)$ is the group of permutations of three elements, for singularities of type A_1 the group $\text{Aut}\,(\psi)$ is trivial.

In the real case, that is when we are considering a real miniversal deformation of a real singularity, E. Looijenga ([232]) proved that the complement of the level bifurcation set of a simple singularity has contractible components. These components are in one to one correspondence with classes of W-conjugate elements of the second order in the adjacent class Wn, where $n \in N/W$ is some element of the second order, W is the corresponding Weyl group and N is its normaliser in the group of all linear transformations. The group N/W is the same as the group of automorphisms of the corresponding Dynkin diagram.

O. V. Lyashko ([234]) described all decompositions of simple singularities, found in the base of its restricted miniversal deformation. Let $f_\lambda(x)$ $(\lambda \in (\mathbb{C}, 0))$ be a deformation of a singularity $f : (\mathbb{C}^n, 0) \to (\mathbb{C}, 0)$. We say that under the deformation f_λ the singularity f decomposes into types

$$X = (X_1, \ldots, X_k),$$

where $X_i = (X_{i1}, \ldots, X_{ij_i})$, if for values of the parameter λ sufficiently close to zero (but different from zero) the function f_λ has (in a small neighbourhood of zero in the space $\mathbb{C}^n$) k distinct critical values $z_1, \ldots, z_k$, the critical value z_i being attained at j_i critical points, at which the function f_λ has singularities of types $X_{i1}, \ldots, X_{ij_i}$.

We shall say that the D-diagram E (with respect to a distinguished basis $\Delta_1, \ldots, \Delta_\mu$) decomposes into the types

$$(E_1, \ldots, E_k)$$

where $E_i = (E_{i1}, \ldots, E_{ij_i})$, if:

(i) $\{E_i\}$ is a decomposition of the set of vertices of the diagram E into disjoint subsets, the vertices of each of the diagrams E_i (basic vanishing cycles) being numbered in the diagram E by successive integers;

(ii) the vertices of the diagram E_i are joined to each other by edges of the same multiplicity as they are joined in the diagram E;

(iii) E_{ij} are the connected components of the diagrams E_i $(j = 1, \ldots, j_i)$.

The decomposition of the singularity f into the types $X = (X_1, \ldots, X_k)$ $(X_i = (X_{i1}, \ldots, X_{ij_i}))$ is said to be compatible with the decomposition of its diagram E into the types $(E_1, \ldots, E_k)$ $(E_i = (E_{i1}, \ldots, E_{ij_i}))$, if for each i the set $E_{i1}, \ldots, E_{ij_i}$ is the set of D-diagrams of the critical points $X_{i1}, \ldots, X_{ij_i}$ corresponding to the ith critical value. It is easy to show that if the singularity f decomposes into the types $X = (X_1, \ldots, X_k)$, then there exists a distinguished basis $\{\Delta_p\}$ such that the D-diagram E of the singularity f decomposes into the types $(E_1, \ldots, E_k)$ compatible with the decomposition of the singularity. O. V. Lyashko proved that for simple singularities we have also a converse assertion: if the diagram E of a simple singularity f with respect to any distinguished basis decomposes into the types $(E_1, \ldots, E_k)$ $(E_i = (E_{i1}, \ldots, E_{ij_i}))$, then there exists a deformation of the singularity f for which the decomposition is compatible with the decomposition of the diagram E.

For a more detailed description of all the decompositions of simple singularities see [234].

3.4   The $\mu = $ constant stratum and the topological type of singularity

Deformations which preserve the multiplicity of a singularity must not change its topology in an essential way.

Indeed, when the number of variables $n \neq 3$, Lê and Ramanujam ([209]) proved that under a deformation which preserves the multiplicity, the topological type of the singular level set (more precisely, of the pair

$$(B_\varrho, f^{-1}(0) \cap B_\varrho),$$

where B_ϱ is a ball of sufficiently small radius ϱ with centre at the critical point) will not change and the differential type of the Milnor fibration will not change. The restriction $n \neq 3$ arises from the fact that the proof makes use of the h-cobordism theorem. At present it is not known whether this result holds for $n = 3$. There were proofs offered in the case $n = 3$, but later gaps were found in them.

Timourian ([353]) proved that under a deformation with constant multiplicity the topological type of the function also does not change. This means the following. Let $F(x, t)$ be a smooth – in $t \in \mathbb{R}^p$ – deformation of the singularity

$$f: (\mathbb{C}^n, 0) \to (\mathbb{C}, 0) \quad (F(x, 0) = f(x))$$

such that for any t the germ $F(\cdot, t)$ has at zero a critical point of one and the same multiplicity $\mu = \mu(f)$ with critical value equal to zero. Then there exist neighbourhoods U of zero in $\mathbb{C}^n \times \mathbb{R}^p$, U_0 of zero in $\mathbb{C}^n$ and D of zero in $\mathbb{R}^p$ and a homeomorphism $\alpha: U \to U_0 \times D$ ($\alpha(0, t) = (0, t)$), giving the commutative diagram

$$
\begin{array}{ccc}
U & \xrightarrow{\alpha} & U_0 \times D \\
\downarrow F & & \downarrow f \times \mathrm{Id} \\
\mathbb{C} & \xleftarrow{\pi} & \mathbb{C} \times D
\end{array}
$$

(here π is projection onto the first factor).

It is easy to show that along the stratum $\mu = $ constant the intersection maxtrix and the monodromy group of the singularity do not change.

Other characteristics of the singularity, however, which have a "more analytic" character, can change. F. Pham showed ([285]) that under a deformation with constant multiplicity the topology of the level bifurcation set of the singularity, more precisely its decomposition into pieces in correspondence with the singularities of the zero level set, can change.

In order to construct such an example, we consider the singularity $f(x, y) = y^3 + x^9$, the multiplicity of which is equal to 16. Its miniversal deformation has a base of dimension 16 and is given by the formula

$$F(x, y, u, v) = y^3 + u(x) y + v(x),$$

where

$$u(x) = u_0 + u_1 x + \ldots + u_7 x^7,$$

$$v(x) = v_0 + v_1 x + \ldots + v_7 x^7 + x^9,$$

$$u = (u_0, u_1, \ldots, u_7), \quad v = (v_0, v_1, \ldots, v_7).$$

Let $X = F^{-1}(0) \subset \mathbb{C}^{18}$ be the zero level set of the deformation F and let $G : X \to \mathbb{C}^{16}_{u,v}$ be its projection onto the base of the deformation. Denoting by $X^{*\alpha}$ the (analytic) set of points $z \in X$ at which the curve $G^{-1}(G(z))$ has degree 3 (that is locally given by an equation belonging to the cube of the maximal ideal) and the order of contact $\geq \alpha$. It is not difficult to see that the set $X^{*\alpha}$ consists of quadruples (x, y, u, v) for which $y = 0$, x is a root of the equation $u(x) = 0$ of multiplicity $\geq 2\alpha$ and a root of the equation $v(x) = 0$ of multiplicity $\geq 3\alpha$. In particular

$$X^{**} = X^{*3} = \{(x, y, u, v) : x = y = 0, \ v = 0, \ u_i = 0 \quad \text{for} \quad i < 6\}.$$

The projection T^{**} of the set X^{**} into the base $\mathbb{C}^{16}_{u,v}$ of the miniversal deformation has dimension equal to 2 and is the stratum $\mu = $ constant in it.

We consider the set $T^{*5/3} = G(X^{*5/3})$. It consists of pairs $(u, v) \in \mathbb{C}^{16}_{u,v}$ such that the polynomials $u(x)$ and $v(x)$ have a common root of multiplicity 4 for $u(x)$ and multiplicity 5 for $v(x)$. The set

$$G^{-1}(T^{*5/3}) \cap X^{*4/3}$$

can be represented as the union of two sets: $X^{*5/3}$ and X'. Here X' is the set of quadruples $(x, y, u, v) \in \mathbb{C}^{18}$, such that $y = 0$, x is a root of the equation $u(x) = 0$ of multiplicity 3 and of the equation $v(x) = 0$ of multiplicity 4 and in addition the polynimials $u(x)$ and $v(x)$ have another common root of multiplicity 4 for $u(x)$ and 5 for $v(x)$. The intersection X'' of these two sets $X^{*5/3}$ and X' (more precisely of their closures) consists of quadruples $(x, y, u, v) \in \mathbb{C}^{18}$ such that $y = 0$, x is a root of multiplicity 7 for the equation $u(x) = 0$ and a root of multiplicity 9 for the equation $v(x) = 0$. From this it follows that $v = (v_0, \ldots, v_7) = 0$, $x = 0$ and $u_0 = \ldots = u_6 = 0$. Thus the set X' lies in the set X^{**} but does not coincide with it. It means that the sets $X^{*\alpha}$ approach different points of the set X^{**} in different fashions and it also means that the level bifurcation set changes along the stratum T^{**} ($\mu = $ constant).

It is not difficult to see that $X^{*4/3}$ is the set of points $z \in X$ in which the curve $G^{-1}(G(z))$ has a singularity of type E_6, $X^{*5/3}$ is the set of points $z \in X$ in which the curve $G^{-1}(G(z))$ has singularity type E_8. Therefore $G(X')$ is the set (more precisely its closure) of those values $(u, v) \in \mathbb{C}^{16}_{u,v}$, for which the curve $G^{-1}(u, v)$

has two singularities of types E_6 and E_8. The constructed reasoning shows that the stratum of the level bifurcation set, consisting of points in which the corresponding function has singularities of types E_6 and E_8 on the zero level set, changes (simply disappears) along the family $\mu=$ constant (T^{**}) of the singularity x^3+x^9.

S. M. Gusein-Zade and N. N. Nekhoroshev ([153]) gave an example of a deformation of constant multiplicity of a homogeneous polynomial of degree 22 in two variables, along which the largest k such that an A_k singularity adjoins the given one changes.

There was a conjecture, that in the base of the restricted miniversal deformation of a singularity the $\mu=$ constant stratum, that is the set of values of parameters for which the corresponding function has a critical point of the same multiplicity as the initial singularity, is a non-singular manifold. A. M. Gabrielov proved ([117]) that the dimension of this set is equal to the modality of the singularity.

The conjecture about the smoothness of the $\mu=$ constant stratum is proved for the case where the number of variables $n=2$. This was first proved, apparently, by J. Wahl, 1971. As regards this see [56], see also the article by Teissier [349]. I. Luengo has shown that this conjecture doesn't take place for $n=3$.

3.5 The resolution of a singularity and some properties of the classical monodromy operator

A useful instrument for studying the topology of a singularity is its resolution.

Let $f:(\mathbb{C}^n,0)\to(\mathbb{C},0)$ be a singularity, that is the germ of a holomorphic function with an isolated critical point at the origin.

Definition. The *resolution* of a singularity f is a proper analytic map

$$\pi:(Y,Y_0)\to(\mathbb{C}^n,0)$$

on a non-singular complex manifold Y such that:

(i) the map $\pi|_{Y\setminus Y_0}$ is an analytic isomorphism: $Y\setminus Y_0\to\mathbb{C}^n\setminus 0$ (or from a neighbourhood of the space Y_0 in Y to a neighbourhood of zero in $\mathbb{C}^n$);

(ii) the subspace $Y_0=\pi^{-1}(0)$ of the space Y is the union of non-singular $(n-1)$-dimensional manifolds (divisors) in Y which are in general position;

(iii) in a neighbourhood of any point of $Y_0=\pi^{-1}(0)$ there exists a local system of coordinates $y_1,\ldots,y_n$ such that

$$f\circ\pi(y_1,\ldots,y_n)=y_1^{k_1}\cdot\ldots\cdot y_n^{k_n};$$

and the Jacobian of the map π is equal to

$$g(y_1, \ldots, y_n) y_1^{m_1} \cdot \ldots \cdot y_n^{m_n},$$

where $g(0, \ldots, 0) \neq 0$.

The existence of a resolution of any singularity is a consequence of a theorem of Hironaka ([158]). In the case when f is a function of two variables, its resolution can be constructed with the help of some successive σ-processes (see [328], and also § 4.3) at singular points.

Many topological characteristics of a singularity (for example its multiplicity, the characteristic polynomial of its classical monodromy operator and others) can be expressed in terms of the topological characteristics of the divisors which are glued in during the resolution of the singularity. Before formulating the corresponding results we introduce several concepts.

The *characteristic polynomial* $P_f(z)$ (of the classical monodromy operator) of a singularity f is

$$\det(z \cdot id - h_* |_{H_{n-1}(V_\varepsilon)})$$

(where V_ε is the non-singular level manifold of the singularity f). The roots of the characteristic polynomial $P_f(z)$ are the eigenvalues of the classical monodromy operator h_* of the singularity.

We can also express, in terms of the characteristic polynomial $P_f(z)$ of the singularity f, the determinant $\det S$ of the intersection form in the homology $H_{n-1}(V_\varepsilon; \mathbb{Z})$ of the non-singular level manifold of the singularity f. It is an invariant, since the determinant of a change of basis of the integer lattice $H_{n-1}(V_\varepsilon; \mathbb{Z})$ is equal to ± 1. The determinant of the intersection form is the same as the determinant of the matrix of the operator

$$i_* : H_{n-1}(V_\varepsilon; \mathbb{Z}) \to H_{n-1}(V_\varepsilon, \partial V_\varepsilon; \mathbb{Z}).$$

If $\det S \neq 0$ then it is equal, modulo the sign, to the order of the group

$$H_{n-1}(V_\varepsilon, \partial V_\varepsilon; \mathbb{Z})/\operatorname{Im} i_*.$$

We have

$$\det S = \det(-\operatorname{Var}^{-1} + (-1)^n (\operatorname{Var}^{-1})^T)$$

$$= (-1)^{\mu n(n+1)/2} \det(-id + (-1)^n \operatorname{Var}(\operatorname{Var}^{-1})^T)$$

$$= (-1)^{\mu n(n+1)/2} \det(-id + h_*) = (-1)^{\mu(n-1)(n-2)/2} P_f(1).$$

Sometimes instead of the characteristic polynomial of the singularity it is more convenient to use what is called the ζ-function of the classical monodromy transformation h of the singularity. Firstly, it usually gives more beautiful solutions and, secondly, the ζ-function of the monodromy is defined also for non-isolated singularities, whilst the characteristic polynomial becomes practically meaningless. Many of the following results hold also for non-isolated critical points, but we shall not specially specify these.

Definition. The ζ-*function of the transformation* $g : X \to X$ of the topological space X (for definiteness a finite CW complex) is the rational function

$$\zeta_g(z) = \Pi_{q \geqslant 0} \{\det [id - zg_* |_{H_q(X; \mathbb{R})}]\}^{(-1)^q}.$$

In this definition the zeroth homology of the space X is taken into account, that is we do not suppose that the homology is reduced modulo a point. The definition also makes sense for a pair of spaces (X, Y) and a transformation $g : X \to X$ carrying the subspace Y into itself. In this case the action of the transformation g on the relative homology $H_q(X, Y; \mathbb{R})$ figures in the formula for the ζ-function.

Definition. The ζ-*function of the monodromy* of the singularity f is the ζ-function of the classical monodromy transformation h of the non-singular level manifold V_ε of the singularity f into itself.

For isolated singularities we have $H_q(V_\varepsilon) = 0$ for $q \neq 0, n-1$. Consequently

$$\zeta_f(z) = (1 - z) (z^\mu P_f(z^{-1}))^{(-1)^{n-1}},$$

from which

$$P_f(z) = z^\mu ((z/(z-1)) \zeta_f(z^{-1}))^{(-1)^{n-1}},$$

where μ is the multiplicity of the singularity. Thus the characteristic polynomial $P_f(z)$ and the ζ-function $\zeta_f(z)$ of a singularity can be derived from each other.

It is easy to see that the degree of the rational function $\zeta_f(z)$ (equal to the degree of the numerator minus the degree of the denominator) is equal to the Euler characteristic $\chi(V_\varepsilon)$ of the non-singular level manifold V_ε.

The following result allows us to derive the ζ-function of a singularity from the topological invariants of the divisors which are glued in during its resolution. Let

$$\pi : (Y, Y_0) \to (\mathbb{C}^n, 0)$$

be the resolution of the singularity f, let S_m be the set of points of the space Y_0, in a neighbourhood of which the function $f \circ \pi$ in some local system of coordinates has the form x_1^m (clearly the intersections of the glued in divisors do not enter into the set S_m).

Theorem 3.10 (N. A'Campo [5]).

$$\mu(V_\varepsilon) = \sum_{m \geq 1} m\chi(S_m),$$

$$\zeta_f(z) = \prod_{m \geq 1} (1 - z^m)^{\chi(S_m)}.$$

For the proof we use the property that if the transformation $h : X \to X$ preserves the subspace Y, then

$$\zeta_{h_X}(z) = \zeta_{h_Y}(z) \cdot \zeta_{h_{(X,Y)}}(z),$$

where $h_X, h_Y, h_{(X,Y)}$ are the transformation h considered on the corresponding spaces (or pair of spaces). There exists a map

$$\phi : V_\varepsilon \to (f \circ \pi)^{-1}(0)$$

(the contraction of the non-singular fibre onto the singular one) for which points from S_m have m preimages, the preimage of the intersection of any k of the glued-in divisors is a fibration whose fibres are $(k-1)$-dimensional tori. The classical monodromy transformation h can be considered to be compatible with this map in the sense that it preserves preimages under the map ϕ of points of the space $(f \circ \pi)^{-1}(0)$, its action on it being trivial. Over the points of the set S_m the transformation h carries out a cyclic permutation of preimages. The ζ-function of a cyclic permutation of m points is equal to $(1 - z^m)$. From this it follows that the ζ-function of the classical monodromy transformation h, restricted to the preimage $\phi^{-1}(S_m)$ of the set S_m is equal to $(1 - z^m)^{\chi(S_m)}$. Over the points of intersection of the glued-in divisors the transformation h is represented as diffeomorphisms of tori which are shears and therefore do not contribute to the ζ-function.

The idea of this construction is due to Clemens ([77]).

From the formula for the ζ-function of a singularity, introduced in Theorem 3.10, it follows that all the eigenvalues of the classical monodromy operator of an isolated singularity are roots of unity of various degrees. Therefore some power N of the classical monodromy operator has all its eigenvalues equal to one. We can take for N a number which is divisible by the multiplicity m of all the divisors which are glued in during the resolution. In this way we obtain

Theorem 3.11. The operator $(h_*^N - \mathrm{id})$ is nilpotent, that is $(h_*^N - \mathrm{id})^k = 0$ for some k.

This theorem is due to Brieskorn ([55]), Katz ([181]) and a number of other authors. Its generalisation to the case of the germ of an analytic function on an analytic space, all level sets of which can have singularities, was obtained in [207].

By considering the resolution of an isolated singularity we can obtain an estimate of the size of the index k, which, as it is not difficult to see, is equal to the maximal dimension of the Jordan blocks of the classical monodromy operator. For this we take the map $\mathbb{C}_u \to \mathbb{C}_z$, mapping u to $z = u^N$ and consider over $\mathbb{C}_u$ the family of manifolds, induced from the family $\{(f \circ \pi)(x) = z\}$ over $\mathbb{C}_z$ via this map. Resolving the fibre over zero, we obtain a gluing in which all the divisors come with multiplicity 1 (that is $S_m = \emptyset$ for $m > 1$). The monodromy operator $\tilde{h}_*$ of this family is equal to h_*^N. Let Z_i be that part of the fibre over zero, which is the union of all i-fold intersections of glued-in non-singular divisors. We have

$$Z_1 = (f \circ \pi)^{-1}(0) \supset Z_2 \supset Z_3 \supset \ldots \supset Z_n \supset Z_{n+1} = \emptyset.$$

As before we have a map from the non-singular fibre into the singular fibre at zero, which is compatible with the monodromy transformation. It can be proved that if a is a cycle in the non-singular fibre which lies in the preimage of the set Z_i, then the cycle $\tilde{h}_* a - a$ is homologous to a cycle lying in the preimage of the set Z_{i+1}. From this it follows that $(\tilde{h}_* - \mathrm{id})^n = 0$, that is that $(h_*^N - \mathrm{id})^n = 0$. In this way we can take as the index k in Theorem 3.11 the number of variables n and consequently the Jordan blocks of the classical monodromy operator h_* have dimension no larger than $n \times n$. Thus we have proved.

Theorem 3.12. The dimensions of the Jordan blocks of the normal form of the classical monodromy operator h_* of the singularity

$$f : (\mathbb{C}^n, 0) \to (\mathbb{C}, 0)$$

of a function of n variables does not exceed $n \times n$.

For example if f is a singularity of a function of two variables, then the maximal possible dimension of the Jordan blocks of its classical monodromy operator is two. For the singularity

$$f(x, y) = (x^3 + y^2)(x^2 + y^3)$$

it is indeed equal to two. We can show that the classical monodromy operator of the singularity is not diagonalisable in the following way. The quadratic form of the singularity f can be found by the method of Chapter 4. In §4.4 it will be shown that it has the following indices of inertia: the positive index of inertia $\mu_+ = 1$, the zero index of inertia $\mu_0 = 1$, the negative index of inertia $\mu_- = \mu - 2 = 9$. From the formula of Picard-Lefschetz it follows that the eigenvectors of the classical monodromy operator h_* of the singularity corresponding to the eigenvalue 1 are elements of the space $H_{n-1}(V_\varepsilon)$, orthogonal in the sense of the intersection form to all the vanishing cycles of the distinguished basis $\{\Delta_i\}$, and, consequently, to all the elements of the space $H_{n-1}(V_\varepsilon)$. In this way for the singularity of the function

$$\hat{f}(x, y, t) = f(x, y) + t^2$$

the subspace of vectors $a \in H_{n-1}(V_\varepsilon)$ satisfying the condition $h_* a = a$ is one-dimensional. For the characteristic polynomial $P_{\hat{f}}(z)$ we have

$$P_{\hat{f}}(z) = \det(z \cdot \mathrm{id} - h_*) = \det(z \cdot \mathrm{id} - L^{-1}L^T)$$

$$= \det(zL + L^T) = \det(zL^T + L)$$

$$= z^\mu(\det(L^T + z^{-1}L)) = z^\mu P_{\hat{f}}(z^{-1}).$$

Consequently $P_{\hat{f}}(z)$ is a reflexive polynomial of degree μ (the coefficients of the monomials z^ν and $z^{\mu-\nu}$ coincide). The multiplicity of unity, as a root of a reflexive polynomial is always even. Therefore the space of elements of the homology group $H_2(V_\varepsilon)$ of the non-singular level manifold of the singularity $\hat{f}(x, y, t)$ associated with the eigenvalue 1 of the classical monodromy operator h_* has even dimension and consequently does not coincide with the space of eigenvectors with eigenvalue 1. In its turn the classical monodromy operator of the singularity $f(x, y)$ is obtained from the classical monodromy operator of the singularity $\hat{f}(x, y, t)$ multiplied by -1 and therefore is also not diagonalisable, namely it has a Jordan block of dimension 2×2 corresponding to the eigenvalue -1.

Steenbrink ([343]) proved that the dimension of the Jordan block of the normal form of the classical monodromy operator of the singularity $f:(\mathbb{C}^n, 0)\rightarrow(\mathbb{C}, 0)$, corresponding to the eigenvalue 1, does not exceed $(n-1) \times (n-1)$. See also § 13.2.5.

For a quasihomogeneous singularity $f:(\mathbb{C}^n, 0)\rightarrow(\mathbb{C}, 0)$ the germ of f belongs to its Jacobian ideal

$$J_f=(\partial f/\partial x_1,\ldots, \partial f/\partial x_n),$$

the classical monodromy operator h_* has finite order, that is it is diagonalisable. Briançon and Skoda ([54]) proved that for an arbitrary singularity $f:(\mathbb{C}^n, 0) \rightarrow(\mathbb{C}, 0)$ in n variables the nth power of the germ f belongs to the Jacobian ideal J_f. As we made clear above, the dimension of the Jordan blocks of the classical monodromy operator h_* does not exceed $n \times n$. Arising from these considerations, Scherk ([316]) conjectured that if the kth power of the germ f belongs to the Jacobian ideal J_f, then the dimension of the Jordan blocks of the classical monodromy operator of the singularity f does not exceed $k \times k$. In [317] he proved this conjecture. In this regard see also Theorem 14.19 in § 14.3.5.

There is a way of constructing the resolution of a singularity using its Newton diagram (the so-called toral resolution). For almost all functions with a given Newton diagram it indeed leads to a resolution of the singularity. The construction of this resolution can be found in [358]. See also Chapter 8.

Arising from this and Theorem 3.10, a description is obtained in [359] of the ζ-function (or the characteristic polynomial) of the classical monodromy operator of a singularity from its Newton diagram.

Let $\Gamma \subset \mathbb{N}^n$ be a Newton diagram ($\mathbb{N}$ is the set of non-negative integers). The ζ-function of the diagram Γ is the function

$$\zeta_\Gamma(z)= \prod_{l=1}^{n} (\zeta^l(z))^{(-1)^{l-1}},$$

where the polynomials $\zeta^l(z)$ are defined below (they are defined by the intersections of the diagram Γ with all possible l-dimensional coordinate planes in the space $\mathbb{R}^n$).

If L is an l-dimensional affine subspace in the space $\mathbb{R}^n$ such that $L \cap \mathbb{N}^n$ is an l-dimensional integer lattice, then we shall adopt the convention that the l-dimensional volume of the parallelepiped spanned by any basis in $L \cap \mathbb{N}^n$ is equal to one. For a set $I \subset \{1,\ldots,n\}$ with the number of elements $\#I=l$ we put

$$L_I=\{k \in \mathbb{R}^n : k_i=0 \quad \text{for} \quad i \notin I\}.$$

Let $\Gamma_j(I)$ $(j=1,\dots,j(I))$ be all $(l-1)$-dimensional faces of the polyhedron $L_I \cap \Gamma$ and let $L_j(I)$ be the $(l-1)$-dimensional affine subspaces in which they lie. The quotient group of the lattice $\mathbb{N}^n \cap L_I$ by the subgroup generated by the vectors of $N^n \cap L_j(I)$ is cyclic. Let us denote its order by $m_j(I)$. Let $V_j(I)$ be the $(l-1)$-dimensional volume of the face $\Gamma_j(I)$ in the space $L_j(I)$. We remark that $m_j(I)(l-1)!V_j(I)$ is equal to the product by $l!$ of the l dimensional volume of the cone over $\Gamma_j(I)$ with vertex at the origin.

Let us put

$$\zeta^l(z) = \prod_{I:\#I=l} \prod_{j=1}^{j(I)} (1-z^{m_j(I)})^{(l-1)!V_j(I)}.$$

Theorem 3.13. For almost all functions $f:(\mathbb{C}^n,0)\to(\mathbb{C},0)$ with Newton diagram Γ the ζ-function of the classical monodromy operator of the singularity f coincides with the ζ-function of the diagram Γ.

The condition identifying the set of functions for which the assertion of Theorem 3.13 holds can be expressed by means of the coefficients which enter in the expansion of the germ of f in monomials lying in the Newton diagram Γ (see Chapter 8). Generally speaking, it can happen that all functions with Newton diagram Γ have non-isolated singularities. Nevertheless Theorem 3.13 remains correct.

Since the degree of the ζ-function of the monodromy operator of a singularity coincides with the Euler characterstic of its non-singular level manifold there follows from Theorem 3.13 a result of A. G. Kushnirenko ([195]), expressing the multiplicity of a singularity in terms of its Newton diagram.

3.6   The monodromy group and distinguished bases of simple singularities

A most effective description of the monodromy group exists for simple singularities, that is for singularities which do not have a continuous modulus (see Volume 1, Chapter 15). Just as for any singularity, we have here two distinct cases: singularities with an odd number of variables and singularities with an even number of variables. In the first case the Picard-Lefschetz operator is reflection in a hyperplane, orthogonal in the sense of the intersection form to the corresponding vanishing cycle; the monodromy group is a group generated by reflections. In the second case the description of the Picard-Lefschetz operator is more unusual.

As is well known, the monodromy group of a singularity is defined by its intersection matrix with respect to a weakly distinguished basis. For simple singularities we get the following result.

Theorem 3.14. For the simple singularities $A_k (x^{k+1} + \Sigma t_i^2)$, $D_k (x^2 y + y^{k-1} + \Sigma t_i^2)$, $E_6 (x^3 + y^4 + \Sigma t_i^2)$, $E_7 (x^3 + xy^3 + \Sigma t_i^2)$, $E_8 (x^3 + y^5 + \Sigma t_i^2)$, there exist distinguished bases of vanishing cycles, in which the D-diagrams coincide with the classical diagrams of the corresponding Lie algebras of the same names (figure 31). The monodromy groups of these singularities with an odd number of variables are finite and isomorphic to the Weyl groups of the corresponding algebras.

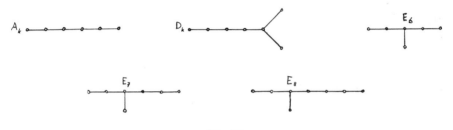

Fig.31.

For a description of the classical Weyl groups see [53].

For a singularity of type A_k Theorem 3.14 was proved in §2.9. For the rest of the simple singularities the proof, based on the fact that all these singularities are stably equivalent to singularities of functions of two variables, will be constructed in §4.1.

A method of constructing D-diagrams of simple singularities with an odd number of variables directly from the monodromy group was found by McKay ([253]).

We denote by $M_f^{(o)}$ and $M_f^{(e)}$ the monodromy groups of singularities, stably equivalent to f, with an odd and an even number of variables respectively. The generators of the groups $M_f^{(o)}$ and $M_f^{(e)}$ are transformations of the integer lattice, defined by the formula of Picard-Lefschetz. These transformations coincide modulo two. Consequently, the corresponding groups of transformations of the homology $H_{n-1}(V_\varepsilon; \mathbb{Z}_2)$ of the non-singular level manifold with coefficients in $\mathbb{Z}_2$ ($H_{n-1}(V_\varepsilon; \mathbb{Z}_2) \approx (\mathbb{Z}_2)^\mu$) are identical. Therefore there is one monodromy group $M_f^{\mathbb{Z}_2}$ of the singularity modulo two, acting on the binary lattice $(\mathbb{Z}_2)^\mu$. There are defined natural epimorphisms

$$M_f^{(o)} \to M_f^{\mathbb{Z}_2} \quad \text{and} \quad M_f^{(e)} \to M_f^{\mathbb{Z}_2},$$

induced by the homomorphism $\mathbb{Z} \to \mathbb{Z}_2$. Consequently, the monodromy group $M_f^{\mathbb{Z}_2}$ in the homology with coefficients in $\mathbb{Z}_2$ is a quotient group of the group $M_f^{(o)}$ and also of the group $M_f^{(e)}$.

For simple singularities a description of the group $M_f^{\mathbb{Z}_2}$ follows from the fact that the kernel of the homomorphism

$$M_f^{(o)} \to M_f^{\mathbb{Z}_2}$$

is either trivial (in which case $M_f^{\mathbb{Z}_2} \approx M_f^{(o)}$) or contains $\pm \mathrm{id}$, where id is the identity transformation (in this case $M_f^{\mathbb{Z}_2} \approx M_f^{(o)}/\mathbb{Z}_2$). The kernel of the map

$$M_f^{(o)} \to M_f^{\mathbb{Z}_2}$$

coincides with the group $\mathbb{Z}_2$ if and only if the monodromy group $M_f^{(o)}$ of the singularity f with an odd number of variables contains the transformation which is multiplication by -1.

For simple singularities with an even number of variables the monodromy group $M_f^{(e)}$ is described by the following result.

Theorem 3.15 (A. N. Varchenko, S. V. Chmutov [72]; see also [6], [397]). For simple singularities with an even number of variables the monodromy group coincides with the group of all linear operators g on the integral lattice

$$\mathbb{Z}^\mu = H_{n-1}(V_\varepsilon; \mathbb{Z}),$$

satisfying the following three conditions:

(i) the operator g preserves the (skew-symmetric) intersection form of the singularity;

(ii) the restriction of the operator g to the kernel of the intersection form (that is to the set of vectors orthogonal to all the elements of the lattice $\mathbb{Z}^\mu$) is the identity transformation;

(iii) the operator g, reduced modulo 2, belongs to the monodromy group $M_f^{\mathbb{Z}_2}$ in the homology with coefficients in the group $\mathbb{Z}_2$.

The necessity of satisfying conditions (i)–(iii) for any (not necessarily simple) singularity is obvious. For a singularity f of a function of two variables the dimension of the kernel K of the intersection form is equal to $r-1$ where r is the number of irreducible components of the germ of the curve $\{f=0\}$. Therefore for simple singularities we have: $\dim K = 0$ for the singularities A_{2s}, E_6, E_8, $\dim K = 1$ for the singularities A_{2s+1}, D_{2s+1}, E_7 and $\dim K = 2$ for the singularities D_{2s}.

In [74] a generalisation was obtained of Theorem 3.15 to the case of an arbitrary singularity. By Frobenius' theorem (see [243]) we can choose in the integral lattice

$$\mathbb{Z}^\mu = H_{n-1}(V_\varepsilon; \mathbb{Z})$$

with antisymmetric intersection form an (integral) basis

$$e_1, \ldots, e_n, f_1, \ldots, f_n, g_1, \ldots, g_l$$

$(2n+l=\mu(f))$ such that

$$(e_i \circ f_i) = \lambda_i, \ (e_i \circ f_j) = 0 \quad \text{for} \quad i \neq j,$$
$$(e_i \circ e_j) = (f_i \circ f_j) = (g_i \circ g_j) = (e_i \circ g_j) = (f_i \circ g_j) = 0,$$

the integer λ_i dividing the number λ_{i-1} $(i=2,\ldots,n)$. The sequence of numbers $\lambda_1, \ldots, \lambda_n$ is defined uniquely.

Analogous to the monodromy group $M_f^{\mathbb{Z}_2}$ in the homology $H_{n-1}(V_\varepsilon; \mathbb{Z}_2)$ with coefficients in $\mathbb{Z}_2$, we can define the monodromy group $M_f^{\mathbb{Z}_k}$ in the homology $H_{n-1}(V_\varepsilon; \mathbb{Z}_k)$ with coefficients in the cyclic group $\mathbb{Z}_k$. In the following result a special role will be played by the case $k = 2\lambda_1$, where the integer λ_1 was defined in the previous paragraph.

Theorem 3.16. The monodromy group $M_f^{(e)}$ of an isolated singularity of a function with an even number of variables coincides with the group of all linear operators g on the integral lattice

$$\mathbb{Z}^\mu = H_{n-1}(V_\varepsilon; \mathbb{Z}),$$

satisfying conditions (i) and (ii) of Theorem 3.15 and also the following condition:

(iii)' the operator g, reduced modulo $2\lambda_1$ belongs to the monodromy group $M_f^{\mathbb{Z}_{2\lambda_1}}$ in the homology with coefficients in $\mathbb{Z}_{2\lambda_1}$.

For all singularities of functions of two variables $\lambda_1 = 1$ ([74]).

W. Janssen obtained a generalization of this result of Chmutov for complete intersections ([172], [173]) and gave a classification of skew-symmetric vanishing lattices. G. Ilyuta has transferred the results of Chmutov to the case of boundary singularities ([171]).

In §3.3 it was indicated that the problem of describing all decompositions of simple singularities, encountered in the base of the restricted miniversal deformation, reduces to the problem of describing all of its D-diagrams with respect to distinguished bases. From this the problem naturally arises of describing all distinguished bases of a singularity.

Let $f: (\mathbb{C}^n, 0) \to (\mathbb{C}, 0)$ be an arbitrary germ of a function, with an isolated critical point at zero, let $\Delta_1, \ldots, \Delta_\mu$ be a distinguished basis of vanishing cycles in the homology group

$$H_{n-1}(V_\varepsilon; \mathbb{Z}) \approx \mathbb{Z}^\mu$$

of the non-singular level manifold. With respect to such a basis the variation operator Var_f of the singularity f is represented by an upper triangular matrix (§ 2.5). The classical monodromy operator h_* of the singularity f is the product $h_1 \circ \ldots \circ h_\mu$ of the Picard-Lefschetz operators h_i

$$(h_i(a) = a + (-1)^{n(n+1)/2}(a \circ \Delta_i) \Delta_i),$$

corresponding to the vanishing cycles $\Delta_1, \ldots, \Delta_\mu$ (ibid.).

It can be shown that for simple singularities there is also a converse result.

Theorem 3.17 ([151]). Let $\Delta_1, \ldots, \Delta_\mu$ be a basis of the homology group

$$H_{n-1}(V_\varepsilon; \mathbb{Z}) \approx \mathbb{Z}^\mu$$

of the non-singular level manifold of a simple singularity f. We suppose that with respect to the basis $\Delta_1, \ldots, \Delta_\mu$ the variation operator Var_f of the singularity f is represented by an upper triangular matrix. Then $\Delta_1, \ldots, \Delta_\mu$ is a distinguished basis of vanishing cycles.

Theorem 3.18. Let $\Delta_1, \ldots, \Delta_\mu$ be a basis of vanishing cycles in the homology of the non-singular level manifold of a simple singularity f. Let us suppose that the classical monodromy operator h_* of the singularity f is the product $h_1 \circ \ldots \circ h_\mu$ of Picard-Lefschetz operators corresponding to the vanishing cycles $\Delta_1, \ldots, \Delta_\mu$. Then $\Delta_1, \ldots, \Delta_\mu$ is a distinguished basis of vanishing cycles.

The proof of Theorem 3.18 is contained in letters of P. Deligne to E. Looijenga (1980, not published). For the proof of this, and other assertions, the following result is used which is of interest in its own right.

Let $f: (\mathbb{C}^n, 0) \to (\mathbb{C}, 0)$ be a simple singularity in an odd number of variables n, let $\tilde{f}: U \to \mathbb{C}$ be a small Morse perturbation of it, defined in a neighbourhood U of zero in the space $\mathbb{C}^n$, let $z_1, \ldots, z_\mu$ be the critical values of the function $\tilde{f}$, let z_0 be a non-critical value, such that

$$|z_i| < |z_0| \quad (i = 1, \ldots, \mu)$$

and let Δ be an arbitrary vanishing cycle in the homology of the non-singular level set F_{z_0} of the function $\tilde{f}$.

Lemma 3.3 ([151]). The cycle Δ is a vanishing cycle along some non-self-intersecting path u, joining some critical value z_i of the function $\tilde{f}$ with the non-critical value z_0 and lying wholly inside the circle $|z| < |z_0|$ (with the exception of the end coinciding with the non-critical value z_0).

In the formulation of this result an essential condition is that f is a function of an odd number of variables and also that the path u lies inside the circle $|z| < |z_0|$. In the case when f is a singularity of a function of an even number of variables, the assertion of Lemma 3.3 is not true. If we relax the requirement that the path u lies inside the circle $|z| < |z_0|$ then the lemma becomes trivial (true for any singularity in any number of variables) and without content. Lemma 3.3 is equivalent to the following assertion.

Lemma 3.4. There exists a distinguished basis of vanishing cycles $\Delta_1, \ldots, \Delta_\mu$ in the homology of the non-singular level manifold of a singularity f with the first element Δ_1 coinciding with the vanishing cycle Δ.

It is not known whether there are analogous theorems to Theorems 3.17 and 3.18 and to Lemma 3.3 for singularities which are not simple.

3.7   The polar curve and the intersection matrix of a singularity

Results were obtained in the work [119] relating the intersection matrix of an isolated singularity $f: (\mathbb{C}^n, 0) \to (\mathbb{C}, 0)$ with the intersection matrix of one of the singularities $f + z^2$ or $f|_{z=0}$ and the invariants of the polar curve of the singularity f relative to the linear function z. They allow us to determine the intersection matrices for a large number of singularities and are also useful for proofs of assertions about intersection matrices of singularities, using induction on the dimension n (such an inductive argument was used, for example, in [74] for the proof of Theorem 3.16 of §3.6). The present section contains a brief exposition of the results of the work [119]. The polar curve of a singularity arises also in other problems. A more detailed exposition of the theory of polar curves can be found in [350].

Let $f: (\mathbb{C}^n, 0) \to (\mathbb{C}, 0)$ be the germ of a holomorphic function with an isolated singularity at zero, let $z: \mathbb{C}^n \to \mathbb{C}$ be a linear function. We consider the germ of the map

$$(f, z): (\mathbb{C}^n, 0) \to (\mathbb{C}^2, 0).$$

The set of critical points of this map is the germ of an analytic space. We denote this by $\Gamma_z(f)$. It is not difficult to see that $\Gamma_z(f)$ is the germ of a curve, that is $\dim \Gamma_z(f) = 1$. This follows, for example, from the fact that the critical points of the map (f, z) are critical points of all functions of the form $f - \varepsilon z$ ($\varepsilon \in \mathbb{C}$), and each of the functions $f - \varepsilon z$ has (in a neighbourhood of zero in the space $\mathbb{C}^n$) a finite number of critical points.

Definition. The curve $\Gamma_z(f)$ is called the *polar curve* of the singularity f relative to the linear function z.

Another (equivalent) way of describing the polar curve $\Gamma_z(f)$ is the following: the curve $\Gamma_z(f)$ consists of all points $x \in \mathbb{C}^n$ in which the tangent space of the level set of the function f (passing through this point) is parallel to the fixed hyperplane $\{z = 0\}$, that is all points $x \in \mathbb{C}^n$ in which the differential df is proportional to the differential dz.

Let $\Gamma_z(f) = \cup_i \Gamma_i$ be the decomposition of the germ of the curve $\Gamma_z(f)$ into irreducible components. As we have seen, the critical points of the function $(f - \varepsilon z)$ lie on the curve $\Gamma_z(f)$ for all ε. Let μ_i be the number of critical points of the function $f - \varepsilon z$ ($\varepsilon \neq 0$) lying in the component Γ_i (counted with their multiplicities). It is clear that $\mu(f) = \Sigma_i \mu_i$.

The critical points of the function $f|_{z=\varepsilon}$ (on the hyperplane $\{z = \varepsilon\} \subset \mathbb{C}^n$) also lie on the polar curve $\Gamma_z(f)$. Let $\Gamma_i \not\subset \{z = 0\}$, let ν_i be the number of critical points of the function $f|_{z=\varepsilon}$ ($\varepsilon \neq 0$) lying in the component Γ_i (counted with their multiplicities). If the function $f|_{z=0}$ has an isolated critical point at zero, then no one of the components Γ_i lies in the hyperplane $\{z = 0\}$ and $\mu(f|_{z=0}) = \Sigma_i \nu_i$.

If $\Gamma_i \not\subset \{z = 0\}$, then $f|_{\Gamma_i} \not\equiv 0$. Indeed, if $f|_{\Gamma_i} \equiv 0$, then

$$df|_{\Gamma_i} = 0, \quad dz|_{\Gamma_i} = 0, \quad z|_{\Gamma_i} \equiv 0.$$

On the curve Γ_i the function f can be expanded in a series of (fractional) powers of z (this expansion coincides with the Puiseux expansion of the image of the curve Γ_i under the map

$$(f, z): (\mathbb{C}^n, 0) \to (\mathbb{C}^2, 0),$$

see §4.3). Let $a_i z^{\alpha_i}$ be the first term of this expansion ($a_i \neq 0$). We have $\alpha_i > 1$. This follows from the fact that as $\varepsilon \to 0$ the roots of the equation $f'_z = \varepsilon$ (defining the critical points of the function $(f - \varepsilon z)$) must tend to zero. If $\Gamma_i \subset \{z = 0\}$ (in this case the function $f|_{z=0}$ has a non-isolated critical point), then $f|_{\Gamma_i} \equiv 0$. In this case we shall suppose that $\alpha_i = 1$.

Lemma 3.5. If $\alpha_i > 1$ then

$$\mu_i = v_i(\alpha_i - 1).$$

Proof. If $\alpha_i > 1$, then $\Gamma_i \not\subset \{z = 0\}$ and

$$f = a_i z^{\alpha_i} + \ldots \qquad (a_i \neq 0).$$

We suppose that the components Γ_i enter into the polar curve $\Gamma_z(f)$ with multiplicity one. This means that the critical points of the function $f - \varepsilon z$ ($\varepsilon \neq 0$), lying in the component Γ_i, are non-degenerate. In the opposite case the proof must be altered somewhat. We write down the Puiseux expansion of the image of the Γ_i under the map $(f, z): (\mathbb{C}^n, 0) \to (\mathbb{C}^2, 0)$ in the form

$$z = z(t) = t^k, \qquad f = f(t) = a_i t^m + \ldots$$

(a series of integer powers of the variable t), where t is a uniformizing parameter. We have $\alpha_i = m/k$. The number μ_i is equal to the number of (non-zero) roots of the equation $f'_t - \varepsilon z'_t$, tending to zero as $\varepsilon \to 0$. The number v_i is equal to the number of roots of the equation $z(t) = \varepsilon$. It is clear that $\mu_i = m - k$, $v_i = k$, and hence

$$\mu_i / v_i = (m - k)/k = \alpha_i - 1,$$

which is what we were trying to prove.

Knowledge of the polar curve of a singularity allows us to observe the behaviour of the critical points and critical values for perturbations of a special form. For example we consider a small perturbation

$$F_\varepsilon = f + (z - \varepsilon)^2$$

of the function $f + z^2$. The critical points of the function F_ε lie on the polar curve $\Gamma_z(f)$ of the singularity f. We consider those critical points which lie on the component Γ_i and tend to zero as $\varepsilon \to 0$. They can be determined from the equation $(F_\varepsilon)'_z|_{\Gamma_i} = 0$. We have

$$a_i \alpha_i z^{\alpha_i - 1} + 2(z - \varepsilon) + o(z^{\alpha_i - 1}) = 0.$$

For $\alpha_i > 2$ it follows from this that

$$z = \varepsilon + O(\varepsilon^{\alpha_i - 1}), \qquad F_\varepsilon = a_i \varepsilon^{\alpha_i} + o(\varepsilon^{\alpha_i}).$$

The number of such critical points is equal to ν_i.
    For $\alpha_i = 2$, $a_i \neq -1$ we have

$$z = \varepsilon / (a_i + 1) + o(\varepsilon), \qquad F_\varepsilon = \varepsilon^2 a_i / (a_i + 1) + o(\varepsilon^2).$$

The number of such critical points is also equal to ν_i.
    For $\alpha_i < 2$ we have to a first approximation $f'_z = 2\varepsilon$, which is the same as the equation defining the critical points of the function $(f - 2\varepsilon z)$. The number of such critical points is equal to μ_i. We have

$$z = (2\varepsilon / (a_i \alpha_i))^{1/(\alpha_i - 1)} + \ldots = o(\varepsilon).$$

Therefore $F_\varepsilon = \varepsilon^2 + o(\varepsilon^2)$.
    From this follows

Lemma 3.6. If $a_i \neq -1$ for $\alpha_i = 2$, then the Milnor number $\mu(f + z^2)$ of the singularity $f + z^2$ is equal to

$$\sum_{i : \alpha_i < 2} \mu_i + \sum_{i : \alpha_i \geqslant 2} \nu_i.$$

Corollary.

$$\mu(f + z^2) \leqslant \mu(f|_{z=0});$$

$\mu(f + z^2) = \mu(f|_{z=0})$ if and only if $\alpha_i \geqslant 2$ for all i.

It is not hard to show that if $f \in m^k$, that is if the Taylor expansion of the germ f does not have terms of degree $< k$, and if the linear function z is chosen in general position then $\alpha_i \geqslant k$ for all i.

In order to determine the intersection matrix of the singularity f, we need to know the intersection matrix of the singularity $(f + z^2)$ with respect to a distinguished basis of a special form. We describe this basis. Let A be the set of all distinct values of the index α_i. From the asymptotics of the critical values of the perturbation

$$F_\varepsilon = f + (z - \varepsilon)^2$$

of the function $f + z^2$ described before Lemma 3.6, it follows that for sufficiently small $\varepsilon \neq 0$ we can choose positive numbers r'_α and r''_α for $\alpha = 2$ and for $\alpha \in A$, $\alpha > 2$ with $r''_\alpha < r'_\beta$ for $\alpha > \beta$, such that the critical values of the function F_ε at the critical points belonging to the component Γ_i with $\alpha_i = \alpha > 2$ are contained in the annulus

$$\{u : r'_\alpha < |u| < r''_\alpha\},$$

and the critical values of the function F_ε at the critical points belonging to the component Γ_i with $\alpha_i = \alpha \leqslant 2$, are contained in the annulus

$$\{u : r'_2 < |u| < r''_2\}.$$

Let $\sigma(r)$ be a continuous monotonic decreasing function such that $\sigma(r) = \alpha - 1$ for $r'_\alpha \leqslant r \leqslant r''_\alpha$. We let

$$V_m = \{u : \arg u + 2\pi\sigma(|u|) \geqslant \pi(2m - 1)\},$$

where $m = 1, 2, \ldots$, $-\pi \leqslant \arg u \leqslant \pi$. We suppose that $(-a_i)^{q_i} \notin \mathbb{R}_+$, where $\alpha_i = p_i/q_i$, $(p_i, q_i) = 1$ ($\mathbb{R}_+$ is the positive half-axis). In this case the critical values of the function F_ε do not belong to the half-axis $\mathbb{R}_-$, nor to the boundaries of the regions V_m. We consider a system of paths joining the critical values of the function F_ε with the non-critical value 0, which defines the distinguished basis of vanishing cycles in the homology of the non-singular level manifold of the singularity $f + z^2$. We require of this system of paths that all paths intersect the half-axis $\mathbb{R}_-$ only at zero and that all paths leading from those critical values of the function F_ε which belong to the region V_m are themselves contained in V_m. In particular, we can choose, as such a system of paths, a system of line segments joining the critical values of the function F_ε with zero. One of the principal results of the work [119] is the following.

Theorem 3.19. Let $\{\varDelta_j\}$ be the distinguished basis of vanishing cycles for the singularity $f+z^2$, defined by the system of paths described above. Then:
1. There exists a distinguished basis of vanishing cycles $\{\varDelta_j^m\}$ for the singularity f with the following intersection numbers:

$$(\varDelta_j^m \circ \varDelta_{j'}^m) = (\varDelta_j \circ \varDelta_{j'}),$$

$$(\varDelta_j^m \circ \varDelta_j^{m'}) = (m'-m)^{n-1} \quad \text{for} \quad |m'-m|=1,$$

$$(\varDelta_j^m \circ \varDelta_{j'}^{m'}) = (-1)^n (\varDelta_j \circ \varDelta_{j'}) \quad \text{for} \quad |m'-m|=1, \ (m'-m)(j'-j)<0,$$

$$(\varDelta_j^m \circ \varDelta_{j'}^{m'}) = 0 \quad \text{for} \quad |m'-m|>1 \quad \text{or} \quad (m'-m)(j'-j)>0.$$

Here the pair (m,j) is admissible (that is there corresponds to it a vanishing cycle $\varDelta_j^m$) if and only if the cycle $\varDelta_j$ vanishes along a path contained in V_m.
2. $\varDelta_j^m = h_* \varDelta_j^{m-1}$ for $m>1$, where h_* is the classical monodromy operator of the singularity f.
3. The condition of admissibility of the pair (m,j) can be reformulated in the following way: for each i the first μ_i pairs from (m,j) are admissible, where the cycle $\varDelta_j$ vanishes at a critical point belonging to the component $\varGamma_i$ of the polar curve $\varGamma_z(f)$.

There are analogous links between the intersection matrices of the singularities f and $f|_{z=0}$ in the case when $f|_{z=0}$ has an isolated singularity. The function $G_\varepsilon = f|_{z=\varepsilon}$ is a small perturbation of the singularity $f|_{z=0}$. The critical values of the function G_ε at the critical points belonging to the component $\varGamma_i$, are equal to

$$a_i \varepsilon^{\alpha_i} + o(\varepsilon^{\alpha_i})$$

as $\varepsilon \to 0$. Consequently, for sufficiently small ε we can choose positive numbers r_α' and r_α'' for all $\alpha \in A$ in such a way that

$$r_\alpha' < r_\alpha'' \quad \text{for all} \quad \alpha \in A,$$

$$r_\alpha'' < r_\beta' \quad \text{for} \quad \alpha > \beta$$

and the critical values of the function G_ε at the critical points belonging to the curve $\varGamma_i$ with $\alpha_i = \alpha$ are contained in the annulus

$$\{u : r_\alpha' < |u| < r_\alpha''\}.$$

We define the function $\sigma(r)$, the regions V_m and the system of paths joining the critical values of the function G_ε with the non-critical value 0 in exactly the same way as for the function F_ε, the only difference being that we include all $\alpha \in A$, and not only $\alpha \geq 2$. Let $\{\tilde{\Delta}_j'\}$ be a distinguished basis of vanishing cycles for the singularity $f|_{z=0}$ defined by such a system of paths, let $\{\tilde{\Delta}_j\}$ be the corresponding distinguished basis of vanishing cycles (defined by the same system of paths) for the singularity $f|_{z=0}+z^2$, which is stably equivalent to the singularity $f|_{z=0}$. The connection between the intersection matrices of the singularities $f|_{z=0}$ and $f|_{z=0}+z^2$ with respect to the distinguished bases $\{\tilde{\Delta}_j'\}$ and $\{\tilde{\Delta}_j\}$ respectively is given by Theorem 2.14 of §2.8.

Theorem 3.20. Suppose that the germ $f|_{z=0}$ has an isolated singularity at zero. In this case in theorem 3.19 the singularity $f+z^2$ and the distinguished basis of vanishing cycles $\{\Delta_j\}$ can be changed into $f|_{z=0}+z^2$ and $\{\tilde{\Delta}_j\}$ respectively.

Theorem 3.20 reduces the problem of calculating the intersection matrix of the singularity f of a function of n variables to the problem of calculating the intersection matrix of the singularity $f|_{z=0}$ of a function of $n-1$ variables with respect to a distinguished basis of a special type and indices α_i for the components Γ_i of the polar curve $\Gamma_z(f)$. The results of the calculation for the majority of the singularities classified in Chapter 15 of Volume 1 are summarised in tables displayed below. Here the intersection matrix of the singularity $f|_{z=0}$ with respect to the distinguished basis $\{\tilde{\Delta}_j'\}$ and consequently also of the singularity $f|_{z=0}+z^2$ with respect to the distinguished basis $\{\tilde{\Delta}_j\}$ is defined by one of the following D-diagrams:

$$\underset{1 \quad 2}{\circ\!\!-\!\!\!-\!\!\bullet}$$

for a singularity f from the series J and E;

$$\underset{1 \quad 3 \quad 2}{\circ\!-\!\circ\!-\!\circ}$$

for a singularity f from the series X, Y, Z and W;

for a singularity f from the series Q, S, T and U.

In the right-hand column of the tables are displayed numbers $M_1, \ldots, M_{\mu'}$ ($\mu' = \mu(f|_{z=0})$), where the natural number M_j is defined so that in the distinguished basis $\{\Delta_j^m\}$ of the singularity f there are cycles Δ_j^m with $1 \leqslant m \leqslant M_j$.

Singularity	$M_1, \ldots, M_{\mu'}$
$J_{k,i}$	$3k+i-1,\ 3k-1$
E_{6k}	$3k,\ 3k$
E_{6k+1}	$3k+1,\ 3k$
E_{6k+2}	$3k+1,\ 3k+1$
$X_{k,p}$	$4k-1,\ 4k-1,\ 4k+p-1$
$Y_{r,s}^k$	$4k+r-1,\ 4k+s-1,\ 4k-1$
$Z_{i,p}^k$	$4k-1,\ 4k+3i-p-1,\ 4k+3i-1$
$Z_{12k+6i-1}^k$	$4k+3i,\ 4k-1,\ 4k+3i$
Z_{12k+6i}^k	$4k+3i+1,\ 4k-1,\ 4k+3i$
$Z_{12k+6i+1}^k$	$4k+3i+1,\ 4k-1,\ 4k+3i+1$
W_{12k}	$4k\ 4k,\ 4k$
W_{12k+1}	$4k+1,\ 4k,\ 4k$
$W_{k,i}$	$4k+1,\ 4k+1,\ 4k+i+1$
$W_{k,2q-1}^{\neq}$	$4k+q+1,\ 4k+q,\ 4k+1$
$W_{k,2q}^{\neq}$	$4k+q+1,\ 4k+q+1,\ 4k+1$
W_{12k+5}	$4k+2,\ 4k+1,\ 4k+2$
W_{12k+6}	$4k+2,\ 4k+2,\ 4k+2$
$Q_{k,i}$	$2,\ 2,\ 3k-1,\ 3k+i-1$
Q_{6k+4}	$2,\ 2,\ 3k,\ 3k$
Q_{6k+5}	$2,\ 2,\ 3k+1,\ 3k$
Q_{6k+6}	$2,\ 2,\ 3k+1,\ 3k+1$
S_{12k-1}	$2,\ 4k-1,\ 4k-1,\ 4k-1$
S_{12k}	$2,\ 4k,\ 4k-1,\ 4k-1$
$S_{k,i}$	$2,\ 4k,\ 4k,\ 4k+i$
$S_{k,2q-1}^{\neq}$	$2,\ 4k+q,\ 4k+q-1,\ 4k$
$S_{k,2q}^{\neq}$	$2,\ 4k+q,\ 4k+q,\ 4k$
S_{12k+4}	$2,\ 4k+1,\ 4k,\ 4k+1$
S_{12k+5}	$2,\ 4k+1,\ 4k+1,\ 4k+1$
$T_{p,q,r}$	$p-1,\ q-1,\ r-1,\ 2$
U_{12k}	$3k,\ 3k,\ 3k,\ 3k$
$U_{k,2q-1}$	$3k+q,\ 3k+q,\ 3k,\ 3k+1$
$U_{k,2q}(q>0)$	$3k+q+1,\ 3k+q,\ 3k,\ 3k+1$
U_{12k+4}	$3k+1,\ 3k+1,\ 3k+1,\ 3k+1$

For singularities of corank 2 (that is for singularities from the series J, E, X, Y, Z and W) analogous diagrams can be easily obtained by the methods of Chapter 4.

3.8   Intersection forms of unimodal and bimodal singularities

In Chapter 15 of Volume 1 we derived the classification of unimodal and bimodal singularities of functions. Here we shall derive results on their quadratic forms.

The D-diagrams of unimodal singularities and their monodromy groups were calculated by A. M. Gabrielov [118]. They can be obtained using the results given in § 3.7. We denote by μ_+, μ_0 and μ_- the positive, zero and negative indices of inertia of the quadratic form of the singularity, that is the number of positive, zero and negative diagonal elements in a diagonalisation of the intersection form of a singularity stably equivalent to the given one and depending on $n \equiv 3 \bmod 4$ variables $(\mu_+ + \mu_0 + \mu_- = \mu)$.

Theorem 3.21. The D-diagrams of the parabolic singularities P_8, X_9 and J_{10} with respect to some weakly distinguished bases have the form

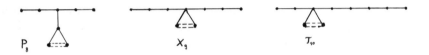

For these singularities $\mu_+ = 0$, $\mu_0 = 2$.

The D-diagrams of the hyperbolic singularities $T_{p,q,r}$ have the form

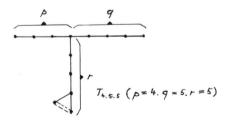

For these singularities $\mu_+ = \mu_0 = 1$.

The D-diagrams of the 14 exceptional unimodal singularities have the form

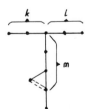

Here (k, l, m) are the so-called Gabrielov numbers of the singularity (see the introduction to Part II of Volume 1), $\mu = k + l + m$. For these singularities $\mu_+ = 2$, $\mu_0 = 0$.

Between the Gabrielov numbers (GN) and the Dolgachev numbers (DN; see the introduction to Part II of Volume 1) of the exceptional unimodal singularities there is a "strange duality", expressed by the fact that the GN of each singularity is the same as the DN of some (generally speaking different) singularity, and the GN of the latter is the same as the DN of the former. An explanation of this duality was given by Dolgachev and Pinkham ([97], [290]). They showed that the DN of a quasihomogeneous unimodal singularity in some sense are its GN at infinity and vice-versa.

The quadratic forms and D-diagrams of bimodal singularities in two variables and their indices of inertia can be easily obtained by the methods of Chapter 4. For bimodal singularities of three variables the D-diagrams were obtained by A. M. Gabrielov (see §3.7) but they are not convenient for the computation, for example, of the indices of inertia of their quadratic forms. The indices of inertia of quadratic forms can be obtained by using two general results which are due to Steenbrink. One of these gives a way of calculating the inertia indices of the quadratic forms for quasihomogeneous singularities.

Let $f : (\mathbb{C}^n, 0) \to (\mathbb{C}, 0)$ $(n \equiv 3 \bmod 4)$ be a quasihomogeneous singularity with weights $w_1, \ldots, w_n$ and degree 1 (this means that

$$f(x_1, \ldots, x_n) = \Sigma a_{\beta_1 \ldots \beta_n} x_1^{\beta_1} \ldots x_n^{\beta_n},$$

where $\Sigma_{i=1}^n w_i \beta_i = 1$), having an isolated critical point at the origin. We set

$$l(\alpha) = \sum_{i=1}^n (\alpha_i + 1) w_i \qquad (\alpha = (\alpha_1, \ldots, \alpha_n)).$$

Let the monomials $x^{\alpha^{(j)}}$ $(j = 1, 2, \ldots, \mu)$ generate a basis of the local ring $J = {}_n O / (\partial f / \partial x_i)$ of the singularity f. We denote by $[c]$ the integral part of c.

Theorem 3.22 ([341]).

$$\mu_+ = \text{the number of } \alpha^{(j)} : l(\alpha^{(j)}) \notin \mathbb{Z}, \ [l(\alpha^{(j)})] \text{ is even};$$

$$\mu_- = \text{the number of } \alpha^{(j)} : l(\alpha^{(j)}) \notin \mathbb{Z}, \ [l(\alpha^{(j)})] \text{ is odd};$$

$$\mu_0 = \text{the number of } \alpha^{(j)} : l(\alpha^{(j)}) \in \mathbb{Z}.$$

This assertion in the form of a conjecture was previously formulated by V. I. Arnold (see [150]).

Among the bimodal singularities some (but not all) have a quasihomogeneous representation. For these singularities Theorem 3.22 gives $\mu_+ = 2$, $\mu_0 = 0$. To calculate the indices of inertia of the quadratic forms of the other bimodal singularities we use the following facts.

Theorem 3.23 ([356]). Let $f_t(x)$ be a continuous deformation of the singularity

$$f_0 : (\mathbb{C}^n, 0) \to (\mathbb{C}, 0) \qquad (t \in [0, 1]),$$

with $\mu(f_0) = \mu$, $\mu(f_t) = \mu'$ for $0 < t \leqslant 1$. Then $\mu \geqslant \mu'$ and the homology group $H_{n-1}(V_{t,\lambda}; \mathbb{Z})$ of the non-singular level set $V_{t,\lambda}$ of the germ of the function f_t near zero has a natural inclusion in the homology group $H_{n-1}(V_{0,\lambda}; \mathbb{Z})$ of the non-singular level set $V_{0,\lambda}$ of the germ f_0. Moreover a distinguished basis of the group $H_{n-1}(V_{t,\lambda}; \mathbb{Z})$ can be expanded to a distinguished basis of the group $H_{n-1}(V_{0,\lambda}; \mathbb{Z})$.

Theorem 3.24 ([342]). For any singularity $\mu_+ + \mu_0$ is even.

From these results we infer

Theorem 3.25. The quadratic forms of all bimodal singularities have the following indices of inertia: $\mu_+ = 2$, $\mu_0 = 0$.

Chapter 4

The intersection matrices
of singularities of functions
# of two variables

The method of calculating the intersection matrix of a singularity of a function of two variables described in this chapter is due to S. M. Gusein-Zade ([147], [148]) and N. A'Campo ([7], [8]). It applies to all singularities of two variables. Using it allows us substantially to simplify many calculations connected with the quadratic form of a singularity (for example, the calculation of its signature).

4.1   Intersection matrices of real singularities

The intersection matrix of a real singularity of a function f of two variables can be determined by the (real) zero level curve of a perturbation $\tilde{f}$ of the function of a special type.

Let $f(x, y)$ be the germ of a real (that is taking real values on $\mathbb{R}^2 \subset \mathbb{C}^2$) holomorphic function $(\mathbb{C}^2, 0) \to (\mathbb{C}, 0)$ which has an isolated (in the space $\mathbb{C}^2$) critical point at zero. We suppose that there exists a real perturbation $\tilde{f}$ of the function f such that all its critical points (into which the critical point 0 of the function f bifurcates) are real and non-degenerate, and that the values of the function $\tilde{f}$ at all the saddle points are zero. It is not difficult to see that in this case the values of the function $\tilde{f}$ at all minima are negative, and at all maxima are positive. If such a perturbation exists, then the intersection matrix of the singularity f can be determined from the real curve $\{\tilde{f}(x, y) = 0\}$ in the plane $\mathbb{R}^2$. In order to formulate the corresponding results, we introduce some definitions.

Let us be given in the (open) disk D in the plane $\mathbb{R}^2$ a real curve l (closed in the topological sense), which has as singularities only a finite number of simple double self-intersections and which approaches transversely the boundary circle of D. The curve l gives rise to symmetric and antisymmetric integral bilinear forms on a lattice according to the rules described below.

Each connected component of the complement of the curve l is a curvilinear polygon. In such a polygon some pairs of vertices can coincide (as in figure 32). We divide the set of components of the complement of the curve l into two classes (the first and the second) so that two components with a common side are in different classes. Such a division into two classes is possible and unique modulo interchange of the two classes. We assign to each point p_j of self-intersection of the curve l the formal generator Δ_j^1, to each relatively compact in D (that is having no boundary points in common with the complement of D) component of the complement of the curve l from the first class (U_i^0) the formal generator Δ_i^0, and from the second class (U_k^2) the formal generator Δ_k^2. We denote by $n_{10}(j, i)$ (respectively $n_{21}(k, j)$) the number of vertices of the curvilinear polygon U_i^0 (respectively U_k^2) coinciding with the point p_j. The numbers $n_{10}(j, i)$ and $n_{21}(k, j)$ can take the values 0, 1 or 2. We denote by $n_{20}(k, i)$ the number of common edges of the curvilinear polygons U_k^2 and U_i^0. So, for example, in figure 32, $n_{20} = 1$, $n_{10} = 2$, $n_{21} = 1$ ($i = j = k = 1$).

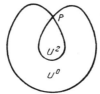

Fig. 32.

We construct on the generators $\{\Delta_m^s\}$ an integer lattice and will suppose that the basis Δ_k^2, Δ_j^1, Δ_i^0 is a distinguished basis of this lattice (the order of the elements Δ_m^σ with the same σ is immaterial). The division of the components of the complement of the curve l into two classes is used only for fixing the order of the elements Δ_m^σ in the distinguished basis.

Definition. The *quadratic form corresponding to the curve l* is defined by the following table of scalar products of the generators:

$$(\Delta_m^\sigma \circ \Delta_{m'}^\sigma) = -2\delta_{mm'},$$

$$(\Delta_i^0 \circ \Delta_j^1) = n_{10}(j, i),$$

$$(\Delta_j^1 \circ \Delta_k^2) = n_{21}(k, j),$$

$$(\Delta_i^0 \circ \Delta_k^2) = -n_{20}(k, i).$$

Definition. The *antisymmetric bilinear form corresponding to the curve l* is defined by the following table of scalar products (the fact that the notation is the same as in the previous definition will not create confusion):

$$(\Delta_m^{\sigma} \circ \Delta_{m'}^{\sigma}) = 0,$$

$$(\Delta_j^1 \circ \Delta_i^0) = n_{10}(j, i),$$

$$(\Delta_k^2 \circ \Delta_j^1) = n_{21}(k, j),$$

$$(\Delta_i^0 \circ \Delta_k^2) = n_{20}(k, i).$$

(here if we interchange the arguments the scalar product changes sign.)

By the D-diagram of the curve l we shall mean the D-diagram of the corresponding quadratic form. Its vertices correspond to the self-intersections of the curve l and the relatively compact in D components of the complement of the curve l. The rule for joining vertices follows from the table of intersection numbers. For example, the D-diagram of the curve l in figure 32 is depicted in figure 33. We remark that this diagram is not the D-diagram of any singularity.

Fig. 33.

Let $\tilde{f}$ be the perturbation of the singularity f described above, that is such that all the critical points of the function $\tilde{f}$ (into which the singularity f bifurcates) are real, and the values of the function $\tilde{f}$ at all saddle points equal zero. In this case the real curve $\{\tilde{f} = 0\}$ (in a small disc D with centre at zero) has only simple double-intersections. We shall say that a component of the complement of this curve is of the first class if the function $\tilde{f}$ takes negative values on it, and of the second class otherwise. The critical values of the function $\tilde{f}(x, y)$ (or, which is the same thing, the function of three variables $\tilde{f}(x, y) + t^2$) by assumption lie on the real axis in the plane $\mathbb{C}$ of values of the function $\tilde{f}$. We choose a z_0 such that $\text{Im } z_0 > 0$, and fix a system of paths joining the non-critical value z_0 with the critical values of the function $\tilde{f}$, subject to the condition that these paths lie in their entirety in the upper half-plane $\text{Im } z > 0$ except for the ends which coincide with the critical values. We remark that the critical points of the function $\tilde{f}$ are in one-one correspondence with the self-intersections of the real curve $\{\tilde{f} = 0\}$ and the relatively compact components of its complement, because in each such component there is exactly one critical point of the function $\tilde{f}$ (maximum or minimum). Having fixed a system of paths in this way, we define distinguished bases of vanishing cycles in the homology of the non-singular level manifold of the singularities $f(x, y)$ and $f(x, y) + t^2$. We shall denote these bases also by $\{\Delta_m^\sigma\}$. Here the cycles Δ_j^1 vanish at the saddle points p_j of the function $\tilde{f}$, the cycles Δ_i^0 at the minima, lying in the components U_i^0 and the cycles Δ_k^2 at the maxima, lying in the components U_k^2.

Theorem 4.1. The intersection form in the homology of the non-singular level manifold of the singularity $f(x, y)$ with respect to the above-described distinguished basis $\{\Delta_m^\sigma\}$ coincides with the antisymmetric bilinear form corresponding to the real curve $\{\tilde{f} = 0\}$, and the intersection form of the singularity of three variables $f(x, y) + t^2$ with the quadratic form corresponding to the same curve.

Examples.

(i) Let $f(x, y) = x^m + y^n$. The multiplicity of this singularity is equal to $(m - 1) \cdot (n - 1)$. Put

$$\tilde{f}(x, y) = \lambda^{mn}(T_m(\lambda^{-n}x) + 2^{n-m}T_n(\lambda^{-m}2^{(m-n)/n}y)).$$

Here $T_n(x) = 2^{1-n}\cos(n \cdot \arccos x)$ are the Chebyshev polynomials. It is not diffcult to see that the perturbation $\tilde{f}$ of the singularity f satisfies the conditions of Theorem 4.1. The curve $\{(x, y) \in \mathbb{R}^2 : \tilde{f}(x, y) = 0\}$ and its D-diagram are depicted in figure 34 ($m = 6$, $n = 5$). For $m = k + 1$, $n = 2$ we get, as in §2.9, the classical Dynkin diagram A_k, *but with the vertices in a different order.*

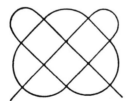 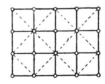

Fig. 34.

(ii) $f(x, y) = x(x^{k-2} - y^2)$ is the singularity D_k, $k \geqslant 4$. Its multiplicity is equal to k. It is not difficult to choose a perturbation $\tilde{f}$ of the function f, the zero level line of which (for $k = 9$) is depicted in figure 35 (for this we can take the perturbation of the function $x^{k-2} - y^2$, similar to the one described in the previous example, and multiply it by $(x - 2\lambda^2)$). Its D-diagram is the classical diagram D_k.

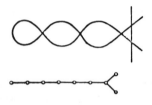

Fig. 35.

(iii) $f(x, y) = x^3 + xy^3 = x(x^2 + y^3)$ is the singularity E_7. A perturbation, satisfying the conditions of Theorem 4.1 is

$$\tilde{f}(x, y) = (x + \lambda^{3/2}/3)(x^2 + y^3 - \lambda^2 y - 2\lambda^3/3\sqrt{3}).$$

The curve $\{\tilde{f} = 0\}$ and its D-diagram are shown in figure 36.

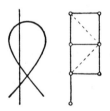

Fig. 36.

A shortcoming of Theorem 4.1 is that generally it is hard to choose a perturbation $\tilde{f}$ satisfying its conditions. In addition the diagrams of such singularities as, for example,

$$E_6(x^3 + y^4), \quad E_7 \quad \text{and} \quad E_8(x^3 + y^5)$$

for the most natural choices of perturbation turn out to be different from their classical forms and require a transformation. It turns out that it is simpler to construct not a perturbation of the function but a perturbation of its zero level line. A description of the corresponding procedure will be given in the following sections. The result formulated below allows us in some degree to remove the second of the above shortcomings.

Let the curve *l* be the same as before. If we divert our attention from the self-intersections, then the curve *l* consists of several circles and intervals. This means that there exists a proper non-degenerate map $\chi: L \to D$ of a one-dimensional smooth manifold L to the disc D such that $\operatorname{Im} \chi = l$, and χ maps L to l bijectively outside the singular points of the curve *l* (the self-intersections). Let $\chi_t : L \to D$ ($t \in [0, 1]$) be a homotopy of the map χ ($\chi = \chi_0$) in the class of proper non-degenerate maps, constant on the preimage of the boundary circle of the disc D. If χ_t is a homotopy of general form, then the type of the curve $\operatorname{Im} \chi_t$ (as a one-dimensional smooth submanifold of the disc D with simple self-intersections) will change for a finite number of values of the parameter t. For these values of the parameter t we will get one of three types of bifurcation:

1) two points of self-intersection of the curve $\operatorname{Im} \chi_t$ come together and vanish (figure 37, at the exceptional value of the parameter the two branches of the curve $\operatorname{Im} \chi_t$ simply touch);

Fig.37.

2) there appear two new points of self-intersection (this type of bifurcation can be transformed into the previous one by changing the direction of the parameter t);

3) three points of self-intersection of the curve $\operatorname{Im} \chi_t$ come together and then separate again (figure 38, for the exceptional value of the parameter there appears on the curve $\operatorname{Im} \chi_t$ a point of threefold intersection).

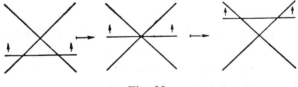

Fig. 38.

Other types of bifurcation of codimension one are absent, in view of the fact that a non-degeneracy condition (its differential not mapping to zero) has been imposed on the function χ_t. For the first two types of bifurcation the total number of points of self-intersection of the curve $\operatorname{Im} \chi_t$ is not preserved, and therefore the dimension of the integer lattice corresponding to it changes. For a bifurcation of the third type the dimension of the lattice is preserved. It can be shown that if all bifurcations of the curve $\operatorname{Im} \chi_t$ correspond to the third type, then the bilinear form corresponding to the curve $\operatorname{Im} \chi_t$ also does not change. We get only a change of basis of the lattice on which it is defined.

Definition. The homotopy χ_t is said to be *admissible*, if for each value of the parameter $t \in [0, 1]$ there do not exist points $h_1 \neq h_2$ of L for which

$$\chi_t(h_1) = \chi_t(h_2),$$

$$\operatorname{Im} d\chi_t(h_1) = \operatorname{Im} d\chi_t(h_2)$$

($d\chi_t$ is the differential of the curve χ_t) and in addition the curve $\operatorname{Im} \chi_1$ has only simple double self-intersections.

It can be shown that an admissible homotopy is a homotopy for which all the bifurcations of the curve $\operatorname{Im} \chi_t$ are of the third type (and $t = 1$ is not an exceptional value of the parameter, that is there is not a bifurcation of the curve there). If χ_t is an admissible homotopy, then the curves $\operatorname{Im} \chi_0 = l$ and $\operatorname{Im} \chi_1$ have the same number of self-intersections (and also the same number of components of the complement). Therefore the integral lattices corresponding to these curves have the same dimension.

Theorem 4.2. Let $\chi_t : L \to D$ be an admissible homotopy. Then the symmetric and antisymmetric bilinear forms, corresponding to the curve $\operatorname{Im} \chi_1$ can be obtained from the forms, corresponding to the curve $\operatorname{Im} \chi_0 = l$ with the help of the operations of change of distinguished basis.

For the proof we can display explicitly the change of basis which corresponds to one bifurcation of the curve $\operatorname{Im} \chi_t$ of the third type. We transform only four vanishing cycles, corresponding to the three points of self-intersection of the curve which come together in this bifurcation, and the curvilinear triangle with vertices at these points, which is a connected component of the complement of the curve $\operatorname{Im} \chi_t$. An explicit form for such a change of basis can be obtained for one example in which the corresponding homotopy χ_t can be realised as a deformation of the perturbation $\tilde{f}$ of the singularity (for example, for $\tilde{f}_t(x, y) = xy(x + y + t)$; the exceptional value of the parameter being $t = 0$). The set of paths defining the distinguished basis in this case were described in the formulation of Theorem 4.1. The parameter t going out into the complex region and going half way round the exceptional value $t=0$ allows us to observe the transformation of the corresponding system of paths.

Let $t=t_0$ be the value of the parameter for which there is a bifurcation of the curve $\operatorname{Im}\chi_t$ of the third type. For definiteness we shall assume that for $t<t_0$ the triangle with vertices at the three converging points belongs to the first class of components of the complement of the curve $\operatorname{Im}\chi_t$ (corresponding to the components on which the perturbation $\tilde{f}$ takes negative values). For $t>t_0$ the analogous triangle will belong to the second class. Let Δ_m, Δ_{m+1}, Δ_{m+2} be the vanishing cycles corresponding to the three points of self-intersection, let Δ_{m+3} be the cycle corresponding to the triangle with vertices at these points. Then the sequence of operations

$$\beta_{m+3}, \beta_{m+2}, \beta_{m+1}, \beta_{m+1}, \beta_{m+2}, \beta_{m+3}, \beta_{m+3}, \beta_{m+2}, \beta_{m+1}$$

is equivalent to the above bifurcation of the curve $\operatorname{Im}\chi_t$.

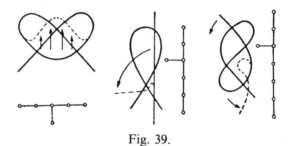

Fig. 39.

As examples of the application of Theorem 4.2 we show in figure 39 the reduction of the D-diagrams of the singularities

$$E_6(x^3+y^4), \quad E_7(x^3+xy^3) \quad \text{and} \quad E_8(x^3+y^5)$$

(see examples (i) and (iii)) to classical form with the help of an admissible homology.

Thus we have completed the proof of Theorem 3.14 of § 3.6.

A small modification allows us to adapt the above method of calculating the intersection matrix of a singularity of two variables for a boundary singularity (see Volume 1, § 17.4) and prove that the D-diagrams of the singularities B_k, C_k and F_4 are the classical diagrams of the corresponding Lie algebras, and their monodromy groups (for the case of an odd number of variables) are the corresponding classical Weyl groups (§ 5.2).

4.2   Germs of complex curves and singularities of functions of two variables

Let $f: (\mathbb{C}^2, 0) \to (\mathbb{C}, 0)$ be the germ of a holomorphic function with an isolated critical point at zero. The germ of the curve

$$M(f) = \{\bar{x} \in \mathbb{C}^2 : f(\bar{x}) = 0\}$$

may be reducible. Let

$$M(f) = \cup_{i=1}^r M_i$$

be its representation as a union of irreducible components. For each of the germs of the irreducible complex curves M_i there exists the germ of a mapping

$$\phi_i : (\mathbb{C}_i, 0) \to (\mathbb{C}^2, 0)$$

(the uniformization) such that $\mathrm{Im}\, \phi_i = M_i$ and ϕ_i is an isomorphism of the curves $\mathbb{C}_i$ and M_i away from zero. The germ ϕ_i is defined modulo germs of holomorphic isomorphisms

$$(\mathbb{C}_i, 0) \to (\mathbb{C}_i, 0)$$

(a change of uniformizing parameter). For small perturbations $\tilde{\phi}_i$ of the maps ϕ_i in general position the complex curve

$$\cup_{i=1}^r \mathrm{Im}\, \tilde{\phi}_i$$

(in a neighbourhood of zero) has as singularities only simple double points. Their number, which, of course, does not depend on the choice of perturbation, we denote by $s = s\{\phi_i\} = s(f)$.

Lemma 4.1. The multiplicity of the singularity f is equal to

$$\mu(f) = 2s(f) - (r - 1).$$

The problem of calculating the intersection matrix of an arbitrary singularity of a function of two variables can be reduced to the case of a real singularity with the help of the following result.

Theorem 4.3. For any set of germs of maps $\{\phi_i\}$ ($\phi_i : (\mathbb{C}_i, 0) \to (\mathbb{C}^2, 0)$, ϕ_i maps $\mathbb{C}_i$ in a one-one fashion onto its image and $\operatorname{Im} \phi_i \neq \operatorname{Im} \phi_j$ for $i \neq j$) there exists a set of real germs of maps $\{\psi_i\}$, lying in the same connected component as ϕ_i of the set $s = \text{const}$ in the space of all sets of r maps $(\mathbb{C}_i, 0) \to (\mathbb{C}^2, 0)$.

The mappings (or curves) $\{\phi_i\}$ and $\{\psi_i\}$ have the same Puiseux pairs and the same pairwise orders of tangency, which gives us a way of constructing the maps $\{\psi_i\}$.

Corollary. For any singularity of a function of two variables there exists a real singularity, lying in the same connected component of the set $\mu = \text{const}$ in the space of all germs of functions $(\mathbb{C}^2, 0) \to (\mathbb{C}, 0)$ and having therefore the same intersection matrix.

For singularities of a greater number of variables analogous results have not been proved.

If all the maps ϕ_i are real and there exist real perturbations $\tilde{\phi}_i$ of them such that the curve

$$\cup_{i=1}^r \operatorname{Im} \tilde{\phi}_i$$

has only simple double points and all these s points are real, then the D-diagram of the real curve

$$(\cup_{i=1}^r \operatorname{Im} \tilde{\phi}_i) \cap \mathbb{R}^2$$

is the D-diagram of the singularity f, corresponding to the set of maps $\{\phi_i | i = 1, \ldots, r\}$. It follows from this that the perturbation $\tilde{f}$ of the singularity f, corresponding to the perturbations $\{\tilde{\phi}_i\}$ of the set of mappings $\{\phi_i\}$, satisfies the conditions of Theorem 4.1.

Theorem 4.4. For any set of germs of real maps $\{\phi_i\}$ ($\phi_i : (\mathbb{C}_i, 0) \to (\mathbb{C}^2, 0)$) there exist real perturbations $\{\tilde{\phi}_i\}$ for which the curve

$$\cup_{i=1}^r \operatorname{Im} \tilde{\phi}_i$$

has only simple double points and all these points are real.

We remark that both existing proofs of this theorem have a constructive character, that is they contain ways of constructing such a perturbation. One of these ways (more convenient for exposition but not more effective) will be described in § 4.3.

From Theorem 4.4 it follows that if $f: (\mathbb{C}^2, 0) \to (\mathbb{C}, 0)$ is a real singularity such that the curve $\{f = 0\}$ is real (that is all its irreducible components are real), then there exists a real perturbation $\tilde{f}$ of it, satisfying the conditions of Theorem 4.1, that is having only non-degenerate real critical points with the same critical value at all the saddle points.

Almost in the same way this result can be proved for any real singularity of two variables.

4.3   The resolution of singularities of functions of two variables and the construction of their real perturbations

A resolution of the singularity of a function $f: (\mathbb{C}^2, 0) \to (\mathbb{C}, 0)$ (or of the curve $\{(x, y): f(x, y) = 0\}$) can be constructed with the help of a sequence of σ-processes.

We consider the complex vector space $\mathbb{C}^n$ and the point 0 in it. The σ-process with centre at the point 0 is the map

$$\sigma: \Pi^n \to \mathbb{C}^n$$

of an n-dimensional complex manifold Π^n, which is constructed in the following manner: outside the preimage of the point $0 \in \mathbb{C}^n$ the map σ is an analytic isomorphism, the preimage $\sigma^{-1}(0)$ of the point 0 is an $(n-1)$-dimensional complex projective space $\mathbb{C}P^{n-1}$ (projectivisation of the space $\mathbb{C}^n$), which is glued to the complement

$$\Pi^n \setminus \sigma^{-1}(0) \approx \mathbb{C}^n \setminus 0$$

so that the line in the space $\mathbb{C}^n$, passing through zero is pasted to that point of the projectivisation $\mathbb{C}P^{n-1}$ of the space $\mathbb{C}^n$ to which it corresponds. Thus, the manifold Π^n is obtained from the space $\mathbb{C}^n$ except that in place of the point 0 there is an $(n-1)$-dimensional projective space $\mathbb{C}P^{n-1}$.

The σ-process with centre at zero in the space $\mathbb{C}^n$ can be described in the following manner. Let

$$(\mathbb{C}^n \setminus 0) \to \mathbb{C}P^{n-1}$$

be the projectivisation map, mapping each non-zero vector in the space $\mathbb{C}^n$ to the line generated by it. We consider the graph of this map as a subspace of the product $\mathbb{C}^n \times \mathbb{C}P^{n-1}$. It is not a closed submanifold of the space $\mathbb{C}^n \times \mathbb{C}P^{n-1}$. However it can be shown that its closure Π^n is a non-singular n-dimensional closed submanifold of the product $\mathbb{C}^n \times \mathbb{C}P^{n-1}$. The natural projection

$$\Pi^n \hookrightarrow \mathbb{C}^n \times \mathbb{C}P^{n-1} \to \mathbb{C}^n$$

is a one-one mapping away from zero of the space $\mathbb{C}^n$. The preimage of zero is the projective space $0 \times \mathbb{C}P^{n-1}$. The map $\Pi^n \to \mathbb{C}^n$ is a σ-process with centre at zero in the space $\mathbb{C}^n$.

Another (coordinate) description of the manifold Π^n is the following. Let $x_1, \ldots, x_n$ be the coordinates in the space $\mathbb{C}^n$, and let $u_1 : \ldots : u_n$ be the corresponding homogeneous coordinates in the complex projective space $\mathbb{C}P^{n-1}$. Let Π^n be the subspace of the product $\mathbb{C}^n \times \mathbb{C}P^{n-1}$ given by the equations

$$x_i u_j = u_i x_j \ (1 \leqslant i, j \leqslant n),$$

$(x_1, \ldots, x_n) \in \mathbb{C}^n$, $(u_1 : \ldots : u_n) \in \mathbb{C}P^{n-1}$. We shall show below that Π^n is an n-dimensional manifold. We denote by

$$\sigma : \Pi^n \hookrightarrow \mathbb{C}^n \times \mathbb{C}P^{n-1} \to \mathbb{C}^n$$

the projection onto the first factor. If $x = (x_1, \ldots, x_n) \neq 0$, then the preimage $\sigma^{-1}(x)$ of the point $x \in \mathbb{C}^n$ consists of the one point

$$(x_1, \ldots, x_n; x_1 : \ldots : x_n).$$

Therefore outside the preimage of the point $0 \in \mathbb{C}^n$ the map σ is an isomorphism

$$\Pi^n \backslash \sigma^{-1}(0) \to \mathbb{C}^n \backslash 0.$$

The preimage of the point $0 \in \mathbb{C}^n$ is the space $0 \times \mathbb{C}P^{n-1}$, which is isomorphic to the space $\mathbb{C}P^{n-1}$.

Let L be the line in the space $\mathbb{C}^n$ passing through the points 0 and $(x_1^0, \ldots, x_n^0)$. It consists of points of the type $(tx_1^0, \ldots, tx_n^0)$ $(t \in \mathbb{C})$. The preimage $\sigma^{-1}(L \backslash 0)$ of the line L without the point 0 consists of points of the product $\mathbb{C}^n \times \mathbb{C}P^{n-1}$ of the form

$$(tx_1^0, \ldots, tx_n^0; x_1^0 : \ldots : x_n^0) \ (t \neq 0).$$

Therefore the closure of the space $\sigma^{-1}(L \setminus 0)$ is the line

$$\{(tx_1^0, \ldots, tx_n^0; x_1^0 : \ldots : x_n^0)\} \subset \mathbb{C}^n \times \mathbb{C}P^{n-1}.$$

This line passes through the point $(0, \ldots, 0; x_1^0 : \ldots : x_n^0)$ of the preimage $\sigma^{-1}(0) = 0 \times \mathbb{C}P^{n-1}$, corresponding to the line L.

In order to show that the space Π^n is a non-singular complex manifold, we consider it at a neighbourhood of the point

$$(0, \ldots, 0; u_1^0 : \ldots : u_n^0) \in \mathbb{C}^n \times \mathbb{C}P^{n-1}.$$

Let, for example, $u_1^0 \neq 0$. Since $u_1 : \ldots : u_n$ are homogeneous coordinates in the space $\mathbb{C}P^{n-1}$, we can assume that $u_1^0 = 1$. Let $\mathbb{C}_1^{n-1} \subset \mathbb{C}P^{n-1}$ be the affine part of the projective space $\mathbb{C}P^{n-1}$ given by the condition $u_1 = 1$. In the part $\mathbb{C}^n \times \mathbb{C}_1^{n-1}$ of the product $\mathbb{C}^n \times \mathbb{C}P^{n-1}$ the space Π^n can be given by the equations

$$x_j = x_1 u_j \quad (j = 2, \ldots, n).$$

From this it can be seen that the space

$$\Pi^n \cap (\mathbb{C}^n \times \mathbb{C}_1^{n-1})$$

is isomorphic to an n-dimensional complex vector space with coordinates $x_1, u_2, \ldots, u_n$ and is therefore non-singular. From the equations defining the space Π^n it follows that in the part defined by the condition $u_1 = 1$, the coordinates $u_2, \ldots, u_n$ are expressed in terms of the coordinates $x_1, \ldots, x_n$ in the space $\mathbb{C}^n$ by the formulae

$$u_j = x_j / x_1 \quad (j = 2, \ldots, n).$$

It is not difficult to show that the above construction does not depend on the choice of coordinates $x_1, \ldots, x_n$ in the space $\mathbb{C}^n$ and is therefore applicable to any complex analytic space and a non-singular point of it. To prove this we need to verify that any local complex analytic isomorphism $(\mathbb{C}^n, 0) \to (\mathbb{C}^n, 0)$ lifts to a complex analytic isomorphism $\Pi^n \to \Pi^n$ in a neighbourhood of the preimage $\sigma^{-1}(0) = \mathbb{C}P^{n-1}$.

Now let $n = 2$, let $f : (\mathbb{C}^2, 0) \to (\mathbb{C}, 0)$ be the germ of a holomorphic function with an isolated critical point at zero and let

$$M = \{(x, y) : f(x, y) = 0\}$$

be the germ of a complex analytic curve in the space $\mathbb{C}^2$. Under the σ-process $\sigma: \Pi^2 \to \mathbb{C}^2$ with centre at zero, the projective line $\mathbb{C}P^1 \subset \Pi^2$ with coordinates $(u:v)$ is glued in place of the point $0 \in \mathbb{C}^2$. In this case the composition $f \circ \sigma$ is a holomorphic function on a neighbourhood of the space $\sigma^{-1}(0)$ in the manifold Π^2. The function $f \circ \sigma$ will, generally speaking, have non-isolated critical points. It is zero on the glued-in projective line $\mathbb{C}P^1 \subset \Pi^2$, moreover the line $\mathbb{C}P^1$ lies in the divisor $\{f \circ \sigma = 0\}$ with multiplicity equal to the degree m of the germ $f: (\mathbb{C}^2, 0) \to (\mathbb{C}, 0)$. This means that in the neighbourhood of the point

$$(1:0) \in \mathbb{C}P^1 \subset \Pi^2,$$

given by the condition $u = 1$, the function $f \circ \sigma$ has the form

$$f \circ \sigma = x^m \cdot g_1,$$

where g_1 is not identically zero on the glued-in projective line $\mathbb{C}P^1$. The degree m of the germ f is the least of the degrees of the monomials occurring in the expansion of f with non-zero coefficients. We have an analogous relation $(f \circ \sigma = y^m \cdot g_2)$ in the neighbourhood of the point $(0:1) \in \mathbb{C}P^1 \subset \Pi^2$, given by the condition $v = 1$. Therefore we need to consider as singularities of the function $f \circ \sigma$ not all the points at which its differential equals zero. There are too many such points and among them the majority are such that in a neighbourhood of them the function $f \circ \sigma$ is equivalent to the function x^m. We need to consider only the points at which the function g_1 (or g_2) takes the value zero. However g_1 (or g_2) is a function on the part of the manifold Π^2 defined by the condition that one of the coordinates is not equal to zero. To define it on the manifold Π^2 in an invariant way is not possible (without considering it as a section of bundle).

In order to avoid difficulties of this sort, we shall consider not the singularity of the function $f: (\mathbb{C}^2, 0) \to (\mathbb{C}, 0)$, but the singularity of the germ of the curve $M = \{f = 0\}$. In accordance with §4.2, we can replace the problem of constructing a real perturbation of the function f, necessary for the definition of its intersection matrix, by the problem of constructing a real perturbation of the curve M. Let

$$M = \cup_{i=1}^r M_i$$

be the decomposition of the germ M into irreducible components. Each of the curves M_i can be given by an equation $f_i = 0$ of degree m_i (where

$$f = \Pi_{i=1}^r f_i, \ m = \Sigma_{i=1}^r m_i).$$

We consider temporarily one of the irreducible curves M_i (the index i will for the present be omitted). As we said above, there exists a germ of a map

$$\phi : (\mathbb{C}, 0) \to (\mathbb{C}^2, 0),$$

the image of which is the germ M and which away from zero is an isomorphism $\mathbb{C} \backslash 0 \to M \backslash 0$. The map ϕ is given by the formulae

$$x = x(t) = at^m + \dots, \qquad y = y(t) = bt^m + \dots,$$

where the dots denote the sum of terms of higher degree (t is a coordinate on the line $\mathbb{C}$, x and y are coordinates in the plane $\mathbb{C}^2$; either $a \neq 0$ or $b \neq 0$). It can be shown that the natural number m is the same as the degree of the germ of the function defining the curve M. A linear change of coordinates in the space $\mathbb{C}^2$ allows us to take away the term of degree m from the series $y(t)$, that is to suppose that

$$x(t) = at^m + \dots, \; y(t) = bt^n + \dots$$

where $n > m$, $a \neq 0$. After this the (local) change of coordinates

$$\tilde{t} = \sqrt[m]{x(t)} = \sqrt[m]{a} \cdot \sqrt[m]{(t^m + \dots)} = \sqrt[m]{a}(t + \dots)$$

on the line $\mathbb{C}$ allows us to suppose that

$$x(t) = t^m, \qquad y(t) = bt^n + \dots \; (n > m).$$

If n is divisible by m ($n = km$), then the change

$$\tilde{x} = x, \qquad \tilde{y} = y - bx^k$$

eliminates the term of degree n from the series $y(t)$. Therefore after change of coordinates in the source $\mathbb{C}$ and in the target $\mathbb{C}^2$ we can suppose that the map ϕ is given by the formulae

$$x(t) = t^m, \qquad y(t) = \sum_{k \geq n} a_k t^k \; (n > m, \, a_n \neq 0),$$

where n is not divisible by m. Moreover the highest common factor of m, n and those k for which $a_k \neq 0$ is equal to 1. It is not difficult to see that in this case the curve M touches the coordinate line $y = 0$. The equation of the curve M can be written in the form

$$y = \sum_{k \geqslant n} a_k x^{k/m}.$$

The series

$$\sum_{k \geqslant n} a_k x^{k/m}$$

of fractional powers of the variable x is called the *Puiseux series* of the curve M. The pair of natural numbers (n, m) we call the *principal Puiseux indices* of the curve M.

For a fuller description of the germ M the set of so-called *characteristic Puiseux pairs* is used. We take the ratio n/m as a fraction in lowest terms n_1/m_1. The pair (n_1, m_1) is called the first characteristic Puiseux pair of the germ of the curve M. If the highest common factor of the numbers n and m (which is equal to m/m_1) is equal to 1 (in this case, of course, $m_1 = m$), then this exhausts the set of all characteristic Puiseux pairs. If $m/m_1 > 1$, then let

$$k_2 = \min\{k : a_k \neq 0, \ k \text{ is not divisible by } (m/m_1)\}.$$

We put the ratio $k_2/(m/m_1)$ in the form of a fraction in lowest terms n_2/m_2. The pair (n_2, m_2) is called the second characteristic Puiseux pair of the germ of the curve M. If $m/m_1 m_2 = 1$, then this second pair exhausts the set of characteristic Puiseux pairs. If $m/m_1 m_2 > 1$, then let

$$k_3 = \min\{k : a_k \neq 0, \ k \text{ is not divisible by } (m/m_1 m_2)\},$$

$$k_3/(m/m_1 m_2) = n_3/m_3, \ldots.$$

In the end we obtain a sequence of coprime pairs of natural numbers

$$(n_1, m_1), \ (n_2, m_2), \ldots, (n_g, m_g),$$

called the *characteristic Puiseux pairs* of the curve M. There are the relations:

$$m_1 \cdot \ldots \cdot m_g = m, \ n_{i-1} m_1 < n_i.$$

It can be shown that the characteristic Puiseux pairs give a sufficiently detailed description of the topology of the germ of the curve M. In particular, the germs of curves with the same characteristic Puiseux pairs are topologically equivalent to each other and are the zero level manifolds of singularities of germs of functions which are in one family $\mu = \text{const}$.

Let us observe what happens to the curve M under the action of the σ-process $\sigma : \Pi^2 \to \mathbb{C}^2$. In the affine part of the manifold Π^2, defined by the condition $u = 1$, the coordinates are x and y/x ($(u:v)$ are homogeneous coordinates in the projective line $\mathbb{C}P^1 \subset \Pi^2$). From the formulae

$$x(t) = t^m, \ y(t) = t^n + \ldots$$

it follows that in terms of the coordinates (x, v) on the manifold Π^2 we have

$$x(t) = t^m, \ v(t) = t^{n-m} + \ldots$$

From this it follows that the preimage $\sigma^{-1}(M \backslash 0)$ of the curve M without zero under the σ-process extends to the curve $\sigma^{-1}(M)$ in the manifold Π^2. The curve $\sigma^{-1}(M)$ intersects the glued-in projective line $\mathbb{C}P^1$ at the point $(1:0)$. From the parametric equations defining the curve $\sigma^{-1}(M)$ it follows that they have in some sense a smaller degree than the equations of the curve M. If $m < n < 2m$, then the degree of the curve $\sigma^{-1}(M)$ (equal to $n - m$) is strictly less than the degree of the curve M. If $n > 2m$, then the degree of the curve $\sigma^{-1}(M)$ (equal to m) is the same as the degree of the curve M, but the first of the principal Puiseux indices ($n - m, m$) of the curve $\sigma^{-1}(M)$ is less than that for the curve M.

In this way the singularity of the curve

$$\tilde{M} = \sigma^{-1}(M)$$

is simpler in the above sense than the singularity of the curve M. Carrying out a σ-process on the surface Π^2 with centre at the singular point of the curve $\tilde{M} = \sigma^{-1}(M)$, we obtain a surface $\tilde{\Pi}^2$ ($\tilde{\sigma} : \tilde{\Pi}^2 \to \Pi^2$) and a curve

$$\tilde{\tilde{M}} = \tilde{\sigma}^{-1}(\tilde{M})$$

on it, the singularity of which will be even simpler. The complex projective line on the surface $\tilde{\Pi}^2$, which is glued in during this σ-process will intersect transversely the projective line, glued in during the first σ-process, at one point only. Repeating this process the requisite number of times, we arrive in the end with the preimage of the curve M being non-singular.

Suppose now that the curve M is not necessarily irreducible

$$(M = \cup_{i=1}^r M_i).$$

Carrying out the σ-processes at zero and then at the singular points of the preimages of the curves M_i, we arrive at a position in which all r preimages of the curves M_i are non-singular. This means that we obtain an analytic map

$$\pi : (Z, Z_0) \rightarrow (\mathbb{C}^2, 0)$$

(a composition of σ-processes) from a non-singular analytic surface Z into the space $\mathbb{C}^2$ such that:
   (i) the restriction of the map π to $Z \backslash Z_0$ is an isomorphism

$$Z \backslash Z_0 \rightarrow \mathbb{C}^2 \backslash 0;$$

   (ii) the subspace $Z_0 = \pi^{-1}(0)$ is the union of complex projective lines on the surface Z which are in general position;
   (iii) the preimages $\pi^{-1}(M_i)$ of the curves $M_i \subset \mathbb{C}^2$ (the closures of $\pi^{-1}(M_i \backslash 0)$) are non-singular curves on the surface Z. The fact that the projective lines from which the subspace Z_0 is constructed are in general position means that they only intersect each other in pairs, two projective lines either not intersecting at all or intersecting transversely at one point.
   The curves $\pi^{-1}(M_i)$ can touch each other and also the glued-in complex projective lines. It is not difficult to see that a σ-process at a point of tangency of two non-singular curves reduces the degree of tangency, and a σ-process at a point of their transverse intersection separates them, that is gives no inter-section. Therefore, by carrying out a sufficient number of σ-processes, we can suppose that the curves $\pi^{-1}(M_i)$ do not intersect each other and intersect the preimage Z_0 of zero transversely at its non-singular points (that is not at points of intersection of the glued-in projective lines).
   We shall call such a map

$$\pi : (Z, Z_0) \rightarrow (\mathbb{C}^2, 0)$$

a *resolution* of the curve

$$M = \{(x, y) : f(x, y) = 0\}.$$

It can be proved that this map is a resolution of the singularity of the function f in the sense of §3.5.
   Let, for example, the curve M be irreducible and given by the parametric equations

$$x = t^2, \ y = t^5.$$

The σ-process with centre at zero reduces it to a curve of the form

$$x=t^2, \ y=t^3.$$

The σ-process at the singular point of this curve reduces it to a non-singular curve, touching the glued-in projective line, this tangency being simple. The σ-process at the point of tangency reduces to the situation where the preimage of the curve M intersects two glued-in projective lines at the point of their intersection. Finally, the σ-process at this point of intersection at last reduces the preimage of the curve we started with and all the glued-in projective lines to general position (figure 40; for clarity the curve M and its preimage are depicted with heavier lines and the glued-in projective lines are numbered in the order of their introduction).

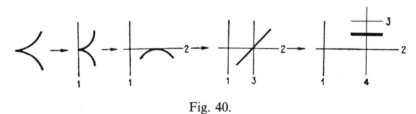

Fig. 40.

Another example (for the singularity of the function $f(x,y)=x(x^2+y^3)$, $M=M_1\cup M_2$, $M_1=\{x=0\}$, $M_2=\{x^2+y^3=0\}$) leads to figure 41.

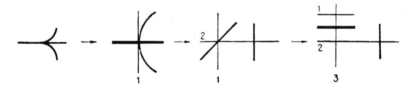

Fig. 41.

It is not difficult to prove that the multiplicity and consequently the intersection matrix of a singularity of a function corresponding to the curve

$$M=\cup_{i=1}^r M_i$$

is determined by which of the glued-in projective lines and preimages of the curves M_i intersect each other and does not depend on the specific points at which they (transversely) intersect. From this it follows that as far as the multiplicity of the singularity and its intersection matrix are concerned there is no harm in supposing that all the glued-in projective lines and preimages of the curves M_i are real (that is that all σ-processes are carried out only at real points of the corresponding surfaces, and the curves M_i themselves are real). In the case of figures 40 and 41 we depicted the real parts of the glued-in projective lines and the curves M_i.

The construction described above is in essence the proof of Theorem 4.3.

The contraction of one of the glued-in projective lines (the last in the order of introduction) reduces us to the situation when several *non-singular* curves (glued-in projective lines and preimages of the curves M_i) intersect at one point, pairwise transversely. By perturbation it is possible to arrange that these curves do not have more than simple intersections, and that the simple pairwise intersections are all real (figure 42).

By contracting in this way all the glued-in projective lines (in the oppsite order to their introduction in the resolution of the singularity) and at each step getting rid, by perturbation, of more than pairwise intersections, we arrive at a perturbation of the initial curves M_i which have only simple double intersections (both with themselves and with each other), all these intersections being real. In this way we construct the perturbation of the curves M_i, required for the definition of the intersection matrix of the singularity in accordance with §4.2.

Fig.42.

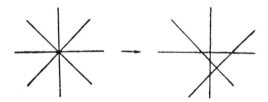

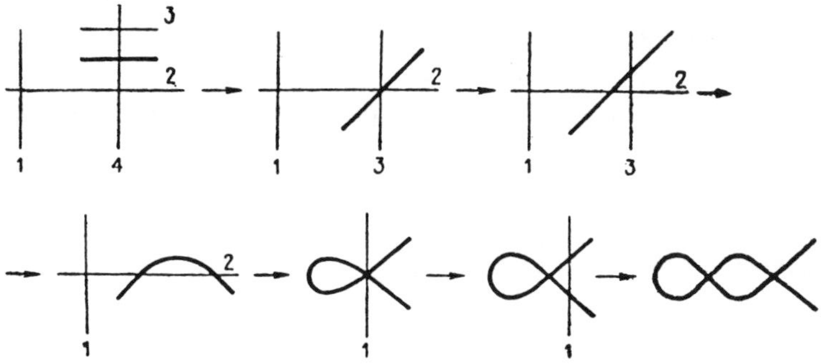

Fig. 43.

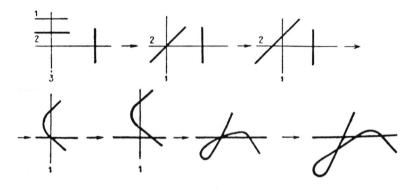

Fig. 44.

For the singularities, the resolutions of which were depicted in figures 40 and 41, the corresponding construction leads to figures 43 and 44.

The above-described way of constructing a real perturbation has the shortcoming that to carry it out we need to construct the resolution of the singularity with the help of a sequence of σ-processes, which is rather a long procedure. For example, for the singularity $x^m + y^n$, discussed in § 4.1 (Example 1), this method requires numerous (although simple) constructions. A more effective way is that based on induction on Puiseux pairs of the singularity, although it is more difficult to describe.

4.4   Partial diagonalisation of the quadratic form of a singularity

We have already said that the possibility of representing the D-diagram of a singularity of a function of two variables in the form of the D-diagram of a real curve allows us to simplify significantly some calculations connected with the corresponding quadratic forms (for example, the calculation of the inertia indices). The intersection numbers of the (formal) vanishing cycles, corresponding to the real curve l (see § 4.1) are connected with each other by relations of a special form.

Lemma 4.2. The following equality holds:

$$2n_{20}(k, i) = \Sigma_j n_{21}(k, j) n_{10}(j, i).$$

The proof can be obtained from simple geometrical considerations.

We embed the integer lattice with basis

$$\Delta_i^0, \Delta_j^1, \Delta_k^2$$

in the real linear space $\mathbb{R}^\mu$ with the same basis and we extend the quadratic form $(x \circ y)$ to the whole space $\mathbb{R}^\mu$. We define in the space $\mathbb{R}^\mu$ a new basis

$$\bar{\Delta}_i^0, \bar{\Delta}_j^1, \bar{\Delta}_k^2$$

by the formulae

$$\bar{\Delta}_i^0 = \Delta_i^0 + \tfrac{1}{2}\Sigma_j(\Delta_i^0 \circ \Delta_j^1)\Delta_j^1 ,$$

$$\bar{\Delta}_j^1 = \Delta_j^1 ,$$

$$\bar{\Delta}_k^2 = \Delta_k^2 + \tfrac{1}{2}\Sigma_j(\Delta_k^2 \circ \Delta_j^1)\Delta_j^1 .$$

Using Lemma 4.2 it is not difficult to verify that

$$(\bar{\Delta}_i^0 \circ \bar{\Delta}_j^1) = (\bar{\Delta}_k^2 \circ \bar{\Delta}_j^1) = (\bar{\Delta}_k^2 \circ \bar{\Delta}_i^0) = 0.$$

In addition $(\bar{\Delta}_j^1 \circ \bar{\Delta}_{j'}^1) = -2\delta_{jj'}$. Thus we have obtained a partial diagonalisation of the intersection matrix.

It is not difficult to see that

$$(\bar{\Delta}_i^0 \circ \bar{\Delta}_i^0) = -2 + \tfrac{1}{2}\sum_j [n_{10}(j, i)]^2,$$

$$(\bar{\Delta}_k^2 \circ \bar{\Delta}_k^2) = -2 + \tfrac{1}{2}\sum_j [n_{21}(k, j)]^2.$$

From this it follows that if among the regions bounded by the curve l (or a curve obtained from it with the help of an admissible homotopy), there is even one quadrilateral, then its quadratic form is not negative definite, and that if among them there is a polygon with more than four edges then it is not even negative semi-definite.

We shall demonstrate one application of this construction. For quadratic forms on the lattice $\mathbb{Z}^\mu$ it makes sense to talk about the determinant, since the determinant of a change of basis in the integer lattice is equal to ± 1. We denote by $D(f)$ the determinant of the quadratic form, corresponding to the singularity f. The determinant $D(f)$ is equal to zero if the map

$$i_* : H_{n-1}(V_\varepsilon) \to H_{n-1}(V_\varepsilon, \partial V_\varepsilon)$$

of homology groups induced by the inclusion $V_\varepsilon \hookrightarrow (V_\varepsilon, \partial V_\varepsilon)$ (for the singularity $f(x, y) + t^2$) is not a monomorphism and otherwise is equal, up to sign, to the order of the cokernel

$$H_{n-1}(V_\varepsilon, \partial V_\varepsilon)/\operatorname{Im} i_*$$

of the map i_* (the matrix of the map i_* is also the intersection matrix of the singularity).

Theorem 4.5. Let $f: (\mathbb{C}^2, 0) \to (\mathbb{C}, 0)$ be a singularity of a function of two variables and let r be the number of irreducible factors into which the germ of the function f can be decomposed (that is the number of irreducible components of the curve $\{f = 0\}$). Then $D(f)$ is divisible by 2^{r-1}.

In particular, if the function f is reducible, that is if $r > 1$, then the map i_* (for the singularity $f(x, y) + t^2$) cannot be an isomorphism.

Proof. In accordance with Theorems 4.3 and 4.4 (§ 4.2) we construct a set of real maps $\tilde{\phi}_i : \mathbb{C}_i \to \mathbb{C}^2$ (defined in a neighbourhood of zero) such that the D-diagram of the curve

$$\left(\cup_{i=1}^r \operatorname{Im} \tilde{\phi}_i\right) \cap \mathbb{R}^2$$

is the same as the D-diagram of the singularity f. Let $\{\Delta_m^\sigma\}$ be the basis of the lattice on which the quadratic form $(x \circ y)$ is defined, corresponding to this curve. The number of self-intersections s of the curve

$$\left(\cup_{i=1}^r \operatorname{Im} \tilde{\phi}_i\right) \cap \mathbb{R}^2$$

is equal to the number of basic elements Δ_j^1. The number of remaining elements of the basis (Δ_i^0 and Δ_k^2) is equal to $s - (r - 1)$ (compare with Lemma 4.1 of § 4.2). The transition matrix from the basis $\{\Delta_m^\sigma\}$ of the integer lattice to the basis $\{\bar{\Delta}_m^\sigma\}$ of the space $\mathbb{R}^\mu$, described above, has determinant equal to $+1$. Therefore $D(f)$ is equal to the determinant of the matrix of the quadratic form of the singularity f with respect to the basis $\{\bar{\Delta}_m^\sigma\}$. This determinant factors into the product of three determinants:

$$|(\bar{\Delta}_i^0 \circ \bar{\Delta}_{i'}^0)|, \quad |(\bar{\Delta}_k^2 \circ \bar{\Delta}_{k'}^2)|, \quad \text{and} \quad |(\bar{\Delta}_j^1 \circ \bar{\Delta}_{j'}^1)|.$$

The last of these is equal to $(-2)^s$. It is easy to see that the intersection numbers $(\bar{\Delta}_i^0 \circ \bar{\Delta}_{i'}^0)$ and $(\bar{\Delta}_k^2 \circ \bar{\Delta}_{k'}^2)$ belong to the set of half-integers $\frac{1}{2}\mathbf{Z}$. Therefore

$$|(\bar{\Delta}_i^0 \circ \bar{\Delta}_{i'}^0)| \cdot |(\bar{\Delta}_k^2 \circ \bar{\Delta}_{k'}^2)| \in 2^{-s+(r-1)} \cdot \mathbf{Z},$$

from which it follows that

$$D(f) = \pm 2^s \cdot |(\bar{\Delta}_i^0 \circ \bar{\Delta}_{i'}^0)| \cdot |(\bar{\Delta}_k^2 \circ \bar{\Delta}_{k'}^2)|$$

belongs to $2^{r-1}\mathbf{Z}$, which is what we were required to prove.

For example, for the singularity A_k $(f(x,y) = x^{k+1} + y^2)$ the determinant $D(f)$ is equal to $(-1)^k(k+1)$.

We give one more example of the application of the partial diagonalisation of the quadratic form of a singularity of a function of two variables. We consider the singularity of the function

$$f(x,y) = x^5 + x^2 y^2 + y^5.$$

It is equivalent to the singularity of the function

$$(x^3 + y^2)(x^2 + y^3).$$

A real perturbation of the curve $\{f(x,y) = 0\}$, which has only real simple double self-intersections, is depicted in figure 45.

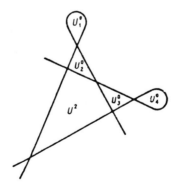

Fig. 45.

The multiplicity of the singularity f is equal to 11. With respect to the basis $\bar{\Delta}^\sigma_m$, defined above, the quadratic form of the singularity f decomposes into the direct sum of three quadratic forms. One of these, on a six-dimensional subspace, generated by the elements $\bar{\Delta}^1_j$, is negative definite. The second, on the one-dimensional subspace generated by the element $\bar{\Delta}^2$, which in figure 45 corresponds to the quadrilateral U^2, is zero. Finally the third, on the four-dimensional subspace generated by the elements $\bar{\Delta}^0_1, \ldots, \bar{\Delta}^0_4$, which in figure 45 correspond to the "monangles" and triangles $U^0_1, \ldots, U^0_4$.

We have

$$(\bar{\Delta}^0_1 \circ \bar{\Delta}^0_1) = (\bar{\Delta}^0_4 \circ \bar{\Delta}^0_4) = -3/2,$$

$$(\bar{\Delta}^0_2 \circ \bar{\Delta}^0_2) = (\bar{\Delta}^0_3 \circ \bar{\Delta}^0_3) = -1/2,$$

$$(\bar{\Delta}^0_1 \circ \bar{\Delta}^0_2) = (\bar{\Delta}^0_2 \circ \bar{\Delta}^0_3) = (\bar{\Delta}^0_3 \circ \bar{\Delta}^0_4) = 1/2$$

(see figure 46, where we have shown the self-intersection numbers and inter-section numbers of the basic elements $\bar{\Delta}^0_i$, multiplied, for convenience, by two).

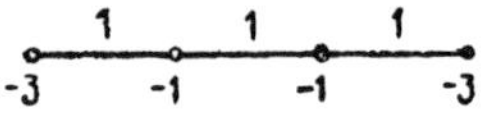

Fig. 46.

Such a form is easily diagonalised and has its positive inertia index equal to 1 and its negative one equal to 3. From this it follows that the quadratic form of the singularity f has its positive inertia index $\mu_+ = 1$, its zero one $\mu_0 = 1$ and its negative one $\mu_- = \mu - 2 = 9$. Therefore this singularity is hyperbolic in the sense of Volume 1.

Chapter 5

The intersection forms of boundary singularities and the topology of complete intersections

In this chapter we shall give a short exposition of some generalizations concerned, principally, with the concepts of the intersection form and vanishing cycles for singularities of functions on manifolds with boundary, for complete intersections, ...

5.1 Singularities with the action of finite groups

Some of the concepts and results which we discussed in the previous chapters can be generalised to the case when we consider the germ of a function

$$f: (\mathbb{C}^n, 0) \to (\mathbb{C}, 0)$$

which is invariant relative to the linear action of a finite group G on the space $\mathbb{C}^n$. In addition, there will arise on the way, in a natural manner, singularities corresponding to the Lie algebras

$$B_k, C_k, F_4 \quad \text{and} \quad G_2,$$

the root systems of which contain vectors of different lengths.

Let us suppose that we are given a linear representation of a finite group G on the complex vector space $\mathbb{C}^n$. The transformation of the space $\mathbb{C}^n$, corresponding to an element g of the group G, we shall denote by T_g. Let us suppose that $f: (\mathbb{C}^n, 0) \to (\mathbb{C}, 0)$ is the germ of a function, invariant under the action of the group G, that is, such that

$$f(T_g x) = f(x)$$

for all $g \in G$. In this case the group G acts on the non-singular level manifold

$$V_\varepsilon = f^{-1}(\varepsilon) \cap \bar{B}_\varrho$$

of the function f near the critical point, and thus on its homology group $H_{n-1}(V_\varepsilon)$ with coefficients in the groups $\mathbb{Z}$, $\mathbb{R}$ or $\mathbb{C}$.

In the case when f is an ordinary singularity of a function (that is when the group G is trivial), the multiplicity of the singularity f is defined to be the dimension of the space $H_{n-1}(V_\varepsilon; \mathbb{R})$ (or $H_{n-1}(V_\varepsilon; \mathbb{C})$). The natural analogue of dimension for the case when the group G is not trivial is the G-module

$$[H_{n-1}(V_\varepsilon; \mathbb{R})] \text{ (or } [H_{n-1}(V_\varepsilon; \mathbb{C})])$$

as an element of the ring $R_\mathbb{R}(G)$ (respectively $R_\mathbb{C}(G)$) of real (respectively complex) representations of the group G (see [327]). We shall denote by $[H]$ the element $[H_{n-1}(V_\varepsilon; \mathbb{C})]$ of the ring $R(G) = R_\mathbb{C}(G)$ of complex representations of the group G.

Let $_n\mathcal{O}$ be the ring of germs at zero of holomorphic functions of n variables and let $(\partial f/\partial x_1, \ldots, \partial f/\partial x_n)$ be the Jacobian ideal of the germ f, that is the ideal generated by the partial derivatives of the function f. The dimension of the quotient ring Q_f of the ring $_n\mathcal{O}$ by the Jacobian ideal $(\partial f/\partial x_1, \ldots, \partial f/\partial x_n)$ (as a complex vector space) is the same as the multiplicity of the singularity f. The action of the group G on the space $\mathbb{C}^n$ defines its representation in the ring $_n\mathcal{O}$. In the case when the germ f is invariant relative to the action of the group G, this representation in a natural way defines a representation of the group G on the vector space Q_f. As we said above, the dimensions of the vector spaces $H = H_{n-1}(V_\varepsilon; \mathbb{C})$ and Q_f are the same. The relationship between $[H]$ and $[Q_f]$ as elements of the group $R(G)$ was revealed in [399].

The group G acts linearly on the space $\mathbb{C}^n$, on which the germ of the function f is defined. Therefore it acts on its nth exterior power $\lambda^n\mathbb{C}^n$, which is a one-dimensional vector space. The action of the element $g \in G$ on the space $\lambda^n\mathbb{C}^n$ is the same as multiplication by the determinant $\det T_g$ of the operator T_g.

The representation of the group G on the vector space V defines a corresponding representation on the dual vector space V^*. For example, the representation of the group G on the space $H = H_{n-1}(V_\varepsilon; \mathbb{C})$ of the homology of the non-singular level manifold has the corresponding dual representation on the space $H^* = H^{n-1}(V_\varepsilon; \mathbb{C})$ of cohomology of the non-singular level manifold.

Theorem 5.1 ([399]). The G-modules (that is vector spaces with representations of the group G)

$$H \quad \text{and} \quad Q_f^* \otimes_\mathbb{C} \lambda^n\mathbb{C}^n$$

are isomorphic.

The isomorphism whose existence is stated in Theorem 5.1 is not defined canonically.

In § 3.5 we mentioned that instead of calculating the multiplicity of a singularity, that is the dimension of the homology group of the non-singular level manifold V_ε, it is frequently more convenient to calculate its Euler characteristic $\chi(V_\varepsilon)$. If X is a topological space (for definiteness a finite CW complex) on which a group G acts, then the equivariant Euler characteristic $\chi_G(X)$ of the space X is the element

$$\Sigma_q(-1)^q[H_q(X;\mathbb{C})]$$

of the ring $R(G)$ of complex representations of the group G, where $[H_q(X;\mathbb{C})]$ is the element of the ring $R(G)$ defined by the qth homology group of the space X with the corresponding representation of G. The equivariant Euler characteristic $\chi_G(V_\varepsilon)$ of the non-singular level manifold V_ε is equal to

$$[\mathbb{C}]+(-1)^{n-1}[H],$$

where $[\mathbb{C}]$ is the element of the ring $R(G)$ defined by the one-dimensional space $\mathbb{C}$ with the trivial representation of the group G (the zeroth homology group $H_0(V_\varepsilon,\mathbb{C})$ of the non-singular level manifold is one-dimensional and the representation of the group G on it is trivial).

If the action of the group G preserves the CW-complex structure of the space X then there is defined on the vector space $C_q(X;\mathbb{C})$ of q-dimensional CW chains on the space X with coefficients in the field $\mathbb{C}$ a natural representation of the group G. We can show that by analogy with the relation

$$\chi(X)=\Sigma_q(-1)^q \dim C_q(X;\mathbb{C})$$

for the ordinary Euler characteristic, we get the formula

$$\chi_G(X)=\Sigma_q(-1)^q[C_q(X;\mathbb{C})]$$

for the equivariant Euler characteristic.

It is well-known (see, for example [327]) that an element $[V]$ of the ring $R(G)$ of complex representations of the group G is defined by its character

$$[V](g)=\operatorname{tr} T_g|_V$$

as a function on the group G ($\operatorname{tr} T_g|_V$ is the trace of the operator $T_g|_V$). If X is a CW complex with the group action of G preserving the CW-complex structure, then the value $\chi_G(X)(g)$ of the character of $\chi_G(X) \in R(G)$ on $g \in G$ is the same as the (usual) Euler characteristic $\chi(X_g)$ of the set X_g of fixed points of the action of the element g on the space X. If follows from this that outside the set X_g the element g moves cells and therefore these cells give zero contribution in

$$\mathrm{tr}\,(g_*|C_q(V_\varepsilon;\mathbb{C})).$$

The cells lying in X_g stay in place and give in $\mathrm{tr}\,(g_*|C_q(V_\varepsilon;\mathbb{C}))$ a contribution of one.

From this follows

Theorem 5.2. The character of the equivariant Euler characteristic of the non-singular level manifold of the singularity f is defined by the formula

$$\chi_G(V_\varepsilon)(g) = 1 + (-1)^{d_g - 1}\mu_g,$$

where d_g is the dimension of the subspace of the space $\mathbb{C}^n$ on which the element $g \in G$ acts trivially, and μ_g is the multiplicity of the restriction of the function f to this subspace.

The multiplicity μ_g is defined, since if the function f has an isolated critical point at zero in the space $\mathbb{C}^n$, then in the subspace fixed by the element $g \in G$ it also has an isolated critical point.

Corollary. The character of the natural representation of the group G in the space of the homology $H = H_{n-1}(V_\varepsilon;\mathbb{C})$ of the non-singular level manifold is defined by the formula

$$[H](g) = (-1)^{n - d_g}\mu_g.$$

If G is a finite subgroup of the unitary group $U(n)$, generated by reflections, then $\mathbb{C}^n/G \cong \mathbb{C}^n$. The germ $f:(\mathbb{C}^n, 0) \to (\mathbb{C}, 0)$, invariant under the action of the group G, defines the germ f_* on $(\mathbb{C}^n/G, 0) \cong (\mathbb{C}^n, 0)$.

Theorem 5.3 ([399]). The multiplicity $\mu(f_*)$ is equal to

$$(1/|G|) \sum_{g \in G} (-1)^{n - d_g}\mu_g,$$

where $|G|$ is the number of elements in the group G.

Example. Let $G = \mathbb{Z}_2$ be the group with two elements acting on the space $\mathbb{C}^n$ with coordinates $x_1, \ldots, x_n$ according to the formula

$$\sigma(x_1, x_2, \ldots, x_n) = (-x_1, x_2, \ldots, x_n)$$

($\sigma \in \mathbb{Z}_2$ is the non-identity element of the group). An isomorphism of the space $\mathbb{C}^n / \mathbb{Z}_2$ with the space $\mathbb{C}^n$ with coordinates $y_1, \ldots, y_n$ is defined by the formulae

$$y_1 = x_1^2, \quad y_i = x_i \quad \text{for} \quad 2 \leqslant i \leqslant n.$$

In this case $|G| = 2$, the subspace of the space $\mathbb{C}^n$, fixed by the action of the transformation σ is the $(n-1)$-dimensional space $\{x_1 = 0\}$. Theorem 5.3 gives

$$\mu(f_*) = \tfrac{1}{2}(\mu(f) - \mu(f|\{x_1 = 0\})),$$

where $\mu(f|\{x_1 = 0\})$ is the multiplicity of the restriction of the germ f to the subspace $\{x_1 = 0\}$. It is not difficult to see that

$$f|\{x_1 = 0\} = f_*|\{y_1 = 0\}.$$

Therefore

$$\mu(f) = 2\mu(f_*) + \mu(f_*|\{y_1 = 0\}).$$

We consider the action of the group $\mathbb{Z}_2$ on the homology group $H_{n-1}(V_\varepsilon; \mathbb{R})$ of the non-singular level manifold of the function f. The group $\mathbb{Z}_2$ has two irreducible real representations: the trivial one and multiplication by -1. Therefore the homology group $H_{n-1}(V_\varepsilon; \mathbb{R})$ decomposes into the direct sum

$$H^+ \oplus H^-,$$

where H^+ is the space of cycles which are invariant under the involution σ, and H^- is the anti-invariant one:

$$\dim H^+ + \dim H^- = \mu(f).$$

The corollary of Theorem 5.2 gives

$$[H](\sigma) = \dim H^+ - \dim H^- = -\mu(f|\{x_1 = 0\}) = -\mu(f_*|\{y_1 = 0\}).$$

Since

$$\dim H^+ + \dim H^- = 2\mu(f_*) + \mu(f_*|\{y_1 = 0\}),$$

then

$$\dim H^+ = \mu(f_*), \qquad \dim H^- = \mu(f_*) + \mu(f_*|\{y_1 = 0\}).$$

These formulae can also be proved directly (without the use of Theorems 5.2 and 5.3). We leave this as an exercise for the reader.

The classification of singularities of functions invariant under such an action of the group $\mathbb{Z}_2$ is the same as the classification of singularities of functions on manifolds with boundary (see Volume 1, Chapter 17). The classification of singularities of functions of small codimension, invariant relative to the action of the group $(\mathbb{Z}_2)^q$, (interpreted as the classification of singularities of functions on manifolds with "corners") was considered, in particular, in [336].

5.2   Singularities of functions on manifolds with boundary

We shall give here a brief exposition of the analogues of some of the above concepts for singularities of functions on manifolds with boundary. A more detailed exposition and a motivation of the corresponding concepts can be found in [18].

Let f be a singularity of a function on a manifold with boundary (see Volume 1, Chapter 17). This means that f is the germ of a holomorphic function $(\mathbb{C}^n, 0) \to (\mathbb{C}, 0)$ on the complex vector space $\mathbb{C}^n$ in which the hyperplane $\mathbb{C}^{n-1}$ is fixed, the function f having an isolated critical point at zero both in the space $\mathbb{C}^n$ and on the subspace $\mathbb{C}^{n-1}$ (or more generally not having a critical point at zero in the space $\mathbb{C}^n$). We can suppose that the hyperplane $\mathbb{C}^{n-1}$ is given by the equation $x_1 = 0$, where $x_1, \ldots, x_n$ are the coordinates in the space $\mathbb{C}^n$.

Let $\hat{\mathbb{C}}^n$ be the double covering of the space $\mathbb{C}^n$, branching along the hyperplane $\mathbb{C}^{n-1}$. If $\hat{x}_1, \ldots, \hat{x}_n$ are the coordinates in the space $\hat{\mathbb{C}}^n$, then the branched covering $\hat{\mathbb{C}}^n \to \mathbb{C}^n$ is defined by the formulae

$$x_1 = \hat{x}_1^2, \; x_2 = \hat{x}_2, \ldots, \; x_n = \hat{x}_n.$$

On the space $\hat{\mathbb{C}}^n$ there is the natural involution

$$(\hat{x}_1, \hat{x}_2, \ldots, \hat{x}_n) \mapsto (-\hat{x}_1, \hat{x}_2, \ldots, \hat{x}_n),$$

i.e. an action of the cyclic group $\mathbb{Z}_2$ of order two. The germ f induces the germ

$$\hat{f}(\hat{x}_1, \hat{x}_2, \ldots, \hat{x}_n) = f(\hat{x}_1^2, \hat{x}_2, \ldots, \hat{x}_n)$$

of a function on the space $\hat{\mathbb{C}}^n$ which is invariant under the action of the group $\mathbb{Z}_2$. A locally analytic automorphism of the space $\mathbb{C}^n$, preserving the subspace $\mathbb{C}^{n-1}$ induces a locally analytic automorphism, commuting with the action of the group $\mathbb{Z}_2$, of the space $\hat{\mathbb{C}}^n$. In this way, a singularity of a function on a manifold with boundary can be considered in a natural way as the germ of a function $\hat{f}(\hat{x}_1, \hat{x}_2, \ldots, \hat{x}_n)$, invariant under the involution

$$(\hat{x}_1, \hat{x}_2, \ldots, \hat{x}_n) \mapsto (-\hat{x}_1, \hat{x}_2, \ldots, \hat{x}_n).$$

Conversely each such germ induces a singularity of a function on the manifold $\mathbb{C}^n$ with boundary $\mathbb{C}^{n-1}$.

A singularity f of a function on a manifold with boundary can define a non-singular level manifold in two ways. To the function itself corresponds its non-singular level manifold

$$V_\varepsilon = \{x \in \mathbb{C}^n : f(x) = \varepsilon, \|x\| \leqslant \varrho\},$$

which is an $(n-1)$-dimensional complex manifold with boundary (understood in the usual real sense). To the boundary $\mathbb{C}^{n-1}$ corresponds an $(n-2)$-dimensional complex submanifold

$$V_\varepsilon' = \{x \in \mathbb{C}^{n-1} : f(x) = \varepsilon, \|x\| \leqslant \varrho\},$$

which is the non-singular level manifold of the restriction of the function f to the hyperplane $\mathbb{C}^{n-1}$.

From the exact homology sequence for the pair $(V_\varepsilon, V_\varepsilon')$

$$\ldots \to H_{n-1}(V_\varepsilon') \to H_{n-1}(V_\varepsilon) \to H_{n-1}(V_\varepsilon, V_\varepsilon')$$
$$\to H_{n-2}(V_\varepsilon') \to H_{n-2}(V_\varepsilon) \to \ldots$$

in which

$$H_k(V_\varepsilon) = 0 \quad \text{for} \quad k \neq n-1$$
$$H_k(V_\varepsilon') = 0 \quad \text{for} \quad k \neq n-2,$$

it follows that

$$H_k(V_\varepsilon, V_\varepsilon') = 0 \quad \text{for} \quad k \neq n-1,$$

and $H_{n-1}(V_\varepsilon, V_\varepsilon')$ is a free abelian group, the rank of which equals the sum of the multiplicities of the critical point of the function f on the space $\mathbb{C}^n$ and the critical point of its restriction to the space $\mathbb{C}^{n-1}$. From general theorems of homotopy topology it follows (at least, when the number of variables $n > 2$, when the space $V_\varepsilon/V_\varepsilon'$ is simply connected; for $n = 2$ the proof is still simpler) that the quotient space $V_\varepsilon/V_\varepsilon'$ has the homotopy type of a bouquet of spheres. The number $\mu = \mu(f|x_1)$ of these spheres is called the *multiplicity* of the boundary singularity f (the notation identifying the function itself and the coordinate mapping to zero on the boundary $\mathbb{C}^{n-1}$). As we have shown, the multiplicity $\mu(f|x_1)$ of a boundary singularity is equal to the sum

$$\mu(f) + \mu(f|\{x_1 = 0\})$$

of the multiplicities of singular points of the function f on the spaces $\mathbb{C}^n$ and $\mathbb{C}^{n-1}$.

A basis in the homology group $H_{n-1}(V_\varepsilon, V_\varepsilon')$ can be constructed in the following way. Let $\tilde{f}$ be a perturbation of the singularity f of general form. This last expression means that the function $\tilde{f}$ on the space $\mathbb{C}^n$ and its restriction $\tilde{f}|\mathbb{C}^{n-1}$ to the boundary $\mathbb{C}^{n-1} = \{x_1 = 0\}$ are Morse and in addition the critical values of the functions $\tilde{f}$ and $\tilde{f}|\{x_1 = 0\}$ are different. In particular the function $\tilde{f}$ does not have critical points lying on the hyperplane $\{x_1 = 0\}$. Let

$$\mu(f) = \mu_0, \quad \mu(f|\{x_1 = 0\}) = \mu_1,$$

let $z_1, \ldots, z_{\mu_0}$ be the critical values of the function $\tilde{f}$ in a neighbourhood of zero in the space $\mathbb{C}^n$, let $z_1', \ldots, z_{\mu_1}'$ be the critical values of the restriction of the function $\tilde{f}$ to the subspace $\mathbb{C}^{n-1} = \{x_1 = 0\}$, and let z_0 be a non-critical value of the functions $\tilde{f}$ and $\tilde{f}|\{x_1 = 0\}$. Let

$$F_{z_0} = \{x \in \mathbb{C}^n : \tilde{f}(x) = z_0, \|x\| \leqslant \varrho\}$$

and let

$$F_{z_0}' = F_{z_0} \cap \{x_1 = 0\}.$$

As for ordinary singularities of functions under natural restrictions there is a diffeomorphism between the pair of manifolds (F_{z_0}, F_{z_0}') and the pair $(V_\varepsilon, V_\varepsilon')$. If u is a path joining some critical value z_i of the function $\tilde{f}$ with the non-critical value

z_0 (and not passing through the critical values of the functions $\tilde{f}$ and $\tilde{f}|\{x_1=0\}$), then, as before, there corresponds to it a vanishing cycle Δ in the homology group $H_{n-1}(F_{z_0})$ of the non-singular level manifold of the function $\tilde{f}$, which is isomorphic to the homology group $H_{n-1}(V_\varepsilon)$ of the non-singular level manifold of the function f. Moreover, since the path u does not pass through the critical values of the function $\tilde{f}|\{x_1=0\}$, a lifting of the homotopy $t\to u(t)$ to the homotopy of the fibre can be chosen so that it preserves the submanifold

$$F_{u(t)}\cap\{x_1=0\}.$$

In this case the cycle Δ lies entirely outside the submanifold $F_{z_0}\cap\{x_1=0\}$.

From the exact sequence for the pair $(V_\varepsilon, V_\varepsilon')$ it follows that the natural homomorphism

$$i_*:H_{n-1}(V_\varepsilon)\to H_{n-1}(V_\varepsilon, V_\varepsilon'),$$

induced by the inclusion $V_\varepsilon\hookrightarrow(V_\varepsilon, V_\varepsilon')$, is a momomorphism. Therefore the path u gives rise to a vanishing cycle in the relative homology

$$H_{n-1}(V_\varepsilon, V_\varepsilon')$$

of the non-singular level manifold modulo the submanifold $\{x_1=0\}$.

If u is a path joining the critical value z_j' of the function $\tilde{f}|\{x_1=0\}$ (which is not critical for the function $\tilde{f}$) with the non-critical value z_0 ($u(0)=z_j'$, $u(1)=z_0$) and not passing through the critical values of the functions $\tilde{f}$ and $\tilde{f}|\{x_1=0\}$, then there corresponds to it a vanishing "hemicycle"

$$\Delta'\in H_{n-1}(V_\varepsilon, V_\varepsilon'),$$

defined in the following fashion. Let $p_j'\in\mathbb{C}^{n-1}$ be a critical point of the function $\tilde{f}|\{x_1=0\}$, and let $\tilde{f}(p_j')=z_j'$. It can be shown that in a neighbourhood of the point p_j', by a local change of coordinates preserving the hyperplane $\{x_1=0\}$, the function $\tilde{f}$ can be reduced to the form

$$\tilde{f}=-x_1+\sum_{k=2}^{n}x_k^2+z_j'.$$

Without loss of generality we can suppose that $z_j'=0$ and that for t small $u(t)=t$. In this case for small $t>0$ there is on the non-singular level manifold $F_{u(t)}=\{\tilde{f}=t\}$ an $(n-1)$-dimensional real submanifold $D^{n-1}(t)$ with boundary defined by the relations

$$\text{Im } x_k = 0 \quad (k = 1, 2, \ldots, n)$$

$$\sum_{k=2}^{n} x_k^2 \leqslant t,$$

$$x_1 = \sum_{k=2}^{n} x_k^2 - t \leqslant 0$$

(see figure 47 for $n=2$). The manifold $D^{n-1}(t)$ is diffeomorphic to an $(n-1)$-dimensional ball. Its boundary (an $(n-2)$-dimensional sphere) lies on the submanifold

$$F_{u(t)} \cap \{x_1 = 0\}.$$

Changing t from 0 to 1 defines a continuous family of $(n-1)$-dimensional discs $D^{n-1}(t) \subset F_{u(t)}$, for which

$$S^{n-2}(t) = \partial D^{n-1}(t) \subset F_{u(t)} \cap \{x_1 = 0\}$$

for all values of $t \in [0, 1]$. Here

$$D^{n-1}(t) \cap \{x_1 = 0\} = \partial D^{n-1}(t).$$

The disk $D^{n-1}(1) \subset F_{z_0}$, with boundary $\partial D^{n-1}(1) = S^{n-2}(1)$, lying in the submanifold $F_{z_0} \cap \{x_1 = 0\}$, defines a relative cycle Δ' in the homology group

$$H_{n-1}(F_{z_0}, F_{z_0} \cap \{x_1 = 0\}),$$

isomorphic to the group $H_{n-1}(V_\varepsilon, V_\varepsilon')$.

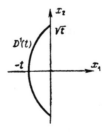

Fig. 47.

Let $u_1, \ldots, u_{\mu_0}$ (respectively $u'_1, \ldots, u'_{\mu_1}$) be a system of paths joining the critical values $z_1, \ldots, z_{\mu_0}$ (respectively $z'_1, \ldots, z'_{\mu_1}$) of the function $\tilde{f}$ (respectively $\tilde{f}|\{x_1 = 0\}$) with the non-critical value z_0 and defining in the homology of the non-singular level manifold of the function $\tilde{f}$ (respectively $\tilde{f}|\{x_1 = 0\}$) a distinguished basis of vanishing cycles. Moreover we shall suppose that the paths $u_1, \ldots, u_{\mu_0}, u'_1, \ldots, u'_{\mu_1}$ do not pass through the critical values $z_1, \ldots, z_{\mu_0}$, $z'_1, \ldots, z'_{\mu_1}$ of the functions $\tilde{f}$ and $\tilde{f}|\{x_1 = 0\}$ (for $t \neq 0$). As we explained above, such a system of paths determines a set of vanishing cycles

$$\Delta_1, \ldots, \Delta_{\mu_0}$$

and a set of vanishing hemicycles

$$\Delta'_1, \ldots, \Delta'_{\mu_1}$$

in the relative homology group $H_{n-1}(V_\varepsilon, V'_\varepsilon)$. The boundary homomorphism

$$H_{n-1}(V_\varepsilon, V'_\varepsilon) \to H_{n-2}(V'_\varepsilon)$$

of the pair $(V_\varepsilon, V'_\varepsilon)$ maps the vanishing hemicycle Δ'_j into the vanishing cycle in the homology of the non-singular level manifold of the function $\tilde{f}|\{x_1 = 0\}$ corresponding to the path u'_j. From this and from the exact sequence of the pair $(V_\varepsilon, V'_\varepsilon)$ it follows that the set of elements

$$\Delta_1, \ldots, \Delta_{\mu_0}, \Delta'_1, \ldots, \Delta'_{\mu_1}$$

is a basis in the relative homology group $H_{n-1}(V_\varepsilon, V'_\varepsilon)$.

It can be shown that the rank of the homology group $H_{n-1}(V_\varepsilon, V'_\varepsilon)$ is the same as the dimension of the base of the miniversal deformation of the boundary singularity $f|x_1$, which is equal to

$$\dim_{\mathbb{C}} {_n\mathcal{O}}/(x_1 \partial f/\partial x_1, \partial f/\partial x_2, \ldots, \partial f/\partial x_n),$$

where ${_n\mathcal{O}}$ is the ring of germs of holomorphic functions at zero in the space $\mathbb{C}^n$.

Everything that we have stated above suggests that the relative homology group $H_{n-1}(V_\varepsilon, V'_\varepsilon)$ must play the same role for boundary singularities as the absolute homology group $H_{n-1}(V_\varepsilon)$ does for ordinary singularities. It turns out, however, that it is impossible to define an intersection form on the group $H_{n-1}(V_\varepsilon, V'_\varepsilon)$ in an invariant manner and to obtain an analogue of the Picard-Lefschetz formula. This forces us to consider, in place of it, another group, also isomorphic to the integer lattice

$$\mathbb{Z}^\mu = \mathbb{Z}^{\mu_0 + \mu_1}$$

of dimension $\mu = \mu(f|x_1)$, but not having a canonically defined isomorphism with the group $H_{n-1}(V_\varepsilon, V'_\varepsilon)$. This group is defined in the following manner.

To the boundary singularity f we can associate one more non-singular level manifold – the level manifold $\hat{V}_\varepsilon$ of the corresponding function $\hat{f}$, which is invariant under the group $\mathbb{Z}_2$ on the space $\hat{\mathbb{C}}^n$. On this level manifold is defined the action of the involution, induced from its action on the space $\hat{\mathbb{C}}^n$. The quotient space of the space $\hat{V}_\varepsilon$ by the action of this involution is the same as the level manifold V_ε of the function f. The manifold $\hat{V}_\varepsilon$ is a ramified cover of the manifold V_ε with branching along the submanifold V'_ε. We have

$$\dim H_{n-1}(\hat{V}_\varepsilon) = 2 \dim H_{n-1}(V_\varepsilon) + \dim H_{n-2}(V'_\varepsilon)$$

$$= \dim H_{n-1}(V_\varepsilon) + \dim H_{n-1}(V_\varepsilon, V'_\varepsilon)$$

(see the example in § 5.1). In the homology group $H_{n-1}(\hat{V}_\varepsilon, \mathbb{Z})$ there are distinguished two subspaces H^+ and H^-, corresponding to the two possible irreducible real representations of the group $\mathbb{Z}_2$. The subspace H^+ consists of homology classes which are invariant relative to the action of the involution σ (that is such that $\sigma_* a = a$), and the subspace H^- consists of the antiinvariant ones ($\sigma_* a = -a$). In the example in § 5.1 it was shown that

$$\dim H^+ = \dim H_{n-1}(V_\varepsilon),$$

$$\dim H^- = \dim H_{n-1}(V_\varepsilon) + \dim H_{n-2}(V'_\varepsilon) = \dim H_{n-1}(V_\varepsilon, V'_\varepsilon).$$

The group H^-, isomorphic to the integer lattice of dimension $\mu(f|\{x_1 = 0\})$ plays the role for the boundary singularities that the group $H_{n-1}(V_\varepsilon)$ plays for ordinary singularities. In particular the intersection form is defined on it (as on a subgroup of the homology group $H_{n-1}(\hat{V}_\varepsilon)$ of the non-singular manifold).

A basis of the group H^- can be constructed in the following manner. Let

$$\Delta_1, \ldots, \Delta_{\mu_0}, \Delta'_1, \ldots, \Delta'_{\mu_1}$$

be a basis of vanishing cycles and hemicycles of the group $H_{n-1}(V_\varepsilon, V'_\varepsilon)$, constructed above by a system of paths $u_1, \ldots, u_{\mu_0}, u'_1, \ldots, u'_{\mu_1}$, joining the critical values $z_1, \ldots, z_{\mu_0}, z'_1, \ldots, z'_{\mu_1}$, of the functions $\tilde{f}$ and $\tilde{f}|\{x_1 = 0\}$ with the non-critical value z_0. The preimage of the cycle Δ_i ($i = 1, \ldots, \mu_0$) under the map $\hat{V}_\varepsilon \to V_\varepsilon$ (which is a double ramified cover, branching along the submanifold V_ε') consists of two cycles $\varDelta_i^{(1)}$ and $\varDelta_i^{(2)}$, each of which projects isomorphically onto the cycle $\varDelta_i$. Their difference

$$\hat{\varDelta}_i = \varDelta_i^{(1)} - \varDelta_i^{(2)}$$

is an antiinvariant cycle in the homology of the manifold $\hat{V}_\varepsilon$. The preimage of the hemicycle $\varDelta_j'$ also consists of two hemicycles $\varDelta_j'^{(1)}$ and $\varDelta_j'^{(2)}$, projecting isomorphically onto $\varDelta_j'$, but the hemicycles $\varDelta_j'^{(1)}$ and $\varDelta_j'^{(2)}$ have a common boundary (lying on the branch manifold). Therefore their difference

$$\hat{\varDelta}_j' = \varDelta_j'^{(1)} - \varDelta_j'^{(2)}$$

is also an absolute antiinvariant cycle in the homology of the manifold $\hat{V}_\varepsilon$. The cycles

$$\hat{\varDelta}_1, \ldots, \hat{\varDelta}_{\mu_0}, \hat{\varDelta}_1', \ldots, \hat{\varDelta}_{\mu_1}'$$

form a basis in the group H^- of the antiinvariant homology classes.

They can also be described in the following manner. We consider the function

$$\hat{\tilde{f}}(\hat{x}_1, \ldots, \hat{x}_n) = \tilde{f}(\hat{x}_1^2, \hat{x}_2, \ldots, \hat{x}_n).$$

It has $\mu_0 + \mu_1$ critical values

$$z_1, \ldots, z_{\mu_0}, z_1', \ldots, z_{\mu_1}'.$$

The critical values $z_1, \ldots, z_{\mu_0}$ are doubly degenerate and are taken at two separate critical points. To each of these critical values (z_i) there correspond two vanishing cycles $\hat{\varDelta}_i^1$ and $\hat{\varDelta}_i^2$. Moreover we can suppose that $\sigma_* \hat{\varDelta}_i^1 = \hat{\varDelta}_i^2$. To the critical value z_j' there corresponds one vanishing cycle $\hat{\varDelta}_j'$. It is not hard to see that $\sigma_* \hat{\varDelta}_j' = -\hat{\varDelta}_j'$. The cycles $\hat{\varDelta}_i = \hat{\varDelta}_i^1 - \hat{\varDelta}_i^2$ and also the cycles $\hat{\varDelta}_j'$ are antiinvariant; the cycles $\hat{\varDelta}_i^1 + \hat{\varDelta}_i^2$ are invariant relative to the action of the involution σ. The set of cycles

$$\{\hat{\varDelta}_i, \hat{\varDelta}_j'\}$$

is the same as that described above.

In this way we define an integer lattice H^- for the boundary singularities with an intersection form defined on it. A basis of the lattice H^-, constructed from a system of paths $\{u_i, u_j'\}$ joining the critical values z_i and z_j' of the function $\hat{f}$ with the non-critical value z_0, contains μ_0 "long" vanishing cycles $\hat{\Delta}_i$ $(=\Delta_i^{(1)}-\Delta_i^{(2)})$ and μ_1 "short" vanishing cycles $\hat{\Delta}_j'$. If the number of variables $n \equiv 3 \bmod 4$, then

$$(\hat{\Delta}_i \circ \hat{\Delta}_i) = -4, \quad (\hat{\Delta}_j' \circ \hat{\Delta}_j') = -2.$$

The monodromy group acts on the lattice H^- as the image of the natural representation

$$\pi_1(\mathbb{C}\setminus\{z_i, z_j'\}) \to \mathrm{Aut}\, H^-$$

of the fundamental group of the complement of the set of critical values of the function $\tilde{f}$. The monodromy group is generated by the monodromy operators arising from the simple loops τ_i and τ_j', corresponding to the paths u_i and u_j'. To the simple loop τ_j' there corresponds the usual Picard-Lefschetz operator

$$h_j'(a) = a + (-1)^{n(n+1)/2}(a \circ \Delta_j')\Delta_j',$$

to the loop τ_i there corresponds the Picard-Lefschetz operator

$$h_i(a) = a + (-1)^{n(n+1)/2}(a \circ \Delta_i)\Delta_i/2.$$

We remark that the intersection number $(a \circ \Delta_i)$ of an antiinvariant cycle $a \in H^-$ with a long vanishing cycle Δ_i is always even. When the number of variables $n \equiv 1 \bmod 2$ the operators h_i and h_j' are reflections in hyperplanes orthogonal (in the sense of the intersection form) to the vanishing cycles Δ_i and Δ_j' respectively.

For boundary singularities, as for ordinary ones, we define miniversal deformations and level and function bifurcation sets. The miniversal deformation of a boundary singularity $f|x_1$ can be given in the form

$$F(x, \lambda) = f(x) + \sum_{i=1}^{\mu} \lambda_i \phi_i(x)$$

$(x \in \mathbb{C}^n, \lambda = (\lambda_1, \ldots, \lambda_\mu) \in \mathbb{C}^\mu)$, where the germs $\phi_1, \ldots, \phi_\mu$ form a basis of the vector space

$$_n\mathcal{O}/(x_1 \partial f/\partial x_1, \partial f/\partial x_2, \ldots, \partial f/\partial x_n).$$

The set (more precisely its germ at zero) of values of the parameters $\lambda = (\lambda_1, \ldots, \lambda_\mu)$ in the base of the miniversal deformation, for which either the corresponding function $F(\cdot, \lambda)$ or its restriction to the boundary $\mathbb{C}^{n-1} = \{x_1 = 0\}$ has zero as a critical value, is called the *level bifurcation set* of the boundary singularity $f|x_1$ (and denoted by Σ). The condition specifying the values of the parameters $\lambda \in \Sigma$ can be changed to an equivalent condition, that zero is a critical value of the function

$$\hat{F}(\hat{x}_1, \hat{x}_2, \ldots, \hat{x}_n, \lambda) = F(\hat{x}_1^2, \hat{x}_2, \ldots, \hat{x}_n, \lambda).$$

The level bifurcation set of a boundary singularity is reducible. It is the union of two components. The first consists of those values of the parameter λ for which the function $F(\cdot, \lambda)$ has zero as a critical value, and the second consists of those values λ for which its restriction to the boundary $\mathbb{C}^{n-1}$ has zero as a critical value. Sometimes it is more convenient to say that the second component consists of those values λ for which the hypersurface

$$\{x : F(x, \lambda) = 0\} \subset \mathbb{C}^n$$

is not transverse to the boundary $\mathbb{C}^{n-1}$. This formulation does not need a special definition for the case $n = 1$.

From the fact that the level bifurcation set of an ordinary singularity is irreducible (Theorem 3.2 of § 3.2), it follows that each of the above components of the level bifurcation set of a boundary singularity is irreducible. Just as in Theorem 3.4 of §3.2, it follows from this that the monodromy group of a boundary singularity acts transitively on the sets of short and long vanishing cycles (not mixing them, of course, with each other).

For the simple boundary singularities B_k, C_k, F_4 (Volume 1, Chapter 17) the level bifurcation sets can be obtained in the way described in § 3.3 for ordinary singularities of functions. This means that it is biholomorphically equivalent to the variety of nonregular orbits of the corresponding group, generated by reflections, acting on the complexification of Euclidean space. Two types of mirrors (orthogonal to the long and the short roots respectively) generate the two components of the level bifurcation set of a simple boundary singularity.

Let us show how to check this for simple singularities of types $B_k (f(x_1) = x_1^k, n = 1)$ and $C_k (f(x_1, x_2) = x_1 x_2 + x_2^k, n = 2)$. Their miniversal deformation can be given in the form

$$F = x_1^k + \lambda_1 x_1^{k-1} + \ldots + \lambda_k \quad \text{for} \quad B_k,$$

$$F = x_1 x_2 + x_2^k + \lambda_1 x_2^{k-1} + \ldots + \lambda_k \quad \text{for} \quad C_k.$$

In both the first and the second case zero is not a critical value of the function $F(\cdot, \lambda)$ if the polynomial

$$x^k + \lambda_1 x^{k-1} + \ldots + \lambda_k$$

does not have multiple roots, and the local level manifold of the function $F(\cdot, \lambda)$ is transverse to the boundary $\{x_1 = 0\}$ if zero is not a root of this polynomial. In this way the level bifurcation sets of the singularities B_k and C_k are identified with the space of polynomials of the form

$$x^k + \lambda_1 x^{k-1} + \ldots + \lambda_k$$

which have multiple or zero roots.

The Weyl groups B_k and C_k are the same. They consist of the transformations of the space $\mathbb{R}^k$ (or its complexification $\mathbb{C}^k$) which are permutations of the coordinates with arbitrary changes of sign. The mirrors are the hyperplanes

$$z_i = 0 \quad \text{and} \quad z_i = \pm z_j.$$

In one case the first of these corresponds to the short cycles and the second to the long ones; in the other conversely. The space of orbits of the action of the Weyl group on the complexification $\mathbb{C}^k$ is identified with the space of polynomials of degree k of type

$$x_{\mu}^k + \lambda_1 x^{k-1} + \ldots + \lambda_k$$

(with complex coefficients), the identification taking the point $(z_1, \ldots, z_k) \in \mathbb{C}^k$ to the polynomial with roots $z_1^2, \ldots, z_k^2$. The space of polynomials of degree k is isomorphic to a k-dimensional complex vector space. The space of non-regular orbits (that is the image of the union of the mirrors under the factorisation map) consists of polynomials with multiple or zero roots, that is it is the same as the level bifurcation sets of the singularities B_k and C_k.

The level bifurcation sets of the singularities B_2 and C_2 consist of two curves

$$\lambda_2 = 0 \quad \text{and} \quad 4\lambda_2 = \lambda_1^2$$

(figure 48).

A description of bases of vanishing cycles and intersection forms for simple boundary singularities (and for other boundary singularities of two variables) can be obtained by the method of Chapter 4. It can be shown that for boundary

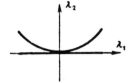

Fig. 48.

singularities there are Theorems analogous to 4.1, 4.2 and 4.3, but we shall not dwell on this. For the simple singularities B_k ($\pm x^k \pm y^2$), C_k ($xy \pm y^k$) and F_4 ($\pm x^2 + y^3$) the corresponding equivariant germs of functions on the cover $\hat{\mathbb{C}}^2$ of the space $\mathbb{C}^2$ are given by the formulae

$$\hat{f}(\hat{x}, \hat{y}) = \hat{x}^{2k} + \hat{y}^2 \quad \text{for} \quad B_k,$$

$$\hat{f}(\hat{x}, \hat{y}) = \hat{y}(\hat{x}^2 - \hat{y}^{k-1}) \quad \text{for} \quad C_k$$

and

$$\hat{f}(\hat{x}, \hat{y}) = \hat{x}^4 + \hat{y}^3 \quad \text{for} \quad F_4$$

(the choice of sign is made deliberately to ensure that the curve $\{\hat{f}=0\}$ will be real). As ordinary germs of functions they have singularities of types A_{2k-1}, D_{k+1} and E_6 respectively. It is easy to construct perturbations $\tilde{\hat{f}}$ of the germs of functions $\hat{f}$ (or the germs of curves $\{\hat{f}=0\}$) which would satisfy the conditions of Theorem 4.1 of § 4.1 and would be invariant under the involution acting on the space $\hat{\mathbb{C}}^2$. In fact the perturbations of the singularities A_{2k-1}, D_{k+1} and E_6 used in Chapter 4 will possess these properties. The corresponding real curves $\{\tilde{\hat{f}}=0\}$ are depicted in figure 49. The line $x=0$ is drawn with dashes. The basis of the

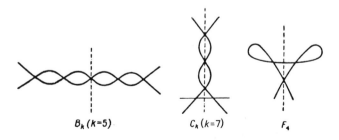

$B_k\,(k=5)$ $C_k\,(k=7)$ F_4

Fig. 49.

homology group of the local level manifold of the singularity

$$\hat{f}(\hat{x}, \hat{y}) + \hat{t}^2,$$

described in Theorem 4.1 of §4.1 is invariant relative to the involution

$$\sigma : (\hat{x}, \hat{y}, \hat{t}) \to (-\hat{x}, \hat{y}, \hat{t}),$$

acting on the space $\hat{\mathbb{C}}^3$ in the sense that $\sigma_* \varDelta = -\varDelta$ if the basic vanishing cycle $\varDelta$ corresponds to a critical point of the function $\hat{f}(\hat{x}, \hat{y})$ lying on the line $\hat{x} = 0$ and $s_* \varDelta_1 = \varDelta_2$ if $\varDelta_1$ and $\varDelta_2$ are basic vanishing cycles corresponding to critical points arranged symmetrically relative to the line $\hat{x} = 0$. From this it followes at once that the D-diagrams of the singularities B_k, C_k and F_4 are as shown in figure 50. The rules for reading these diagrams are somewhat different from the rules in §2.8. The arrows on the edges point from the vertices corresponding to long vanishing cycles (with self-intersection number -4) to vertices corresponding to short vanishing cycles (with self-intersection number -2). The intersection number of vanishing cycles corresponding to vertices joined by edges of multiplicity k is equal to $2k$ if both cycles are long and otherwise to k (if both are short or one of them is long and the other one is short). For the diagrams of the singularities B_k, C_k and F_4, shown in figure 50, this means that the angle between vanishing cycles corresponding to vertices joined by edges of multiplicity 1 is equal to $2\pi/3$, and the angle between vanishing cycles, corresponding to vertices joined by arcs of multiplicity 2 is equal to $3\pi/4$.

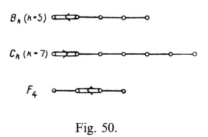

Fig. 50.

I. G. Shcherbak proved that the transition from the function $f(x, y)$ with boundary $\{x = 0\}$ to the function $f(x, y) + zx$ with boundary $\{z = 0\}$ defines on the set of classes of stably equivalent boundary singularities an involution transposing the singularity in the non-boundary sense and its restriction to the boundary.

5.3   The topology of complete intersections

Let

$$f = (f_1, \ldots, f_p) : (\mathbb{C}^n, 0) \to (\mathbb{C}^p, 0)$$

be the germ of an analytic map defining a complete intersection with isolated singularity at zero ($n \geqslant p$, $f_i : (\mathbb{C}^n, 0) \to (\mathbb{C}, 0)$). This means that at all points of the germ of the analytic space $\{f = 0\}$ (that is $\{x \in \mathbb{C}^n : f_1(x) = \ldots = f_p(x) = 0\}$), except zero, the map f has rank equal to p, that is its differential is surjective, or, which is the same thing, $rk(\partial f_i/\partial x_j) = p$. From this it follows that away from zero the space $\{f = 0\}$ is a non-singular $(n - p)$-dimensional complex manifold.

By analogy with Lemma 2.1 of §2.1 it is not hard to show that there exists a $\varrho > 0$ such that for all $0 < r \leqslant \varrho$ the sphere $S_r \subset \mathbb{C}^n$ of radius r with centre at zero intersects the manifold $\{f = 0\}$ transversely. In this case for sufficiently small $z = (z_1, \ldots, z_p) \in \mathbb{C}^p$ ($\|z\| \leqslant \varepsilon_0$) the space $\{f = z\}$ will intersect the sphere S_ϱ transversely. The space $\{f = z\}$, generally speaking, will not be non-singular for $z \neq 0$. The set (more precisely, the germ) Σ of those $z \in \mathbb{C}^p$ ($\|z\| < \varepsilon_0$), for which the space $\{f = z\}$ has a singular point inside the ball B_ϱ of radius ϱ with centre at zero is called the discriminant set of the map f. It is not hard to see that for $z \in \Sigma$ the analytic space $\{f = z\}$ has only isolated singularities inside the ball B_ϱ. It follows from Sard's theorem that the complement of the set Σ is everywhere dense in the ball $\{z : \|z\| \leqslant \varepsilon_0\} \subset \mathbb{C}^p$. For $z \notin \Sigma$ ($\|z\| \leqslant \varepsilon_0$) the space

$$F_z = \{f = z\} \cap B_\varrho = \{x \in \mathbb{C}^n : \|x\| \leqslant \varrho, f(x) = z\}$$

is a non-singular $(n - p)$-dimensional complex manifold with boundary

$$\{f = z\} \cap S_\varrho.$$

For all $z \notin \Sigma$, $\|z\| \leqslant \varepsilon_0$ the manifolds F_z are diffeomorphic to each other. They are called the *non-singular level manifolds* of the map f.

We get the result, analogous to Theorem 2.1 of §2.1:

Theorem 5.4 ([156]). If the germ of the map

$$f : (\mathbb{C}^n, 0) \to (\mathbb{C}^p, 0)$$

defines a complete intersection with isolated singularity at zero then the non-singular level manifold F_z of the map f is homotopically equivalent to a bouquet of spheres of dimension $n - p$.

We outline the principal ideas of the proof of this theorem.

For $p = 1$ the statement of Theorem 5.4 follows from Theorem 2.1 of § 2.1 and is therefore proved. Let us suppose that it has already been proved for maps

$$(\mathbb{C}^n, 0) \to (\mathbb{C}^{p-1}, 0)$$

defining a complete intersection of dimension $n - p + 1$ with isolated singularity at zero. The singular points of the map $f : (\mathbb{C}^n, 0) \to (\mathbb{C}^p, 0)$, that is the points $x \in \mathbb{C}^n$ for which $rk(\partial f_i/\partial x_j) < p$, form the germ of an analytic space S. The discriminant set

$$(\Sigma, 0) \subset (\mathbb{C}^p, 0)$$

is the image of the germ S under the map f. The preimage of zero for the map

$$f|_S : S \to \mathbb{C}^p$$

consists of the one point $0 \in \mathbb{C}^n$. Consequently, the map $f|_S$ is proper (in a sufficiently small neighbourhood of the point $0 \in \mathbb{C}^n$) and therefore the germ of the discriminant set Σ is the germ of an analytic subspace of dimension (at least no more than) $p - 1$ in the space $(\mathbb{C}^p, 0)$. From this it follows that for almost all lines $l \subset \mathbb{C}^p$ passing through zero the intersection $l \cap \Sigma$ has zero as an isolated point. We fix one such line. We can suppose that ε_0 is chosen só small that inside the ball $\{\|z\| \leq \varepsilon_0\}$ in the space $\mathbb{C}^p$ the intersection $l \cap \Sigma$ consists of the one point $0 \in \mathbb{C}^p$. After making, if necessary, a linear change in the system of coordinates $z_1, \ldots, z_p$ in the space $\mathbb{C}^p$ we can suppose that it is chosen so that the line l coincides with the coordinate axis

$$z_1 = \ldots = z_{p-1} = 0.$$

In this case, the subspace $f^{-1}(l) \subset (\mathbb{C}^n, 0)$, which is the same as the zero level manifold of the map

$$f' = (f_1, \ldots, f_{p-1}) : (\mathbb{C}^n, 0) \to (\mathbb{C}^{p-1}, 0),$$

is a complete intersection of dimension $n - p + 1$ with an isolated singular point at zero. We have shown that almost all linear changes of the system of equations

$$f_1 = \ldots = f_p = 0,$$

defining complete intersection of dimension $n - p$ with isolated singularity at zero, reduce to a system of equations such that the first $p-1$ of them ($f_1 = \ldots$ $= f_{p-1} = 0$) also define a complete intersection (of dimension $n-p+1$) with isolated singularity at zero.

From the induction hypothesis, for almost all sufficiently small

$$z' = (z_1, \ldots, z_{p-1}) \in \mathbb{C}^{p-1}$$

the set

$$F' = F'_{z'} = (f')^{-1}(z') \cap B_\varrho$$

is a non-singular $(n-p+1)$-dimensional complex manifold with boundary and homotopically equivalent to a bouquet of spheres of dimension $n-p+1$. It can be shown that for sufficiently small z' all non-triviality of the homotopy type of the space $F'_{z'}$ is concentrated in a small neighbourhood of zero. More precisely it means that, in particular, for $\|z'\| \ll \varepsilon_p^0$ the space $F' = F'_{z'}$ is homotopically equivalent to its own subspace

$$F'' = F' \cap f_p^{-1}(\{|z_p| \leqslant \varepsilon_p^0\}) = \{x \in F' : f_p(x) \leqslant \varepsilon_p^0\}.$$

To prove this some care is needed in carrying out the induction, but we shall not dwell on this.

The restriction $f_p|_{F'}$ of the function f_p to F' defines the map of the manifold F' to the complex line $\mathbb{C}$, possessing the properties described in §1.1 (with the disc $\{|z_p| \leqslant \varepsilon_p^0\}$ as the region U). The function $f_p|_{F'}$, generally speaking, can have degenerate critical points. Changing the function $f_p|_{F'}$ to a small perturbation $\tilde{f}_p$ of it, we obtain a function on the manifold F', also possessing the indicated properties. We can suppose that on the space

$$\tilde{F}'' = \tilde{f}_p^{-1}(\{|z_p| \leqslant \varepsilon_p^0\})$$

the function $\tilde{f}_p$ has only non-degenerate critical points, v_0 in number, with distinct critical values $z^{(1)}, \ldots, z^{(v_0)}$, lying inside the disc $\{|z_p| < \varepsilon_p^0\}$. The space $\tilde{F}''$ is diffeomorphic to the space F'' and therefore has the homotopy type of a bouquet of spheres of dimension $n-p+1$. The non-singular level manifold

$$\tilde{f}_p^{-1}(z_p)$$

of the function $\tilde{f}_p$ on the manifold F' is diffeomorphic to the non-singular level manifold of the function $f_p|_{F'}$ (for $|z_p| \leqslant \varepsilon_p^0$), which is the same as the non-singular level manifold of the map f.

We arrive at a situation very similar to that which was considered in the proof of Theorem 2.1 of § 2.1. The role of the function $\tilde{f}$ on the ball B_ϱ in the space $\mathbb{C}^n$ is played by the function $\tilde{f}_p$ on the manifold F'. Over the complement of the set

$$\{z^{(1)}, \ldots, z^{(v_0)}\}$$

of its critical values in the disc $\{|z_p| \leqslant \varepsilon_p^0\}$ the function $\tilde{f}_p$ defines a locally trivial fibration, the fibre of which is a non-singular level manifold of the map f. One difference is that the preimage of the disc $\{|z_p| \leqslant \varepsilon_p^0\}$ under the map $\tilde{f}_p$ is not contractible, but is homotopically equivalent to a bouquet of μ_1 spheres of dimension $n - p + 1$. Let $z^{(0)}$ be a non-critical value of the function $\tilde{f}_p$ with $|z^{(0)}| = \varepsilon_p^0$,

$$\tilde{f}_p^{-1}(z^{(0)}) = F, \quad \tilde{f}_p^{-1}(\{|z_p| \leqslant \varepsilon_p^0\}) = \tilde{F}''$$

(the space $\tilde{F}''$ is homotopically equivalent to a bouquet of μ_1 spheres of dimension $n - p + 1$).

We choose a system of paths u_i $(i = 1, \ldots, v_0)$, joining the critical values $z^{(1)}, \ldots, z^{(v_0)}$ of the function $\tilde{f}_p$ with the non-critical value $z^{(0)}$ and satisfying the conditions formulated in the definition of distinguished bases (§ 1.2). This means that the paths are not self-intersecting and do not have common points other than the chosen non-critical value $z^{(0)}$. As above, such a system of paths defines a set of v_0 vanishing cycles

$$\Delta_1, \ldots, \Delta_{v_0}$$

in the homology group $H_{n-p}(F; \mathbb{Z})$ of the non-singular level manifold F of the map f. Carrying out the same argument as in the proof of Theorem 2.1 of § 2.1, we obtain:

(i) the space

$$\tilde{F}'' = \tilde{f}_p^{-1}(\{|z_p| \leqslant \varepsilon_p^0\})$$

is homotopy equivalent to its own subspace

$$X = \tilde{f}_p^{-1}(\cup_{i=1}^{v_0} \{u_i(t)\}),$$

which is the preimage of the union of the paths u_i;

(ii) the quotient space

$$X/F = X/\tilde{f}_p^{-1}(z^{(0)})$$

is homotopy equivalent to a bouquet of v_0 spheres of dimension $n - p + 1$;
  (iii) under the action of the boundary homomorphism

$$H_{n-p+1}(X/F) = H_{n-p+1}(X, F) \rightarrow H_{n-p}(F)$$

of the pair (X, F) cycles corresponding to these spheres map into the homology classes of the vanishing cycles $\varDelta_1, \ldots, \varDelta_{v_0}$.

From this follows the simple-connectedness of the non-singular level manifold F of the map f (for $n - p > 1$; for $n - p \leqslant 1$ the proof of Theorem 5.4 is even simpler). From general results about complex submanifolds in the space $\mathbb{C}^n$ it follows that the homology groups $H_i(F; \mathbb{Z})$ of the space F are zero for $i > n - p$, and the group $H_{n-p}(F; \mathbb{Z})$ is free Abelian. This follows, for example, from the fact that any non-singular complex submanifold of the ball B_ϱ in the space $\mathbb{C}^n$ with (complex) dimension m is homotopy equivalent to a finite CW complex of (real) dimension m. This theorem can be proved in exactly the same way as the theorem of Andreotti and Frankel in [255], the difference being that the complex manifold considered was not in the ball B_ϱ but in the whole space $\mathbb{C}^n$.
  From the exact homology sequence of the pair (X, F):

$$\ldots \rightarrow H_{i+1}(X) \rightarrow H_{i+1}(X, F) \rightarrow H_i(F) \rightarrow H_i(X) \rightarrow \ldots$$

it follows that $H_i(F) = 0$ for $i \neq n - p$ and there is a short exact sequence

$$0 \rightarrow H_{n-p+1}(X) \rightarrow H_{n-p+1}(X, F) \rightarrow H_{n-p}(F) \rightarrow 0,$$

in which all the groups are free Abelian. From this it follows that the non-singular level manifold F of the map f is homotopy equivalent to a bouquet of $v_0 - \mu_1$ spheres of dimension $n - p$, where μ_1 is the rank of the $(n - p + 1)$th homology group of the non-singular level manifold F' of the map

$$f' = (f_1, \ldots, f_{p-1})$$

and v_0 is the number of critical points of the function f_p on the manifold F' (counted with their multiplicities). Theorem 5.4 is thereby proved.

By analogy with the case of an ordinary singularity of a function

$$(\mathbb{C}^n, 0) \to (\mathbb{C}, 0)$$

the number $\mu_0 = (v_0 - \mu_1)$, equal to the rank of the $(n-p)$th homology group of the non-singular level manifold F of the germ of the map $f: (\mathbb{C}^n, 0) \to (\mathbb{C}^p, 0)$, is called the *Milnor number* of the isolated singularity of the germ f.

In the proof of Theorem 5.4 we constructed a short exact sequence

$$0 \to \mathbb{Z}^{\mu_1} \to \mathbb{Z}^{v_0} \to H_{n-p}(F; \mathbb{Z}) \to 0,$$

where μ_1 is the Milnor number of the isolated singularity of the germ

$$f' = (f_1, \ldots, f_{p-1}): (\mathbb{C}^n, 0) \to (\mathbb{C}^{p-1}, 0),$$

obtained from the germ f by forgetting one of the components (f_p), v_0 is the number of critical points (counting multiplicities) of the function f_p of the non-singular level manifold of the germ of the singularity f' near the critical point. A basis of the group $\mathbb{Z}^{v_0}$ is formed by the (formal) vanishing cycles

$$\Delta_1, \ldots, \Delta_{v_0},$$

corresponding to the paths $u_1, \ldots, u_{v_0}$, described above. Their intersection numbers on the non-singular level manifold F of the germ f define a bilinear form on the group $\mathbb{Z}^{v_0}$. The group $\mathbb{Z}^{\mu_1}$ is the same as the group of linear relations between the vanishing cycles $\Delta_1, \ldots, \Delta_{v_0}$ in the homology group $H_{n-p}(F)$ of the non-singular level manifold of the map f. It, of course, lies in the kernel of the form defined on the group $\mathbb{Z}^{v_0}$ by the intersection numbers.

The natural numbers μ_1 and v_0, naturally, depend on the choice of system of coordinates in the space $\mathbb{C}^p$. It is not difficult to see, however, that for systems of coordinates $z_1, \ldots, z_p$ in general position the numbers μ_1 and v_0 depend only on the germ $f: (\mathbb{C}^n, 0) \to (\mathbb{C}^p, 0)$ itself. In this way, for isolated singularities of germs of complete intersections, the short exact sequence

$$0 \to \mathbb{Z}^{\mu_1} \to \mathbb{Z}^{v_0} \to H_{n-p}(F)(= \mathbb{Z}^{\mu_0}) \to 0,$$

is defined invariantly.

In an analogous manner for the map

$$f': (\mathbb{C}^n, 0) \to (\mathbb{C}^{p-1}, 0)$$

there is defined the short exact sequence

$$0 \to \mathbb{Z}^{\mu_2} \to \mathbb{Z}^{\nu_1} \to \mathbb{Z}^{\mu_1} \to 0,$$

and further the short exact sequences

$$0 \to \mathbb{Z}^{\mu_3} \to \mathbb{Z}^{\nu_2} \to \mathbb{Z}^{\mu_2} \to 0,$$

.
.
.

$$0 \to \mathbb{Z}^{\mu_{p-1}} \to \mathbb{Z}^{\nu_{p-2}} \to \mathbb{Z}^{\mu_{p-2}} \to 0,$$

$$0 \to \mathbb{Z}^{\nu_{p-1}} \widetilde{\to} \mathbb{Z}^{\mu_{p-1}} \to 0.$$

All together they give a long exact sequence (resolvent)

$$0 \to \mathbb{Z}^{\nu_{p-1}} \to \mathbb{Z}^{\nu_{p-2}} \to \ldots \to \mathbb{Z}^{\nu_2} \to \mathbb{Z}^{\nu_1} \to \mathbb{Z}^{\nu_0} \to H_{n-p}(F; \mathbb{Z}) \to 0,$$

consisting of free Abelian groups. Here ν_i is the number of critical points (counted with their multiplicities) of the function f_{p-i} on the non-singular level manifold of the germ of the map

$$(f_1, \ldots, f_{p-i-1}) : (\mathbb{C}^n, 0) \to (\mathbb{C}^{p-i-1}, 0)$$

(for a choice of system of coordinates in general position in the space $(\mathbb{C}^p, 0)$).

For the germ of a map, defining a complete intersection with an isolated singularity, we can define its intersection matrix as the intersection matrix of the vanishing cycles $\Delta_1, \ldots, \Delta_{\nu_0}$ defined above, and also the corresponding diagram (for $n - p \equiv 2 \bmod 4$). Remember that the cycles $\Delta_1, \ldots, \Delta_{\nu_0}$ generate the $(n - p)$th homology group $H_{n-p}(F; \mathbb{Z})$ of the non-singular level manifold, but do not form a basis in it. Therefore everywhere (except the trivial case when the complete intersection defined by the map $f' = (f_1, \ldots, f_{p-1})$ happens to be non-singular, in which case the complete intersection, defined by the map f, is isomorphic to the germ of a hypersurface) the intersection matrix of the germ of such a map is degenerate, and the number of vertices in the corresponding diagram is greater than the Milnor number of the singularity.

5.4   Singularities of projections onto a line

A *projection onto a line* (or here simply a projection) is a triple

$$E \hookrightarrow (\mathbb{C}^n, 0) \to (\mathbb{C}, 0)$$

where E is the germ of a complete intersection of codimension p in the space $\mathbb{C}^n$, with an isolated singular point at zero,

$$\pi : (\mathbb{C}^n, 0) \to (\mathbb{C}, 0)$$

is a linear projection along the hyperplane $\mathbb{C}^{n-1} = \pi^{-1}(0) \subset \mathbb{C}^n$. The two projections $E_1 \hookrightarrow (\mathbb{C}^n, 0) \to (\mathbb{C}, 0)$ and $E_2 \hookrightarrow (\mathbb{C}^n, 0) \to (\mathbb{C}, 0)$ are considered equivalent if there exists a commutative diagram

$$
\begin{array}{ccc}
E_1 \hookrightarrow (\mathbb{C}^n, 0) \to (\mathbb{C}, 0) \\
\downarrow \qquad \downarrow \qquad \downarrow \\
E_2 \hookrightarrow (\mathbb{C}^n, 0) \to (\mathbb{C}, 0)
\end{array}
$$

in which all vertical arrows are isomorphisms in a neighbourhood of zero.

As always, the singularity of the projection $E \to (\mathbb{C}^n, 0) \to (\mathbb{C}, 0)$ is said to be simple if among small perturbations of it there are a finite number of projections which are distinct relative to the above equivalence. The simple singularities of projection were described in [132]. They exist for $p = 1$ and $p = 2$. For a description of their representatives we choose in the space $\mathbb{C}^n$ a system of coordinates $(x_1, \ldots, x_n)$ such that the projection π maps the point $(x_1, \ldots, x_n) \in \mathbb{C}^n$ to $x_1 \in \mathbb{C}$. For $p = 1$ the simple singularities of projection exist for all $n \geq 2$. They are given by the equations $f_1 = 0$ with the following functions $f_1(x_1, \ldots, x_n)$ (here $q = x_3^2 + \ldots + x_n^2$):

$A_0 : f_1 = x_2$;

$X_\mu : f_1 = x_1 + X_\mu$, where X_μ is one of the simple singularities of functions
of the $n - 1$ variables $x_2, \ldots, x_n$ (A_μ, D_μ or $E_\mu, \mu > 0$);

$B_\mu : f_1 = x_1^\mu + x_2^2 + q \quad (\mu \geq 2)$;

$C_\mu : f_1 = x_1 x_2 + x_2^\mu + q \quad (\mu \geq 3)$;

$F_4 : f_1 = x_1^2 + x_2^3 + q$.

For $p = 2$ simple singularities of projection exist for $n = 3$. They are given by the equations $f_1 = f_2 = 0$ with the following functions f_1 and f_2:

$C_{k+l}^{k,l}$:	$f_1 = x_2 x_3,$	$f_2 = x_1 + x_2^k + x_3^l$	$(2 \leq k \leq l)$;
F_{2k+1}:	$f_1 = x_2^2 + x_3^3,$	$f_2 = x_1 + x_3^k$	$(k \geq 2)$;
F_{2k+4}:	$f_1 = x_2^2 + x_3^3,$	$f_2 = x_1 + x_2 x_3^k$	$(k \geq 1)$.

If we fix the linear projection

$$\pi : (\mathbb{C}^n, 0) \rightarrow (\mathbb{C}, 0)$$

along a hyperplane $\mathbb{C}^{n-1} \subset \mathbb{C}^n$, then the germs of the complete intersections E_1 and E_2, entering into the equivalent singularites of projection

$$E_i \hookrightarrow (\mathbb{C}^n, 0) \rightarrow (\mathbb{C}, 0)$$

($i = 1, 2$) are obtained one from the other under the action of a local analytic isomorphism $(\mathbb{C}^n, 0) \rightarrow (\mathbb{C}^n, 0)$ mapping the hyperplane $\mathbb{C}^{n-1}$ into itself. In this manner, to the singularity of projection

$$E \hookrightarrow (\mathbb{C}^n, 0) \rightarrow (\mathbb{C}, 0)$$

corresponds the germ of a complete intersection E with isolated singularity at zero, considered modulo local analytic diffeomorphism $(\mathbb{C}^n, 0) \rightarrow (\mathbb{C}^n, 0)$ mapping the hyperplane $\mathbb{C}^{n-1} \subset \mathbb{C}^n$ to itself. Therefore the singularities of projections are generalisations and in some sense are mixtures of boundary singularities and singularities of germs of complete intersections, considered, respectively, in § 5.2 and § 5.3.

In accordance with § 5.2 and § 5.3 for boundary singularities (that is for isolated singularities of germs of functions or hypersurfaces in the space $(\mathbb{C}^n, 0)$ considered modulo local analytic isomorphisms of the space $(\mathbb{C}^n, 0)$, preserving the hyperplane $\mathbb{C}^{n-1} \subset \mathbb{C}^n$) and for isolated singularities of germs of complete intersections in the space $(\mathbb{C}^n, 0)$, respectively, as also for ordinary singularities of functions, we can define an integral lattice with integral bilinear form and chosen (distinguished) sets of elements generating it. The difference from the usual singularities of functions consists in the first case of the fact that among these elements there are both "short" and "long" vanishing cycles (for $n \equiv 3 \bmod 4$ their self-intersection numbers equal -2 and -4 respectively), and in the second case the difference consists of the fact that the set of these elements is redundant in the sense that their number is more than the dimension of the lattice: they are linearly dependent in it. Arising from the considerations described in § 5.2 and § 5.3, there is defined also for the singularity of projection

$$E \hookrightarrow (\mathbb{C}^n, 0) \rightarrow (\mathbb{C}, 0)$$

an integer lattice with an integral bilinear form and a chosen set of elements generating it. These sets possess both the above-mentioned differences from the case of ordinary singularities of functions: they include both short and long vanishing cycles and the number of elements in them is more than the dimension of the lattice.

The precise construction of the lattice and the distinguished sets of elements in it is as follows. Let the complete intersection $E \hookrightarrow (\mathbb{C}^n, 0)$ be defined by the set of equations

$$f_1 = f_2 = \ldots = f_p = 0$$

(dim $E = n - p$). Having made a linear change of general form to this system of equations, we can suppose that the system of equations

$$f_1 = f_2 = \ldots = f_{p-1} = 0$$

defines a complete intersection of dimension $n - p + 1$ with isolated singularity at zero (see § 5.3). The non-singular level manifold

$$F' = F'_z = \{f_1 = z_1, \ldots, f_{p-1} = z_{p-1}\} \cap B_\varrho$$

has the homotopy type of a bouquet of spheres of dimension $n - p + 1$, a number which does not depend on the concrete choice of linear change of the system of equations of general form. The intersection of this level manifold with the hyperplane $\mathbb{C}^{n-1} \subset \mathbb{C}^n$ is a non-singular $(n - p)$-dimensional submanifold in the manifold F' (again for a general choice of system of equations), and the function f_p defines a function on the manifold F', with isolated critical point. Changing, if necessary, the function f_p to a small perturbation of it, we can suppose that it has, on the manifolds F' and $F' \cap \mathbb{C}^{n-1}$, only non-degenerate critical points with distinct critical values. For the pair of manifolds $(F', F' \cap \mathbb{C}^{n-1})$ and the function f_p on it we can realise the construction described in § 5.2 for the pair $(\mathbb{C}^n, \mathbb{C}^{n-1})$. This means that we ought to consider the double cover $\hat{F}'$ of the manifold F', branching along the submanifold $F' \cap \mathbb{C}^{n-1}$ and the function $\hat{f}_p$, obtained from f_p by lifting to the covering space. On the level manifold of the function $\hat{f}_p$ acts the involution, interchanging the sheets of the cover. In the integral homology group of the non-singular level manifold of the function $\hat{f}_p$ there is picked out a subgroup H^- consisting of the homology classes, antiinvariant under this involution. The subgroup H^- is the integer lattice, associated to the singularity of projection

$$E \hookrightarrow (\mathbb{C}^n, 0) \to (\mathbb{C}, 0).$$

To a system of paths joining the critical values of the function $\hat{f}_p$ with the non-critical value (and satisfying the conditions applying to systems of paths defining distinguished bases) there corresponds a system of "ordinary" vanishing cycles in the homology of the non-singular level manifold of the function $\hat{f}_p$ on the manifold $\hat{F}'$: two for each critical value of the function $f_p|_{F'}$ and one for each critical value of the function $f_p|_{F' \cap \mathbb{C}^{n-1}}$. The involution interchanges the first of these and acts on the second by multiplication by -1. The differences of cycles corresponding to the critical values of the function $f_p|_{F'}$ and cycles corresponding to the critical values of the function $f_p|_{F' \cap \mathbb{C}^{n-1}}$ generate the group H^- of antiinvariant cycles in the homology of the non-singular level manifold of the function $\hat{f}_p$ on the manifold $\hat{F}'$. This set of cycles must be considered as distinguished in the integer lattice H^-.

Since the manifold $\hat{F}'$ (as distinct from the space $\mathbb{C}^n$) has, generally speaking, non-trivial homology, the number of these cycles is greater than the dimension of the lattice H^- (their number is equal to the sum of the dimensions of the lattice H^- and the lattice consisting of antiinvariant homology classes of the manifold $\hat{F}'$). The intersection numbers of the above cycles allows us, using the same rules as in §5.2, to define the corresponding D-diagram. Here we must bear in mind the fact that these rules define the diagram according to the intersection numbers of the vanishing cycles for $n - p \equiv 2$ mod 4. Therefore for $p = 2$ ($n - p = 1$) we must formally perform the same count of intersection matrices, which occurs with the addition of a quadratic function of one new variable to the usual singularity (see §2.8) and further construct the diagram as if it were true that $n - p \equiv 2$ mod 4. In the given case this means that the ith and jth vertices are joined by an edge of multiplicity $(\varDelta_i \circ \varDelta_j)$ (if even one of the cycles $\varDelta_i$ and $\varDelta_j$ is short) or $(\varDelta_i \circ \varDelta_j)/2$ (if both of the cycles are long) for $i < j$ (in accordance with the usual order of cycles $\{\varDelta_k\}$ in the distinguished set); the arrows on the edges are directed from the vertices corresponding to the long vanishing cycles to the vertices corresponding to the short vanishing cycles.

It is not difficult to see that the simple singularities of projection onto a line with $p = 1$ have diagrams coinciding with the diagrams of the same name of ordinary singularities of functions or boundary singularities. The diagrams of the singularities of projection $C_{k+l}^{k,l}$ ($2 \leqslant k \leqslant l$) and F_μ ($\mu \geqslant 5$) have the form depicted in figure 51 ([132]; the numbering of the vertices is omitted).

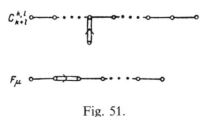

Fig. 51.

Part II

Oscillatory integrals

This part is devoted to an investigation of asymptotic oscillatory integrals, that is integrals of the form

$$I(\tau) = \int_{\mathbb{R}^n} e^{i\tau f(x)} \phi(x) dx_1 \ldots dx_n,$$

for large values of the real parameter τ. Here f and ϕ are smooth functions. The function f is called the *phase*, the function ϕ is called the *amplitude*. In accordance with the principle of stationary phase the main contribution in the asymptotics is given by neighbourhoods of the critical points of the phase. In this part we discuss the connection between asymptotics and different characteristics of the critical points of the phase (resolution of singularities, Newton polyhedra) and explain the methods for calculating asymptotics. In Part III we discuss the connection between asymptotics and the monodromy and mixed Hodge structures of critical points.

In the last ten years the theory of singularities has been exceptionally closely linked with the investigation of oscillatory integrals. On one hand a great many reasonable problems of the theory of singularities arose from attempts to understand the nature of the behaviour of integrals. On the other hand much of the study of critical points has found direct application in the study of asymptotics. As a first example, remember that the classification of simple critical points of functions arose as a by-product of the calculation of asymptotics of the simplest oscillatory integrals [11, 12]. As a second example we mention the connection between asymptotic integrals and the mixed Hodge structure of critical points (see Part III).

Chapter 6
Discussion of results

6.1   Examples and definitions

6.1.1   Oscillatory integrals and shortwave oscillations

The problems of optics, acoustics and quantum mechanics, the theory of partial differential equations, probability theory and number theory lead to the need to study oscillatory integrals with large values of the parameter.

Example. We consider a surface in three-dimensional space. We suppose that each point of the surface radiates a spherical wave of fixed frequency and fixed wavelength. We suppose that the wavelength is small in comparison with the size of the surface and with the rate of change of the amplitude of the wave with change of the point on the surface.

The total oscillatory behaviour at the point y of the space is given by the function

$$e^{2\pi i\omega t} \int_{S} \frac{e^{2\pi i\|x-y\|/\lambda}}{\|x-y\|} \phi(x)dx,$$

where t is the time, ω is the frequency, λ is the wavelength, S is the surface radiating the wave, ϕ is the amplitude, and dx is an element of surface area. In this way the complex oscillation is given by an oscillatory integral in which the reciprocal of the wavelength plays the rôle of the large real parameter, and the distance function from the point on the surface to the fixed point of the space serves as the phase. The principal contribution to the complex oscillation (that is to the oscillatory integral) is given by neighbourhoods of the critical points of the phase. If all the critical points of the phase are non-degenerate, then the contribution to the complex oscillation from each of them is proportional to the wavelength. If the phase has degenerate critical points then the contribution of their small neighbourhoods in the complex oscillation is still bigger, namely, the order of the contribution is proportional to the wavelength to some power less than one. As a rule the function on the surface, equal to the distance from the fixed point of the space, has only non-degenerate critical points. The points of the space are called *caustic* or *focal* if the function on the surface, equal to the distance from the point, has a degenerate critical point. The caustic points form in the space a new surface called the *caustic*. At points of the caustic the complex oscillation has exceptionally large magnitude. If the surface radiates light waves, then the caustic is the surface of exceptionally bright points. It can be seen on a wall, illuminated by rays reflected from a concave surface (for example, the surface of a cup). Caustics can be defined in another way. The caustic is the set of critical values of the exponential map of the space of the normal bundle of the radiating surface. We racall the definition of the exponential map. A point of the space of the normal bundle is a pair, consisting of a point on the surface and a vector, based at it, which is perpendicular to the surface. The exponential map maps such a pair to the point of space which is the endpoint of the vector. Finally there is a third description of the caustic. On the normal to the radiating surface we mark out the principal radii of curvature. The surface of the endpoints of all these segments is the caustic (see [145]).

We give one more example of the appearance of oscillatory integrals.

One of the classical problems of the theory of linear partial differential equations is the problem of constructing the solution, asymptotic in a parameter, of the Cauchy problem with rapidly oscillating initial conditions. Asymptotic methods (see [244–246]) reduce in this problem to the following result. For any natural number N in a small neighbourhood of any point y^0 the solution of the Cauchy problem can be represented in the form of a finite sum of oscillatory integrals

$$\int e^{i\tau F(y,x)}\phi(y, x, (i\tau)^{-1})dx$$

and a remainder term of order $o(\tau^{-N})$ as $\tau \to +\infty$. In this integral F is a real function, τ is the large parameter of the problem, x are real parameters, the function ϕ has compact support in x and is a polynomial in $(i\tau)^{-1}$. Therefore the calculation of the asymptotic solution of the Cauchy problem is reduced to the calculation of asymptotic oscillatory integrals.

For a multitude of examples of physical problems in which the need arises of studying asymptotic integrals, see the works of M. Berry and J. Nye cited in the bibliography. We note also the interesting articles [33, 122, 283].

6.1.2. The principle of stationary phase states: the principal contribution in oscillatory integrals is given by a neighbourhood of a critical point of the phase.

Theorem 6.1. Let the amplitude of an oscillatory integral have compact support. Let the phase of the oscillatory integral not have critical points on the support of the amplitude. Then as the parameter of the oscillatory integral tends to $+\infty$, the integral tends to zero more rapidly than any power of the parameter.

Proof. First let the integral be one-dimensional. Integrating it by parts:

$$\int_{-\infty}^{+\infty} e^{i\tau f(x)}\varphi(x)\,dx = \frac{-1}{i\tau} \int_{-\infty}^{+\infty} e^{i\tau f(x)}(\varphi(x)/f'(x))'\,dx.$$

Repeating the integration a sufficient number of times, we obtain the theorem. The many-dimensional case reduces to the one-dimensional case with the help of a partition of unity and a change to new variables of integration, in which the phase function is one of the variables.

6.1.3 Fresnel integrals

An oscillatory integral the phase of which has only non-degenerate critical points is called a *Fresnel integral*.

Example. We consider a one-dimensional oscillatory integral, the phase of which is the function x^2. In figure 52 we have depicted the graph of

$$y = \cos{(\tau x^2)}\phi(x),$$

which is the real part of the integrand of the oscillatory integral. It is clear that for large values of the parameter τ the integral is proportional to the area under the first loop of the graph, that is proportional to $\phi(0)\tau^{-1/2}$. Exact calculation shows

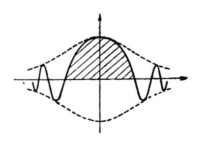

Fig. 52.

that as the parameter τ tends to $+\infty$ the oscillatory integral can be represented in the form

$$\phi(0)\sqrt{(\pi/\tau)}\exp(i\pi/4)$$

and a remainder term of order $O(\tau^{-3/2})$ (see [110]).

We consider the many-dimensional Fresnel integral

$$\int_{\mathbb{R}^n}\exp(i\tau f(x))\varphi(x)dx_1\ldots dx_n.$$

Theorem 6.2 [109, 110]. We suppose that the phase of this integral has a non-degenerate critical point at the origin, and that the support of the amplitude is compact and does not contain any other critical point of the phase. Then as the parameter of the integral tends to $+\infty$ the integral can be represented in the form

$$\varphi(0)(2\pi/\tau)^{n/2}\exp(i\tau f(0)+(i\pi/4)\text{ sign } f''_{xx}(0))|\det f''_{xx}(0)|^{-1/2}$$

$$+O(\tau^{-n/2-1}),$$

where sign $f''_{xx}(0)$ is the signature of the matrix of second derivatives of the phase at the origin and det $f''_{xx}(0)$ is the determinant of the matrix of second derivatives of the phase at the origin.

Proof. By the Morse lemma the phase has the form

$$y_1^2+\ldots+y_k^2-y_{k+1}^2-\ldots-y_n^2$$

with respect to a suitable system of coordinates in a neighbourhood of the critical point. Therefore it is sufficient to prove the theorem in this case. This case easily reduces, with the help of Fubini's theorem, to the assertion of the previous example. The theorem is proved.

6.1.4   Caustics

In applications, as a rule, the phase and amplitude of oscillatory integrals depend on additional parameters. We consider such integrals.

Let us suppose that the phase is a general family of functions, depending on additional parameters (a propos of this see Part II of Volume 1). In this case the integral is a Fresnel integral for almost all values of the parameters and for these values has order $\tau^{-n/2}$ (Theorem 6.2). The set of values of the parameters for which the phase has degenerate critical points forms a hypersurface in the space of parameters. This hypersurface is called the *caustic*. For caustic values of the parameters the order with which the integral tends to zero is determined by the degenerate critical points of the phase.

6.1.5 Asymptotic oscillatory integrals near caustics

Let us suppose that for a given value of the additional parameters the phase of an oscillatory integral has a unique critical point and the phase, considered as a family of functions depending on parameters, is a family of functions in general position. In this case the caustic in a neighbourhood of the given value of the parameter is said to be *elementary*.

Examples of elementary caustics, occurring when the number of parameters is two and three are depicted in figures 53–57, where near each part of the caustic is a label indicating the type of degenerate critical point occurring for these caustic values of the parameters. For example $A_2 + A_2$ means that the phase has two

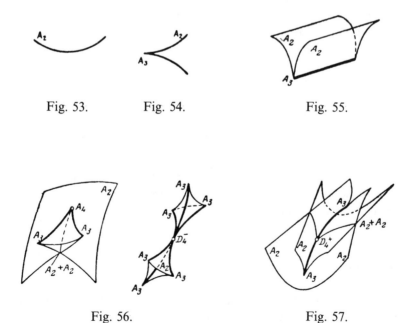

Fig. 53.   Fig. 54.   Fig. 55.

Fig. 56.   Fig. 57.

critical points of type A_2 and the other critical points of the phase are non-degenerate. Each degenerate critical point of the phase gives in the integral a contribution of order $\tau^{\beta-n/2}$. The number β for the critical points of types A_k, D_k is equal, respectively, to $(k-1)/(2k+2)$, $(k-2)/(2k-2)$ (see Theorem 6.4 below).

In accordance with the results of Part III of Volume 1 for a phase depending in a general way on two or three parameters, each elementary caustic is locally diffeomorphic to one of the caustics depicted in figures 53–57. The integrals of

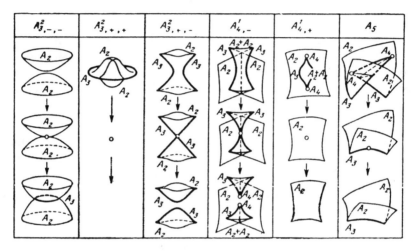

Fig. 58.

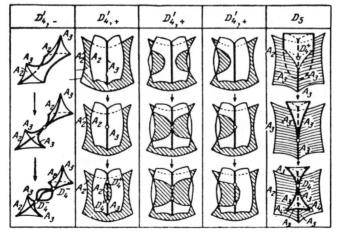

Fig.59.

equal orders correspond to the values of the parameters transforming one into the other under local diffeomorphism.

Let us suppose that there are four additional parameters and one of the parameters is chosen; we shall call it time. Then depending on the time the caustic will be reconstructed. For a family of functions in general position all possible reconstructions are shown in figures 58, 59. The classification of the reconstructions of the caustic was carried out by V. M. Zakalyukin (see Part III of Volume 1). Each reconstruction has its own designation. The families corresponding to these designations were given in § 22.3 of Volume 1.

Notice that figure 58, (the reconstruction A_3^2, $_+$, $_+$) depicts the unique reconstruction, for which for positive time the caustic is absent but for which for negative time it exists. According to V. M. Zakalyukin this reconstruction, possibly, illustrates the phenomenon of the disappearance of "flying saucers".

6.1.6   Oscillatory integrals in a halfspace

We return to the example of § 6.1.1. Let us suppose that the radiating surface is opaque for the emitted waves. Then at a given point of the space there arrive waves from the visible parts of the surface only. Therefore the complex oscillation at a point of the space is expressed as a sum of oscillatory integrals each of which is taken over a part of the surface. In this way it is useful in the study of short-wave radiation to know how to calculate asymptotic oscillatory integrals along a region with boundary. We investigate the case of a smooth boundary.

Let us consider an oscillatory integral on that part of the space $\mathbb{R}^n$ given by the condition that the first coordinate is positive. Moreover we shall suppose that the phase and the amplitude of the integral are smooth functions on the whole space.

Theorem 6.1'. Let the amplitude of the oscillatory integral on the halfspace have compact support. Let the phase of the oscillatory integral on the halfspace not have critical points on the support of the amplitude in the region of integration. Let the restriction of the phase to the boundary of the halfspace not have critical points on the support of the amplitude. Then as the parameter of the oscillatory integral tends to $+\infty$ the integral tends to zero more rapidly than any power of the parameter.

Proof. It is sufficient to carry out the proof in the case when the support of the amplitude is concentrated in a small neighbourhood of a boundary point of the halfspace. By changing the variables of integration we can transform to the case in which the halfspace of integration is given by the condition that the first variable is positive, but the phase is the second variable. After integrating by parts with respect to the second variable a sufficient number of times, we obtain the theorem.

Let us suppose that the phase of an oscillatory integral in a halfspace does not have critical points on the boundary of the halfspace. Let us suppose that all its critical points inside the halfspace are non-degenerate and the critical points of its restriction to the boundary of the halfspace of integration are also non-degenerate. An oscillatory integral with such a phase we shall call a *Fresnel integral on a halfspace*.

Theorem 6.2′. Let us consider a Fresnel integral on the halfspace in which the first coordinate is positive. Let us suppose that the origin is not a critical point of the phase, but is a non-degenerate critical point of the restriction of the phase to the boundary of the halfspace. Let us suppose that the support of the amplitude is compact, does not contain a critical point of the phase and does not contain other critical points of the restriction of the phase to the boundary of the halfspace. Then as the parameter of the integral tends to $+\infty$ the integral can be represented in the form

$$\varphi(0)\,(i\tau)^{-1}(2\pi/\tau)^{(n-1)/2}\exp\left(i\tau f(0)+(i\pi/4)\,\text{sign}\,\tilde{f}''_{x'x'}(0)\right)\times$$

$$\times\,|\det\tilde{f}''_{x'x'}(0)|^{-1/2}+O(\tau^{-(n+1)/2}),$$

where sign $\tilde{f}''_{x'x'}(0)$ is the signature of the matrix of second derivatives at the critical point of the restriction of the phase to the boundary, and det $\tilde{f}''_{x'x'}(0)$ is the determinant of the matrix of second derivatives at the critical point of the restriction of the phase to the boundary.

Proof. In a neighbourhood of the origin we change the first variable so that the halfspace of integration, as before, satisfies the condition that it is positive, and the phase of the integral takes the form

$$x_1+h(x_2,\ldots,x_n).$$

Then Theorem 6.2′ reduces to Theorem 6.2 by integrating by parts with respect to the first coordinate.

Let us suppose that the phase and amplitude of an oscillatory integral in a halfspace depend on additional parameters. Let us suppose that the phase, considered as a family of functions depending on parameters, is a family of functions in general position (see Part II of Volume 1). In this case the integral is a Fresnel integral for almost all values of the parameters. Those values of the parameters for which the integral is not a Fresnel integral form a hypersurface in the space of parameters called the *caustic*.

Let us suppose that for the given values of the parameters the phase has a unique critical point on the boundary of the halfspace of integration. A caustic in a neighbourhood of such values of the parameters is called *elementary*.

Examples of elementary caustics, occuring when the number of parameters is 2 or 3 are depicted in figures 60–64.

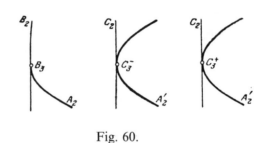

Fig. 60.

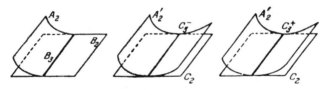

Fig. 61.

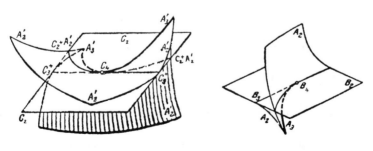

Fig. 62   Fig. 63.

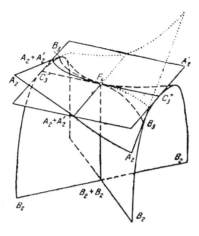

Fig. 64.

For caustic values of the parameters either the phase has a degenerate critical point in the halfspace of integration or the phase has a critical point on the boundary or the restriction of the phase to the boundary has a degenerate critical point. On the diagrams near each part of the caustics is given the designation of these critical points. For the normal forms of the critical points marked on the caustics, see Chapter 17 of Volume 1 and also [16]. Each indicated critical point gives in the integral a contribution of order $\tau^{\beta - n/2}$. The number β for the critical points of types

$$A_k', \ D_k', \ B_k, \ C_k^{\pm}, \ F_4$$

are equal, respectively, to

$$-1/(k+1), \ -1/(2k-2), \ (k-1)/2k, \ 0, \ 1/6$$

(see Theorem 8.9 below).

In accordance with [18] (see also Chapter 17 of Volume 1) for the phase of an oscillatory integral in a halfspace, depending in a general way on two or three parameters, each elementary caustic is locally diffeomorphic to one of the caustics depicted in figures 60–64. Values of the parameters which are related by a local diffeomorphism give integrals of the same order.

6.1.7 Light, dark and twilight zones (according to P. K. Mandrykin)

Let us suppose that the phase and amplitude of an oscillatory integral depend on additional parameters. Let us consider the space of parameters and a caustic located in it. Let us consider an arbitrary value of the parameters away from the caustic. The phase of the oscillatory integral, corresponding to this value of the parameters, has only non-degenerate critical points or, has no critical points at all. In the first case the oscillatory integral has order $\tau^{-n/2}$, where n is the dimension of the space of integration; in the second case as $\tau \to +\infty$ the oscillatory integral tends to zero more rapidly than any power of the parameter τ. Corresponding to these cases the region away from the caustic is called
– a *light zone*, if to the values of the parameters from this region there correspond oscillatory integrals, the phases of which have at least one critical point;
– a *dark zone*, if to the values of the parameters from this region there correspond oscillatory integrals, the phases of which do not have even one critical point.

Example. Figures 53–57 depict caustics corresponding to critical points of types

$$A_2, A_3, A_4, D_4^{\pm}.$$

In the pictures of the caustics corresponding to

$$A_2, A_4, D_4^{+},$$

the dark zones lie under the caustics. The other regions away from the caustics depicted in the figure are light zones.

Let us suppose now that our oscillatory integral is an integral on a halfspace. Let us consider an arbitrary value of the parameters away from the caustic. There are three possibilities:

(i) The phase of the oscillatory integral, corresponding to the given value of the parameters, has at least one critical point in the halfspace of integration. In this case the integral has order $\tau^{-n/2}$.

(ii) The phase of the oscillatory integral, corresponding to the given value of the parameters, does not have a critical point in the halfspace of integration, but the restriction of the phase to the boundary of the halfspace of integration has at least one critical point. In this case the integral has order $\tau^{-(n+1)/2}$ (Theorem 6.2').

(iii) The phase of the oscillatory integral, corresponding to the given value of the parameters does not have a critical point in the halfspace of integration, and also its restriction to the boundary of the halfspace of integration does not have a critical point. In this case as $\tau \to +\infty$ the integral tends to zero more rapidly than any power of the parameter τ.

Corresponding to these three possibilities, the region away from the caustic is called a *light zone*, a *twilight zone* or a *dark zone*, respectively.

Example. Figures 61–64 depict caustics, corresponding to critical points of types

$$B_3, \ C_3^{\pm}, \ C_4, \ B_4, \ F_4.$$

We shall indicate the dark and twilight zones on these pictures, the other regions away from the caustics being light zones.

In figure 61a the twilight zone is above the caustic. In figure 61b the twilight zone is under the caustic. In figure 61c the dark zone is above the caustic; the twilight zone is between the two leaves of the caustic. In figure 62 the twilight zone is on one side of the plane of the caustic. In figure 63 the twilight zone is to the right over the plane of the caustic. In figure 64 the dark zone is over all the caustic; the twilight zone is behind the caustic under the plane of the caustic; there is also a twilight zone to the right between the plane of the caustic and the ruled surface (the third part of the caustic).

It is interesting to ask if there exists a critical point, away from the caustic of which there are two dark zones. Probably the dark zones possess certain convexity properties.

6.1.8.

Theorem 6.3 (on asymptotic expansions, see [32, 46, 47, 239, 358].

Let us consider the oscillatory integral

$$\int\limits_{\mathbb{R}^n} \exp(i\tau f(x))\, \varphi(x)dx_1 \ldots dx_n. \tag{1}$$

Let the phase be an analytic function in a neighbourhood of its critical point x^0. Then the oscillatory integral can be expanded in an asymptotic series

$$\exp(i\tau f(x^0)) \sum_{\alpha} \sum_{k=0}^{n-1} a_{k,\alpha}(\varphi)\tau^{\alpha}(\ln \tau)^k \quad \text{for} \quad \tau \to +\infty \tag{2}$$

if the support of the amplitude is concentrated in a sufficiently small neighbourhood of this critical point of the phase. Here the parameter α runs through a finite set of arithmetic progressions, depending only on the phase, and consisting of negative rational numbers. The numerical coefficients $a_{k,\alpha}$ are generalised functions of the amplitude. The support of each such generalised function lies in the critical set of the phase.

Example [110]. Let us consider a Fresnel integral. Let us suppose that the phase of the integral has a non-degenerate critical point at the origin and that the support of the amplitude is compact and does not contain other critical points of the phase. Then as the parameter tends to $+\infty$ the integral can be expanded in the asymptotic series

$$\exp{(i\tau f(0))}\tau^{-n/2} \sum_{j=0}^{\infty} a_j\tau^{-j}.$$

The number a_j is equal to a linear combination of the $(2j)$th mixed derivatives of the amplitude at the origin. The number a_0 is indicated in Theorem 6.2.

Remark. In Theorem 6.3 the condition that the phase is analytic is practically always satisfied: the phase is a polynomial in suitable coordinates in a neighbourhood of a finite-multiplicity critical point. Infinite-multiplicity critical points are very rare: the coefficients of the Taylor series of an infinite-multiplicity critical point satisfy an infinite set of independent algebraic relations.

We give two proofs of Theorem 6.3. One, based on Hironaka's theorem on the resolution of singularities, is given in Chapter 7. For the other proof, which uses complex analytic reasoning, see Chapter 11.

In the asymptotic series of oscillatory integrals the phase and the amplitude do not enjoy the same status: the phase determines the indices for the powers of the parameter, but the amplitude determines the coefficients for the powers of the parameter. The dependence on the phase is more important. As a rule investigating oscillatory integrals we fix the phase, but we allow the amplitude to change. In the example in § 6.1.1 on a surface radiating waves, the phase depends on the geometry of the radiating surface, but the amplitude depends on the intensity of the radiation.

6.1.9   The oscillation index and the singular index

The fundamental characteristics of the asymptotic series of an oscillatory integral are: the index of the power of the parameter in the leading term of the series;
the power of the logarithm of the parameter in the leading term of the series;
the numerical coefficient of the leading term of the series;
finally, the set of all the indices of the powers of the parameter occurring in the series.

Definitions. The *index set* of an analytic phase at a critical point is the set of all numbers α possessing the property: for any neighbourhood of the critical point there is an amplitude with support in the neighbourhood for which in the asymptotic series (2) there is a number k with the property that the coefficient $a_{k,\alpha}$ is not equal to zero. The *oscillation index* of an analytic phase at a critical point is the maximal number in the index set. The oscillation index will be denoted by β. The *multiplicity of the oscillation index* of an analytic phase at a critical point is the maximal number k possessing the property: for any neighbourhood of the critical point there is an amplitude with support in this neighbourhood for which in the asymptotic series (2) the coefficient $a_{k,\beta}$ is not equal to zero. The multiplicity of the oscillation index will be denoted by K.

Example. The index set of a phase in n variables at a non-degenerate critical point is the set of all numbers of the form $-n/2 - l$, where $l = 0, 1, \ldots$ The oscillation index of this critical point is equal to $-n/2$, its multiplicity is equal to zero.

The oscillation index and its multiplicity satisfy the following simple property. Let

$$f(x_1, \ldots, x_n), \quad g(y_1, \ldots, y_l)$$

be analytic functions with critical points at the origin. Then for the function

$$f(x_1, \ldots, x_n) + g(y_1, \ldots, y_l)$$

the oscillation index at the origin and its multiplicity are equal, respectively, to the sum of the oscillation indices and the sum of the multiplicities of the oscillation indices at the critical points of the functions f and g:

$$\beta(f+g) = \beta(f) + \beta(g), \quad K(f+g) = K(f) + K(g)$$

(this is a corollary of Fubini's theorem, see Chapter 9). The additivity of the oscillation index and its multiplicity motivate the following definition.

Definition. The *singular index* of an analytic phase in n variables at a critical point is the oscillation index at this critical point, increased by $n/2$. The *multiplicity of the singular index* is the multiplicity of the oscillation index.

The singular index and its multiplicity are equal to zero at a non-degenerate critical point of the phase. The singular index and its multiplicity are equal at stably equivalent critical points.

6.1.10   Tables of singular indices

In this part we calulate (in the cases enumerated below) the principal characteristics of critical points of phases of oscillatory integrals: the oscillation index, its multiplicity, the index set. We describe amplitudes such that the principal term of the asymptotic series is different from zero. The results proved in this part allow us to calculate the singular indices and their multiplicities for all critical points classified in Part II of Volume 1. Namely for all simple, unimodal and bimodal critical points, for all critical points of multiplicity less than 16, for all critical points existing in classes of codimension less than 10, see Chapter 15 of Part 1.

The results of the calculations follow in tables 1–5. In the first row of the tables is the designation of the type of critical point of the phase. The normal forms of the critical points corresponding to these designations are shown in §§ 15.1 and 17.1 of Volume 1. In the second row of the tables are the singular indices. The meaning of our tables is this: if in a neighbourhood of the critical point the phase is reduced to the tabulated form by a diffeomorphism of the space, then its singular index is equal to the singular index of the tabulated function.

For critical points of types

$$J_{10+k},\ X_{9+k},\ Y_{r,s},\ P_{8+k}^{-},\ R_{l,m},\ \tilde{R}_{m}^{+,-},$$

$$\tilde{R}_{m}^{-,+},\ T_{p,q,r}\,(p^{-1}+q^{-1}+r^{-1}<1),\ X_{1,p}\,(p>0),\ Y_{r,s}^{1},$$

$$Z_{i,0}^{3},\ Z_{35+6i}^{3},\ Z_{36+6i}^{3},\ Z_{37+6i}^{3},\ Z_{i,p}^{3}$$

the multiplicity of the singular index is equal to 1, for all other critical points in tables 1–5 the multiplicity of the singular index is equal to 0. For all the critical

Table 1. Simple singularities.

A_k	D_k	E_6	E_7	E_8
$\dfrac{k-1}{2k+2}$	$\dfrac{k-2}{2k-2}$	$\dfrac{5}{12}$	$\dfrac{4}{9}$	$\dfrac{7}{15}$

Table 2. Unimodal singularities.

$$P_8, X_9, J_{10}, J_{10+k}, X_{9+k}, Y_{r,s}, \bar{Y}_r, P_{8+k}, R_{l,m}, \bar{R}_m, T_{p,q,r}, \bar{T}_{p,m}$$

$$\frac{1}{2}$$

E_{12}	E_{13}	E_{14}, Q_{10}	Z_{12}	Z_{13}, Q_{11}	W_{12}	W_{13}, S_{11}	Q_{12}	S_{12}	U_{12}
$\dfrac{11}{21}$	$\dfrac{8}{15}$	$\dfrac{13}{24}$	$\dfrac{6}{11}$	$\dfrac{5}{9}$	$\dfrac{11}{20}$	$\dfrac{9}{16}$	$\dfrac{17}{30}$	$\dfrac{15}{26}$	$\dfrac{7}{12}$

Table 3. Bimodal singularities.

$\begin{array}{c}J_{3,0}\\J_{3,p}\end{array}$	$\begin{array}{c}Z_{1,0}, E_{19}\\Z_{1,p}\end{array}$	$\begin{array}{c}W_{1,0}, W_{1,2q-1}^{\#}, Q_{2,0}, Z_{17}\\W_{1,p}, W_{1,2q}^{\#}, Q_{2,p}\end{array}$	$\begin{array}{c}S_{1,0}, S_{1,2q-1}^{\#}, W_{17}\\S_{1,p}, S_{1,2q}^{\#}, Q_{17}\end{array}$
$\dfrac{5}{9}$	$\dfrac{4}{7}$	$\dfrac{7}{12}$	$\dfrac{3}{5}$

S_{17}	$U_{1,0}, U_{1,2q-1}, U_{1,2q}$	E_{18}	E_{20}	Z_{18}	Z_{19}	W_{18}	Q_{16}	Q_{18}	S_{18}	U_{16}
$\dfrac{5}{8}$	$\dfrac{11}{18}$	$\dfrac{17}{30}$	$\dfrac{13}{24}$	$\dfrac{10}{17}$	$\dfrac{16}{27}$	$\dfrac{17}{28}$	$\dfrac{25}{42}$	$\dfrac{29}{48}$	$\dfrac{21}{34}$	$\dfrac{19}{30}$

Table 4. Singularities of corank 2 with non-zero 4-jet.

$J_{k,0}, J_{k,i}$	E_{6k}	E_{6k+1}	E_{k+2}	$X_{k,0}, X_{k,p}, Y_{r,s}^k$
$\dfrac{2k-1}{3k}$	$\dfrac{6k-1}{9k+3}$	$\dfrac{4k}{6k+3}$	$\dfrac{6k+1}{9k+6}$	$\dfrac{3k-1}{4k}$

$\left.\begin{array}{l} Z_{i,0}^k, Z_{i,p}^k, Z_{12k+6i-1}^k, \\ Z_{12k+6i}^k, Z_{12k+6i+1}^k \end{array}\right\}$ for $k>2$	$\begin{array}{c} Z_{i,0}^2 \\ Z_{i,p}^2 \end{array}$
$\dfrac{3k-1}{4k}$	$\dfrac{2i+5}{3i+8}$

Z_{23+6i}^2	Z_{24+6i}^2	Z_{25+6i}^2	$Z_{i,0}, Z_{i,p}$	Z_{6i+11}	Z_{6i+12}	Z_{6i+13}
$\dfrac{6i+17}{9i+27}$	$\dfrac{4i+12}{6i+19}$	$\dfrac{6i+19}{9i+30}$	$\dfrac{2i+2}{3i+4}$	$\dfrac{6i+8}{9i+15}$	$\dfrac{4i+6}{6i+11}$	$\dfrac{6i+10}{9i+18}$

W_{12k}	W_{12k+1}	$W_{k,0}, W_{k,i}, W_{k,2q-1}^\#, W_{k,2q}^\#$	W_{12k+5}	W_{12k+6}
$\dfrac{12k-1}{16k+4}$	$\dfrac{9k}{12k+4}$	$\dfrac{12k+2}{16k+8}$	$\dfrac{9k+3}{12k+8}$	$\dfrac{12k+5}{16k+12}$

Table 5. Singularities of corank 3 with reduced 3-jet and 3-jet x^2y.

$Q_{k,0}, Q_{k,i}$	Q_{6k+4}	Q_{6k+5}	Q_{6k+8}	S_{12k-1}	S_{12k}
$\dfrac{4k-1}{6k}$	$\dfrac{12k+1}{18k+6}$	$\dfrac{8k+2}{12k+6}$	$\dfrac{12k+5}{18k+12}$	$\dfrac{12k-3}{16k}$	$\dfrac{18k-3}{24k+2}$

$\begin{array}{c} S_{k,0}, S_{k,2q-1}^\# \\ S_{k,i}, S_{k,2q}^\# \end{array}$	S_{12k+4}	S_{12k+5}	U_{12k}	$\begin{array}{c} U_{k,2q} \\ U_{k,2q-1} \end{array}$	U_{12k+4}	$\begin{array}{c} V_{1,0}, V_{1,2q-1}^\# \\ V_{1,p}, V_{1,2q}^\# \end{array}$
$\dfrac{6k}{8k+2}$	$\dfrac{18k+3}{24k+10}$	$\dfrac{12k+3}{16k+8}$	$\dfrac{15k-1}{18k+6}$	$\dfrac{10k+1}{12k+6}$	$\dfrac{15k+4}{18k+12}$	$\dfrac{5}{8}$

points in the tables (except

$$P_{8+k}^{+}, \ \tilde{R}_m^{+,+}, \ \tilde{R}_m^{-,-}, \ \tilde{T}_{p,m})$$

the multiplicity of the singular index is equal to 1, for all other critical points in tables 1–5 the multiplicity of the singular index is equal to 0. For all the critical points in the tables (except

$$P_{8+k}^{+}, \ \tilde{R}_m^{+,+}, \ \tilde{R}_m^{-,-}, \ \tilde{T}_{p,m},$$

the same assertion is true about the imaginary part of the coefficient $a_{K,\beta}$.

For the proof of these results see Chapter 9.

6.2   Formulation of the results

The main results of this part are formulated in terms of the Newton polyhedra of the Taylor series of the critical points of the phase. The Newton polyhedron is that convex polyhedron formed by the indices of the monomials occurring in the Taylor series. We consider the class of critical points with fixed Newton polyhedron. We prove that almost all critical points of the class have the same oscillation index. We prove the formula expressing this common oscillation index in terms of the geometry of the Newton polyhedron. The exceptions consist of the critical points, the coefficients of whose Taylor series satisfy a finite set of explicit algebraic conditions.

The class of critical points with fixed Newton polyhedron is a useful thing to consider in a study of discrete invariants of critical points. As a rule an invariant has a single value for almost all points of the class, and this common value can be simply expressed in terms of the geometry of the Newton polyhedron (see §§ 6.2.4, 3.5, and also [31, 44, 45, 65–67, 76, 89, 106, 159–165, 183, 195, 196, 200, 223, 314, 343, 358, 359, 380–382, 386]).

6.2.1   The Newton polyhedron

Let us consider the positive orthant of the space $\mathbb{R}^n$, that is the set of points with non-negative coordinates. We define the Newton polyhedron of an arbitrary subset of the orthant consisting of points with integer coordinates. At each point of the subset we take a parallel positive orthant. The *Newton polyhedron* is the convex hull in $\mathbb{R}^n$ of the union of all the orthants constructed above. The Newton polyhedron is a convex polyhedron with vertices at points with non-negative integer coordinates. Together with each point it contains the positive orthant parallel translated to this point. The *Newton diagram of a subset* is the union of all the compact faces of its Newton polyhedron.

We consider the power series

$$f = \Sigma a_k x^k$$

with real or complex coefficients, where

$$k = (k_1, \ldots, k_n),\ x^k = x_1^{k_1} \ldots x_n^{k_n}.$$

The *support of the series* is the set of indices of all the monomials occurring in the series with non-zero coefficients. The support of the series is a subset of the positive orthant, consisting of points with non-negative integer coordinates. We remove from the support the origin (if it lies in the support). The set obtained is called the *reduced* support of the series. The *Newton polyhedron of a power series* is the Newton polyhedron of its reduced support. The *Newton diagram of a power series* is the Newton diagram of its reduced support.

The Newton polyhedron is denoted by Γ, the Newton diagram is denoted by Δ.

Example. For the functions

$$f = (x_1^2 + x_2^2)^2 + x_1^5,$$

$$g = (x_1^2 - x_2^3)^2,$$

$$h = (x_1 + x_2)^2 x_1^2 + x_1^5 + x_2^5$$

the Newton polyhedra and Newton diagrams of the Taylor series at the origin are depicted in figure 65.

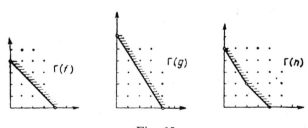

Fig. 65.

For each face γ of the Newton polyhedron of a power series the γ-part of this power series is the power series consisting of monomials the indices of which lie in the face γ; moreover each monomial occurs with the same coefficient as in the original power series. If the face γ is compact, then the γ-part is a polynomial. The *principal part* of a power series is a polynomial consisting of monomials, the indices of which lie in the Newton diagram of the power series; moreover the monomials occur with the same coefficients as in the original power series. The γ-part of the series f is denoted by f_γ, the principal part of the series is denoted by f_Δ.

Example. For the functions in the previous example the principal parts of the Taylor series are the polynomials

$$f_\Delta = (x_1^2 + x_2^2)^2,$$
$$g_\Delta = (x_1^2 - x_2^3)^2,$$
$$h_\Delta = (x_1 + x_2)^2 x_1^2 + x_2^5$$

6.2.2   Nondegeneracy of the principal part

We define the concept of nondegeneracy of the principal part of a power series. In the sequel we shall see that functions, the Taylor series of which have nondegenerate principal parts, have good properties: their discrete characteristics can be simply expressed in terms of the geometry of their Newton polyhedra, see § 6.2.4.

Definitions [195, 196]. The principal part of the power series f with real coefficients (respectively power series with complex coefficients) is $\mathbb{R}$-*nondegenerate* (resp. $\mathbb{C}$-*nondegenrate*) if for every compact face γ of the Newton polyhedron of the series the polynomials

$$\partial f_\gamma / \partial x_1, \ldots, \partial f_\gamma / \partial x_n$$

do not have common zeros in $(\mathbb{R} \backslash 0)^n$ (respectively in $(\mathbb{C} \backslash 0)^n$).

Example. All principal parts in the previous example are $\mathbb{C}$-degenerate, the principal part of f_Δ is $\mathbb{R}$-non-degenerate, the principal parts of g_Δ, h_Δ are $\mathbb{R}$-degenerate.

Remark. For any compact face γ the γ-part polynomial is quasihomogeneous. By the Theorem of Euler on homogeneous functions, the common zeroes in $(\mathbb{R}\setminus 0)^n$ of all the first partial derivatives of the γ-part polynomial lie on the zero level manifold of the γ-part polynomial.

The following lemma shows that series with degenerate principal parts are rare.

Lemma 6.1 (see [195, 196]). The set of $\mathbb{R}$-degenerate (respectively $\mathbb{C}$-degenerate) principal parts is a proper semialgebraic (resp. constructive) subset in the space of all principal parts corresponding to a given Newton polyhedron, the complement of which is everywhere dense.

Proof. For a fixed compact face γ of the Newton polyhedron we prove that, in the space of polynomials which are γ-parts, the semialgebraic subset corresponds to those polynomials for which the zero level manifold has a singular point in $(\mathbb{R}\setminus 0)^n$ and the complement of the subset is everywhere dense.

The Theorem of Tarski-Seidenberg (see [130, 325]) guarantees that the subset is semialgebraic. Let us prove that the complement is dense. The zero level manifold is given by the equation

$$\sum_{k\in\gamma} c_k x^k = 0.$$

We pick out one of these monomials. Then the zero level manifold in $(\mathbb{R}\setminus 0)^n$ can be given by the equation

$$c_{k_0} = - \sum_{k\in\gamma\setminus k_0} c_k x^{k-k_0}.$$

By the Theorem of Bertini-Sard only a finite set of values of the coefficient c_{k_0} (where the other coefficients are fixed) correspond to singular zero level manifolds. The lemma is proved.

6.2.3 The distance to a polyhedron and the remoteness of a polyhedron

To study oscillatory integrals we use geometrical characteristics of the Newton polyhedron, called the distance to a polyhedron and the remoteness of a polyhedron. Let us consider the bisector of the positive orthant in $\mathbb{R}^n$, that is the line consisting of points with equal coordinates. The bisector intersects the boundary of the Newton polyhedron in exactly one point. This point is called the *centre of the boundary* of the Newton polyhedron. The coordinate of the centre is called the *distance to the Newton polyhedron*. The *remoteness of the Newton polyhedron* is the reciprocal of the distance, taken with a minus sign.

Example. For the functions f, g, h of the example on page 189, the distances to the Newton polyhedra are equal, respectively, to

2, 12/5, 2,

whilst the remoteness of the Newton polyhedra are equal, respectively, to

$-1/2$, $-5/12$, $-1/2$.

The further from the origin the Newton polyhedron is, the larger is its remoteness. We call the Newton polyhedron *remote* if its remoteness is greater than -1. In other words the Newton polyhedron is remote if it does not contain the point $(1, \ldots, 1)$.

We consider the open face which contains the centre of the boundary of the Newton polyhedron. The codimension of this face, less one, is called the *multiplicity* of the remoteness. In particular, if the indicated face is a vertex of the polyhedron, then the multiplicity is equal to $n-1$, if the indicated face is an edge of the polyhedron then the multiplicity is equal to $n-2$, and so on.

Example. For the functions f, g, h of the example on page 189 the multiplicities of the remoteness of the Newton polyhedra equal, respectively, 0, 0, 1.

6.2.4 Formulation of the main results

The main result of this part is the following: the oscillation index of a critical point of the phase is determined by the remoteness of the Newton polyhedron of its Taylor series (under conditions formulated in the following two theorems).

Theorem 6.4 [358]. Let the phase be an analytic function in a neighbourhood of its critical point. Let us suppose that the principal part of the Taylor series of the phase at this critical point is $\mathbb{R}$-nondegenerate, and that the Newton polyhedron of this series is remote. Then the oscillation index of the critical point of the phase is equal to the remoteness of the Newton polyhedron.

Example 1. The degenerate critical point at the origin of the phase $x_1^{k_1} + x_2^{k_2}$ satisfies the conditions of the theorem. Its oscillation index is equal to $-1/k_1 - 1/k_2$.

Example 2. The critical point at the origin of the phase f of the example on page 189 satisfies the conditions of the theorem. Its oscillation index is equal to $-1/2$.

The following assertions supplement the theorem.

(i) If the conditions of the theorem are satisfied then the multiplicity of the oscillation index of the critical point of the phase is equal to the multiplicity of the remoteness of the Newton polyhedron of the Taylor series of the phase at this critical point.

(ii) If the principal part of the Taylor series of the critical point of the phase is $\mathbb{R}$-nondegenerate then the oscillation index of the critical point is not more than the remoteness of the Newton polyhedron of the Taylor series.

(iii) Let us consider the critical point at the origin of the phase

$$x_4^2 + x_1^9 + x_2^9 + x_3^9 + (x_4 - (x_1^2 + x_1^4 + x_2^2 + x_3^2))x_5.$$

Then the principal part of the Taylor series of the critical point is $\mathbb{R}$-non-degenerate; the remoteness of the Newton polyhedron of the Taylor series is less than -1; the oscillation index of the critical point is less than the remoteness of the Newton polyhedron.

(iv) If the Newton polyhedron of the Taylor series of the critical point of the phase is remote then the oscillation index of the critical point of the phase is not less than the remoteness of the polyhedron.

(v) If the Newton polyhedron of the Taylor series of the critical point of the phase is remote, and this critical point has finite multiplicity, then the coefficient of the principal term of the asymptotic series of the oscillatory integral (the coefficient $a_{K,\beta}$ of the series (2) on page 181) is equal to the value of the amplitude at the critical point of the phase, multiplied by a non-zero constant, depending only on the phase.

(vi) Let the principal part of the Taylor series of the critical point of the phase be $\mathbb{R}$-nondegenerate and the remoteness of the Newton polyhedron be equal to -1. Then the oscillation index of the critical point of the phase is equal to -1 if at least one of the following two conditions is satisfied:

– the open face which contains the centre of the boundary of the Newton polyhedron has dimension less than $n-1$;

– the closure γ of the open face which contains the centre of the boundary of the Newton polyhedron is compact and the γ-part of the Taylor series has a zero in $(\mathbb{R} \setminus 0)^n$.

(vii) If the hypotheses of supplement (vi) are satisfied then the multiplicity of the oscillation index of a critical point of the phase is equal to the multiplicity of the remoteness of the Newton polyhedron of the Taylor series or to one less than the multiplicity of the remoteness.

Theorem 6.4 and supplements (i), (ii), (iv), (v), (vi), (vii) will be proved in Chapter 8, supplement (iii) will be proved in Chapter 9.

According to Theorem 6.4, the oscillation index of a critical point of the phase can be expressed in terms of the Newton polyhedron of its Taylor series if the principal part of the Taylor series is $\mathbb{R}$-nondegenerate and the Newton polyhedron of the Taylor series is remote. A system of coordinates in which the Taylor series possesses these properties does not always exist. For example, it does not do so for the critical point at the origin of the function g of the example on page 189. None the less, for critical points of functions of two variables conditions on the existence of the indicated system of coordinates can be omitted.

Let the phase be an analytic function in a neighbourhood of its critical point.

The *remoteness of the critical point* of the phase is the upper bound of the remotenesses of the Newton polyhedra of the Taylor series of the phase in all systems of local analytic coordinates with origin at the critical point.

A local analytic coordinate system with origin at the critical point of the phase is called *adapted* to the critical point if the remoteness of the Newton polyhedron of the Taylor series of the phase in this system of coordinates has the greatest possible value, equal to the remoteness of the critical point.

Theorem 6.5 (see [358]). Let the phase be an analytic function of two variables in a neighbourhood of its critical point. Then the oscillation index of the critical point is equal to its remoteness.

The following assertions supplement the theorem.

(i) Under the conditions of the theorem there exists a system of coordinates, adapted to the critical point.

(ii) If the critical point of a two-dimensional phase has finite multiplicity then the coefficient of the principal term of the asymptotic series of the oscillatory integral (the coefficient $a_{K,\beta}$ of series (2) on page 181) is equal to the value of the amplitude at the critical point of the phase, multiplied by a non-zero constant depending only on the phase.

(iii) The oscillation index is greater than the remoteness for the critical point at the origin of the phase

$$(-x_1^2 + x_1^4 + x_2^2 + x_3^2)^2 + x_1^9 + x_2^9 + x_3^9$$

of three variables.

Theorem 6.5 and supplements (i), (ii) will be proved in § 8.4, supplement (iii) will be proved in Chapter 9 (see also [358]).

In [358] there is described an algorithm for the creation of a system of coordinates adapted to the critical point of a phase of two variables. The following lemma is useful for the creation of adapted coordinates.

Lemma 6.2 (see [358]). A local system of coordinates with origin at the critical point of a phase of two variables is adapted to the critical point if at least one of the following conditions is satisfied.

(i) The centre of the boundary of the Newton polyhedron of the Taylor series of the phase with respect to this system of coordinates is a vertex of the polygon.

(ii) The centre of the boundary of the Newton polyhedron lies on a non-compact edge of the polygon.

(iii) The centre of the boundary of the Newton polyhedron lies on a compact edge of the polygon and neither the tangent nor the cotangent of the angle formed by the edge and the first coordinate axis in $\mathbb{R}^2$ is equal to an integer (we remark that interchanging the axes changes tangents to cotangents and does not influence the truth of the formulated condition).

Example. Let us consider the functions g and h of the example on page 189. The system of coordinates x_1, x_2 is adapted to the critical points of these functions (because of sections (iii) and (i) of Lemma 6.2 respectively). By Theorem 6.5 the oscillation indices are equal, respectively, to $-5/12$, $-1/2$.

Remark 1. The assertion of the fact that the coefficient of the leading term of the asymptotic series is proportional to the value of the amplitude at a critical point (see the supplements of Theorems 6.4, 6.5), can be used to solve the following problem of integral geometry, posed by I. M. Gelfand in Amsterdam in 1954.

Problem. Let ϕ be a smooth function with support concentrated in a small neighbourhood of the critical point of a smooth function f. Knowing the integrals of the function ϕ over every level hypersurface of the function f find the value of the function ϕ at the indicated critical point.

For the solution of this problem it is sufficient to take the coefficient of the leading term of the asymptotic series of the oscillatory integral with phase f and amplitude ϕ if the critical point of the function f has finite multiplicity and its remoteness is greater than -1. For more details see § 7.3.

Remark 2. The singular index is non-negative for critical points of phases of one and two variables (see Theorems 6.4, 6.5), for critical points satisfying the conditions of supplement (iv) of Theorem 6.4. Using supplement (iv) of Theorem 6.4, it can be proved that the singular index is non-negative for critical points of phases of three variables. Apparently the singular index is always non-negative. This means that the order of complex shortwave oscillation at a caustic point is, apparently, always greater than the order of complex shortwave oscillation at a non-caustic point (see §§ 6.1.1, 6.1.4). In particular a light caustic, apparently, is always distinguished by its brightness.

In Chapter 13 we shall define the complex singular index of a critical point of a holomorphic function. The complex singular index is always non-negative, see § 13.3. The proof of this fact uses the connection between asymptotic integrals and mixed Hodge structures.

6.3   The resolution of a singularity

The proof of Theorems 6.3–6.5 uses the resolution of the singularity of the critical point of the phase.

Let us consider a function $f : \mathbb{R}^n \to \mathbb{R}$, analytic in a neighbourhood of its critical point x. Let us suppose that the value of the function at this point is equal to 0. The *resolution of the singularity of the critical point* is an n-dimensional analytic manifold Y and an analytic map

$$\pi : Y \to \mathbb{R}^n$$

possessing the following properties.
1. At each point of the preimage of the critical point x there are local coordinates with respect to which the function $f \circ \pi$ and the Jacobian map of π are equal to monomials modulo multiplication by a function which does not take the value zero.

2. In a small neighbourhood of the critical point x there is a proper analytic subset outside which in this neighbourhood the map π has an analytic inverse.
3. The preimage of any compact subset of a small neighbourhood of the point x is compact.

Remark 1. In particular, from the first condition it follows that in a neighbourhood of the preimage of the point x the zero level hypersurface of the function $f \circ \pi$ is locally structured as the union of coordinate hyperplanes.

Remark 2. Sometimes the requirements of the map π are strengthened, property 2 being replaced by property $2'$ or even by property $2''$ and property 4 being added.
$2'$. In a small neighbourhood of the point x the map π is invertible outside the zero level hypersurface of the function f.
$2''$. In a small neighbourhood of the point x the map π is invertible outside the critical set of the function f.
4. In a small neighbourhood of the preimage of the point x the zero level hypersurface of the function $f \circ \pi$ is the union of non-singular $(n-1)$-dimensional submanifolds.

Theorem 6.6 (Hironaka, see [158, 32]). There exists a resolution of the singularity (with properties 1, $2''$, 3, 4) of the critical point of an analytic function.

This theorem was formulated in [32]. It is a special case of a general theorem of Hironaka on resolutions of singularities [158].

Remark 3. The concept of resolution of singularities has a natural complex analogue. We consider the function $f: \mathbb{C}^n \to \mathbb{C}$, analytic in a neighbourhood of its critical point x. The resolution is an n-dimensional complex analytic manifold Y and an analytic map

$$\pi: Y \to \mathbb{C}^n$$

which satisfies the properties formulated above. Also in this case, the theorem of Hironaka is true.

The theorem of Hironaka leads the investigation of oscillatory integrals with analytic phase to the investigation of sums of oscillatory integrals the phase of each of which is a monomial. It is necessary for this to make a change of variables in the integral with the help of the map π. The oscillatory integrals with monomial phase are called *elementary*. The elementary integrals are studied in Chapter 7. For these it is not hard to find the oscillation index, its multiplicity and its index set. Therefore in investigating oscillatory integrals with analytic phase, it is important to know the resolution of the singularity of the phase, to be able to see how the asymptotic series of the integral under investigation is added up from asymptotic series of elementary integrals, to see whether or not the leading term cancels out. The result of such analysis is an expression for the oscillation index and analogous characteristics in terms of the resolution of the singularity of the phase (Theorem 7.5). Theorems 6.4 and 6.5 are reformulations of the properties of resolutions of singularities, which arise in this analysis, in terms of Newton polyhedra.

6.4   Asymptotics of volumes

The asymptotics of oscillatory integrals are closely connected with the asymptotics of the volume of the set of points at which the phase takes values less than a given number, as this number changes and tends to the critical value of the phase.

6.4.1 The Gelfand-Leray form

For the study of oscillatory integrals it is very useful to know the following method, which reduces many-dimensional oscillatory integrals to one-dimensional ones. The method consists of applications of Fubini's Theorem. Namely, let us consider the oscillatory integral:

$$\int_{\mathbb{R}^n} e^{i\tau f(x)} \varphi(x) dx_1 \ldots dx_n.$$

Using Fubini's Theorem, we reduce the integral to another in which we first integrate along a level hypersurface of the phase, and then with respect to the remaining variable, the value of the phase. To do this, we change to new variables, one of which is the phase.

We make two remarks. Firstly the phase can be taken as a variable only away from its critical points. Therefore we cut out of consideration the union of the critical level hypersurfaces of the phase. The union of these hypersurfaces has zero measure and has no effect on the integral. Secondly, for the integration along the level hypersurfaces we do not need to know each of the remaining new variables. It is sufficient to know on the level hypersurfaces the density $(n-1)$-form, which after multiplication by the differential of the phase becomes the volume form of the space. This density form is called the *Gelfand-Leray form* and is denoted by

$$dx_1 \wedge \ldots \wedge dx_n/df.$$

And so the oscillatory integral is reduced to the form

$$\int_{-\infty}^{\infty} e^{i\tau t} \left(\int_{f=t} \varphi \, dx_1 \wedge \ldots \wedge dx_n/df \right) dt.$$

In this representation the oscillatory integral is the Fourier transform of the function given by the inside integral. The function of one variable, defined by the inside integral, is called the *Gelfand-Leray function*.

The Gelfand-Leray function is smooth outside the critical values of the phase. In a neighbourhood of the critical value of the phase the Gelfand-Leray function can be expanded in an asymptotic series of the form

$$\sum_{\alpha} \sum_{k=0}^{n-1} a_{k,\alpha}(t-t_0)^{\alpha}(\ln(t-t_0))^k.$$

Knowing the asymptotic series of the Gelfand-Leray function one can determine the asymptotic series of the oscillatory integral and conversely the asymptotics of the oscillatory integral give information about the asymptotics of the Gelfand-Leray function. These properties of the Gelfand-Leray function will be proved in Chapter 7.

6.4.2 The volume of an infralevel set

Let us suppose that the phase has an isolated minimum and that the minimal value of the phase is equal to zero. Let us suppose also that the amplitude in a neighbourhood of the minimum point is identically equal to 1. Let us denote by J the Gelfand-Leray function and let us consider the new function

$$V(t) = \int_0^t J(s) \, ds.$$

It is clear that for negative values of the argument this function is equal to zero, and for small positive values of the argument this function is equal to the volume of the set of points in which the phase takes values less than the given one (an "infralevel set"). In this way the asymptotics of the volume function of the infralevel set determine the asymptotics of the oscillatory integral in the case when the phase has an isolated minimum and the amplitude is equal to a constant in a neighbourhood of the minimum point of the phase.

Let us find the rate of convergence to zero of the volume of the infralevel set for simple isolated minimum points. The classification of critical minimum points, not removed by a small perturbation from the family of functions depending on at most 16 parameters, was produced by V. A. Vasilev in [386], and the asymptotics of the volume of infralevel sets were evaluated in the same place. According to this classification the only minimum points not removed by small perturbations from the family of functions depending on at most 5 parameters are minimum points in which the function can be reduced, by a diffeomorphism of the space, to the form

$$A_s : x_1^{s+1} + x_2^2 + \ldots + x_n^2,$$

where $s = 1, 3, 5$. For small positive t the leading term of the asymptotics of the volume of the infralevel set has the form

$$\mathrm{const} \cdot t^{-\beta + n/2}$$

where β is equal, respectively, to 0, 1/4, 1/3.

We formulate a general theorem on the evaluation of the rate at which the volume of the infralevel set tends to zero.

Theorem 6.7 (cf. [386]). Let us suppose that an analytic function has an isolated minimum and that the minimal value is equal to zero. Then as $t \to +0$ the volume function V of the infralevel set can be expanded in the asymptotic series

$$\sum_\alpha \sum_{k=0}^{n-1} a_{k,\alpha} t^\alpha (\ln t)^k.$$

Here the variable α runs through a finite number of arithmetic progressions, consisting of positive rational numbers. If in addition it is known that the Taylor series of the function at the minimum point has a $\mathbb{R}$-nondegenerate principal part, then the index α of the maximal term of the asymptotic series is equal to minus the remoteness of the Newton polyhedron of the Taylor series.

The theorem will be proved in § 8.3.3.

Remark 1. If the volume form of the space is changed (that is if it is multiplied by a positive function) then the order of the leading term of the above asymptotic series does not change.

Remark 2. The above asymptotic series converges for small positive t.

6.4.3 The area of a level surface

In § 8.3.3. we shall formulate a theorem on calculating the asymptotic of area of a compact level surface as the level tends to the critical one.

6.4.4 The set of points with small gradient

Yet one more characteristic of critical points, similar to those considered above, is the rate at which the volume of those points at which the length of the gradient is less than a given number tends to zero as the given number tends to zero.

Let us suppose that in the space there is given a Riemannian metric. This metric (with the help of the matrix inverse to the matrix of the metric) defines a metric on the cotangent bundle of the space. In this metric we calculate the square of the length of the gradient

$$df = (\partial f/\partial x_1, \ldots, \partial f/\partial x_n)$$

of the function f under consideration. In a neighbourhood of the chosen critical point of the function we consider for each small positive t the volume $V(t)$ of the set of those points in the neighbourhood for which the square of the length of the gradient is less than t. We shall be interested in the asymptotic of the volume as $t \to +0$. Since all the metrics in a neighbourhood of the point are mutually bounded, the order of the leading term of the asymptotic series will not depend on the choice of metric.

Example. For critical points of types

$$A_\mu, D_4, D_\mu(\mu > 4), E_6, E_7, E_8,$$

the leading term of the asymptotic series of the function V has the form, as $t \to +0$,

$$\text{const} \cdot t^{-\alpha + n/2} (\ln t)^k,$$

where (α, k) equal, respectively, $((\mu - 1)/2\mu, 0), (1/2, 0), (1/2, 1), (7/12, 0), (5/8, 0),$ $(5/8, 0)$.

To calculate the asymptotics of the volume of the set of points with small gradient we can use Theorem 6.7, applied to the function (df, df), and we can also use Theorems 6.4 and 6.5.

6.5   Uniform estimates

As well as the asymptotics of individual oscillatory integrals it is often useful to have uniform estimates of oscillatory integrals, depending on additional parameters.

We define the concept of uniform estimate and the uniform oscillation index. Let $f: \mathbb{R}^n \to \mathbb{R}$ be a smooth function. A *deformation* of it is any smooth function

$$F: \mathbb{R}^n \times \mathbb{R}^l \to \mathbb{R},$$

which is equal to the function f when the second argument takes the value zero.

Definition. At the critical point x^0 of the phase f we get a *uniform estimate with index α* if for any deformation F of the phase f there is a neighbourhood in $\mathbb{R}^n \times \mathbb{R}^l$ of the point $x^0 \times 0$ such that for any smooth function ϕ with support in this neighbourhood and for any positive ε there exists a number $C(\varepsilon, \phi)$, for which for all positive τ

$$\left| \int_{\mathbb{R}^n} e^{i\tau F(x,y)} \varphi(x,y) dx_1 \ldots dx_n \right| < C(\varepsilon, \varphi) \tau^{\alpha + \varepsilon}.$$

The lower bound of such numbers α is called the *uniform oscillation index* of the phase at the critical point.

It is clear that the uniform oscillation index is not less than the individual one.

There arises a natural conjecture, formulated by V. I. Arnold in [12, 13, 14], that the uniform oscillation index is equal to the individual index. That is that an oscillatory integral permits a uniform estimate with respect to additional parameters in terms of quantities proportional to the value of the integral for the initial values of the additional parameters.

For a justification of this conjecture it is necessary that the individual oscillation index is upper semicontinuous for a continuous deformation of critical point. Namely it is necessary that the oscillation index of a complex critical point is not less than the oscillation index of a simpler critical point, obtained by decomposing the complex one. An analysis of the tables of singular indices and the known adjacencies of the critical points classified in Part II of Volume 1, shows that such semicontinuity takes place for the critical points classified in Part II of Volume 1.

Theorem 6.8. The uniform oscillation index is equal to the individual one for critical points of functions of one variable (I. M. Vinogradov [391]), for simple critical points (J. J. Duistermaat [100]), for parabolic critical points (Y. Colin de Verdier [80]), for hyperbolic critical points of the series $T_{p,q,r}$ (V. N. Karpushkin [178]), for critical points of functions of two variables (V. N. Karpushkin [176]).

Corollary. For critical points, occurring unavoidably in a family of phases in general position, depending on not more than seven parameters, the uniform oscillation index is equal to the individual one.

According to Theorem 6.8, as we move on a caustic, corresponding to one of the critical points enumerated in the theorem, the intensity of the shortwave oscillation at the limiting point is not less than the intensity of the radiation at a point near the limiting point. Surprisingly, this phenomenon does not take place for all caustics. Namely there are examples of degenerate critical points of phases for which the uniform oscillation index is greater than the individual one (see [358]).

An exposition of these examples will be given in Chapter 9. The critical points of the constructed examples are very degenerate, the codimension of such critical points being of order 80 or more (that is these critical points disappear under small perturbations from a family of functions with less than this number of parameters).

According to the constructed examples there exists a critical point and a deformation of it which has the following property. The oscillation index of the critical point of the deformation for the chosen value of the parameter is less than the oscillation index of the critical point of the deformation for a general value of the parameter, that is the modulus of the oscillatory integral of the deformation for the chosen value of the parameter is substantially less than the modulus of the integral for a general value of the parameters.

It would be interesting to elucidate *whether it is possible to observe the indicated phenomenon physically in the form of a subset of a caustic which is dark in comparison with its neighbourhood.* As we have already mentioned such a phenomenon cannot be observed on caustics in general position in small-dimensional spaces (Theorem 6.8 and its corollary).

Remark. The proof of V. N. Karpushkin of the equality of the uniform and individual oscillation indices for critical points of functions of two variables is based on Theorem 6.5. As we have already mentioned, the equality of the uniform and individual oscillation indices is possible only under the condition of upper semicontinuity of the individual oscillation index under deformations of the critical point. According to Theorem 6.5 this property of semicontinuity can be reformulated for functions of two variables as follows: let us be given an arbitrary family of functions of two variables, depending on a parameter and having a critical point at the origin, then the remoteness of this critical point depends in an upper-semicontinuous way on the parameter. It is very likely that this result is correct for functions of any number of variables. An interesting problem in this case is to express the uniform oscillation index in terms of the other characteristics of the critical point (the Newton polyhedron, the resolution of the singularity etc.). It could be that the uniform oscillation index can be expressed in terms of the remoteness of critical points, stably equivalent to the given one. Another likely candidate for expressing the uniform oscillation index is the complex oscillation index, to be defined in Chapter 13. The complex oscillation index is defined for critical points of holomorphic functions. It is the complex analogue of the oscillation index. B. Malgrange in [239] formulated a conjecture on semicontinuity of the complex oscillation index under defor-mation of the critical point. The complex oscillation index is one of the spectral numbers of a critical point of a holomorphic function (the spectrum will be defined in Chapter 13). In § 14.3 is formulated a conjecture of V. I. Arnold on the semicontinuity of spectra under deformation of the critical point. The conjecture is proven in [371–375, 345].

In this part we study the asymptotics of individual oscillatory integrals. In this section we discussed uniform estimates of them. There is yet another approach to the estimation of integrals – this is the estimate in mean. We formulate the corresponding results.

Let us consider an oscillatory integral depending on additional parameters,

$$I(\tau, y) = \int_{\mathbb{R}^n} e^{i\tau F(x, y)} \varphi(x, y) \, dx_1, \ldots, dx_n.$$

Let us denote by Σ the set of critical points of the phase, that is

$$\Sigma = \{(x, y) | \partial F / \partial x_j(x, y) = 0, j = 1, \ldots, n\}.$$

Theorem 6.9 (see [100]). Let us suppose that Σ is a submanifold, that is that the differentials $d(\partial F / \partial x_j)$, $j = 1, \ldots, n$, are linearly independent at each point of the set Σ. Let us suppose that the support of the amplitude is concentrated in a small neighbourhood of one of the points of the set Σ. Then as $\tau \to \infty$ we get the asymptotic expansion

$$\int |I(\tau, y)|^2 dy \approx \sum_{l=n}^{\infty} a_l(\varphi) \tau^{-l},$$

where the numerical coefficients a_l are generalised functions of the amplitude with support in Σ. In particular the leading coefficient a_n is proportional to the integral of the square of the modulus of the amplitude over the critical set Σ.

This result corresponds to results on the unitariness of the canonical operator of Maslov (see [244–246, 144]) and means that, for individual values of the additional parameters, the asymptotic behaviour of the integral may have complex character but the integral of the square of the modulus of the oscillatory integral behaves as if the phase had only non-degenerate critical points in the variables of integration.

The proof of Theorem 6.9 is based on the fact that the integral of the square of the modulus is an oscillatory integral. Its phase

$$F(x, y) - F(z, y)$$

has critical points on the set

$$\{(x, y, z) | x = z, (x, y) \in \Sigma\}$$

(if x, z are sufficiently close), and the critical points are non-degenerate in a transversal direction to this set.

6.6 The number of integral points in a family of homothetic regions

Let us consider in the space $\mathbb{R}^n$ a bounded region D with smooth boundary. We shall estimate the difference between the volume of the region, stretched out by a factor of λ, and the number $N(\lambda)$ of points with integer coordinates lying in the stretched out region, that is the difference

$$R(\lambda) = \lambda^n V(D) - N(\lambda).$$

The study of this question is motivated by the following considerations (see [80]):

1) The case in which D is an ellipsoid is considered in number theory in connection with the study of the arithmetical properties of quadratic forms (see [37, 169, 175, 218, 240, 392]).

2) If the region D is defined by the condition $\{f \leqslant 1\}$, where $f : \mathbb{R}^n \setminus 0 \rightarrow \mathbb{R}_+$ is a smooth homogeneous function (say a homogeneous polynomial), then the function $N(\lambda)$ is interpreted as the spectral function of the pseudodifferential operator P on the torus $\mathbb{R}^n/(2\pi\mathbb{Z})^n$, given by its spectral decomposition

$$P(\exp(i\langle k, x\rangle)) = f(k)\exp(i\langle k, x\rangle)$$

3) The following problem, arising in numerical integration, is studied in an analogous fashion: let f be a smooth function and

$$N_f(\lambda) = \sum_{x \in \lambda D \cap \mathbb{Z}^n} f(x/\lambda).$$

It is required to estimate the difference

$$R_f(\lambda) = \lambda^n \int_D f \, dx - N_f(\lambda).$$

For the difference $R(\lambda)$ one usually obtains an estimate by the degree of the parameter λ: $R(\lambda) = O(\lambda^\beta)$.

A trivial estimate for any region is obtained if we take $\beta = n - 1$. Indeed the difference $R(\lambda)$ is less than the volume of the neighbourhood of width $2\sqrt{n}$ of the boundary of the blown-up region.

For a ball of radius 1 with centre at the origin $\beta \geqslant n - 2$. More precisely there are arbitrarily large λ for which $\sim \lambda^{n-2}$ points with integer coordinates lie on the sphere $\lambda \, \partial D$. Indeed, let us consider the integer points lying between the spheres $(\lambda + 1)\partial D$ and $\lambda \, \partial D$. Their number is proportional to the volume, that is proportional to

$$(\lambda + 1)^n - \lambda^n \sim \lambda^{n-1}.$$

Between these spheres there are approximately λ spheres with centre at the origin for which the square of the radius is an integer. Therefore λ^{n-1} integer points lie on λ spheres, and so there is a sphere on which lie not less than $\sim \lambda^{n-2}$ points.

The best (least) number β depends on the form of the region. The most studied case has been a region in the plane.

Theorem 6.10 (see [297, 299, 80]). Let us suppose that $n=2$. Let us denote by l the maximal order of vanishing of the curvature of the boundary of the region. Then if $l=0$ or 1 (this is the situation of general position), β can be taken as $2/3$. If $l \geqslant 1$, then β can be taken as $1-1/(l+2)$. Furthermore if $l \geqslant 2$, then, generally speaking, β cannot be taken smaller (for example for $D=\{x^{2k}+y^{2k} \leqslant 1\}$).

In the many-dimensional situation the only cases studied have been that of a strictly convex region and for $n \leqslant 7$ the case of a region with boundary lying in general position.

Theorem 6.11 (see [297, 298]). If the region $d \subset \mathbb{R}^n$ is convex and the second fundamental form of its boundary is non-degenerate then we can take β as $n-2+2/(n+1)$.

Theorem 6.12 (see [80]). Let us suppose that $n \leqslant 7$. Let X be a compact oriented smooth manifold of dimension $n-1$. Then there exists an open, everywhere dense, subset in the space of all embeddings of the manifold X in $\mathbb{R}^n$, which possesses the property: if the embedding belongs to the subset and the image of the embedding bounds a region in $\mathbb{R}^n$, then for this region as the number β we can take $n-2+2/(n-1)$.

As the example of a sphere with centre at the origin shows, the estimate with $\beta=n-2+2/(n-1)$ cannot, generally speaking, be substantially improved.

6.6.1 The Poisson summation formula

We explain how the estimate of the number of integer points is connected with oscillatory integrals.

The number of points on the integer lattice in the blown-up region λD is equal to the number of points on the condensed lattice $\frac{1}{\lambda} \mathbb{Z}^n$ in the original region. We shall suppose for simplicity that λ is a natural number and that the region D lies in a standard n-dimensional cube with edge length 1. In this case we can consider the region as a region on the torus $T^n = \mathbb{R}^n/\mathbb{Z}^n$ and count up on the torus the points on the projection of the lattice $\frac{1}{\lambda}\mathbb{Z}^n$ falling in D. We shall denote by χ the characteristic function of the region D, that is the function equal to 1 on D and equal to 0 outside D. Then

$$R(\lambda) = \lambda^n \int_{T^n} \chi dx - \sum_{0 \leqslant x_1, \ldots, x_n \leqslant \lambda - 1} \chi(x/\lambda).$$

We expand the characteristic function in a Fourier series

$$\chi(x) = \sum_{k \in \mathbb{Z}^n} \hat{\chi}(k) \exp(2\pi i \langle k, x \rangle),$$

and consider the analogous difference for each term of the series:

$$\lambda^n \int \exp(2\pi i \langle k, x \rangle) dx - \sum_x \exp(2\pi i \langle k, x \rangle/\lambda).$$

For $k = 0$, the difference is equal to zero. If $k \neq 0$, then the first term of this difference is equal to zero and it remains to calculate the second term. The second term is the product of sums of n geometric progressions. Summing these we find that the second term of the difference is equal to zero if at least one of the coordinates of the vector k is not divisible by λ. If all the coordinates of the vector k are divisible by λ, then the sum is equal to $-\lambda^n$. This argument shows that

$$R(\lambda) = -\lambda^n \sum_{k \in \lambda\mathbb{Z}^n \setminus 0} \hat{\chi}(k) = -\lambda^n \sum_{k \in \mathbb{Z}^n \setminus 0} \hat{\chi}(\lambda k). \tag{3}$$

This formula is called the *Poisson summation* formula. Unfortunately for characteristic functions it is not correct: in the derivation of the formula we transposed the order of summation with respect to k and summation with respect to the points of the condensed lattice. For the Poisson formula to be correct it is sufficient that the Fourier series be bounded by an absolutely convergent series with constant coefficients. In particular the Poisson formula is true for any smooth finite function χ on $\mathbb{R}^n$.

To study the difference $R(\lambda)$ we first smooth out the characteristic function, then apply the Poisson formula and study its right-hand part (see, for example, [80]). To smooth the characteristic function we convolute with a standard function. The Fourier transform of the convolution is equal to the product of the Fourier transforms of the characteristic function and the standard one. Therefore for studying the right-hand part of the Poisson formula, applied to the smoothed characteristic function it is important to know how the Fourier coefficient $\hat{\chi}(\lambda k)$ of the characteristic function behaves as $\lambda \to \infty$. The Fourier coefficients are oscillatory integrals.

6.6.2 The Fourier transform of a characteristic function

The Fourier coefficient

$$\hat{\chi}(k) = \int_D \exp(-2\pi i \langle k, x \rangle) dx$$

is an oscillatory integral, in which the rôle of the larger parameter is played by the length of the vector k, and the rôle of the phase is played by the function $-\langle \alpha(k), x \rangle$, where $\alpha(k) = k/\|k\|$ is the corresponding vector of unit length. This formula is transformed by Stokes' formula into an oscillatory integral on the boundary of the region. The phase of the new integral, as before, is the function $-\langle \alpha(k), x \rangle$. Consequently the magnitude of the Fourier coefficient $\hat{\chi}(k)$ when the vector k has large length is determined by the critical points of the restriction to the boundary of the linear function $\langle \alpha(k), x \rangle$. For example, if the region is convex and the second fundamental form of the boundary is non-degenerate, then all the critical points of the restriction are non-degenerate and

$$\hat{\chi}(k) \sim \|k\|^{-(n+1)/2}$$

(Theorem 6.2).

Let us analyse in more detail the case of a region in the plane. The critical points of the restriction of the function $\langle \alpha(k), x \rangle$ to the curve ∂D are those points at which the normal vector to the curve is equal to $\pm \alpha(k)$. If at such a point the curvature of the curve is different from zero, then the critical point is non-degenerate and its contribution to the Fourier coefficient has order $\|k\|^{-3/2}$. If at a point of the boundary with normal $\pm \alpha(k)$ the multiplicity of zero of the curvature is equal to l, then the critical point has type A_{l+1} and in this case its contribution to the Fourier coefficient has order

$$\|k\|^{-1-1/(l+2)}.$$

The normal at a point on the boundary at which the curvature is zero can have a gradient with irrational tangent. Such a point of the boundary will not be a critical point for the function $\langle\alpha(k), x\rangle$. The contribution of such a point to the sum (3) is determined by the rate at which the tangent of the gradient of its normal can be approximated by rational numbers. If the tangent of the gradient has a good approximation by a rational number with relatively small numerator and denominator, then the critical points of the restriction of the linear function $\langle\alpha(k), x\rangle$ with vector k of relatively short length will be almost degenerate and will give a large contribution to the sum (3). A curve in general position in the plane has as degeneracies only points of inflection, that is for a curve in general position the multiplicity of zero of the curvature function is not more than 1. Therefore for a curve in general position the critical points of the restriction to the boundary of a linear function are either non-degenerate, or have type A_2. Consequently in general position we can estimate that $R(\lambda) \sim \lambda^{2/3}$. These arguments explain Theorem 6.10.

6.6.3 The estimate averaged over rotations

The principal contribution to the Fourier coefficients of the characteristic function of a region is given by neighbourhoods of those points of the boundary at which the normal has rational direction and the curvature is zero. B. Randol had the idea that after rotating the region such points, in general, would not exist, and that the estimate, averaged over rotations could be better than an individual estimate.

Theorem 6.13 (see [299, 300, 366, 367]). Let us denote by ds the Haar measure on the special orthogonal group SO_n. Let us denote by $R(\lambda, s)$ the difference, corresponding to the region blown up by a factor of λ and then rotated by the transformation $s \in SO_n$. Then

$$\int_{SO_n} |R(\lambda, s)| ds = O(\lambda^{n-2+2/(n+1)}).$$

Theorem 6.14 (see [299, 300, 366, 367]). Let us denote by G the group of all motions of the form st, where $s \in SO_n$ and t is parallel translation of the space $\mathbb{R}^n$. Let $I \subset G$ be the subgroup of all parallel translations by vectors with integer coordinates. Let us denote by H the factor group G/I. H is topologically equivalent to $SO_n \times T^n$, where $T^n = \mathbb{R}^n/\mathbb{Z}^n$ is the n-dimensional torus. Let us denote by dh the Haar measure on H. Then

$$\left(\int_H |R(\lambda,h)|^2\,dh\right)^{1/2} = O(\lambda^{(n-1)/2}).$$

An analogue of Theorem 6.13 for polyhedra in $\mathbb{R}^n$ was proved by M. Tarnopolska-Weiss.

Theorem 6.15 (see [347]). Let D be a polyhedron in $\mathbb{R}^n$ containing the origin and possessing the property: the prolongations of its faces do not pass through the origin. Then

$$\int_{SO_n} |R(\lambda,s)|\,ds = O((\ln\lambda)^{2+\delta}).$$

The proofs of Theorems 6.13 and 6.14 are based on an estimate of the square of the modulus of the Fourier coefficients of the characteristic function of the region.

Theorem 6.16 ([366, 367]). As $\|k\| \to \infty$ the following estimate is correct

$$\int_{SO_n} |\hat{\chi}(k,s)|^2\,ds = O(\|k\|^{-(n+1)}). \tag{4}$$

If the boundary of the region depends in an infinitely differentiable manner on the additional parameters, then this estimate is uniform with respect to the additional parameters under the condition that the parameters differ little from their initial values.

The proof of this theorem is analogous to the proof of Theorem 6.9.

Let us deduce Theorem 6.14 from Theorem 6.16. Each element $h \in H$ has a unique representation in the form st, where $s \in SO_n$, $t \in T^n$. Let us fix s. Then $R(\lambda,h)$ is a function on T^n. Let us expand it in a Fourier series:

$$R(\lambda,st) = \sum_k a(\lambda,s,k)e^{2\pi i\langle k,t\rangle}.$$

A simple, direct calculation shows that

$$a(\lambda,s,0) = 0, \quad a(\lambda,s,k) = (-1)^{n-1}\hat{\chi}(-\lambda k,s)\lambda^n.$$

Using Parseval's equality we obtain

$$\int\limits_{SO_n} \left(\int\limits_{T^n} |R(\lambda, st)|^2 \, dt \right) ds = \int\limits_{SO_n} \sum_{k \in \mathbb{Z}^n \backslash 0} \lambda^{2n} |\hat{\chi}(-\lambda k, s)|^2 \, ds.$$

Then Theorem 6.16 follows from (4).

Remark. For a region, the boundary of which has a non-degenerate second fundamental form, the estimate $\hat{\chi}(k) \sim \|k\|^{-(n+1)/2}$ is true (see § 6.7.2). Therefore for such a region

$$\left(\int\limits_{T^n} |R(\lambda, t)|^2 \, dt \right)^{1/2} = O(\lambda^{(n-1)/2}).$$

6.6.4. The proof of Theorem 6.12 is based on two interesting results on uniform estimates of oscillatory integrals, depending on additional parameters. In these results it is assumed that all critical points of the phase are either simple or parabolic.

Theorem 6.17 (see [80]). Let us consider the oscillatory integral

$$I(\tau, y) = \int\limits_{\mathbb{R}^n} e^{i\tau F(x, y)} \varphi(x, y) \, dx.$$

Let us suppose that for each value of the additional parameters all the critical points of the phase of this integral are either simple or parabolic. Then we get the inequality

$$|I(\tau, y)| \leqslant \text{const} \cdot \tau^{-n/2} \sum_{(x, y) \in \Sigma \cap \text{supp} \varphi} |\det F''_{xx}(x, y)|^{-1/2},$$

where we denote by Σ the set of all critical points of the phase with respect to the variables of integration, and by F''_{xx} the matrix of second derivatives of the phase with respect to the variables of integration.

In order to formulate the following theorem we make several remarks. Let $F: \mathbb{R}^n \times \mathbb{R}^\mu \to \mathbb{R}$ be a minimal versal deformation of a simple or parabolic critical point (μ is the multiplicity of the critical point). Let us denote by W_r the subset of the base of the deformation consisting of all points y for which the function $F(\cdot, y)$ has a critical point of multiplicity r. The set W_r has codimension $r-1$ with the exception of the case when the initial critical point is parabolic. In this case the dimension of the set W_μ is equal to 2. We shall denote by $\beta(\sigma)$ the oscillation index of the critical point of type σ.

Theorem 6.18 (see [80]). Let us suppose that the phase of the oscillatory integral $I(\tau, y)$ is a minimal versal deformation of a simple or parabolic critical point. If the support of the amplitude is concentrated in a sufficiently small neighbourhood of the initial critical point, then the oscillatory integral permits the estimate:

$$|I(\tau, y)| \leqslant \varphi_p(y)\tau^{\beta_p},$$

where $\beta_p = \max\{\beta(\sigma)|\sigma$ is a critical point of multiplicity p adjacent to the initial critical point$\}$ and

$$\varphi_p(y) \leqslant \text{const} \cdot d(y, W_{p+1})^{-\alpha^p_{p+1}} \ldots d(y, W_\mu)^{-\alpha^p_\mu}.$$

In this formula d is the distance with respect to an arbitrary Riemannian metric on the base of the versal deformation, the numbers $\alpha^p_{p+1} \ldots, \alpha^p_\mu$ are positive rational numbers, depending on the initial critical point (see [80] for their definitions).

Remark 1. In this theorem $p = 1, \ldots, \mu$. For $p = \mu$ the theorem asserts the uniform estimate of the oscillatory integral with the uniform index equal to the individual index of the initital critical point.

Remark 2. All simple and parabolic critical points are quasihomogeneous. The base of the versal deformation of a quasihomogeneous critical point has a natural quasihomogeneous structure. Simple and parabolic critical points are distinguished in the class of quasihomogeneous critical points by the condition that the weights of quasihomogeneity of the base of the versal deformation are non-negative. The non-negativity of the weights is the basis of the proofs of Theorems 6.17 and 6.18. The theorems are proved by induction on the multiplicity of the initial critical point. In the basis of the versal deformation is considered a quasisphere. By the induction hypothesis applied to the restriction of the parameters of the deformation to the quasisphere the required estimate is already proved. This estimate can be extended to the whole base with the help of the quasihomogeneous structure. The numbers α_j^p, occurring in the theorem, are constructed from the weights of quasihomogeneity of the base and the oscillation index of the adjacent critical points.

6.7   The greatest singular index

Let us consider critical points, which are not removable by small perturbations from families of functions in n variables, depending on l parameters. The maximum of their singular indices in dependence on l and n has the form

$$\beta_l(n) = 1/2 - 1/N,$$

where the number N for $n \geqslant 3$ is given by the table

l	0	1	2	3	4	5	6	7	8	9	10, $n=3$	11, $n=3$	10, $n>3$
N	$+2$	$+3$	$+4$	$+6$	$+8$	$+12$	∞	∞	-24	-16	-12	-8	-6

All the numbers $\beta_l = \beta_l(n)$ are rational (see § 7.4). For sufficiently large n the number β_l does not depend on n (a corollary of the theorem of Kushnirenko in § 12.7 of Volume 1 and a theorem on selection of squares in § 11.1 of Volume 1).

The calculation of all these rational numbers seems to be a hard problem. Probably, $\beta_l \sim \sqrt{(2\,l)}/6$. It is conjectured that a non-degenerate cubic form in n variables is the critical point with maximal singular index for its codimension (that is for $l = n(n+1)/2$). In other words $\beta_{n(n+1)/2} = n/6$ (see [12]). From Theorem 6.5 it follows that

$$\beta_l(2) \sim 1 - \sqrt{(2/l)}.$$

From Theorem 6.5 it also follows that for $n = 2$ the maximum singular index for given multiplicity μ has asymptotic $1 - 2/\sqrt{\mu}$.

6.8   Arrangement of the material in the next three chapters

In Chapter 7 we define the Gelfand-Leray form and we discuss its properties. We consider the critical point of a monomial and express its oscillation index and index set in terms of the indices of the monomial. We define discrete characteristics of the resolution of singularities of critical point of an analytic phase and express in terms of them the oscillation index and the index set of the critical point.

In Chapter 8 we prove Theorem 6.4. For this we construct in terms of the Newton polyhedron an analytic manifold and its mapping onto $\mathbb{R}^n$. The constructed manifold and its mapping resolve the singularities of any critical point with the given Newton polyhedron under the condition that the principal part of its Taylor series is $\mathbb{R}$-nondegenerate. In Chapter 9 we prove the additivity of the oscillation index and its multiplicity, we make explicit the calculation of the indices of the tabulated functions, and produce examples, demonstrating the absence of semicontinuity of the oscillation index under deformation of the critical point.

Chapter 7

Elementary integrals and the resolution of singularities of the phase

In this chapter we shall study the asymptotics of an oscillatory integral, the phase of which is a monomial. We shall indicate the connection between the asymptotics of an oscillatory integral and the poles of the meromorphic function

$$F(\lambda) = \int f^{\lambda}(x)\phi(x)\,dx,$$

where f is the phase, and ϕ is the amplitude of the oscillatory integral. We shall introduce the discrete characteristics of the resolution of the singularity of a critical point of the phase: the weight of the resolution and the multiplicity set. We shall describe the connection between these characteristics and the basic characteristics of the asymptotic behaviour of the oscillatory integral: the oscillation index, its multiplicity and the index set.

7.1 The Gelfand-Leray form

In the study of integrals of the form

$$\int e^{i\tau f(x)}\phi(x)\,dx, \qquad \int f^{\lambda}(x)\phi(x)\,dx,$$

where τ, λ are parameters, it is convenient to take as one of the variables the function f. In this case the integrals turn into the usual Fourier transform and the Mellina transform of the integral with respect to the remaining variables. The expression under the integral sign in this latter integral is called the Gelfand-Leray form.

Let $f : \mathbb{R}^n \to \mathbb{R}$ be a smooth function, and ω be a smooth differential n-form on $\mathbb{R}^n$. We shall denote by ψ a smooth differential $(n-1)$-form for which

$$df \wedge \psi = \omega. \tag{1}$$

Lemma 7.1. If at a certain point the differential of the function f differs from zero, then in a neighbourhood of the point there exists a form ψ with property (1). The restriction of this form to an arbitrary level manifold of the function is defined uniquely.

The form with the property (1) is called the *Gelfand-Leray form* of the form ω and denoted by ω/df.

For a proof of the lemma it is sufficient to change to coordinates in which the function is one of the coordinates.

Example. Let

$$f(x, y) = y^2 - x^3 - sx$$

(where s is a number) be a function, $\omega = dx \wedge dy$ be a 2-form. Then on the level t curve the Gelfand-Leray form is equal to

$$-dx/2\,y = -dx/2\sqrt{(x^3 + sx + t)}.$$

The integral of such a form is called *elliptic*.

We shall prove two remarkable properties of the Gelfand-Leray form.

Let us orient the level manifold of the function in the standard way.

Lemma 7.2.

1. Let ω be a smooth differential n-form with compact support. Let us suppose that the support of the form does not intersect the critical set of the function f. Then

$$\int_{\mathbb{R}^n} \omega = \int_{-\infty}^{+\infty} \left(\int_{f=t} \omega/df \right) dt. \tag{2}$$

2. Let ψ be a smooth differential $(n-1)$-form with compact support. Let us suppose that the support of the form does not intersect the critical set of the function f. Then

$$\frac{d}{dt} \left(\int_{f=t} \psi \right) = \int_{f=t} d\psi/df. \tag{3}$$

Proof. Property (2) clearly follows from Fubini's theorem. Property (3) is a corollary of Stokes' theorem (see [55, 213]).

Corollary. Let f be a non-constant analytic function, ω be a smooth differential n-form with compact support, then

$$\int_{\mathbb{R}^n} e^{it f(x)}\omega = \int_{-\infty}^{+\infty} e^{itt}\left(\int_{f=t} \omega/df\right) dt. \tag{4}$$

Indeed, the union of the singular level manifolds is of measure zero.

7.2 The asymptotics of integrals of the Gelfand-Leray form

Definition. An *elementary oscillatory integral* is an integral of the form

$$\int_{\mathbb{R}^n} e^{\pm i\tau x_1^{k_1}\dots x_n^{k_n}}|x_1^{m_1}\dots x_n^{m_n}|\phi(x_1,\dots,x_n)\,dx_1\dots dx_n. \tag{5}$$

where $k_1,\dots,k_n, m_1,\dots,m_n$ are non-negative integers, ϕ is a smooth function with compact support, and τ is a real parameter.

Further, we shall denote by f the function $\pm x_1^{k_1}\dots x_n^{k_n}$, and by ω the form

$$|x_1^{m_1}\dots x_n^{m_n}|\phi(x_1,\dots,x_n)\,dx_1 \wedge \dots \wedge dx_n.$$

We shall suppose that $k_1 + \dots + k_n \geqslant 2$.

For non-zero t let us put

$$J(t) = \int_{f=t} \omega/df.$$

J is a smooth function on $\mathbb{R}\setminus 0$, equal to zero outside a sufficiently large interval. The function J is called the *Gelfand-Leray function* of the form ω.

We shall study the asymptotic behaviour of the elementary integral as $\tau \to +\infty$ in the following way. First we make explicit the asymptotics of the Gelfand-Leray function and then, using formula (4) and standard formulae for the asymptotics of one-dimensional oscillatory integrals [110], we obtain the asymptotic expansion of the elementary integral.

We shall need the following theorem.

Theorem 7.1 (see [174]). The Gelfand-Leray function can be expanded in the asymptotic series

$$J(t) \approx \sum_{\alpha} \sum_{k=0}^{n-1} a_{k,\alpha}^{+} t^{\alpha} (\ln t)^{k} \quad \text{as} \quad t \to +0 \tag{6}$$

$$J(t) \approx \sum_{\alpha} \sum_{k=0}^{n-1} a_{k,\alpha}^{-} t^{\alpha} (\ln t)^{k} \quad \text{as} \quad t \to -0 \tag{7}$$

where α runs through some discrete subset of the real numbers, bounded below. These asymptotic expansions can be differentiated term by term.

Theorem 7.1 can be proved without difficulty by induction on n.

In order to describe the asymptotic expansion of the Gelfand-Leray function, we consider the integrals

$$F_{\pm} = \int_{\pm f > 0} (\pm f)^{\lambda} \omega,$$

where λ is a complex parameter. We shall prove that the integrals are meromorphic functions of the parameter. We shall express the coefficients of the series (6) and (7) and the indices α in these series by the poles and Laurent coefficients of the resulting meromorphic functions. Then we shall give these poles and Laurent coefficients explicitly.

7.2.1 Asymptotics of the Gelfand-Leray function and the poles of its Mellina transformation

Let $J: (0, \infty) \to \mathbb{R}$ be a smooth function equal to zero for sufficiently large values of the argument. Let us suppose that there is an asymptotic expansion

$$J(t) \approx \sum_{\alpha} \sum_{k=0}^{l} a_{k,\alpha} t^{\alpha} (\ln t)^{k} \quad \text{as} \quad t \to +0 \tag{8}$$

where α runs through some discrete subset of the real numbers, bounded below. Let us consider the integral

$$F(\lambda) = \int_{0}^{\infty} t^{\lambda} J(t) \, dt,$$

where λ is a complex parameter. The integral is well defined if the real part of the parameter is sufficiently large and under these conditions the integral depends holomorphically on the parameter.

Theorem 7.2 (see [123]). The function F can be analytically continued over the whole complex plane as meromorphic function. The analytic continuation has poles at the points $\lambda = -(\alpha+1)$, where α runs through the same discrete set as in (8). The coefficient of $(\alpha+1+\lambda)^{-(k+1)}$ in the Laurent expansion at the point $\lambda = -(\alpha+1)$ is equal to $(-1)^k k! a_{k,\alpha}$.

7.2.2   Poles and Laurent coefficients of the Mellina transform of the Gelfand-Leray function

Let $f = \pm x_1^{k_1} \ldots x_n^{k_n}$ be a monomial. Let us consider two integrals

$$F_\pm(\lambda) = \int\limits_{\pm f > 0} (\pm f)^\lambda |x_1^{m_1} \ldots x_n^{m_n}| \varphi(x_1, \ldots, x_n) dx_1 \ldots dx_n,$$

where λ is a complex parameter. According to formula (2) on page 216

$$F_+'(\lambda) = \int\limits_0^\infty t^\lambda J(t)dt, \qquad F_-(\lambda) = \int\limits_{-\infty}^0 (-t)^\lambda J(t)dt,$$

where J is the Gelfand-Leray function. Consequently, the integrals depend holomorphically on the parameter for sufficiently large values of its real part. Under analytic continuation over the whole complex plane the integral has as singularities poles arranged on a discrete subset of the real numbers. Theorem 7.2 connects the poles and the Laurent coefficients with the asymptotic expansion of the Gelfand-Leray function. We shall show that the poles of the analytic continuation and the Laurent coefficients can be given explicitly. In this way we can give explicitly the asymptotic expansion of the Gelfand-Leray function.

Lemma 7.3.   1. The functions $F_\pm$ are holomorphic away from the points of the complex plane which belong to the following n arithmetic progressions:

$$-(m_1+1)/k_1, \qquad -(m_1+2)/k_1, \ldots, ;$$
$$-(m_2+1)/k_2, \qquad -(m_2+2)/k_2, \ldots, ;$$

$$\cdots\cdots$$

$$-(m_n+1)/k_n, \qquad -(m_n+2)/k_n, \ldots.$$

At a point belonging to exactly r of these progressions the functions $F_\pm$ have poles of not higher than rth order.

2. All the coefficients of the Laurent expansion of the functions $F_\pm$ at an arbitrary point of the complex plane are generalised functions of the amplitude φ.

Proof. It is sufficient to prove the conclusion for the integral

$$F(\lambda) = \int_0^a \ldots \int_0^a x_1^{k_1\lambda + m_1} \ldots x_n^{k_n\lambda + m_n} \varphi(x) dx_1 \ldots dx_n,$$

where a is a positive number. It is useful to consider a more general integral

$$\tilde{F}(\lambda_1, \ldots, \lambda_n) = \int_0^a \ldots \int_0^a x_1^{k_1\lambda_1 + m_1} \ldots x_n^{k_n\lambda_n + m_n} \varphi(x) dx_1 \ldots dx_n.$$

Let N be a large natural number. Let us make the transformation

$$\tilde{F} = \int_0^a \ldots \int_0^a x^{k\lambda + m} R\varphi dx + \sum_{l_1 = 0}^N \frac{a^{k_1\lambda_1 + m_1 + l_1 + 1}}{k_1\lambda_1 + m_1 + l_1 + 1} \times$$

$$\times \frac{1}{l_1!} \int_0^a \ldots \int_0^a x_2^{k_2\lambda_2 + m_2} \ldots x_n^{k_n\lambda_n + m_n} \frac{\partial^{l_1}\varphi}{\partial x_1^{l_1}} (0, x_2, \ldots, x_n) dx_2 \ldots dx_n,$$

$$(9)$$

where $R\varphi$ is the difference between the function φ and its Taylor polynomial of degree N in x_1.

The first of these integrals does not have a singularity in λ_1 for

$$\mathrm{Re}(k\lambda_1 + m_1 + N + 1) > 0,$$

and the poles in λ_1 of the second term in the right hand side belong to the first of the progressions indicated in the lemma. Repeating successively with each integral in the right hand side the same procedure with respect to the other variables and then putting

$$\lambda_1 = \lambda_2 = \ldots = \lambda_n = \lambda,$$

we obtain the first part of the lemma. This argument allows us to give an explicit analytic continuation of the integral in a neighbourhood of the given point of the complex plane (see [123]). Each coefficient of the Laurent expansion at an arbitrary point of the plane is equal to the sum of the integrals of the function φ and its derivatives over some coordinate subspace. This proves the second part of the lemma.

Lemma 7.4. Let the number λ_0 belong to exactly r of the arithmetic progressions of Lemma 7.3. For definiteness let us suppose that it is the first r progressions, and

$$\lambda_0 = -(m_1 + l_1 + 1)/k_1 = \ldots = -(m_r + l_r + 1)/k_r$$

where $l_1, \ldots, l_r$ are certain non-negative integers. Then the coefficient of $(\lambda - \lambda_0)^{-r}$ in the Laurent expansion at the point λ_0 of the function F is equal to

$$\prod_{j=1}^{r} \frac{1}{k_j} \frac{1}{l_j!} \left(\int_{\substack{x_j = 0 \\ j = 1, \ldots, r}} x_{r+1}^{k_{r+1}\lambda + m_{r+1}} \ldots x_n^{k_n\lambda + m_n} \times \right.$$

$$\left. \times \frac{\partial^{l_1 + \ldots + l_r} \varphi}{\partial x_1^{l_1} \ldots \partial x_r^{l_r}} (0, \ldots, 0, x_{r+1}, \ldots, x_n) dx_{r+1} \ldots dx_n \right)_{\lambda = \lambda_0},$$

where $(\int)_{\lambda = \lambda_0}$ denotes analytic continuation of the integral in parentheses to the point λ_0.

Lemma 7.4 is a corollary of formula (9).

Lemma 7.5.

1. Let $\beta = \max \left\{ -(m_1 + 1)/k_1, \ldots, -(m_n + 1)/k_n \right\}$ be the maximal number in the union of the arithmetic progressions of Lemma 7.3. Let us suppose that the number β belongs to exactly r arithmetic progressions of Lemma 7.3. Let us suppose that the amplitudes φ in the integrals F_+, F_- are non-negative and that their values at the origin are positive. Then the sums of the coefficients of $(\lambda - \beta)^{-r}$ in the Laurent expansions of the functions F_+ and F_- are positive, and each of these coefficients is non-negative.

2. Among the numbers $k_1, \ldots, k_n$ let precisely one be equal to 1. Let this be k_1. Let λ_0 be a number belonging to the first progression of Lemma 7.3 and belonging to no other progression of Lemma 7.3. In particular this means that

$$\lambda_0 = -(m_1 + l + 1),$$

where l is a non-negative integer. Let us denote by a^+, a^- the coefficients of $(\lambda - \lambda_0)^{-1}$ in the Laurent series at the point λ_0 of the functions F_+ and F_- respectively. Then

$$a^+ = (-1)^l a^-.$$

3. Let f have a minimum at the origin, that is

$$f = +x_1^{k_1} \ldots x_n^{k_n},$$

where all the powers are even. Then $F_- (\lambda) \equiv 0$. In addition if β, r, φ are the same as in Section 1, then the coefficient of $(\lambda - \beta)^{-r}$ in the Laurent expansion of the function F_+ is greater than zero.

This lemma is clearly a corollary of Lemma 7.4 and the decomposition of the integrals F_+, F_- into sums of integrals on the coordinate orthants.

7.2.3   Asymptotics of elementary oscillatory integrals

Theorem 7.3 (see [358]).

1. An elementary oscillatory integral (see (5) on page 217) can be expanded, as $\tau \to +\infty$, in the asymptotic series

$$\sum_\alpha \sum_{k=0}^{n-1} a_{k,\alpha}(\varphi) \tau^\alpha (\ln \tau)^k, \tag{10}$$

where the numerical coefficients $a_{k,\alpha}$ are generalised functions of the amplitude φ, and the parameter α runs through the arithmetic progressions of Lemma 7.3. If the number α belongs to exactly r arithmetic progressions of Lemma 7.3, then $a_{k,\alpha} \equiv 0$ for $k \geqslant r$.

2. Let

$$\beta = \max\{-(m_1 + 1)/k_1, \ldots, -(m_n + 1)/k_n\}$$

be the maximal number in the union of the arithmetic progressions of Lemma 7.3. Let r be the number of arithmetic progressions of Lemma 7.3 to which β belongs. Let us suppose that β is not an odd integer. Let us suppose that the amplitude φ is non-negative and that its value at the origin is positive. Then the real part of the numerical coefficient of the leading term of the asymptotic series (that is the real part of the number $a_{r-1,\beta}$) is not equal to zero and has the same sign as the number $\cos(\pi\beta/2)$; in this way the sign of the real part is determined by the number β.

3. Let $k_1 = 1$ and m_1 be even, that is let the hypersurface $x_1 = 0$ not belong to the critical set of the phase of the elementary integral and not belong to the subset on which the expression under the integral sign of the elementary integral is not smooth. Then in the expansion (10) the number α runs through only the arithmetic progressions of Lemma 7.3 with numbers $2, \ldots, n$.

4. Let the phase of the elementary integral have a minimum at the origin, that is let

$$f = +x_1^{k_1} \ldots x_n^{k_n},$$

where all the powers are even. Let the amplitude φ be non-negative and let its value at the origin be positive. Then the numerical coefficient of the leading term of the asymptotic series (that is the number $a_{r-1,\beta}$) is not equal to zero and has the same argument as the number $\exp(-\pi i\beta/2)$, where the numbers β, r were defined in section 1; in this way the argument of the coefficient is determined by the number β.

Theorem 7.3 follows easily from Theorems 7.1, 7.2 and Lemmas 7.3, 7.5 with the help of the following standard formulae. Let $\theta: \mathbb{R} \to \mathbb{R}$ be a smooth function with compact support, identically equal to 1 in a neighbourhood of the origin. Then as $\tau \to +\infty$ modulo infinitesimals of arbitrarily high order

$$\int_0^\infty e^{i\tau t} t^\alpha (\ln t)^k \theta(t) dt \approx \frac{d^k}{d\alpha^k} \frac{\Gamma(\alpha+1)}{(-i\tau)^{\alpha+1}},$$

$$(11)$$

$$\int_{-\infty}^0 e^{i\tau t} (-t)^\alpha (\ln(-t))^k \theta(t) dt \approx \frac{d^k}{d\alpha^k} \frac{\Gamma(\alpha+1)}{(i\tau)^{\alpha+1}}.$$

In these formulae $\arg(\pm i\tau) = \pm \pi/2$, and Γ is the gamma-function (see [110]).

7.2.4 Asymptotics of elementary Laplace integrals

Definition. An *elementary Laplace integral* is an integral

$$\int_{\mathbb{R}^n} e^{-\tau x_1^{k_1} \ldots x_n^{k_n}} |x_1^{m_1} \ldots x_n^{m_n}| \varphi(x_1, \ldots, x_n) dx_1 \ldots dx_n,$$

where $k_1, \ldots, k_n, m_1, \ldots, m_n$ are non-negative integers, $k_1, \ldots, k_n$ are even, $k_1 + \ldots + k_n \geqslant 2$, φ is a smooth function with compact support, τ is a real parameter.

Theorem 7.4.

1. An elementary Laplace integral can be expanded, as $\tau \to +\infty$, in an asymptotic series (10), with the properties indicated in the conclusion of section 1 of Theorem 7.3.

2. Let the amplitude φ be non-negative and its value at the origin be positive. Then in the asymptotic series of the elementary Laplace integral the numerical coefficient of the leading term (that is the coefficient $a_{r-1,\beta}$) is positive, where the numbers r, β were defined in section 2 of Theorem 7.3.

The proof of Theorem 7.4 is the same as the proof of Theorem 7.3, except that the reference to formula (11) must be replaced by a reference to the formula (see [110])

$$\int_0^\infty e^{-\tau t} t^\alpha (\ln t)^k \theta(t)\, dt \approx \frac{d^k}{d\alpha^k} \frac{\Gamma(\alpha+1)}{\tau^{\alpha+1}}.$$

7.3 Asymptotics and the resolution of singularities

7.3.1 The weight of the resolution of a singularity and the multiplicity set

Let us consider a function $f : \mathbb{R}^n \to \mathbb{R}$, analytic in a neighbourhood of its critical point x. Let us suppose that the value of the function at this point is equal to zero. Let us consider the resolution of the singularity of this critical point (see § 6.4). We shall introduce the characteristics of the resolution of the singularity through which we shall express the oscillation index of the critical point, its multiplicity and its index set.

The resolution of a singularity is a manifold Y and a map $\pi : Y \to \mathbb{R}^n$, possessing the properties indicated in § 6.4. In a small neighbourhood of the preimage of the critical point x we consider the decomposition into irreducible components of the zero level hypersurface of the function $f \circ \pi$. To each irreducible component which intersects the preimage of the point x there are associated two non-negative integers: the multiplicities of zero on this component of the function $f \circ \pi$, and of the Jacobian of the map π, respectively. Let us denote these numbers by k, m, respectively. The ordered pair (k, m) is called the *multiplicity of the component*, the number $-(m+1)/k$ is called the *weight of the component*.

Definition. The *multiplicity set* of the resolution of a singularity is the set of all pairwise distinct multiplicities, possessing the properties

$$(k, m) \neq (1, 0), \quad k > 0.$$

Let us denote the multiplicity set by Mu.

Definition. The *weight of the resolution of a singularity* is the maximum weight of the components, the multiplicities of which possess the properties

$$(k, m) \neq (1, 0), \quad k > 0.$$

In this way the weight of the resolution is equal to the number

$$\max \{ -(m+1)/k | (k, m) \in Mu \}.$$

Remark. The multiplicity set is finite by virtue of the properties of the map π.

Example. Let f be a homogeneous polynomial of degree N with a finite-multiplicity critical point at the origin. Let

$$\pi: Y \to \mathbb{R}^n$$

be a σ-process at the origin (see §4.3, and also Chapter 4 of Part II in [328]). This map resolves the singularity at the origin. The multiplicity set of the resolution consists of the one pair $(N, n-1)$. The weight of this resolution is equal to $-n/N$.

Let us define the concept of the multiplicity of a number relative to the resolution of the singularity. To do this we must first define the concept of the multiplicity of a number at a point of the preimage of a critical point.

Let α be a number, let y be a point of the preimage (relative to the map of the resolution) of the critical point x. Let us consider a small neighbourhood of the point y and a decomposition in it of the zero level hypersurface of the function $f \circ \pi$ into irreducible components. The *multiplicity of the number α at the point y* is the number of irreducible components of weight α which intersect at y. The *multiplicity of the number α relative to the resolution of the singularity* is the maximum of the multiplicities of the number α at points of the preimage of the critical point x. It is clear that the multiplicity of a number is an integer, constrained to lie between 0 and n.

Example. Let us consider the critical point and its resolution, indicated in the previous example. Let $n \neq N$. Then the multiplicity of the number $(-n/N)$ relative to the indicated resolution is equal to 1. The multiplicity of the number -1 is equal to zero if the function f is semidefinite and equal to 1 otherwise. The multiplicity of the remaining numbers is equal to 0.

7.3.2 Asymptotic series of oscillatory integrals

Theorem 7.5 (see [358]). Let us consider the oscillatory integral

$$\int_{\mathbb{R}^n} e^{i\tau f(x)} \varphi(x) dx_1 \ldots dx_n.$$

Let us suppose that the phase is an analytic function in a neighbourhood of its critical point. Let us suppose that the value of the phase at this critical point is equal to zero. Let us consider the resolution of the singularity at the critical point. We assert: if the support of the amplitude is concentrated in a sufficiently small neighbourhood of the critical point of the phase then

1. The oscillatory integral can be expanded in an asymptotic series

$$\sum_{\alpha} \sum_{k=0}^{n-1} a_{k,\alpha}(\varphi) \tau^{\alpha} (\ln \tau)^k \quad \text{as} \quad \tau \to +\infty.$$

The numerical coefficients $a_{k,\alpha}$ are generalised functions of the amplitude. The support of each generalised function lies in the critical set of the phase. The parameter α runs through the following arithmetic progressions. One of these is the negative integers and the others are parametrised by the elements of the set of multiplicities of the resolution of the singularity of the critical point of the phase. The pair (k, m) corresponds to the arithmetic progression

$$-(m+1)/k, \quad -(m+2)/k, \ldots.$$

2. Let y be a point of the preimage of the critical point of the phase and $(k_1, m_1), \ldots, (k_n, m_n)$ be the multiplicities at y of the components of the zero level hypersurface of $f \circ \pi$. Let us consider the arithmetic progressions of Lemma 7.3. If for any y the number α is contained in not more than k progressions then the generalised function $a_{k,\alpha}$ is identically equal to zero.

3. If the weight of the resolution of the singularity at the critical point of the phase is greater than -1 then

(i) the oscillation index of the critical point of the phase is equal to the weight of the resolution of the singularity;
(ii) the multiplicity of the oscillation index of the critical point of the phase is equal to one less than than the multiplicity of the number equal to the weight of the resolution of the singularity relative to the resolution of the singularity;
(iii) if the amplitude is non-negative and its value at the critical point of the phase is positive, then the numerical coefficient of the leading term of the asymptotic series of the oscillatory integral (that is the coefficient $a_{K,\beta}$, where β is the oscillation index, and K is the multiplicity of the oscillation index) is non-zero;
(iv) if the critical point of the phase has finite multiplicity then the numerical coefficient of the leading term of the asymptotic series of the oscillatory integral is equal to the value of the amplitude at the critical point of the phase, multiplied by a non-zero constant which depends only on the phase.

4. If the critical point of the phase is a maximum or minimum point then the conclusions (i)–(iv) of section 3 of this theorem are true.

5. Let us denote by π the map of the resolution of the singularity. Let us suppose that there does not exist a point in the preimage of the critical point of the phase at which intersect two or more irreducible components of the zero level hypersurface of the function $f \circ \pi$, the multiplicities of which are equal to $(1, 0)$. (We note that this assumption is satisfied if the phase has finite multiplicity at the critical point (if $n = 2$, we must also exclude the case of a nondegenerate critical point)). Then the parameter α in the asymptotic series of the oscillatory integral runs through only the arithmetic progressions of Section 1 of this theorem, parametrised by the elements of the set of multiplicities of the resolution of the singularity. In particular, the oscillation index of the critical point of the phase is not more than the weight of the resolution of the singularity.

6. Let β be the weight of the resolution of the singularity and let k be the multiplicity of the number β relative to the resolution of the singularity. Let us consider all the points of the preimage (relative to the map of the resolution of the singularity) at which the multiplicity of the number β is equal to k. Let us suppose that this set does not intersect any irreducible component of the zero level hypersurface of the function $f \circ \pi$, the multiplicity of which is equal to $(1, 0)$. Let us suppose that the condition of section 5 is satisfied. Let us suppose that the weight of the resolution of the singularity is not an odd integer. Then conclusions (i)–(iv) of section 3 of this theorem are true.

7. Let us suppose that the weight of the resolution of the singularity is equal to -1 and that the multiplicity of the number -1 relative to the resolution of the singularity is not less than 2. Then the oscillation index of the critical point of the

phase is equal to -1. Furthermore the multiplicity of the oscillation index is equal to the multiplicity of the number -1 relative to the resolution of the singularity or to one less than the multiplicity of the number -1.

Remark 1. This theorem implies Theorem 6.3 on asymptotic expansions.

Remark 2. The resolution of the singularity at the critical point is not uniquely defined. However, if for one resolution the weight is greater than -1 then for any other resolution the weight is also greater than -1 and does not depend on the resolution (Section 3 of Theorem 7.5). It would be interesting to find a purely algebraic proof of this fact.

Remark 3. Practically all critical points have a resolution of the singularity of weight greater than -1. For a sufficient condition for this see Theorem 8.5 and also supplement 1 of Theorem 6.4.

Remark 4. In § 9.2 (see also [358]) we shall cite an example of a critical point and a resolution of its singularity with the properties: the weight of the resolution is less than -1, and the oscillation index is less than the weight of the resolution.

Remark 5. Let us return to the problem of reconstructing the value of the function ϕ at the critical point of the function f in terms of the integrals of the function ϕ over the level hypersurfaces of the function f, see § 6.3. For a critical point of finite multiplicity of the function f, satisfying the conditions of one of Sections 3, 4, 6 of Theorem 7.5, this problem can be solved in the following way. Let us take as density on the level hypersurfaces of the function f the differential $(n-1)$-form of Gelfand-Leray

$$dx_1 \wedge \ldots \wedge dx_n/df.$$

In this way, knowing the Gelfand-Leray function

$$J(t) = \int_{f=t} \phi \, dx_1 \wedge \ldots \wedge dx_n/df,$$

we must reconstruct the value of the function at the critical point. According to formula (4) on page 217 the oscillatory integral is the Fourier transform of the Gelfand-Leray function. Therefore assertion 3(iv) of Theorem 7.5 gives the solution of the problem.

Example. Let f be a homogeneous polynomial of degree N, with a critical point of finite multiplicity at the origin. The resolution of the singularity at this critical point was indicated in the example on page 225. The multiplicity set of the resolution consists of pairs $(N, n-1)$. According to Theorem 7.5 in the asymptotic expansion of the oscillatory integral with phase f the parameter α runs through the arithmetic progression

$$-n/N, \quad -(n+1)/N, \ldots .$$

If $N > n$ (or f is of definite sign), then the oscillation index of the critical point is equal to $-n/N$, and the multiplicity of the oscillation index is equal to 0.

Proof of the theorem. We shall make a change of variables in the oscillatory integral with the help of the map $\pi: Y \to \mathbb{R}^n$ of the resolution of the singularity. Then the integral is transformed into an integral over Y. Using a sufficiently fine partition of unity we transform the latter integral into a sum of elementary integrals (this is possible since π is a resolution of the singularity).

Now Sections 1 and 2 of Theorem 7.5 follow immediately from Section 1 of Theorem 7.3. Analogously, Sections 3, 4, 5, 6 of Theorem 7.5 follow, respectively, from Sections 2, 3, 4, 2 of Theorem 7.3.

We shall prove Section 7. We have

$$\int_{\mathbb{R}^n} e^{itf} \varphi \, dx = \int_0^\infty e^{itt} J(t) \, dt + \int_0^\infty e^{-itt} J(-t) \, dt,$$

where J is the Gelfand-Leray function. Using the resolution of the singularity and Theorem 7.2 we obtain

$$J(\pm t) \approx a_{r,0}^\pm (\ln t)^r + \ldots + a_{0,0}^\pm + \sum_{\alpha > 0} t^\alpha (\ln t)^k a_{k,\alpha}^\pm,$$

where $(r+1)$ is the multiplicity of the number -1 relative to the resolution of the singularity, $a_{k,\alpha}^\pm$ are real numbers. If the amplitude has fixed sign and is different from zero at the critical point of the phase, then according to Lemma 7.7 the numbers $a_{r,0}^\pm$ have one and the same sign, and their sum is different from zero.

Applying formula (11) on page 223, we can convince ourselves that in the asymptotic expansion of an oscillatory integral the real part of the coefficient of $(\ln \tau)^{r-1}/\tau$ is proportional to $a_{r,0}^+ + a_{r,0}^-$ and the coefficient of $(\ln \tau)^r/\tau$ is proportional to $a_{r,0}^+ - a_{r,0}^-$, the constants of proportionality being different from zero. Section 7 is proved.

Remark. The assumptions of Sections 3, 4, 6 of Theorem 7.5 are necessary in order that the principal term of the asymptotic series is not influenced by points on the non-singular part of the zero level hypersurface of the function f.

7.3.3   The asymptotics of the Laplace integral

A *Laplace integral* is an integral of the form

$$\int_{\mathbb{R}^n} e^{-\tau f(x)} \varphi(x) dx_1 \ldots dx_n,$$

where τ is a positive real parameter. The functions f and φ are called the *phase* and *amplitude* respectively.

Let us suppose that the phase has a minimum point and that it is an analytic function in a neighbourhood of the minimum point. Let us suppose that the value of the phase at the minimum point is equal to zero.

Theorem 7.6. If the support of the amplitude is concentrated in a sufficiently small neighbourhood of the minimum point, then as $\tau \to +\infty$ the Laplace integral can be expanded in the asymptotic series

$$\sum_{\alpha} \sum_{k=0}^{n-1} a_{k,\alpha}(\varphi) \tau^\alpha (\ln \tau)^k,$$

for which the conclusions of Sections 1, 2, 4 of Theorem 7.5 are true.

The proof of Theorem 7.6 is obtained from the proof of Theorem 7.5 by replacing the references to Theorem 7.3 by references to Theorem 7.4.

Corollary. For each small positive t let us denote by $V(t)$ the volume of the set of those points at which the value of the phase is less than t. Then as $t \to +0$ the function V can be expanded in the asymptotic series

$$\sum_\alpha \sum_{k=0}^{n-1} a_{k,\alpha} t^{-\alpha} (\ln t)^k.$$

Here the parameter α runs through a finite set of arithmetic progressions, consisting of positive rational numbers. These arithmetic progressions are the progressions of Section 1 of Theorem 7.5. The assertions about the coefficients of the series and the order of the leading term of the series, given in the conclusions of Sections 2 and 4 of Theorem 7.5, are true.

Proof of the corollary. The derivative of the function V is equal to the Gelfand-Leray function of the phase f and an amplitude which is identically equal to 1.

7.4   The rationality of the greatest singular index $\bar{\beta}_l(n)$, defined in § 6.8

Let us consider a polynomial of degree N with indeterminate coefficients and with zero constant and linear terms.

$$f(x, a) = \sum_{\substack{k_1, \ldots, k_n \geq 0 \\ 1 < k_1 + \ldots + k_n \leq N}} a_{k_1, \ldots, k_n} x_1^{k_1} \ldots x_n^{k_n}.$$

For fixed real coefficients a the polynomial defines the function

$$f(\cdot, a) : \mathbb{R}^n \to \mathbb{R},$$

with a critical point at the origin. According to Theorem 7.5 there exists an arithmetic progression containing the index set of this critical point.

Theorem 7.7. There exists one arithmetic progression containing the index set at the origin of the phase $f(\cdot, a)$ for all a.

Theorem 7.7 follows from Lemma 7.8.

Lemma 7.8. Let us suppose that in the space of coefficients of the polynominal f there is chosen a semialgebraic set A. Then there exists a proper semialgebraic subset $B \subset A$ and an arithmetic progression Q with the property: for any $a \in A \setminus B$ the index set of the critical point of the phase $f(\cdot, a)$ at the origin belongs to Q.

Proof. We can suppose that A is non-singular and connected. Let f' be the restriction of the polynominal f to the manifold $\mathbb{R}^n \times A$. Let us consider the resolution of the singularity

$$\pi : Y \to \mathbb{R}^n \times A$$

of the zero level hypersurface of the function f', see [158]. From the Theorem of Sard-Bertini follows the existence of a proper algebraic subset $B \subset A$ with the property: for any $a \in A \setminus B$ the restriction of the map of the resolution of the singularity to the preimage of the set $\mathbb{R}^n \times a$ is a resolution of the singularity of the zero level hypersurface of the function

$$f(\cdot, a) : \mathbb{R}^n \times a \to \mathbb{R}.$$

Furthermore the topology of this resolution and all the multiplicities depend on a locally constantly. Then the Lemma follows from Theorem 7.5.

Chapter 8
Asymptotics and Newton polyhedra

We shall consider the class of critical points of the phase, the Taylor series of which have fixed Newton polyhedron. If the Newton polyhedron is remote, then almost all the critical points of the class have the same oscillation index. This common oscillation index is equal to the remoteness of the Newton polyhedron. A critical point of the class has the typical oscillation index if the principal part of its Taylor series is $\mathbb{R}$-nondegenerate. (Remember that the condition of $\mathbb{R}$-nondegeneracy is an explicitly written-out algebraic condition on a finite set of Taylor coefficients , see § 6.2). This assertion was formulated as Theorem 6.4. Its proof occupies the whole of the present chapter. The proof uses the resolution of the singularity of the critical point of the phase. In the previous chapter we defined a numerical characteristic of the resolution of a singularity, namely the weight, and we proved that if the weight is greater than -1 then the oscillation index of the critical point is equal to the weight (Theorem 7.5). In this section we construct a manifold and a map of it into $\mathbb{R}^n$, which resolves the singularity of almost all the critical points of the class we are considering. We shall show that the weight of the constructed resolution of the singularity is equal to the remoteness of the Newton polyhedron. In this way we shall prove Theorem 6.4.

The resolution of the singularity is constructed in terms of the Newton polyhedron and consists of three stages. In the first stage we use the Newton polyhedron to construct a decomposition of the positive orthant of the space into convex cones, each of which is given by a finite set of linear conditions with rational coefficients. In the second stage the cones are broken into smaller pieces to construct a new decomposition of the positive orthant. The new decomposition is inscribed in the previous one, all of its cones are simicial and their multiplicity is equal to 1 (for the definitions see § 8.1.1). In the third stage we construct from the new decomposition a manifold and a map of it into $\mathbb{R}^n$. The manifold and the map resolve the singularities of almost all the critical points of the class we are considering.

Each stage uses only the results of the previous stage: in the second stage we do not use the Newton polyhedron, in the third stage we do not use the first decomposition. In the first and third stages the initial data uniquely determine the results. The result of the second stage (the new decomposition) is not determined uniquely by the first decomposition. In this way the resolution of the singularity is not uniquely determined by the Newton polyhedron, although it is constructed from the Newton polyhedron.

We shall describe the third stage first. We shall construct a manifold in terms of a set of cones with the properties mentioned above and we shall define its natural projection onto $\mathbb{R}^n$ (see § 8.1.6). Then we shall describe the first two stages and we shall prove that the manifold and map constructed as a result of the three stages resolve the singularities at the critical points of the class where the Taylor series has $\mathbb{R}$-nondegenerate principal part. At the end we shall derive Theorem 6.4 from Theorem 7.5.

On the manifold which will be constructed in this chapter there acts in a natural way a group – an n-dimensional torus (for more detail see § 8.1.4). A manifold with an action of a torus is called toral. For further discussion of the theory of toral manifolds see [88, 184]. The orbits of the action of the torus on the toral manifold are in one-to-one corresponence with a certain collection of convex cones, constructed with the help of the manifold. In its turn this collection of cones uniquely determines the toral manifold. Our third stage of construction of the resolution of the singularity is this standard (in the theory of toral manifolds) transition from a collection of cones to a manifold. Toral manifolds are remarkable in that the majority of analytic and topological constructions on them reduce to linear algebraic constructions on the corresponding collection of cones. See, for example, our calculation in this chapter of the weight of the resolution of a singularity.

In the study of singularities, Newton polyhedra were first applied in [195, 196]. Toral manifolds were first related to Newton polyhedra by A. G. Hovanski (see [45, 159, 160,] and also [358, 359]).

8.1   Construction of the manifold

8.1.1   Cone, skeleton, multiplicity, fan

The *cone generated by the vectors $a_1, \ldots, a_s \in \mathbb{R}^n$* is the cone consisting of linear combinations of these vectors with nonnegative coefficients.

A cone with vertex at the origin is said to be *rational* if it can be generated by a finite set of vectors with integer coordinates.

The *skeleton* of a rational cone is the set of all of its primitive (not multiple) integer vectors in the faces of dimension 1. It is clear that the skeleton of the cone generates the cone itself.

Example. The cone depicted in figure 66 is rational. Its skeleton consists of the vectors $(3, 1)$, $(1, 2)$.

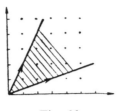

Fig. 66.

A rational cone is said to be *simplicial*, if the vectors making up its skeleton are linearly independent.

The *multiplicity* of a simplicial cone of highest dimension is the index of the sublattice generated by the vectors of the skeleton in the integer lattice of the space. It is clear that the multiplicity of a cone is equal to 1 if and only if its skeleton forms a basis of the integer lattice of the space. Further, if the multiplicity is greater than 1, then there are integer vectors belonging to the cone which are linear combinations of vectors of the skeleton in which all the coefficients are nonnegative, less than 1 and at least one coefficient is not equal to 0.

Example. The cone depicted in figure 66 is simplicial (all two-dimensional rational cones are simplicial). The multiplicity of the cone is equal to 5.

Exercise. Prove that the multiplicity of the cone is equal to the absolute value of the determinant formed from the coordinates of the vectors of the skeleton.

A *fan* is a finite set of rational cones possessing the properties:
(i) each face of a cone from the set also belongs to the set;
(ii) the intersection of any two cones from the set is a face of each of them.
The fan is said to be *simple* if
(iii) all the cones of the fan are simplicial and the skeleton of any cone can be extended to a basis of the integer lattice of the whole space.

8.1.2 Monomial maps

A rational map $h : \mathbb{R}^n \to \mathbb{R}^n$ of the form

$$x_i \circ h = x_1^{a_i^1} \ldots x_n^{a_i^n}, \; i = 1, \ldots, n,$$

where a_i^j is an integer matrix with determinant equal to ± 1 is said to be *monomial*. The region of definition of a monomial map always contains the complement of the union of the coordinate hyperplanes.

It is clear that the inverse map of a monomial map is monomial and is given by the inverse matrix. The region of definition of a monomial map is the whole space if and only if the matrix of the monomial map has non-negative elements.

Let us be given an ordered pair of bases of the integer lattice of the space $\mathbb{R}^n$. A monomial map is said to be *associated* with this pair if it is given by the matrix of the dual operator to the operator which transforms the second basis into the first and which is written down in terms of the second basis. In this way in the columns of the matrix are the coordinates of the vectors of the first basis expressed in terms of the second basis.

Example. Let us be given in the integer lattice of the space $\mathbb{R}^2$ two bases: the first $(1, 0)$, $(1, 1)$, the second $(1, 0)$, $(0, 1)$. Then the monomial map h associated with this pair is given by the formulae

$$x_1 \circ h = x_1^1 x_2^1$$
$$x_2 \circ h = x_1^0 x_2^1$$

Lemma 8.1. If we change the order of the pair then the monomial map associated with the pair will change to its inverse. If we are given three bases then the map associated with the first and third basis is equal to the composition of the map associated with the first and second basis and the map associated with the second and third basis. In other words $h_{1,3} = h_{2,3} \circ h_{1,2}$.

The proof is obvious.

8.1.3 The manifold associated with a simple fan

Let us be given a simple fan. With the help of the fan we shall construct a non-singular n-dimensional real analytic manifold. The construction of the manifold generalises the construction of the standard compactifications of the space $(\mathbb{R} \setminus 0)^n$. Namely, for the fans indicated in figures 67a and 67b, the construction leads to the manifolds $(\mathbb{R} P^1)^n$, $\mathbb{R} P^n$, respectively.

The charts of the manifold are in one-to-one correspondence with the n-dimensional cones of the fan. Each chart is equal to $\mathbb{R}^n$. We introduce an equivalence relation for points of different charts. Then on the set of equivalence

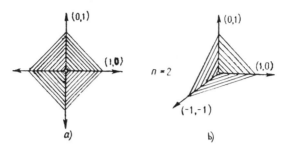

Fig. 67a and b.

classes we introduce a topology and the structure of a manifold. We shall find it useful to order the vectors of the skeleton of each cone of the fan. Let us fix these orders. Later on we shall easily convince ourselves that the manifold which we construct does not depend on this choice of order.

To any ordered pair of charts there is associated a monomial map from the first chart to the second. This is the monomial map associated with an ordered pair of bases of the integer lattice of the space, where the first (respectively second) basis is that basis generated by the skeleton of the n-dimensional cone which corresponds to the first (respectively second) chart. For the future manifold this monomial map will play the role of the transition function from the first chart to the second.

We shall say that a point of the first chart is *equivalent* to a point of the second chart if the monomial map relating these charts is defined at the point of the first chart and maps this point to the point of the second chart.

Example. In the space $\mathbb{R}^2$ let the skeleton of the first cone consist of the vectors $(1, 1)$, $(1, 2)$, and the skeleton of the second cone consist of the vectors $(1, 1)$, $(3, 2)$. Then the monomial map h from the first chart to the second is given by the formulae

$$x_1 \circ h = x_1^1 x_2^4$$
$$x_2 \circ h = x_1^0 x_2^{-1}$$

Therefore, for example, the point $(0, 2)$ of the first chart is equivalent to the point $(0, 1/2)$ of the second chart.

The relation we have introduced on pairs of points will be an equivalence relation if we verify its symmetry. The symmetry follows from Lemma 8.2.

Lemma 8.2. The monomial map relating an ordered pair of charts possesses the following property. If we are given a sequence of points of the first chart, for which the following conditions are satisfied:
(i) the sequence has a finite limit in the first chart;
(ii) the monomial map is defined at the points of the sequence;
(iii) the sequence of images of points of the sequence has a finite limit in the second chart;
then the monomial map is defined and non-degenerate at the limit point of the sequence.

The **proof** of the lemma is based on the fact that the intersection of the cones corresponding to the two charts is a face of each of them. It is sufficient to take the case when the sequence of points lies on a smooth curve and the limit point of the sequence corresponds to the point on the curve the value of whose parameter is zero. The existence of such a curve follows from the curve selection lemma (see [256]).
So let the curve have the form

$$x_j(t) = t^{k_j}(c_j + O(t)), \quad j = 1, \ldots, n$$

where the numbers $c_1, \ldots, c_n$ are all non-zero. Then its image has the form

$$x_j(t) = t^{m_j}(d_j + O(t)), \quad j = 1, \ldots, n$$

where $m_j = \Sigma a_j^i k_i$ and a_j^i is the matrix of the monomial map. By definition, the indices k_j and m_j are non-negative. According to the formula the vector $(m_1, \ldots, m_n)$ is a linear combination of the columns of the matrix a_j^i with coefficients $k_1, \ldots, k_n$. By definition, in the columns are the coordinates of the skeleton of the cone, corresponding to the first chart, expressed in terms of the vectors of the skeleton of the cone corresponding to the second chart. The non-negativity of the numbers m_j means that the indicated linear combination of vectors of the first skeleton belongs to the second cone. However the intersection of the cones is a face of each of them. Therefore if in the linear combination the coefficient k_j is different from zero then in the jth column of the matrix of the monomial map all the elements except one are equal to 0 and the remaining coefficient is equal to 1. The positivity of the coefficient k_j means that the limiting point of the curve as $t \to 0$ lies on the hyperplane $x_j = 0$. According to the above proof the monomial map is defined and non-degenerate at a general point of this hyperplane. The lemma is proved.

So on pairs of points of the charts we are given an equivalence relation. We shall define on the set of equivalence classes a topology and the structure of an analytic manifold.

Each chart is included in a natural way as a subset in the set of equivalence classes. We shall say that a set is open if its intersection with every chart is open. From Lemma 8.2 it follows easily that this definition gives on the set of equivalence classes the structure of a Hausdorff topological space. The inclusion of charts defines a covering of the topological space by open sets and defines a homeomorphism of these sets onto $\mathbb{R}^n$. By construction, the transition functions connected with these homeomorphisms are monomial maps in their regions of definition. In this way on a simple fan we have constructed an analytic manifold. We shall call this manifold the *manifold associated with the simple fan*.

Exercise. Show that $(\mathbb{R}P^1)^n$ and $\mathbb{R}P^n$ are the manifolds associated with the fans depicted in figures 67a and 67b respectively.

8.1.4   A torus acts on the manifold associated with a simple fan

The space $(\mathbb{R}\setminus 0)^n$ together with coordinatewise multiplication forms a group called the n-dimensional torus. The torus acts on itself. Its action extends naturally to $\mathbb{R}^n$.

Let us consider the manifold associated with a simple fan. The torus acts on charts of the manifold. It is easy to see that this action extends to an action of the torus on the whole manifold. We shall describe the orbits of this action.

There is one n-dimensional orbit, isomorphic to the torus. In an arbitrary chart it is $(\mathbb{R}\setminus 0)^n$.

The $(n-1)$-dimensional orbits are in one-to-one correspondence with the one-dimensional cones of the simple fan. Indeed each chart intersects n $(n-1)$-dimensional orbits. Their closure in local coordinates coincides with the coordinate hyperplanes. We put in correspondence with the $(n-1)$-dimensional orbit lying in the hyperplane $x_j = 0$ the jth vector of the skeleton of the cone corresponding to the chart. (Remember that the vectors of the skeletons of the cones of the fan were ordered).

Lemma 8.3.

1. This relation correctly defines a one-to-one correspondence between the set of $(n-1)$-dimensional orbits and the set of one-dimensional cones of the simple fan.

2. The closure of an arbitrary $(n-1)$-dimensional orbit is an $(n-1)$-dimensional submanifold.

Proof. Let us consider two charts of the manifold and the transition function from the first chart to the second.

Let an $(n-1)$-dimensional orbit, lying in the first chart, be identifyed with $(n-1)$-dimensional orbit, lying in the second chart. Let us suppose for simplicity that these orbits, both in the first and in the second chart, lie in the hyperplane $x_1=0$. We shall prove that corresponding to them is one and the same one-dimensional cone. Indeed the transition function is a monomial map. By assumption all the elements in the first column of the monomial map are zero except the first, which is equal to 1. According to the definition of the transition functions this means that in the skeletons of the cones corresponding to the two charts the first vector is the same, which is what we are trying to prove. It can be shown, analogously, that if $(n-1)$-dimensional orbits in different charts correspond to one and the same one-dimensional cone then these orbits are the same.

The assertion of the second part of the lemma is trivial. The lemma is proved.

A one-to-one correspondence analogous to the above, can be established between the set of k-dimensional orbits and $(n-k)$-dimensional cones of a simple fan. The closure of the orbits are submanifolds. In the local charts these are coordinate planes. If one orbit lies in the closure of another orbit, then the cone corresponding to the second orbit is a face of the cone corresponding to the first orbit. For more details see [184].

8.1.5 The map of manifolds associated with simple fans

Let us be given two fans. We shall say that the first fan is *inscribed* in the second fan if for any cone of the first fan there is a cone of the second fan which contains it.

Let us consider two simple fans. Let us suppose that the first fan is inscribed in the second fan. Let us consider the manifold associated with these fans. We shall define an analytic map of the first manifold into the second. To do this we shall define the restriction of the map to each chart of the first manifold. Let us consider an arbitrary chart of the first manifold. A chart corresponds to an n-dimensional cone of the first fan. By assumption there is a cone of the second fan which contains this cone of the first fan. The cone of the second fan is also, of course, n-dimensional. Therefore the cone of the second fan corresponds to a chart of the second manifold. So we have an ordered pair of charts. Let us consider the monomial map connected with these charts from the first chart to the second one (see § 8.1.3). The elements of the matrix of this monomial map are non-negative, since the first cone is inscribed in the second. In this way we have defined an analytic map of an arbitrary chart of the manifold associated with the first fan into one of the charts of the manifold associated with the second fan.

Lemma 8.4. These local maps are in agreement and correctly define an analytic map of the first manifold into the second one.

Lemma 8.4 is a direct corollary of Lemma 8.1.

Remark. On each manifold there is a unique n-dimensional orbit of the action of a torus. The map we have constructed gives an isomorphism of these orbits.

Theorem 8.1. (see [184]). Let us be given two simple fans, with the first fan inscribed in the second. Let us consider the manifolds associated with the fans and the map, constructed above, of the first manifold into the second. Then we assert that if the union of the cones of the first fan contains the union of the cones of the second fan then this map is proper. The converse is also true.

Corollary. Under the conditions of the theorem the first manifold maps *onto* the second manifold.

Indeed the map is proper and invertible on an everywhere dense subset.

The first part of the theorem follows from Lemma 8.5. The converse is analogous.

Lemma 8.5. Let us be given in one of the charts of the second manifold a curve of the form

$$x_j(t) = t^{m_j}(d_j + O(t)), \quad m_j \geqslant 0, \quad j = 1, \ldots, n,$$

where the numbers $d_1, \ldots, d_n$ are all non-zero. Then there exists a chart of the first manifold in which the preimage of the curve has a finite limit as $t \to 0$.

The **proof** is analogous to the proof of Lemma 8.2. We must select a chart of the first manifold and a curve in this chart of the form

$$x_j(t) = t^{k_j}(c_j + O(t)), \quad j = 1, \ldots, n,$$

where the numbers $c_1, \ldots, c_n$ are all non-zero, in such a way that its image coincides with our curve. We shall restrict ourselves to showing how to take the chart. Let us consider a basis forming the skeleton of a cone corresponding to a chart of the second manifold. Let us consider a linear combination of this basis with non-negative coefficients $m_1, \ldots, m_n$. As a result we shall obtain a vector belonging to the cone. By the conditions there exists an n-dimensional cone of the first fan which contains this vector. The n-dimensional cone of the first fan corresponds to a chart of the first manifold. In this chart the preimage of our curve has a finite limit. We shall leave the verification of this fact to the reader.

8.1.6 Important example

Let us consider two simple fans. Let us suppose that the second fan consists of one n-dimensional cone and its faces. Let us suppose that the union of the cones of the first fan coincides with the n-dimensional cone, generating the second fan. According to the construction of § 8.1.3, there is a manifold associated with each fan. The manifold associated with the second fan consists of one chart and is isomorphic to $\mathbb{R}^n$. According to the construction of § 8.1.5, there is a proper analytic map from the manifold associated with the first fan to the manifold associated with the second fan, that is to $\mathbb{R}^n$. This map is invertible outside the union of the coordinate hyperplanes.

   This example will be used in § 8.2 for the construction of the resolution of a singularity.

Exercise. Let $n = 2$ and as the first fan take the fan depicted in figure 68. Prove that the map onto $\mathbb{R}^2$ of the manifold associated with the first fan is the same as a σ-process at the origin.

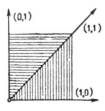

Fig. 68.

8.1.7 Complex analogue

The constructions of manifolds and maps of manifolds, described in this section, have natural complex analytic analogues. Instead of charts, isomorphic to $\mathbb{R}^n$ we need to take charts isomorphic to $\mathbb{C}^n$, and all the maps are given by the same formulae. The modified construction will lead to complex analytic manifolds and complex analytic maps of them. The complex manifolds constructed in this way have natural real parts. The real parts are real manifolds and coincide with the manifolds constructed in this section. The complex analytic maps preserve the real parts. The restrictions of the complex analytic maps to the real parts coincide with the maps constructed in this section.

8.2   Resolution of singularities

8.2.1   The fan associated with a Newton polyhedron

Let us consider a Newton polyhedron, that is a convex polyhedron in $\mathbb{R}^n$ with vertices at points with non-negative integer coordinates, which together with each point contains the positive orthant, parallel translated to this point (see §6.2.1). We shall denote the polyhedron by Γ.

The *supporting function* of a Newton polyhedron is a function on the positive orthant of the space dual to $\mathbb{R}^n$. Its value on the covector a of the positive orthant is equal to

$$\min_{k \in \Gamma} \langle a, k \rangle.$$

The supporting function is denoted by l_Γ.

The *trace* on the Newton polyhedron of the covector a of the positive orthant is the face of the polyhedron distinguished by the condition

$$\{k \in \Gamma \,|\, \langle a, k \rangle = l_\Gamma(a)\}.$$

The *joint trace* of the covectors of the positive orthant is the intersection of their traces.

Two covectors of the positive orthant are said to be *equivalent* relative to a Newton polyhedron if they have the same trace.

Lemma 8.6. The closure of any equivalence class is a rational cone in the space dual to $\mathbb{R}^n$. Furthermore the collection of all these cones forms a fan.

The proof follows easily from the definition of supporting functions. For further details see [159].

The fan formed by the closures of the equivalence classes is called the *fan associated with the Newton polyhedron*. The union of the cones making up this fan coincides with the positive orthant of the space dual to $\mathbb{R}^n$.

Example. Figure 69a depicts a Newton polyhedron in $\mathbb{R}^2$. Figure 69b depicts the fan associated with it.

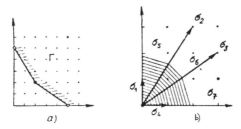

Fig. 69a and b.

8.2.2   A simple subordinate fan

A simple fan in the space dual to $\mathbb{R}^n$ is said to be *subordinate to a Newton polyhedron* if it is inscribed in the fan associated with the polyhedron and if the union of the cones making up this simple fan coincide with the positive orthant.

Lemma 8.7 (see [184]). There exists a simple fan subordinate to a Newton polyhedron.

An algorithm to construct a simple fan, inscribed in a given fan and such that the union of all its cones is the same as before, was given on pages 32–35 in [184]. We shall not reproduce the algorithm in detail. We shall indicate its main features.

The algorithm consists of two stages. In the first stage the cones of the original fan are broken down in an arbitrary way into simplicial cones. In the second stage of the algorithm we must make the multiplicty of all the n-dimensional cones of the decomposition equal to 1. This is done by decreasing induction on the number of cones with maximal multiplicity and then by decreasing induction on the number equal to the maximum multiplicity of the cones of the decomposition. At each step of the second stage we need to break up the simplicial cones of multiplicity greater than one into simplicial cones of smaller multiplicity. It can be done by the addition of a one-dimensional cone, generated by a correctly chosen integer vector. We need to take as this vector integer vector which is a linear combination of the vectors of the skeleton such that all the coefficients are non-negative, less than 1 and at least one coefficient is not equal to 0.

Example. In figure 69b was depicted the fan associated with the Newton polyhedron depicted in figure 69a. All the cones of this fan are simplicial. The multiplicity of the cones σ_5, σ_6, σ_7 are equal, respectively, to $3, 5, 3$. An example of a simple fan, subordinate to the indicated polyhedron, is depicted in figure 70.

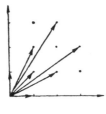

Fig. 70.

8.2.3 Theorem on the resolution of singularities

Let us consider a simple fan, subordinate to a Newton polyhedron. With the simple fan there is associated a manifold. This manifold is called the *manifold subordinate to the Newton polyhedron*. The union of the cones making up the simple fan coincides with the positive orthant, which in particular is a simplicial cone of multiplicity 1. According to the construction of § 8.1.5, the manifold subordinate to the polyhedron projects onto $\mathbb{R}^n$ (see § 8.1.6). This projection is said to be *associated* with the manifold subordinate to the Newton polyhedron. The projection is a proper analytic map.

Theorem 8.2 (on resolutions of singularities, see [45, 159, 358, 359]). Let us consider a convergent power series in n variables without constant term, with real coefficients and with $\mathbb{R}$-nondegenerate principal part. The series gives an analytic function in a neighbourhood of the origin in $\mathbb{R}^n$. Let us consider the manifold, subordinate to the Newton polyhedron of the power series and the projection, associated with the manifold, of the manifold into $\mathbb{R}^n$. We assert that the manifold and the projection resolve the singularity at the origin of the function given by the series.

The proof of the theorem is based on two lemmas formulated below.

8.2.4 Auxiliary lemmas

Let us consider two simplicial n-dimensional cones of multiplicity 1 in the space dual to $\mathbb{R}^n$. Let us suppose that the first cone belongs to the positive orthant, and the second cone coincides with the positive orthant. Let us suppose that the skeletons of the cones are ordered, the skeleton of the second cone being ordered in the standard way:

$$(1, 0, \ldots, 0), \ldots, (0, \ldots, 0, 1).$$

Let us denote by $a^j = (a_1^j, \ldots, a_n^j)$ the jth covector of the ordered skeleton of the first cone. The skeletons of the cones give an ordered pair of bases of the integer n-dimensional lattice. There is a monomial map associated with the pair of bases (see §8.1.2). Let us denote it by h. The matrix (a_i^j) of this monomial map has non-negative elements since the first cone is inscribed in the second. Let us consider a power series f in the variables $x_1, \ldots, x_n$. A monomial map with non-negative matrix induces a transformation of the power series into the power series $f \circ h$.

Lemma 8.8.

1. The maximal power of the variable x_j, by which the power series is divisible after the monomial transformation is equal to the value of the supporting function of the Newton polyhedron of the initial power series evaluated on the jth covector of the ordered skeleton of the first cone.

2. The Jacobian of the monomial map is equal modulo a sign to the monomial in which the power of the variable x_j is equal to one less than the sum of the coordinates of the jth covector of the ordered skeleton of the first cone.

3. The Newton diagram of the power series after the monomial transformation is a point if and only if all the covectors of the interior of the first cone are equivalent relative to the Newton polyhedron of the initial power series.

4. The image of the coordinate hyperplane $x_j = 0$ is contained in the coordinate plane given by the equations $x_i = 0$, $i \in I$, where I is the set of positions of all the non-zero coordinates of the jth covector of the ordered skeleton of the first cone.

The proof is obvious.

Let us suppose that the power series f has real coefficients and converges. We shall denote by the same letter the function given by the series. Let us consider the function $f \circ h$ induced from the function f by the monomial map. With each coordinate hyperplane there are associated two numbers: the multiplicity of zero on the hyperplane of the function $f \circ h$ and the multiplicity of zero on the hyperplane of the Jacobian of the monomial map. Let us denote the first number by k and the second by m. The number $-(m+1)/k$ was called, in Chapter 7, the weight of the hyperplane. Weights play a fundamental role in Theorem 7.5. Let us give a clear geometrical meaning to the weight of a coordinate hyperplane. For definiteness, let this hyperplane be given by the equation $x_1 = 0$.

Let us denote by γ the Newton polyhedron of the initial power series. Let us consider the hyperplane in $\mathbb{R}^n$ given by the equation

$$\langle a^1, x \rangle = l_\Gamma(a^1),$$

where a^1 is the first covector of the ordered skeleton of the first cone, l_Γ is the supporting function of the polyhedron Γ. The intersection of this hyperplane with the polyhedron is the trace of the covector a^1. The hyperplane intersects the bisector of the positive orthant in exactly one point $(t, \ldots, t)$. According to Lemma 8.8, the weight of the coordinate hyperplane $x_1 = 0$ is equal to

$$-(a_1^1 + \ldots + a_n^1)/l_\Gamma(a^1) = -1/t.$$

See figure 71. This remark explains the appearance in Theorem 6.4 of a number equal to the remoteness of the Newton polyhedron.

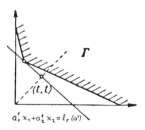

$$a_1' x_1 + a_2' x_2 = l_\Gamma(a^1)$$

Fig. 71.

Let us formulate the second auxiliary lemma.

Lemma 8.9. Let us suppose that the conditions of Lemma 8.8 are satisfied. Let us suppose that under the monomial map h the hyperplane given by the equation $x_1 = 0$ maps into the origin. Let us suppose that the power series f (without constant term) converges and has $\mathbb{R}$-nondegenerate principal part. Let us suppose that all the covectors of the interior of the first cone are equivalent relative to the Newton polyhedron of the series f. Then at each point of the hyperplane $x_1 = 0$ there exist local coordinates in which the function $f \circ h$ and the Jacobian of the map h are equal to monomials modulo multiplication by a function which does not map to zero.

Proof. It is sufficient to prove that the above-mentioned local coordinates exist for points at which the first s Euclidean coordinates are equal to zero and the rest of the Euclidean coordinates are different from zero.

By assumption the series $f \circ h$ can be put in the form

$$x_1^{l_r(a^1)} \ldots x_n^{l_r(a^n)} \, (\text{const} + O(x_1, \ldots, x_n)),$$

where $\text{const} \neq 0$ (see Sections 1 and 3 of Lemma 8.8). Let us rewrite the series $f \circ h$ in the form

$$x_1^{l_r(a^1)} \ldots x_s^{l_r(a^s)} \, (f_0(x_{s+1}, \ldots, x_n) + O(x_1, \ldots, x_s)).$$

It is sufficient to prove that the hypersurface given by the equation $f_0 = 0$ does not have singular points in $(\mathbb{R} \setminus 0)^n$.

All the coordinates of the covector a^1 are positive (Section 4 of Lemma 8.8) therefore f_0 is a polynominal. Let us denote by γ the joint trace of the covectors $a^1, \ldots, a^s$. Now γ is a non-empty compact face of the Newton polyhedron. Let us denote by f_γ the γ-part of the series f. It is clear that

$$f_\gamma \circ h = x_1^{l_r(a^1)} \ldots x_s^{l_r(a^s)} f_0.$$

In view of the $\mathbb{R}$-nondegeneracy of the principal part of the series f the first partial derivatives of the γ-part do not have common zeros in $(\mathbb{R} \setminus 0)^n$. The map h gives a diffeomorphism

$$(\mathbb{R} \setminus 0)^n \to (\mathbb{R} \setminus 0)^n,$$

therefore the polynomials

$$f_0, \ \partial f_0 / \partial x_{s+1}, \ldots, \partial f_0 / \partial x_n$$

do not have common zeros in $(\mathbb{R} \setminus 0)^n$, which is what we were required to prove.

8.2.5 Proof of Theorem 8.2

Let us verify that the three sections of the definition of the resolution of a singularity, listed on page 195–196 are satisfied.

Let us consider the manifold, subordinate to the Newton polyhedron of the series f and the projection associated with it of the manifold onto $\mathbb{R}^n$. According to Theorem 8.1, the projection is a proper analytic map. Consequently, Section 3 of the definition of the resolution of a singularity is satisfied. The projection is invertible away from the union of the coordinate hyperplanes in $\mathbb{R}^n$. Consequently, Section 2 of the definition is satisfied. Finally Section 1 is a direct corollary of Lemma 8.9. The theorem is proved.

Remark 1. The resolution of the singularity in Theorem 8.2 is determined by a simple fan, subordinate to the Newton polyhedron. By changing the fan we can provide a resolution of the singularity with additional properties. Namely we can choose a simple fan, subordinate to the Newton polyhedron, such that the map of the resolution of the singularity, indicated in Theorem 8.2, is invertible away from the zero level hypersurface of the function given by the power series [210]. Invertibility of the map outside the zero level hypersurface means that condition 2′ on page 197 is satisfied.

Lemma 8.10. The resolution of the singularity, indicated in Theorem 8.2, satisfies condition 2′ on page 196 if the simple fan defining the resolution satisfies the following additional property: This fan includes any cone of the fan associated with the polyhedron if this cone is simplicial and its skeleton can be extended to a basis of the integer lattice.

Example. In figures 69 and 70 were depicted a Newton polyhedron, the fan associated with it and a simple fan subordinate to the Newton polyhedron. The simple fan possesses the property indicated in Lemma 8.10.

The lemma is easily proved with the help of Sections 1 and 4 of Lemma 8.8, see also [359].

Lemma 8.11. There exists a simple fan subordinate to the Newton polyhedron and possessing the property indicated in Lemma 8.10.

Such a simple fan can be constructed by the algorithm indicated in § 8.2.2.

Remark 2. Let us formulate the complex analogue of Theorem 8.2.

Theorem 8.2' (see [45, 159, 359]). Let us consider a convergent power series in n variables without constant term, with complex coefficients and with $\mathbb{C}$-nondegenerate principal part. The series gives an analytic function in a neighbourhood of the origin in $\mathbb{C}^n$. Let us consider a complex analytic manifold, subordinate to the Newton polyhedron of the power series and the projection, associated with the manifold, of the manifold onto $\mathbb{C}^n$ (see § 8.1.7). We assert that the manifold and the projection resolve the singularity at the origin of the function given by the series (that is that the function, the manifold and its map onto $\mathbb{C}^n$ possess properties 1–3 on page 195–196).

The proof is the same as the proof of Theorem 8.2.

Theorem 8.2', like Theorem 8.2, admits the condition: there exists a manifold, subordinate to the Newton polyhedron of the power series, for which the resolution of the singularity in Theorem 8.2' possesses the property 2' on page 196 (see [359]).

8.3   Application to oscillatory integrals

8.3.1

Theorem 8.3 (see [358]). Let us consider the oscillatory integral

$$\int_{\mathbb{R}^n} e^{i\tau f(x)} \varphi(x) dx_1 \ldots dx_n.$$

Let us suppose that the phase is an analytic function in a neighbourhood of the origin. Let us suppose that the Taylor series of the phase at the origin has $\mathbb{R}$-nondegenerate principal part. Let us consider the Newton polyhedron of the Taylor series. Let us consider a simple fan, subordinate to this Newton polyhedron. In connection with these objects we claim the following assertions 1–5.

1. The index set of the phase at the origin belongs to the union of the following arithmetic progressions, depending only on the fan and not depending on the coefficients of the Taylor series. One sequence is that of the negative integers. The rest of the progressions are parametrised by the one-dimensional cones of the fan, on which the supporting function of the Newton polyhedron is different

from zero. To such a cone corresponds the arithmetic progression

$$-(a^1 + \ldots + a^n)/l_\Gamma(a), \quad -(1 + a^1 + \ldots + a^n)/l_\Gamma(a), \ldots,$$

where $a = (a^1, \ldots, a^n)$ is the primitive covector generating the cone and l_Γ is the supporting function of the Newton polyhedron.

2. The oscillation index of the phase at the origin is not more than the remoteness of the Newton polyhedron.

3. The oscillation index of the phase at the origin is equal to the remoteness of Newton polyhedron if at least one of the following three conditions is satisfied:

(i) The polyhedron is remote.

(ii) The phase has a maximum or a minimum at the origin.

(iii) Let us denote by γ the closure of the open face of the Newton polyhedron to which the centre of the boundary of the Newton polyhedron belongs (see § 6.2.3); it is required that the γ-part of the Taylor series of the phase at the origin does not have a zero in $(\mathbb{R}\setminus 0)^n$ and the remoteness of the Newton polyhedron was not an odd integer (the condition about the absence of a zero is satisfied, in particular, if γ is a vertex of the polyhedron).

4. If at least one of the conditions (i)–(iii) of section 3 of the theorem is satisfied, then the multiplicity of the oscillation index of the phase at the origin is equal to the multiplicity of the remoteness of the Newton polyhedron (in particular, if the bisector of the positive orthant passes through a vertex of the Newton polyhedron then the multiplicity equals $n-1$, if through an edge then the multiplicity equals $n-2$, etc.). If the support of the amplitude is concentrated in a sufficiently small neighbourhood of the origin, the amplitude is of fixed sign and is different from zero at the origin, then the numerical coefficient of the leading term of the asymptotic series of the oscillatory integral (that is the coefficient $a_{K,\beta}$ of the series (2) on page 181) is different from zero.

5. Let us suppose that at least one of the conditions (i)–(iii) of section 3 of the theorem is satisfied. Let us suppose that the phase has a critical point of finite multiplicity at the origin and that the support of the amplitude is concentrated in a small neighbourhood of the origin. Then the numerical coefficient of the leading term of the asymptotic series of the oscillatory integral (that is the coefficient $a_{K,\beta}$ of the series (2) on page 181) is equal to the value of the amplitude at the origin, multiplied by a non-zero constant, depending only on the phase.

6. Let us suppose that the remoteness of the Newton polyhedron is equal to -1. Then the oscillation index of the phase at the origin is equal to -1 if at least one of the following two conditions is satisfied:

(i) the open face, to which the centre of the boundary of the Newton polyhedron belongs, has dimension less than $n-1$.

(ii) The closure γ of the open face to which the centre of the boundary of the Newton polyhedron belongs is compact and the γ-part of the Taylor series has a zero on $(\mathbb{R}\setminus 0)^n$.

Furthermore in this case the multiplicity of the oscillation index of the phase at the origin is equal to the multiplicity of the remoteness of the Newton polyhedron or one less than the multiplicity of the remoteness. If conditions (i) and (ii) are satisfied simultaneously, then the multiplicity of the oscillation index is equal to the multiplicity of the remoteness.

Remark 1. Theorem 6.4 and its supplements (i), (ii), (vi), (vii) are corollaries of the theorem we have just formulated.

Remark 2. In Section 1 of the theorem we indicated the method of constructing arithmetic progressions. There is a different method for constructing similar progressions. This method uses only the Newton polyhedron, and does not use the simple fan, subordinate to the polyhedron. The method can be used if the function has at the origin a critical point of finite multiplicity and the Taylor series of the phase has $\mathbb{C}$-nondegenerate principal part. The method is based on the following theorem of Malgrange. With each critical point of finite multiplicity of the function there is connected the linear monodromy operator in the vanishing homology at the point (see Part I). With each root λ of the characteristic polynomial of the monodromy operator is connected the arithmetic progression of all the numbers α for which $\exp(2\pi i\alpha) = \lambda$. The theorem of Malgrange (see Chapter 11) asserts: the index set of the critical point is contained in the union of the progressions we constructed. In Theorem 3.13 we indicated a formula expressing the characteristic polynomial of the monodromy operator in terms of the Newton polyhedron of the Taylor series of the critical point.

Proof of the theorem.

Let us suppose that $f(0) = 0$. With the simple fan of the theorem there is associated a manifold subordinate to the Newton polyhedron of the Taylor series. This manifold and the projection associated with it resolve the singularity of the phase at the origin (Theorem 8.2). We apply Theorem 7.5. For the proof of section 1 of the theorem we must indicate the set of multiplicities of the resolution of the singularity, that is we must indicate the multiplicities of the irreducible components of the zero level hypersurface of the phase, lifted to the manifold resolving the singularity. In a local chart of the manifold the lifted phase is equal to the product of a monomial and a function with non-singular zero level hyper-surface (see Lemma 8.9); in addition to which the Jacobian of the resolution is equal to a monomial (Lemma 8.8). Therefore in terms of a local chart the irreducible components with multiplicity not equal to $(1, 0)$ (see §7.3) are those coordinate hyperplanes on which the multiplicity of zero of the lifted phase is greater than 1. Lemma 8.8 expresses the multiplicity of zero of the lifted phase and the Jacobian of the resolution in terms of the corresponding primitive covectors of the one-dimensional cone of a simple fan. Then Section 1 of the theorem follows from Theorem 7.5 and Lemma 8.8.

Let us take notice of the geometrical meaning of the first number of the arithmetic progression corresponding to a primitive covector of a one-dimensional cone (see the remark after Lemma 8.8). This first number is equal to minus the reciprocal of the intersection parameter of the bisector of the positive orthant and the hyperplane defined by the covector and leaning upon the polyhedron. Therefore among the first numbers of the arithmetic progressions, indicated in section 1 of the theorem there is certainly a number equal to the remoteness of the polyhedron. Namely, the remoteness is the first number of the arithmetic progression corresponding to any covector the trace of which contains the centre of the boundary of the polyhedron.

Let us prove Section 2. If the remoteness of the Newton polyhedron is equal to -1 then Section 2 follows from Section 1 of Theorem 7.5. If the remoteness is less than -1 then the Taylor series of the phase is not divisible by any one of the variables. According to Lemmas 8.8, 8.9 this means that the conditions of Section 5 of Theorem 7.5 are satisfied. Then Section 2 of the theorem follows from Section 5 of Theorem 7.5. The case in which the remoteness of the Newton polyhedron is greater than -1 is examined during the proof of Section 3.

Section 3(i) of the theorem follows from Section 3 of Theorem 7.5, since in this case the weight of the resolution is greater than -1 and equal to the remoteness of the Newton polyhedron. Section 3(ii) follows from Section 4 of Theorem 7.5.

Sections 4(i) and 4(ii) will follow from Sections 3, 4 of Theorem 7.5 if it is proved that the multiplicity of the weight of the resolution equals the multiplicity of the remoteness of the Newton polyhedron. For the proof we mention the following obvious fact. Let us consider all the cones of the simple fan, subordinate to the Newton polyhedron, which possesses the property: the traces of all the covectors forming the cone contain the centre of the boundary of the Newton polyhedron. Then the maximum dimension of the indicated cones is equal to one more than the multiplicity of the remoteness of the Newton polyhedron. Then the required result follows from Lemma 8.8.

Sections 5(i) and 5(ii) follow, respectively from Sections 3(iv) and 4(iv) of Theorem 7.5. Sections 3(iii), 4(iii) and 5(iii) follow from Section 6 of Theorem 7.5. Section 6 follows from Section 7 of Theorem 7.5. The theorem is proved.

According to Theorem 8.3, the order of the oscillatory integral is equal to the remoteness of the Newton polyhedron of the phase, if the amplitude is of constant sign and its value at the critical point of the phase is different from zero. We shall formulate an assertion describing the order of the integral in the case when the amplitude is equal to zero at the critical point of the phase.

Let us be given two Newton polyhedra. By the *coefficient of inscription* of the first polyhedron in the second we shall mean the lower bound of the following set of positive numbers. A number belongs to the set if the homothety with centre at the origin and expansion coefficient equal to the number maps the first polyhedron inside the second.

We put into correspondence with the amplitude of an oscillatory integral the Newton polyhedron of its Taylor series multiplied by the product of all the variables. In this way we have two polyhedra: this polyhedron and the Newton polyhedron of the Taylor series of the phase. By the *remoteness of the polyhedra of the phase and the amplitude* we shall mean minus the reciprocal of the coefficient of inscription of the first polyhedron in the second.

Exercise. Prove that the remoteness of the polyhedra of the phase and the amplitude is equal to the remoteness of the Newton polyhedron of the phase if the Taylor series of the amplitude has non-zero constant term.

Example. Let the Taylor series of the phase and the amplitude be equal, respectively, to

$$x_1^6 + x_1^3 x_2^2 + x_2^6 \quad \text{and} \quad x_1 x_2.$$

Then the remoteness of their polyhedra is equal to $-7/9$.

Theorem 8.4. Let us suppose that the phase of an oscillatory integral is an analytic function in a neighbourhood of the origin. Let us suppose that the Taylor series of the phase at the origin has $\mathbb{R}$-nondegenerate principal part. Then:

1. The power of the parameter of the leading term of the asymptotic series of the oscillatory integral is not greater than the remoteness of the polyhedra of the phase and the amplitude.

2. The power of the parameter of the leading term is equal to the remoteness of the polyhedron of the phase and the amplitude if this remoteness is greater than -1 and the polyhedron put in correspondence with the amplitude is congruent to the positive orthant.

The proof is analogous to the proof of Theorem 8.3, except that instead of the references to Theorem 7.5 we must refer directly to Theorem 7.3. We notice that Section 2 of the theorem is analogous to Section 3(i) of Theorem 8.3. There are true analogies to Sections 3(ii), 3(iii), 4, 5.

In conclusion we analyse the case of a degenerate principal part of the Taylor series.

Theorem 8.5 (see [358]). Let the phase be an analytic function in a neighbourhood of the origin. Let us suppose that the Newton polyhedron of the Taylor series of the phase at the origin is remote. Then the oscillation index of the phase at the origin is not less than the remoteness of this polyhedron.

Corollary 1. The weight of the resolution of the singularity of the phase is greater than -1.

See Theorem 7.5.

Corollary 2. The assertion of Section 3 of Theorem 7.5 is true for the phase.

Corollary 3. Let the phase be a function of two variables with a degenerate critical point at the origin. Then the oscillation index of this critical point is equal to the weight of the resolution of its singularity.

Indeed in this case it is easy to select a system of coordinates in which the remoteness of the Newton polyhedron of the Taylor series of the phase is greater than -1.

Corollary 4. The remoteness of the critical point is not greater than the oscillation index if the remoteness is greater than -1 (see the definition in § 6.2.4).

We note that Theorem 6.5 asserts the equality of the remoteness and the oscillation index for all critical points of a phase of two arguments.

Proof of the theorem. Let us consider the manifold X subordinate to the Newton polyhedron and the projection

$$\pi : X \to \mathbb{R}^n$$

associated with it. With the help of the projection we lift the phase to the manifold and we resolve all the singularities of the lifted phase on the preimage of the origin. Namely, by the theorem of Hironaka [158], there exists a new manifold Y and a map $\phi: Y \to X$, possessing the property: the map $\pi \circ \phi$ resolves the singularity of the phase at the origin. The phase, lifted to X, has a component of the zero level hypersurface with weight equal to the remoteness of the Newton polyhedron (see the proof of Theorem 8.3). The preimage of this component on Y has the same weight. Now the theorem follows from Section 3 of Theorem 7.5.

8.3.2   Generalisation of Theorem 8.3 to the Laplace integral

Theorem 8.6. Let us consider the Laplace integral

$$\int_{\mathbb{R}^n} e^{-\tau f(x)} \varphi(x) dx_1 \ldots dx_n.$$

Let us suppose that the phase is an analytic function in a neighbourhood of the origin and has a local minimum at the origin. According to Theorem 7.6 as $\tau \to +\infty$ the Laplace integral can be expanded in the asymptotic series

$$e^{-\tau f(0)} \sum_{k=0}^{n-1} \sum_{\alpha} a_{k,\alpha} \tau^\alpha (\ln \tau)^k.$$

Let us suppose that the Taylor series of the phase at the origin has $\mathbb{R}$-non-degenerate principal part. Let us consider the Newton polyhedron of the Taylor series of the phase. Let us consider a simple fan subordinate to this Newton polyhedron. Then the asymptotics of the Laplace integral possess the properties of the asymptotics of an oscillatory integral with the same phase, indicated in Theorem 8.3 in Sections 1, 3, 4, 5.

The proof of Theorem 8.6 is obtained from the proof of Theorem 8.3 by changing the references to Theorem 7.5 to references to Theorem 7.6.

Corollary of the theorem (compare with the corollary of Theorem 7.6). For each positive t let us denote by $V(t)$ the volume of the set of points in which the value of the phase is less than t. According to Theorem 7.6 the function V as $t \to +0$ can be expanded in the asymptotic series

$$\sum_{\alpha} \sum_{k} a_{k,\alpha} t^\alpha (\ln t)^k.$$

It is asserted that the order α of the maximal term of this series is equal to minus the remoteness of the Newton polyhedron of the Taylor series of the phase at the minimum point (under the condition of $\mathbb{R}$-nondegeneracy of the principal part of the Taylor series).

Remark. The assertion that the leading term of the asymptotic series of the Laplace integral is equal to the remoteness of the Newton polyhedron of the phase was proved by V. A. Vasilev in [386] for the case when the principal part of the Taylor series of the phase is $\mathbb{R}$-nondegenerate and the Newton polyhedron intersects each coordinate axis. The proof of V. A. Vasilev does not use the resolution of singularities.

8.3.3   The area of the level surface of a function

Let us suppose that on the space there is given a Riemannian metric. The Riemannian metric on the space gives rise to a Riemannian metric on the level hypersurface of a function. The Riemannian metric on the hypersurface determines an $(n-1)$-dimensional volume form. Let us calculate the volume of compact level manifolds of a function and let us consider the asymptotic volume as the level tends to the critical value.

Theorem 8.7. Let us suppose that the analytic function f has an isolated minimum point and that the minimal value of the function is equal to zero. Let us suppose that in the space there is given an analytic Riemannian metric. For small positive t let us denote by $V(t)$ the $(n-1)$-dimensional volume of the level t manifold. Then as $t \to +0$ the function V can be expanded in the asymptotic series

$$\sum_{\alpha} \sum_{k=0}^{n-1} a_{k,\alpha} t^{\alpha} (\ln t)^k.$$

in which the parameter α runs through a finite set of arithmetic progressions consisting of positive rational numbers. If in addition it is known that the principal part of the Taylor series of the function f is $\mathbb{R}$-nondegenerate at the minimum point and that the principal part of the Taylor series of the function (df, df) is $\mathbb{R}$-nondegenerate at the minimum point of the function f then the order α of the maximal term of the asymptotic series depends only on the Newton polyhedra of the above Taylor series and is calculated according to the following rule.

Rule. Let us consider the Newton polyhedron of the Taylor series of the function (df, df) at the minimum point of the function f. Let us consider the image of this polyhedron under the action of the homothety with coefficient 1/2 and centre at the origin. Let us move the resulting polyhedron to the vector $(1, \ldots, 1)$. Let us consider a second polyhedron, namely the Newton polyhedron of the Taylor series of the function f at the minimum point. Let us denote by k the coefficient of inscription of the first polyhedron in the second. Then the order α of the maximal term of the asymptotic series is equal to $1/k - 1$.

The proof of the theorem is based on the fact that the number $-1/k$ is the index of the maximal term of the asymptotic series of the Laplace integral

$$\int_{\mathbb{R}^n} e^{-\tau f(x)} \sqrt{(df, df)}\, \varphi(x) dx_1 \ldots dx_n$$

as $\tau \to +\infty$, if the amplitude φ is identically equal to 1 in a small neighbourhood of the minimum point and the support of the amplitude is concentrated in a small neighbourhood of the minimum point.

8.3.4   Oscillatory integrals in a halfspace

Let us consider the oscillatory integral in the halfspace

$$\int_{x_1 \geq 0} e^{i\tau f(x)} \varphi(x) dx_1 \ldots dx_n,$$

where the phase and the amplitude are smooth functions on the whole space. Let us suppose that the phase is an analytic function in a neighbourhood of the origin. If the support of the amplitude is concentrated in a sufficiently small neighbourhood of the origin and the restriction of the phase to the boundary of the halfspace does hot have a critical point at the origin then as $\tau \to +\infty$, the integral decreases faster than any power of the parameter (Theorem 6.1'). Let us suppose that the restriction of the phase to the boundary has a critical point at the origin.

Theorem 8.8. The oscillatory integral on the halfspace can be expanded in the asymptotic series

$$e^{i\tau f(0)} \sum_{\alpha} \sum_{k=0}^{n-1} a_{k,\alpha}(\varphi) \tau^{\alpha} (\ln t)^k \quad \text{as} \quad \tau \to +\infty$$

if the support of the amplitude is concentrated in a sufficiently small neighbourhood of the origin. Here the parameter α runs through a finite set of arithmetic progressions depending only on the phase and consisting of negative rational numbers. The numerical coefficients $a_{k,\alpha}$ are generalised functions of the amplitude. The support of each generalised function lies in the union of the critical sets of the phase and the restriction of the phase to the boundary.

The proof is analogous to the proof of Theorem 7.5. For the proof we need to consider the resolution of the singularity at the origin of the function $x_1 f$ (it is simultaneously a resolution of the singularity of the function f), and then repeat the reasoning of the proof of Theorem 7.5. In the result for an oscillatory integral on a half-space we shall prove the assertions of Sections 1–5 of Theorem 7.5 (a natural addition is needed in Section 5: it is true if the phase is not divisible by x_1).

If the origin is a critical point of the restriction of the phase to the boundary, but is not a critical point of the phase, the analysis of the asymptotic integral on the halfspace reduces to the analysis of the oscillatory integral on the boundary. Indeed by a diffeomorphism preserving the boundary the phase can be reduced to the form

$$x_1 + h(x_2, \ldots, x_n).$$

Then we can integrate by parts with respect to x_1.

Let us suppose that the phase has a critical point at the origin. Let us suppose that the principal part of the Taylor series of the phase at the origin is $\mathbb{R}$-nondegenerate.

Theorem 8.9. Under the above assumptions the asymptotics of the oscillatory integral on the halfspace possess the properties indicated in Sections 1–5 of Theorem 8.3.

The proof is the same as the proof of Theorem 8.3.

Let us consider one more type of oscillatory integral on a halfspace, namely an integral of the type

$$\int_{x_1 > 0} e^{i\tau f(x)} \varphi(x) x_1^{-1/2} \, dx_1 \ldots dx_n.$$

Here, as earlier, the phase and the amplitude φ are smooth functions on the whole space. The analysis of such integrals can be reduced to the analysis of the integrals we considered earlier on the halfspace with the help of the change $x_1 = z^2$. We shall formulate one of the results obtained in this manner.

Theorem 8.10. Let us suppose that the phase is an analytic function in a neighbourhood of the origin. Let us suppose that the principal part of the Taylor series of the phase at the origin is $\mathbb{R}$-nondegenerate. Then the oscillation index of the phase for integrals of the indicated type is determined by the Newton polyhedron of the Taylor series of the phase and is equal to minus the reciprocal of the value of the parameter of the point of intersection of the line

$$2x_1 = x_2 = x_3 = \ldots = x_n = t,$$

where $t \in \mathbb{R}$, and the boundary of the Newton polyhedron, if the value is greater than 1.

All the conclusions about the asymptotics of oscillatory integrals on a halfspace formulated in this section are true for the asymptotics of Laplace integrals on a halfspace.

8.4   The two-variable case

According to Theorem 8.3 the oscillation index of the critical point of the phase is equal to the remoteness of the Newton polyhedron of its Taylor series in some system of coordinates if in this system of coordinates the principal part of the Taylor series is $\mathbb{R}$-nondegenerate and the Newton polyhedron is remote. This theorem applies to an arbitrary critical point of the phase depending on one argument. If the phase depends on two or more arguments then the indicated system of coordinates does not always exist (see § 6.2.4). All the same, we have managed to investigate to the end the case of a phase depending on two arguments and prove the equality of the oscillation index and the remoteness of the Newton polyhedron of the Taylor series of the phase in a correctly chosen system of coordinates. The correctly chosen system of coordinates is said to be *adapted to the phase* and was defined in § 6.2.4.

Theorem 8.11 (see [358]). Let us consider the double oscillatory integral

$$\int_{\mathbb{R}^2} e^{itf(x)} \varphi(x) dx_1 dx_2.$$

Let us suppose that the phase is an analytic function in the neighbourhood of its degenerate critical point. Then
  1. The oscillation index of the critical point is equal to its remoteness.

2. There exists a system of local coordinates adapted to the critical point.

3. The multiplicity of the oscillation index of the critical point is equal to 1 if there exists a system of coordinates, adapted to the critical point, in which the centre of the boundary of the Newton polygon of the Taylor series of the critical point lies at the intersection of two edges of the polygon. Otherwise the multiplicity of the oscillation index is equal to 0.

Remark 1. The assertions of Sections 1 and 2 are true also for non-degenerate critical points; the assertion of Section 3 is not (example: $f = x_1 x_2$).

Remark 2. The assertions about asymptotics, formulated in the theorem, are true also for asymptotic Laplace integrals with phases depending on two arguments.

Remark 3. Theorem 8.11 implies Theorem 6.5 and its supplement (i). Supplement (ii) follows from Corollary 3 of Theorem 8.5 and Section 3 of Theorem 7.5.

Remark 4. In [358] there is given an algorithm to search for an adapted system of coordinates and it is shown how to recognise adapted coordinates. According to one of the signs a system of coordinates is adapted to the critical point if the centre of the boundary of the Newton polygon of the Taylor series of the critical point lies on the intersection of two edges of the polygon, cf. Section 3 of the theorem.

Remark 5. The oscillation index of a degenerate critical point of a phase of two arguments is not less than the remoteness of the critical point according to Corollary 4 of Theorem 8.5.

The proof (as also the proof of Theorem 8.3) depends on the analysis of the resolution of the singularity of the critical point of the phase. The analysis of the resolution of the singularity of the critical point of the phase in the case of two arguments is made simpler by two circumstances. Firstly, in this case there are simple algorithms for the resolution of the singularity by sequences of σ-processes at points. Secondly, in the two-dimensional case the weight of the resolution of a singularity of an arbitrary degenerate critical point is greater than -1 (Corollary 3 of Theorem 8.5.). In accordance with the second remark, the oscillation index is equal to the weight of the resolution of the singularity (Section 3 of Theorem 7.5). In this way it remains to prove that the weight of the resolution of the singularity is equal to the remoteness of the Newton polyhedron of the Taylor series of the phase adapted to the phase of the coordinate system.

Theorem 8.11 is a direct corollary of Theorem 8.12 below.

Theorem 8.12 (see [358]). Let us consider the resolution of the singularity of a degenerate critical point of an analytic function of two arguments. Then

1. The weight of the resolution of the singularity is equal to the remoteness of the critical point.

2. There exists a system of local analytic coordinates in which the remoteness of the Newton polyhedron of the Taylor series of the critical point is equal to the weight of the resolution of the singularity.

3. The multiplicity of the number, equal to the weight, relative to the resolution of the singularity (for the definition see § 7.3.1) is equal to 2 if there exists a system of coordinates, adapted to the critical point, in which the centre of the boundary of the Newton polygon lies at the intersection of two edges of the polygon. Otherwise the multiplicity is equal to 1.

Chapter 9

The singular index, examples

In this chapter we shall prove the additivity of the oscillation index, and describe explicitly the calculation of the singular index in the tables in § 6.1.10. In the second part of the chapter we give an example of the deformation of a critical point. This example illustrates several phenomena. First, the absence of semicontinuity of the oscillation index. Second, the existence of critical points which are complex equivalent but which have distinct singular indices. Third, the existence of a critical point in which the singular index is not equal to the remoteness. Finally, the existence of a critical point in which the principal part of the Taylor series is $\mathbb{R}$-nondegenerate but the remoteness of the Newton polyhedron is greater than the oscillation index.

9.1   The singular index

9.1.1   The additivity of the oscillation index and its multiplicity

Let $f : \mathbb{R}^n \to \mathbb{R}$ and $g : \mathbb{R}^l \to \mathbb{R}$ be smooth functions, and let x and y be their respective critical points. The critical point $x \times y$ of the function

$$f + g : \mathbb{R}^n \times \mathbb{R}^l \to \mathbb{R}$$

is called the *direct sum* of the critical points x and y.

Lemma 9.1. The oscillation index and the multiplicity of the oscillation index are additive.

Proof. Let us denote by β, K, respectively, the oscillation index and the multiplicity of the oscillation index.

It is clear that $\beta(x \times y) \geqslant \beta(x) + \beta(y)$, and if $\beta(x \times y) = \beta(x) + \beta(y)$ then $K(x \times y) \geqslant K(x) + K(y)$. Indeed if the amplitude of the oscillatory integral with phase $f + g$ can be decomposed as the product of two functions, one of which is a function on $\mathbb{R}^n$ and the other is a function on $\mathbb{R}^l$, then the integral itself decomposes into a product of oscillatory integrals with phases, respectively, f and g.

We shall prove the opposite inequality. Let us consider the oscillatory integral with phase $f+g$. Let us suppose that the support of the amplitude is concentrated in a small neighbourhood of the point $x \times y$. In this case the integral can be expanded in the asymptotic series

$$\sum_{\alpha} \sum_{k=0}^{n+l-1} a_{k,\alpha} \tau^{\alpha} (\ln \tau)^k,$$

where the numerical coefficients $a_{k,\alpha}$ are generalised functions of the amplitude. The generalised functions $a_{k,\alpha}$ for $\alpha > \beta(x) + \beta(y)$ are identically equal to zero, since in the space of the amplitudes the linear combinations of amplitudes which can be decomposed into a product of a function on $\mathbb{R}^n$ and a function on $\mathbb{R}^l$ form an everywhere dense set. Therefore

$$\beta(x \times y) = \beta(x) + \beta(y).$$

The equality

$$K(x \times y) = K(x) + K(y)$$

is proved analogously.

Corollary. The singular index and its multiplicity are equal for stably equivalent critical points.

9.1.2 Calculation of the singular index in the tables of § 6.1.10

The singular index and its multiplicity for the critical points of the tables were calculated with the help of Theorems 8.3 and 8.11. Theorems 8.3 and 8.11 can be applied to the critical points classified in Part II of Volume 1, since each of the indicated critical points either has a Taylor series with $\mathbb{R}$-nondegenerate principal part or, if it is a function of two variables, is written down with respect to an adapted system of coordinates. The result formulated may be simply verified in each separate case. After this the calculation leads to the calculation of the remoteness of the Newton polyhedron of the Taylor series.

9.2 Examples

Example 1. Let us put

$$F(x_1, x_2, x_3, y) = (yx_1^2 + x_1^4 + x_2^2 + x_3^2)^2 + x_1^p + x_2^p + x_3^p,$$

where y is a real parameter, $p \geqslant 9$. Let us denote by β_y the oscillation index of the critical point at the origin in $\mathbb{R}^3$ of the function $F(\cdot, y)$.

Theorem 9.1 (see [358]). The family F of functions on $\mathbb{R}^3$ depending on the parameter y has the following properties.

1. The function $F(\cdot, y)$ for all values of the parameter y has a critical point of finite multiplicity at the origin.

2. $\beta_0 = -5/8$.

3. For $y > 0$, $\beta_y = -3/4$.

4. For $y < 0$, $\beta_y > -(1/2 + \gamma(p))$, where the function $\gamma(p)$ tends to zero as $p \to +\infty$.

5. The remoteness of the critical point at the origin of the function $F(\cdot, y)$ for $y \neq 0$ is equal to $-3/4$.

6. There exists a neighbourhood U of the origin in $\mathbb{R}^3$ and a neighbourhood V of the origin in $\mathbb{R}$ such that the oscillation index of the function $F(\cdot, y)$, $y \in V$ at any of its critical points $x \in U \setminus 0$ is less than -1.

Corollary 1. For the critical point at the origin of the function $F(\cdot, 0)$ for sufficiently large p the uniform oscillation index is greater than the individual oscillation index.

From the proof of Theorem 9.1 it follows that such an occurrence has already been observed for $p = 9$.

Corollary 2. The critical points at the origin of the functions $F(\cdot, y)$ and $F(\cdot, -y)$ for $p = 4l$ are complex equivalent (that is they can be transformed into each other by a holomorphic diffeomorphism of the space $\mathbb{C}^3$), however they have a different oscillation index.

Corollary 3. The remoteness of the critical point at the origin of the function $F(\cdot, y)$ for $y < 0$ is less than the oscillation index.

Proof. Assertion 1 can be verified by direct calculation. Assertions 2, 3 follow from Theorem 8.3 in view of the $\mathbb{R}$-nondegeneracy of the principal part of the polynomial $F(\cdot, y)$ for indicated y. Assertion 5 follows from the fact that the remoteness of the critical point at the origin of the function

$$\pm x_1^2 + x_2^2 + x_3^2$$

is equal to $-3/2$.

For the proof of assertion 6 it is sufficient to observe that for small y and for critical points x that are near to zero the coordinates x_2, x_3 are equal to zero. This, as it is easy to see, implies that

$$\partial^2 F/\partial x_2 \partial x_3 (x, y) = 0,$$

$$\partial^2 F/\partial x_2^2 (x, y) \neq 0$$

$$\partial^2 F/\partial x_3^2 (x, y) \neq 0,$$

that is that the rank of the second differential is not less than 2. According to the generalised Morse lemma in a neighbourhood of such a critical point the function reduces to the form

$$\phi(u_1) \pm u_2^2 \pm u_3^2.$$

For this function assertion 6 follows from Theorem 8.3.

The proof of Section 4 is based on Theorem 7.5. A resolution of the singularity of the critical point at the origin is constructed and one can prove that the weight of this resolution is greater than $-(1/2 + \gamma(p))$ where $\gamma(p)$ tends to zero as $p \to +\infty$. In the construction of the resolution first a σ-process is performed at the origin and then a σ-process is performed $p/2$ times with centre on the curve which is the intersection of the preimage of the origin and the proper preimage of the zero level surface of the function F. The component arising last has sufficiently high weight to prove Section 4. For more details see [358].

Example 2. Let

$$f_+ = x_4^2 + x_1^p + x_2^p + x_3^p + (x_4 - (x_1^2 + x_1^4 + x_2^2 + x_3^2)) \cdot x_5,$$

$$f_- = x_4^2 + x_1^p + x_2^p + x_3^p + (x_4 - (-x_1^2 + x_1^4 + x_2^2 + x_3^2)) \cdot x_5.$$

Theorem 9.2. For sufficiently large p the polynomials $f_\pm$ possess the following properties.

1. Their principal parts are $\mathbb{R}$-nondegenerate.

2. Their Newton polyhedra coincide.

3. The oscillation index at the origin of the polynomial f_+ is equal to $-7/4$.

4. The oscillation index at the origin of the polynomial f_- is not less than $-(1/2+1+\gamma(p))$, where $\gamma(p)$ tends to zero as $p \to +\infty$.

5. The remoteness of the Newton polyhedra of the polynomials is less than -1.

Corollary 1. There exists a resolution of the singularity of the critical point at the origin of the polynomial f_+, the weight of which is greater than the oscillation index.

Indeed the resolution of the singularity indicated in Theorem 8.2 is such a resolution (see also Theorem 8.3).

Corollary 2. The remoteness of the Newton polyhedron of the critical point at the origin of the polynomial f_+ is greater than the oscillation index.

Proof. Section 2 is obvious. Sections 1 and 5 can be verified by direct calculation. We shall prove Section 3. Section 4 is proved analogously.

We make the substitution

$$v = x_5 + x_4 + x_1^2 + x_1^4 + x_2^2 + x_3^2$$

then

$$f_+ = (x_1^2 + x_1^4 + x_2^2 + x_3^2)^2 + x_1^p + x_2^p + x_3^p + (x_4 - (x_1^2 + x_1^4 + x_2^2 + x_3^2))v.$$

We make the substitution

$$u = x_4 - (x_1^2 + x_1^4 + x_2^2 + x_3^2),$$

and then $v = s + t$, $u = s - t$, then

$$f = F(x_1, x_2, x_3, 1) + s^2 - t^2,$$

where F is the function from example 1. Then Section 3 follows from Theorem 9.1 and Lemma 9.1.

Part III

Integrals of holomorphic forms over vanishing cycles

The first part of the book was devoted to the topology of a critical point of a holomorphic function; the third part is devoted to its analysis. The basic object of study is the integral of a holomorphic form, defined in a neighbourhood of a critical point, over a cycle lying in the level manifold of the function and vanishing at the critical point. We shall study the change in the integral under continuous deformation of the cycle from one level manifold to another. We shall show that the asymptotic behaviour of such integrals under deformation of the cycles at the critical point contains information about a very varied collection of objects, connected with the critical point.

In Chapter 10 we shall give an account of the simplest properties of the integrals (holomorphic dependence on parameters, expansion in series, connection with the monodromy group). In Chapter 11 we shall describe the interaction of asymptotic integrals over cycles with asymptotic integrals of the saddle point method, for which the holomorphic function serves as the phase (in particular with the asymptotics of oscillatory integrals). In Chapter 12 we shall discuss the differential equations which are satisfied by the functions given by the integral of a holomorphic form over a cycle depending continuously on parameters. Chapter 13 is dedicated to a discussion of properties of coefficients of asymptotic expansions of integrals of holomorphic forms over cycles, depending continuously on parameters. In Chapter 13 we shall define the mixed Hodge structure of a finite-multiplicity critical point of a holomorphic function. In Chapter 14 we shall discuss the interaction of the mixed Hodge structure of a critical point with the other characteristics of the critical point. In Chapter 15 we shall construct with the help of integrals maps from the base of a versal deformation of a critical point into the cohomology vanishing at the point. These maps carry the structures found in the cohomology to the base of the versal deformation.

Chapter 10

The simplest properties
of the integrals

In this chapter we shall prove the holomorphic dependence of the integral on the parameters; we shall explain the connection between branches of integrals and the monodromy group in homology; we shall prove that the integral can be expanded in a series in a neighbourhood of the given value of the parameter.

We mention the importance of the concepts, defined in §§ 10.3.1 and 10.3.4, on the specialisation of the unfolding of a deformation of the germ of a holomorphic function and the Milnor fibration of the deformation of the germ of a holomorphic function.

10.1 Example

Let us consider in $\mathbb{C}^2$ the level lines of a polynomial. If the level value is not critical then the level line is a non-singular Riemann surface. As an example we can consider the level lines of the polynomial

$$f(x, y) = y^2 + x^3,$$

all of them with the exception of the zero level line are surfaces of genus 1 from which one infinitely distant point has been deleted. Let us suppose that on $\mathbb{C}^2$ we are given a polynomial differential 1-form, for example, the form $\omega = y\,dx$. Finally let us suppose that on one of the non-singular level lines is chosen a closed curve. Let us consider the integral of the form along the curve. We shall study the change of the integral under continuous deformation of the curve from one non-singular level line to another. In our chosen case the integral is called elliptic and in this way we intend to study the dependence on the parameter of the period of the elliptic integral.

We shall prove that outside a finite set of exceptional values of the parameter the integral is a many-valued holomorphic function of the parameter. We shall explain the connection between the branches of the integral and the topology of the fibration which is given by the polynomial restricted to the complement of the union of the exceptional level lines. We shall prove that in a neighbourhood of an arbitrary value of the parameter the integral can be expanded into a series of fractional powers of the parameter and integral powers of the logarithm of the parameter in which the powers of the parameter and the powers of the logarithm of the parameter are determined by the Jordan structure of the monodromy operator in the one-dimensional homology of a level line, corresponding to going round the exceptional value of the parameter.

10.1.1 Holomorphic dependence on parameters

We begin with an important remark: a form, restricted to a level line, is closed. Indeed on a complex curve there are no non-zero holomorphic 2-forms. This remark has two consequences. First, the integral does not change as the curve of integration changes to a homologous curve (Stokes' theorem). Second, the integral of a form along a curve, deformed into a neighbouring non-singular level line, is well-defined. Indeed, two different continuous deformations into neighbouring level lines lead to homologous curves. In this way the integral of a form along a curve, deformed into level lines near to the chosen one, defines a function relating the integral to the value of the polynomial.

Theorem 10.1. The integral is an analytic function of the value of the polynomial.

We have a function in a neighbourhood of a chosen point of a complex line, we shall prove its holomorphicity. First we shall prove that this function is smooth.

In a neighbourhood of the initial curve of integration the polynomial does not have critical points, therefore in a neighbourhood of the initial curve of integration the map $f : \mathbb{C}^2 \to \mathbb{C}^1$ is a smooth locally trivial fibration. We choose its smooth trivialisation and with the help of the trivialisation we carry the initial curve of integration into a neighbouring level lines of the polynomial. Let us consider the integral of the form along one curve of the constructed family. It is equal to the integral along the initial curve of integration of the form carried with the help of a diffeomorphism of the trivialisation to the initial level line. In this way on the initial level line there is a curve and a differential 1-form smoothly depending on a complex parameter. By a standard theorem of analysis the integral of the form along a curve depends smoothly on the parameter.

Let us denote by $\sigma(s)$ a curve on the s-level line constructed with the help of the trivialisation.

To complete the proof let us represent our integral in the form of the integral of a meromorphic 2-form on a real surface. As our 2-form let us take the form

$$df \wedge \omega/(f-t)$$

where t is the value of the polynomial on the initial level line. Let us describe the surface. On the complex line let us consider a small path γ going round the number t anticlockwise. Let us denote by Γ the surface in $\mathbb{C}^2$, formed by the union of the curves:

$$\Gamma = \cup_{s \in \gamma} \sigma(s).$$

Lemma 10.1. The integral

$$(1/2\pi i) \int_\Gamma df \wedge \omega/(f-t)$$

does not depend on the choice of path γ and equals $\int_{\sigma(t)} \omega$.

The theorem follows from the lemma, since the integral represented in the lemma depends holomorphically on t.

Proof of the lemma. The non-dependence on the path follows from Stokes' theorem in view of the fact that 2-forms are closed. Let us prove the equality of the integrals.

$$\frac{1}{2\pi i} \int_\Gamma \frac{df \wedge \omega}{f-t} = \frac{1}{2\pi i} \int_\gamma \left(\int_{\sigma(s)} \omega \right) \frac{ds}{s-t} =$$

$$= \frac{1}{2\pi i} \int_\gamma \left(\int_{\sigma(t)} \omega \right) \frac{ds}{s-t} + \frac{1}{2\pi i} \int_\gamma \left(\int_{\sigma(s)} \omega - \int_{\sigma(t)} \omega \right) \frac{ds}{s-t}.$$

The first integral on the right hand side is equal to our integral. The expression under the integral sign of the second integral depends smoothly on s. Therefore the second integral tends to zero, if we take as the path a circle of radius tending to zero. In this way the second integral is equal to zero. The lemma is proved.

10.1.2 Branching of an integral

We shall give an account of the global properties of functions given as the integral of a polynomial differential 1-form along a closed curve lying in the level line of a polynomial.

According to standard theorems (see [357, 390]) we can remove from $\mathbb{C}^2$ a finite set of level lines of a polynomial so that the restriction of the polynomial to their complement is a locally trivial fibration. Let us denote these level values by $t_1, \ldots, t_N$. In our chosen example it is sufficient to delete the zero level line.

Let us return to the integral $\int\limits_{\sigma(s)} \omega$. The integral can be continued in a well-defined manner along any path in $\mathbb{C}^1 \backslash \{t_1, \ldots, t_N\}$ beginning at the point t. It is necessary for this to deform continuously the curve $\sigma(t)$ to the level lines over points of the path; the homology class of the curve, lying in the level line corresponding to the final point of the path, is determined by the path and does not depend on the way it was deformed. If two paths with the same end-point are homotopic in $\mathbb{C}^1 \backslash \{t_1, \ldots, t_N\}$, then the values of the continued integrals at the final point are the same. In this way the integral is a many-valued holomorphic function on $\mathbb{C}^1 \backslash \{t_1, \ldots, t_N\}$.

The branching of the integral is defined by a monodromy transformation of the one-dimensional homology of the level t line. To each homotopy class of closed paths in $\mathbb{C}^1 \backslash \{t_1, \ldots, t_N\}$ beginning at the point t corresponds a linear automorphism – the monodromy – of the one-dimensional homology of the level t line; it is defined by continuous deformation of cycles in the level lines over the points of the path (see Part I). If M_γ is the monodromy automorphism corresponding to the path γ then the continuation of the integral along the path γ by definition equals

$$\int\limits_{M_\gamma \sigma(t)} \omega,$$

where $\sigma(t)$ is the homology class determined by the curve $\sigma(t)$.

For our chosen example the restriction of the polynomial to the complement of the zero level line is a smooth locally trivial fibration. Therefore the elliptic integral $\int y\, dx$ under consideration is a many-valued holomorphic function on $\mathbb{C}^1 \backslash 0$. All the closed paths in $\mathbb{C}^1 \backslash 0$ are a multiple of one, going round the origin once anticlockwise. Therefore to understand the branching of the considered integral we need to understand the structure of the linear monodromy transformation, corresponding to the indicated path. In our example the non-singular level line is a torus with one point missing. Its first homology group is a two-dimensional vector space.

Let us represent the level t line itself, that is the curve

$$\{(x, y) \in \mathbb{C}^2 \,|\, y^2 + x^3 = t\},$$

as a double cover of the x-axis with branching at the cube roots of t. The projection onto the x-axis of cycles giving a basis (a, b) of the first homology is depicted in figure 72a. The monodromy diffeomorphism, corresponding to taking t round zero once, can be given by the formula

$$(x, y) \mapsto (\exp(2\pi i/3)x, \exp(2\pi i/2)y).$$

The image of the cycles is depicted in figure 72b. Therefore the monodromy is given by

$$Ma = a - b, \qquad Mb = a.$$

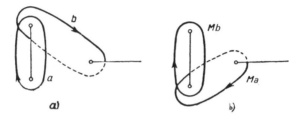

Fig. 72a, b.

To describe the branching of the integral of a holomorphic form along a cycle of the family generated by the cycle $\sigma(t)$, it is sufficient to know the expression of the homology class given by the cycle $\sigma(t)$ in terms of a basis and to know the integral of the form along the basis classes.

Remark. For such a simple form as $y\,dx$ we can describe the branching without knowing the monodromy transformation in the homology. The map

$$(x, y) \mapsto (t^{1/3}x, t^{1/2}y)$$

gives a diffeomorphism of the level 1 line to the level t line. The restriction of the form $y\,dx$ to the level t line mapped with the help of this diffeomorphism to the level 1 line after multiplying by $t^{-5/6}$ is equal to the restriction of the form $y\,dx$ to the level 1 line. Therefore the function given by an integral of the form $y\,dx$ along a cycle of the family is equal to const $\cdot t^{5/6}$ where const is the integral of the form along a cycle of the family lying on the level 1 line.

10.1.3 Expansion of the integral in series

We have explained that the integral of a polynomial differential form along a cycle depending on a parameter is a many-valued holomorphic function on $\mathbb{C}^1 \setminus \{t_1, \ldots, t_N\}$. Consequently each branch of this function in a neighbourhood of an arbitrary value of the parameter which is not exceptional can be expanded in a Taylor series. We shall prove that in a neighbourhood of an exceptional value of the parameter the integral can be expanded as a series also. However this is now a series in fractional powers of the parameter, the coefficients of which are polynomials in the logarithm of the parameter. Such a series converges in each sector of a small neighbourhood of the exceptional value of the parameter (convergence of the series is considered in sectors but not in the full neighbourhoods because of the presence of logarithms). In the description of the series an important role is placed by the monodromy transformation, corresponding to going round the exceptional value of the parameter. So the powers of the parameter are the logarithms of the eigenvalues of the monodromy transformation, divided by $2\pi i$, and the degree of any polynomial in the logarithms which are the coefficients of the series are always less than the maximum dimension of the Jordan blocks associated with the corresponding eigenvalue.

Let us formulate a theorem. To keep the notation simple we shall assume that the exceptional value of the parameter is equal to zero. Let us choose in a neighbourhood of zero the sector $a \leqslant \arg t \leqslant b$ and for each non-zero number t from the sector let us choose a basis

$$\sigma_1(t), \ldots, \sigma_\mu(t)$$

of the one-dimensional integral homology of the level t line, continuously depending on t. Let us denote by M the monodromy transformation corresponding to the parameter going round zero anticlockwise.

Theorem 10.2 (see [55, 93, 138, 181, 239, 262]). In the indicated sector the vector function

$$I(t) = \left(\int_{\sigma_1(t)} \omega, \ldots, \int_{\sigma_\mu(t)} \omega \right)$$

can be expanded in the series

$$\sum_{\alpha, k} a_{k,\alpha} t^\alpha (\ln t)^k.$$

The series converges if the modulus of the parameter is sufficiently small. The coefficients of the series are vectors in the space $\mathbb{C}^{\mu}$. The real parts of all the numbers α are greater than some constant. Each number α possesses the property: $\exp(2\pi i\alpha)$ is an eigenvector of the operator M. A coefficient $a_{k,\alpha}$ of this series is equal to zero if the Jordan form of the operator M does not have a block of dimension $k+1$ or more associated with the eigenvalue $\exp(2\pi i\alpha)$.

Remark. According to Theorems 3.11 and 3.12 the eigenvalues of the monodromy operator M are roots of unity and the dimensions of their Jordan blocks do not exceed two. Therefore in the above series all the numbers α are rational and the power of the logarithm is not greater than 1.

Example. Let $y^2 + x^3$ be the polynomial on the level lines of which lie the cycles, and $\omega = y\,dx$. Then in a neighbourhood of zero

$$\int_{\sigma(t)} \omega = \text{const} \cdot t^{5/6}$$

(see remark in §10.1.2). In §10.1.2 we described the monodromy transformation, corresponding to going round zero, for this case. Its eigenvalues are distinct and equal to $\exp(\pm\pi i/3)$. Therefore according to the theorem for any polynomial form ω its integral in a neighbourhood of zero can be expanded in the series

$$\sum_l a_l t^{5/6+l} + \sum_l b_l t^{7/6+l},$$

where the l are integers, amongst which there is only a finite set of negatives. Actually, as we shall see in §10.1.4, all the indices of the terms of the series are positive.

The proof of the theorem on expansion in series is based on the following important theorem that the integral of a polynomial form in each sector of a neighbourhood of the exceptional value of the parameter grows no faster than a suitable degree of the parameter.

Theorem 10.3 (see [239, 262, 263]). There exists a natural number N for which the following inequality holds in the indicated sector.

$$\left| \int_{\sigma_j(t)} \omega \right| \leqslant \text{const} \cdot |t|^{-N}, \quad j=1,\ldots,\mu.$$

We shall not prove this theorem, but we shall indicate one of the methods of proving it. We resolve the singularity of the exceptional level line. Then with respect to suitable coordinates in a neighbourhood of an arbitrary point of the resolved level line the polynomial becomes a monomial, the polynomial form becomes a holomorphic one and the integral along the part of the cycle lying in the neighbourhood can be used to estimate the required form. We note that it is necessary to resolve the singularities at infinitely remote points of the exceptional level line, in order to estimate the integral along a part of the cycle diverging to infinity. In a neighbourhood of the resolution of such points the form reduces to a meromorphic one but for meromorphic forms the required estimate is not hard to obtain (see [364, 387]). For an elementary proof, not involving resolution of singularities, see [262, 263].

To derive Theorem 10.2 from Theorem 10.3 we must be able to take the logarithm of a non-degenerate linear transformation.

Lemma 10.2. Let A be a non-degenerate $\mu \times \mu$ matrix. Then there exists a $\mu \times \mu$ matrix B for which $\exp B = A$ $(\exp B = \Sigma B^n/n!)$.

Before proving the lemma we give some definitions. A linear operator is *semisimple* if the space on which it acts has a basis consisting of eigenvectors of the operator. A linear operator is *unipotent* if all its eigenvalues are equal to 1. It is well known that for any non-degenerate linear operator M there exists a unique pair of commuting operators, a semisimple operator M_s and a unipotent operator M_u, for which $M = M_u M_s$ (the operator M_s is that operator which on the eigenspace of the operator M acts as multiplication by the corresponding eigenvalue, see [327]). The operators M_u, M_s are called, respectively, the *unipotent* and *semisimple parts* of the operator M.

Proof of Lemma 10.2. It is sufficient to prove the lemma for matrices in Jordan form and furthermore for matrices consisting of a single block. Let λ be the eigenvalue of the block. Then the block matrix can be decomposed into the product of the matrix $\lambda \cdot \text{Id}$ and a matrix which has all its eigenvalues equal to 1. The first matrix is the semisimple part and the second is the unipotent part. The matrices of the semisimple and unipotent parts commute. Therefore it is sufficient to take logarithms of each of them. The logarithm of the first is equal to $\ln \lambda \cdot \text{Id}$. The logarithm of the second is given by the formula

$$\ln C = \ln (\text{Id} + (C - \text{Id})) = \sum (-1)^{s+1} (C - \text{Id})^s/s.$$

Remark. The matrix $\ln C$ is nilpotent, that is all its eigenvalues are equal to zero.

Proof of Theorem 10.2. In a small neighbourhood of zero we extend by continuity a basis of the integral one-dimensional homology to values of the parameter with arbitrary argument. Integrating the form over the basis extends the vector function I to a many-valued vector function in a small punctured neighbourhood of zero. As the parameter goes round zero anticlockwise the vector $I(t)$ changes into the vector $I(t) \cdot A$, where A is the matrix of the monodromy transformation M, with respect to the basis

$$\sigma_1(t), \ldots, \sigma_\mu(t).$$

Let us consider in a punctured neighbourhood of zero the many-valued holomorphic matrix function

$$J(t) = \exp\left(-\ln(t) \cdot \ln(A)/2\pi i\right),$$

where $\ln(A)$ is one of the possible values of the logarithm of the matrix A. As the parameter goes round zero anticlockwise the matrix $J(t)$ changes into the matrix $A^{-1}J(t)$. Therefore the vector function $t \mapsto I(t)J(t)$ is a single-valued function in a punctured neighbourhood of zero.

We shall prove that the function $I \cdot J$ is meromorphic at zero, that is we shall prove that its coordinates grow no faster than a suitable power of the parameter as the parameter tends to zero. According to Theorem 10.3 it is sufficient to prove the analogous result about the coordinates of the matrix J.

Let us explain how to find the elements of the matrix J. To do this it is sufficient to explain how to find the elements of the matrix J in the cases in which the matrix A is diagonal or unipotent (see Lemma 10.2). In the first case the matrix J is also diagonal, with powers of the variable t on the diagonal. In the second case the elements of the matrix are polynomials in $\ln t$, the degrees of the polynomials being less than the dimensions of the Jordan blocks (see the remark after Lemma 10.2). Therefore for arbitrary matrices A each element of the matrix $J(t)$ has the form of a finite sum

$$\sum_\alpha t^\alpha P_\alpha(\ln t).$$

In this sum each number α possesses the property: $\exp(-2\pi i\alpha)$ is an eigenvalue of the matrix A. For each α the coefficient P_α is a polynomial in $\ln t$, with degree less than the maximum dimension of the Jordan blocks of the matrix A associated with the eigenvalue $\exp(-2\pi i\alpha)$.

This result means that the coefficients of the matrix J grow sufficiently slowly. Therefore the vector function $I \cdot J$ is meromorphic at zero. Consequently it can be expanded in a Laurent series in which there are a finite set of negative degrees. Then multiplying this series by J^{-1} we obtain Theorem 10.2.

10.1.4   More precise specifications of Theorems 10.2 and 10.3

If under the deformation of a cycle into an exceptional level line the cycle remains in a bounded part of the space, then the integral along the cycle of a polynomial form remains bounded during the deformation process. Furthermore if the cycle, having been deformed into the exceptional level line, is homologous to zero in it, then the integral of the polynomial form along the cycle tends in the deformation process to zero. This result was proved by B. Malgrange [239]. Let us give an exact formulation of it. We shall assume, as before, that the exceptional value is equal to zero.

Let us denote by X_t the level t line. Let us denote by $X_{t,R}$ the set of points on the level t line, remote from the origin by a distance no greater than R. For small real positive t let us consider a family of integral one-dimensional homology classes $\sigma(t) \in H_1(X_t, \mathbb{Z})$, continously depending on t. We say that for $t \to 0$ this family is a family of homologies *bounded* by the ball of radius R if for all sufficiently small t the class $\sigma(t)$ lies in the image of a natural homomorphism

$$\gamma_R : H_1(X_{t,R}, \mathbb{Z}) \to H_1(X_t, \mathbb{Z}).$$

It is clear that if a family is bounded by a ball of radius R then it is bounded also by any ball of larger radius.

We say that as $t \to 0$ a family of bounded homologies is a family of *vanishing* homologies if there exist $R > 0$, $\delta > 0$, with the properties
(i) the family is bounded by a ball of radius R;
(ii) for each $t \in (0, \delta)$ there exists a class $\sigma_R(t) \in H_1(X_{t,R}, \mathbb{Z})$ which is mapped by the homomorphism γ_R into $\sigma(t)$ and which belongs to the kernel of the natural homomorphism

$$H_1(X_{t,R}, \mathbb{Z}) \to H_1(X, \mathbb{Z})$$

where $X = \cup_{t \in [0, \delta)} X_{t,R}$.

Theorem 10.3′ (see [239]). If σ is a family of bounded homologies then as $t \to +0$ there exists a finite limit for the integral $\int\limits_{\sigma(t)} \omega$. This limit is equal to zero if the family is a family of vanishing homologies.

Corollary of Theorems 10.2 and 10.3'. Let σ be a family of bounded homologies. In a small neighbourhood of zero we extend it by continuity to a parameter with arbitrary arguments. We obtain a many-valued family of integral homology classes continuously depending on the parameter. According to Theorem 10.2 the function given as the integral of a polynomial form over the classes of the family can be expanded in a neighbourhood of zero in the series

$$\sum a_{k,\alpha} t^\alpha (\ln t)^k.$$

According to Theorem 10.3' in this series all the α are non-negative and all the coefficients $a_{k,0}$ for $k > 0$ are equal to zero. Furthermore if σ is a family of vanishing homologies then all the α are positive.

Example. Let $y^2 + x^3$ be the polynomial on the level lines of which lie cycles. In a neighbourhood of zero any family of integral one-dimensional homologies, continuously depending on a parameter is a family of vanishing one-dimensional homologies (see the remark in § 10.1.2). Therefore in the series of the example on page 276 all the indices are positive.

In the following sections of this chapter we shall generalise in three directions the results we have obtained. Firstly we shall increase the dimension of the space. Secondly we shall allow the polynomial to depend holomorphically on additional parameters. Thirdly we shall replace the polynomial by a holomorphic function.

10.2   Holomorphic dependence on parameters

Let us consider a holomorphic function, the variables of which are divided into two groups:

$$F(x_1, \ldots, x_n, y_1, \ldots, y_k).$$

This function gives a family of holomorphic functions on $\mathbb{C}^n$, holomorphically depending on the parameters $y = (y_1, \ldots, y_k)$. Let us denote by $X_{(t,y)}$ the level t hypersurface of the function $F(\cdot, y)$:

$$X_{(t,y)} = \{x \in \mathbb{C}^n | F(x, y) = t\}.$$

Let us suppose that on the non-critical level t_0 hypersurface of the function $F(\cdot, y_0)$ there is chosen an $(n-1)$-dimensional cycle. By continuously deforming this cycle from one level hypersurface to another, we obtain a continuous family of cycles, on the level hypersurfaces near to the chosen level, of functions the parameters of which are near to the chosen one. Let us denote the cycle lying on the hypersurface $X_{(t,y)}$ by $\sigma(t, y)$.

10.2.1   Integrals of holomorphic $(n-1)$-forms

Let us suppose that on the space $\mathbb{C}^n$ we are given a holomorphic differential $(n-1)$-form, holomorphically depending on the parameters y:

$$\omega = \sum_{i=1}^{n} h_i(x_1, \ldots, x_n, y_1, \ldots, y_k) dx_1 \wedge \ldots \widehat{dx_i} \ldots \wedge dx_n,$$

where $\{h_i\}$ is a holomorphic function.

Let us consider the integral

$$I(t, y) = \int_{\sigma(t,y)} \omega(y).$$

The integral is defined as a function of the parameters t and y in a neighbourhood of the point (t_0, y_0) in the space $\mathbb{C} \times \mathbb{C}^k$.

The restriction of the form to the level hypersurface is closed (since on an $(n-1)$-dimensional holomorphic manifold there are no non-zero holomorphic n-forms). Consequently the integral is determined by the class of the cycle in the $(n-1)$st homology group of the hypersurface and does not depend on the cycle representing the class. In addition, the integral of the form along the cycle deformed into a neighbouring level hypersurface does not depend on the deformation since different deformations lead to homologous cycles.

Theorem 10.4 (see [55, 239]). The integral depends holomorphically on (t, y).

The proof of the theorem is practically the same as the proof of Theorem 10.1. In the construction of the proof we need the concept of the coboundary operator of Leray, which we have already implicitly used in the formulation of Lemma 10.1.

Let M be a holomorphic manifold, and let N be a holomorphic submanifold of it of codimension 1. Then for any l there is a natural homomorphism

$$H_l(N) \to H_{l+1}(M \setminus N),$$

called the *coboundary operator of Leray*. It is defined as follows. We choose a
tubular neighbourhood of the submanifold in the manifold. We consider the
projection onto the submanifold of the boundary of the tubular neighbourhood.
The projection is a locally trivial fibration with fibre a circle. We consider a
homology class on the submanifold and a cycle representing it. We consider the
preimage of the cycle on the boundary of the tubular neighbourhood. This is a
cycle, the dimension of which is one greater than the dimension of the initial
cycle. By definition this cycle, considered as a cycle in the complement of the
submanifold, represents the image of the initial homology class under the
coboundary operator of Leray. It is easy to see that the homology class in the
complement does not depend on the choice of cycle representing the initial
homology class on the submanifold or on the choice of tubular neighbourhood.

Proof of Theorem 10.4. A standard theorem of analysis says: if we are given a
chain and a form which depends holomorphically on parameters then the
integral of the form over the chain depends holomorphically on the parameters.
To apply this theorem we need to represent the integrals of the given form along
cycles of the family lying in different hypersurfaces as the integral of a new form
on one and the same chain. The role of the new form is played by the initial form
multiplied by

$$dF(\cdot, y)/(F(\cdot, y) - t).$$

The role of the common chain is played by the cycle representing simultaneously
the images of all the homology classes of cycles of the family under the
coboundary operator of Leray of the pair consisting of the level hypersurface
and $\mathbb{C}^n$ (cf. Lemma 10.1).

Let us consider the boundary of the tubular neighbourhood of the hyper-
surface $X_{(t_0, y_0)}$ in $\mathbb{C}^n$. Let us consider on the boundary an n-dimensional cycle $\partial\sigma$,
representing the image of the homology class of the cycle $\sigma(t_0, y_0)$ on the
hypersurface under the coboundary operator of Leray. The same cycle
represents the image of the homology class of the cycle $\sigma(t, y)$ on the
hypersurface $X_{(t, y)}$ under the coboundary operator of Leray if the point (t, y) is
sufficiently close to (t_0, y_0).

There is an integral representation

$$I(t, y) = \frac{1}{2\pi i} \int_{\partial\sigma(t, y)} dF(\cdot, y) \wedge \omega(y)/(F(\cdot, y) - t), \tag{1}$$

where we denote by $\partial\sigma(t, y)$ the image of the homology class of the cycle $\sigma(t, y)$ on

$X_{(t,y)}$ under the coboundary operator of Leray. (Notice that the expression under the integral sign on the right-hand side is closed in $\mathbb{C}^n \backslash X_{(t,y)}$ so the integral on the right does not depend on the cycle representing the class $\partial \sigma (t, y)$). The proof of the formula cited here is the same as the proof of Lemma 10.1.

Therefore for all (t, y) near to (t_0, y_0) there is an integral representation

$$I(t, y) = \frac{1}{2\pi i} \int_{\partial \sigma} dF(\cdot, y) \wedge \omega(y)/(F(\cdot, y) - t).$$

Now applying standard theorems of analysis we obtain Theorem 10.4.

Further we shall prove the holomorphic dependence on the parameters of the integrals of the Gelfand-Leray form over a continuous family of cycles.

10.2.2   The Gelfand-Leray form in the complex case is defined in the same way as in the real case; see § 7.1

Let f be a holomorphic function in a region of the space $\mathbb{C}^n$. Let η be a holomorphic differential n-form in the same region. Let us consider a point of the region which is not a critical point of the function.

Lemma 10.3. There exists a holomorphic differential $(n-1)$-form ψ, given in a neighbourhood of the point, for which

$$\eta = df \wedge \psi.$$

The restriction of this form to the non-critical level hypersurface of the function is defined invariantly (that is it does not depend on the choice of form ψ satisfying the preceding equation).

The restriction of the form ψ to the non-critical level hypersurface of the function is called the *Gelfand-Leray form* of the form η and is denoted by η/df.

For the proof of the lemma see § 7.1.

10.2.3   Integrals of the Gelfand-Leray form

Let us return to the situation described at the beginning of the section. Let us suppose that on the space $\mathbb{C}^n$ we are given a holomorphic differential n-form η, holomorphically depending on the parameters y:

$$\eta = h(x_1, \ldots, x_n, y_1, \ldots, y_k) dx_1 \wedge \ldots \wedge dx_n.$$

In this way for fixed values of the parameters there is given on the space a holomorphic function and a holomorphic form of the top dimension. Consequently, on each non-critical level hypersurface there is given a Gelfand-Leray form. Let us consider its integral along a cycle, chosen on the level hypersurface:

$$I(t, y) = \int_{\sigma(t, y)} \eta(y)/d_x F(\cdot, y),$$

where d_x is the differential with respect to the variable x.

Theorem 10.5 (see [55, 239]). This integral depends holomorphically on (t, y).

Theorem 10.5 follows from the integral representation

$$I(t, y) = \frac{1}{2\pi i} \int_{\partial\sigma(t, y)} \eta(y)/(F(\cdot, y) - t) \tag{2}$$

(compare with (1)), its proof is analogous to the proof of Lemma 10.1.

10.2.4 Derivatives of functions given by integrals

It is useful to represent these in the form of an integral along the same cycle. Such formulae are given by the integral representations (1), (2).

For brevity of description we shall suppose in the following formulae that the values of the parameters of the forms are the same as the values of the parameters of the cycles. The form will have the lower index j to denote that the coefficients of the form have been differentiated with respect to y_j.

According to (1), (2) we have

$$\frac{\partial}{\partial t} \int_{\sigma(t, y)} \omega = \frac{1}{2\pi i} \int_{\partial\sigma(t, y)} dF \wedge \omega/(F - t)^2 =$$

$$= \frac{1}{2\pi i} \int_{\partial\sigma(t, y)} d\omega/(F - t) = \int_{\sigma(t, y)} d_x\omega/d_x F, \tag{3}$$

$$\frac{\partial}{\partial y_j} \int_{\sigma(t,y)} \omega = \frac{1}{2\pi i} \int_{\partial\sigma(t,y)} (dF \wedge \omega_j/(F-t) +$$

$$+ dF_j \wedge \omega/(F-t) - F_j dF \wedge \omega/(F-t)^2) =$$

$$= \frac{1}{2\pi i} \int_{\partial\sigma(t,y)} (dF \wedge \omega_j/(F-t) - F_j d_x\omega/(F-t)) =$$

$$= \int_{\sigma(t,y)} (\omega_j - F_j d_x\omega/d_x F). \tag{4}$$

The second equation in these formulae is proved by changing the form to a cohomologous one.

10.3   Branching of integrals and expansion of integrals in series

In order to study the analytic continuation of holomorphic functions given by integrals we must know how to deform continuously a cycle along which is integrated a form in hypersurfaces of remote non-critical levels of a function. For this it is sufficient that the aggregate of non-critical level hypersurfaces form a locally trivial fibration. We shall not investigate the most general situation for which this property takes place (compare § 1.1). We shall restrict ourselves to the local case which we shall investigate below.

10.3.1   The Milnor fibration of a deformation of a critical point of a holomorphic function

Let us be given the germ of a deformation of an isolated critical point of a holomorphic function. (Usually we shall be interested in the two extreme cases, in which the deformation is miniversal or trivial). A *Milnor fibration* is a fibration the fibres of which are local non-singular level hypersurfaces of the functions forming the deformation. Let us give a more precise definition.
    Let

$$f : (\mathbb{C}^n, 0) \to (\mathbb{C}, 0)$$

be the germ of a holomorphic function with a finite-multiplicity critical point. Let $F : (\mathbb{C}^n \times \mathbb{C}^k, 0 \times 0) \to (\mathbb{C}, 0)$ be a deformation of it, that is the germ of a holomorphic function with the property:

$$F(\cdot, 0) = f.$$

Let us consider the *unfolding* of the deformation, that is the germ of the map

$$G: (\mathbb{C}^n \times \mathbb{C}^k, 0 \times 0) \to (\mathbb{C} \times \mathbb{C}^k, 0 \times 0),$$

given by the formula $(x, y) \mapsto (F(x, y), y)$.

Let F, G be representatives of the germs F, G.

For small $\varrho > 0$, $\eta > 0$, $\delta > 0$ let us consider in the spaces $\mathbb{C}^n$, $\mathbb{C}$, $\mathbb{C}^k$ balls of corresponding radii:

$$B_\varrho^n = \{x \in \mathbb{C}^n | \, \|x\| < \varrho\},$$

$$B_\eta^1 = \{u \in \mathbb{C} | \, |u| < \eta\},$$

$$B_\delta^k = \{y \in \mathbb{C}^k | \, \|y\| < \delta\}.$$

Put

$$S = B_\eta^1 \times B_\delta^k,$$

$$X = (B_\varrho^n \times B_\delta^k) \cap G^{-1}(S),$$

$$X_s = X \cap G^{-1}(s) \quad \text{for} \quad s \in S,$$

see figure 73.

Let us choose the number ϱ so small that for all r, $0 < r \leqslant \varrho$, the boundary ∂B_r^n of the ball of radius r transversely intersects the zero level hypersurface of the

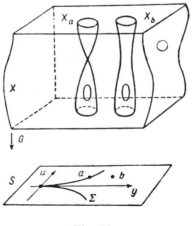

Fig. 73.

function $F(\cdot, 0)$. Let us choose the numbers η, δ so small in comparison with the number ϱ that for any point $(u, y) \in S$ the boundary ∂B_ϱ^n of the ball of radius ϱ transversally approaches the level u hypersurface of the function $F(\cdot, y)$. A triple of numbers ϱ, η, δ with the indicated properties is said to be *admissible*.

If a triple of numbers is admissible, then the map $X \to S$, which is the restriction of the map G, is called a *specialisation of the unfolding* of the deformation. Further restriction of a map will be denoted by the same letter as the map itself.

Let us denote by Σ the set of critical values (*discriminant*) of the map $G : X \to S$, that is the set $\{s \in S \mid X_s \text{ is singular}\}$. If a triple of numbers ϱ, η, δ is admissible then the restriction of the map $G : X \to S$ to the complement of the preimage of the set of critical values is a smooth locally trivial fibration (see § 1.1), that is the map

$$G : X' \to S'$$

where $S' = S \setminus \Sigma$, $X' = X \setminus G^{-1}(\Sigma)$ is a smooth locally trivial fibration. This fibration is called the *Milnor fibration* of the deformation (more precisely, the Milnor fibration of the specialisation of the unfolding of the deformation).

The fibre of the Milnor fibration is an $(n-1)$-dimensional complex analytic manifold with boundary (and correspondingly a $(2n-2)$-dimensional real manifold). According to Milnor's theorem [256] (see also § 2.1) the fibre is homotopy equivalent to a bouquet of spheres of the middle dimension, the number μ of these spheres equal to the multiplicity of the initial critical point. Therefore

$$X_s \sim \underbrace{S^{n-1} \vee \ldots \vee S^{n-1}}_{\mu \text{ spheres}} \quad \text{for} \quad s \in S'$$

The differential type of the fibre of the Milnor fibration does not depend on the choice of admissible triple. For different admissible triples the germs of the sets Σ and S' at the origin of the space S are the same. If the first number of the first triple is less than the first number of the second triple, then for all points of the base near the origin the fibre of the first fibration is included in the fibre of the second fibration. Furthermore the first fibre is a deformation retract of the second fibre. Therefore the inclusion of the first fibre in the second induces an isomorphism of homology and cohomology groups of the fibres of the Milnor fibrations, corresponding to the different admissible triples. In this sense the homology and the cohomology of the fibre of the Milnor fibration are defined invariantly.

We shall be interested in the $(n-1)$st homology and cohomology groups of the fibre of the Milnor fibration. They are the same as for a bouquet of spheres and are free modules over the coefficient ring. Their dimension is equal to the multiplicity of the original critical point if $n-1>0$ and to 1 more than the multiplicity in the case $n-1=0$ (a bouquet of μ zero-dimensional spheres is $\mu+1$ points). We shall always (without further reminder) consider the reduced* homology and cohomology groups. The rank of the reduced (co)homology group equals the multiplicity of the critical point, independently of the value of n. For $n>1$ the reduced $(n-1)$st (co)homology is the same as the ordinary one.

The cohomology and homology of the fibre of the Milnor fibration will be called *vanishing* at the original critical point.

10.3.2   Branching of integrals

Let us be given the germ $f:(\mathbb{C}^n,0)\to(\mathbb{C},0)$ of a holomorphic function with an isolated critical point. Let us be given a deformation

$$F:(\mathbb{C}^n\times\mathbb{C}^k,0\times0)\to(\mathbb{C},0)$$

of the germ f. Let us choose an admissible triple of numbers ϱ, η, δ and let us consider the specialisation corresponding to it of the unfolding and the Milnor fibration.

Let us suppose that in one of the fibres of the Milnor fibration there is chosen an $(n-1)$st integral homology, class. Continuous deformation of a cycle representing it defines a family of $(n-1)$st integral classes in the fibre of the Milnor fibration which depends continuously on the point of the base of the fibration. This dependence is many-valued. The many-valuedness of the classes of the family are described by the monodromy group in the $(n-1)$st homology of the fibre of the fibration: if in the base there is chosen a closed path and over the initial point of the path there is chosen one of the homology classes of the family then as this class is extended to the fibres over the points of the path, the homology class corresponding to the final point of the path is obtained from the homology class corresponding to the initial point of the path by the action of the monodromy transformation corresponding to the path.

Let us suppose that in a neighbourhood of the origin in $\mathbb{C}^n\times\mathbb{C}^k$ there is given a holomorphic differential $(n-1)$-form. Let us suppose that this neighbourhood is so large that it contains the space of the Milnor fibration. Let us consider the integral of the form along a cycle of the family. According to Theorem 10.4 this

---

* Remember that the *reduced k*th homology group of a topological space is the kernel of the map of the kth homology group of the space into the kth homology group of a point induced by the map of the space into the point. Analogously, the *reduced k*th cohomology group is the cokernel of the map of the kth cohomology group of a point into the kth cohomology group of the space.

integral gives a many-valued holomorphic function on the base of the Milnor fibration.

The branching of this function is described by the monodromy group in the $(n-1)$st homology of the fibre of the Milnor fibration. If ω is the form, $\sigma(u, y)$ is the homology class of the family, lying in the fibre over the point $(u, y) \in S'$, γ is a closed path in the base of the fibration with initial and final points at (u, y), M_γ is the monodromy transformation corresponding to it, then the value of the analytic continuation of the branch of the function corresponding to the integral $\int_{\sigma(u,y)} \omega$ is equal to $\int_{M_\gamma \sigma(u,y)} \omega$. In this way the branching of the holomorphic function given by integration along the homology class vanishing at the critical point is determined by the topology of the Milnor fibration of the considered deformation and, in the final analysis, by the topology of the Milnor fibration of the versal deformation of the initial critical point.

Remark 1. The monodromy group in the $(n-1)$st homology of the fibre of the Milnor fibration of the versal deformation of the critical point is called the *monodromy group of the critical point*. Part I was devoted to the monodromy groups of critical points. Any result about monodromy groups can be considered as a result about the branching of the corresponding integrals.

Remark 2. Let us suppose that in a neighbourhood of the origin in $\mathbb{C}^n \times \mathbb{C}^k$ we are given a holomorphic differential n-form η (instead of the $(n-1)$-form as before). The form η defines on each fibre of the Milnor fibration a holomorphic $(n-1)$-form $\eta/d_x F$ (we must restrict the form η and the function F to a subspace in X' of the form $y = \text{const}$, which is a region in $\mathbb{C}^n$, then divide the form by $d_x F$). Let us consider the integral of the form $\eta/d_x F$ along the class of the family indicated above. These integrals give many-valued holomorphic functions on the base of the Milnor fibration (Theorem 10.5). The branching of this function, like the branching of the functions given by the integrals of $(n-1)$-forms is defined by the monodromy group in the $(n-1)$st homology of the fibre of the Milnor fibration. We could begin not with an n-form, but with an $(n+k)$-form and divide it not by $d_x F$ but by

$$d_x F \wedge dy_1 \wedge \ldots \wedge dy_k.$$

The result will be that on each fibre there will be a well-defined holomorphic $(n-1)$-form. Its integral along the class of the family defines a holomorphic many-valued function on the base of the Milnor fibration (Theorem 10.5). The branching of this function like the previous one is determined by the monodromy group.

Remark 3. If an $(n-1)$-form ω is defined in a neighbourhood of the origin in $\mathbb{C}^n \times \mathbb{C}^k$, not containing the space of the Milnor fibration for the given admissible numbers ϱ, η, δ, then, generally speaking, it is impossible to integrate along an arbitrary homology class of the family. However it is possible to integrate along a class of the family lying in the fibres over points lying sufficiently near the origin in $\mathbb{C} \times \mathbb{C}^k$. This is explained by the fact that such classes have representing classes lying in a sufficiently small neighbourhood of the origin in $\mathbb{C}^n \times \mathbb{C}^k$. More precisely, the region of definition of the form (however small it was) contains the space of a Milnor fibration defined with the help of sufficiently small admissible numbers, now the inclusion of the fibre of the smaller Milnor fibration in the fibre of the larger Milnor fibration induces an isomorphism of the homology and cohomology groups (see § 10.3.1).

10.3.3   Expansion of the integral in series

Let us suppose that in the space S (the base of the specialisation) there is chosen a complex line. Let us suppose that the line is not contained in the set of critical values Σ. In this case the intersection of the line with the set of critical values is discrete. Let us consider one of the intersection points. We shall prove that the restriction to the line of a many-valued holomorphic function given by the integral of a holomorphic form along a cycle of a continuous family (see § 10.3.2), in a neighbourhood of the chosen point of intersection of the line with Σ can be decomposed into a series of fractional powers of the parameter on the line, the coefficients of which are polynomials in the logarithm of the parameter on the line. This result is a direct analogue of Theorem 10.2.

Let us denote by t a local holomorphic coordinate on the line with origin at the chosen intersection point. In a small neighbourhood of the intersection point let us choose the sector $a \leqslant \arg t \leqslant b$. For each non-zero value of the parameter t belonging to the sector let us choose a basis $\sigma_1(t), \ldots, \sigma_\mu(t)$ of the $(n-1)$st integral homology of the fibre of the Milnor fibration, lying over the corresponding point of the line, so that this basis depends continuously on the point of the line. Let us denote by M the monodromy transformation in homology, corresponding to the parameter of the line going along a small path round zero anticlockwise. (Remember that according to Theorems 3.11 and 3.12 the eigenvalues of this operator are roots of unity and the dimension of its Jordan blocks is not greater than n.) Let ω be a holomorphic differential $(n-1)$-form, given in a neighbourhood of the origin in $\mathbb{C}^n \times \mathbb{C}^k$, containing the space of the preimage of the specialisation of the deformation (see the definition in § 10.3.1). Then in the sector indicated above there is defined a vector function

$$I(t) = \left(\int_{\sigma_1(t)} \omega, \ldots, \int_{\sigma_\mu(t)} \omega \right).$$

Theorem 10.6 (see [55, 93, 138, 181, 239, 262]). This vector function can be expanded in the series

$$\sum_{\alpha,k} a_{k,\alpha} t^\alpha (\ln t)^k.$$

The series converges if the modulus of the parameter is sufficiently small. The coefficients of the series are vectors in the space $\mathbb{C}^\mu$. The numbers α are non-negative rational numbers. All the coefficients $a_{k,0}$, for $k > 0$, are equal to zero. Each number α possesses the property: $\exp(2\pi i\alpha)$ is an eigenvalue of the operator M. The coefficients $a_{k,\alpha}$ are equal to zero at any time that the Jordan form of the operator M has no blocks of dimension $k+1$ or more associated with the eigenvalue $\exp(2\pi i\alpha)$.

Remark. Let us suppose that the line in the base S (about which we were speaking in Theorem 10.6) is a line of values of the function F for fixed y and transversally intersects the discriminant at a point corresponding to a non-degenerate critical point of the function F with y fixed. In this case the expansion, mentioned in Theorem 10.6, can be made more precise; see Lemma 12.2 on page 321.

Theorem 10.6 depends on Theorem 10.7, which is formulated below, and follows from it in exactly the same way as Theorem 10.2 follows from Theorem 10.3.

Theorem 10.7 (see [239, 262, 263]). There exists a natural number N for which in the sector indicated above we have the inequality

$$\left| \int_{\sigma_j(t)} \omega \right| \leqslant \text{const} \cdot |t|^{-N}, \quad j = 1, \ldots, \mu.$$

Furthermore if we choose in the sector a ray $\arg t = \text{const}$, then there exists a finite limit of the vector function as the parameter tends to zero along the ray.

We shall not prove Theorem 10.7 (see [239] and the remark relating to Theorem 10.3). Theorem 10.7 has a more precise specification (compare with Theorem 10.3′) and we shall formulate it.

First let us define the concept of a family of homologies, vanishing at the chosen intersection point of the line and the discriminant Σ (compare with the definition in § 10.1.4). Let us denote by X_t the fibre of the Milnor fibration lying over the point of the line at which the parameter equals t. Let us assume that on the line is chosen a ray $\arg t = \text{const}$, and for each sufficiently small t, belonging to the ray, there is chosen a homology class

$$\sigma(t) \in H_{n-1}(X_t, \mathbb{Z}),$$

continuously depending on t. We shall say that this family of homologies is a family of *vanishing* homologies as the parameter tends to zero along the ray, if for each sufficiently small t, the class $\sigma(t)$ belongs to the kernel of the natural homomorphism

$$H_{n-1}(X_t, \mathbb{Z}) \to H_{n-1}(\cup_{s \in [0,t]} X_s, \mathbb{Z}).$$

If a continuous family of homologies is defined for parameters belonging to the sector then it is easy to show that the property of being a family of vanishing homologies does not depend on the ray of the sector.

Theorem 10.7′ (see [239]). Let us suppose that a continuous family of homologies has been defined for parameters belonging to a sector. Let us suppose that the family is a family of vanishing homologies as the parameter tends to zero along one of the rays of the sector. Then the limit of the integral $\int_{\sigma(t)} \omega$ is equal to zero as the parameter tends to zero along the ray.

Remember that the existence of a finite limit follows from Theorem 10.7.

Corollary to Theorem 10.7′. The integral of the holomorphic form over a family of vanishing homologies can be expanded in each sector in the series

$$\sum a_{k,\alpha} t^\alpha (\ln t)^k,$$

in which each α is positive.

10.3.4   The Milnor fibration of a critical point

Let us consider the germ $f : (\mathbb{C}^n, 0) \to (\mathbb{C}, 0)$ of a holomorphic function with an isolated critical point. The germ is its own trivial deformation. Let us consider for such a deformation the objects introduced in the previous sections. Namely let us consider a specialisation $f : X \to S$ of the germ and the corresponding Milnor fibration $f : X' \to S'$.

For any holomorphic differential $(n-1)$-form on X its integral along the class of a continuous family of integral vanishing homologies defines a holomorphic many-valued function on the punctured disc S'. The branching of this function as the parameter goes round zero is determined by the monodromy transformation of the homology, which in the given case is called the *classical monodromy* (see § 2.1). In each sector $a \leqslant \arg t \leqslant b$ this function can be expanded in the series

$$\sum a_{k,\alpha} t^\alpha (\ln t)^k.$$

In this series all the powers of the parameters are positive. In addition all the powers of the parameters are logarithms of the eigenvalues of the classical monodromy operator divided by $2\pi i$. Each power of the logarithm of the parameter in this series is less than the maximum size of the Jordan block of the classical monodromy operator associated with the corresponding eigenvalue (see Theorems 10.7 and 10.7′).

Let us consider on the space X a holomorphic differential n-form η and a continuous many-valued family of integral vanishing homologies $\sigma(t)$, where $t \in S'$. Let us consider on the base of the Milnor fibration the many-valued holomorphic function $\int_{\sigma(t)} \eta/df$.

Theorem 10.8 (see [239]). In each sector $a \leqslant \arg t \leqslant b$ this function can be expanded in the series

$$\sum_{\alpha, k} a_{k,\alpha} t^\alpha (\ln t)^k.$$

The series converges if the modulus of the parameter is sufficiently small. All the numbers α are rational. Each number α is greater than -1. Each number α possesses the property: $\exp(2\pi i \alpha)$ is an eigenvector of the classical monodromy operator in the homology. The coefficients $a_{k,\alpha}$ are equal to zero at any time that the classical monodromy operator does not have Jordan blocks of dimension $k+1$ or greater associated with the eigenvalue $\exp(2\pi i \alpha)$.

Proof. In view of the Poincaré lemma $\eta = d\omega$, where ω is a holomorphic differential $(n-1)$-form. According to formula (3) on page 284

$$d/dt \int_{\sigma(t)} \omega = \int_{\sigma(t)} \eta/df.$$

Now the theorem follows from Theorem 10.6.

Remark about the application of the Poincaré lemma. It is easy to show that X is a Stein manifold (see the definitions in [113, 146]). Consequently the cohomology of the manifold X can be calculated with the help of holomorphic forms (see [113], [326, Theorem 12.13]). The manifold X is contractible, the form η is closed, therefore there exists on X a holomorphic $(n-1)$-form ω, for which $\eta = d\omega$. Theorem 10.8 can be proved without using the steinness of the manifold X. In view of the standard Poincaré lemma for the given form η there exists a holomorphic $(n-1)$-form ω, given in a sufficiently small neighbourhood of the origin in $\mathbb{C}^n$, for which $\eta = d\omega$. Now formula (5) is true for sufficiently small t. This assertion is sufficient for the proof of Theorem 10.8.

Example. Let

$$f = x_1^2 + \ldots + x_n^2.$$

The space of vanishing homologies is generated by vanishing cycle. Let $\sigma(t)$ be a vanishing cycle over t, continuously depending on the parameter. As the parameter goes round zero

$$\sigma(t) \mapsto (-1)^n \sigma(t).$$

Let ω be a holomorphic differential $(n-1)$-form. Let us consider the holomorphic function $I(t) = \int_{\sigma(t)} \omega$. According to Theorem 10.6

$$I(t) = \sum a_\alpha t^\alpha,$$

where $\alpha = 1, 2, \ldots$ for n even, and $\alpha = 1/2, 3/2, \ldots$ for n odd. Decomposing ω into homogeneous parts

$$\omega_{n-1} + \omega_n + \ldots,$$

and using the homogeneity of the germ, we obtain the result that the integral $\int_{\sigma(t)} \omega_p$ is equal to zero if $p-n$ is odd, and equals $a_{p/2}\,t^{p/2}$ if $p-n$ is even. Let us calculate the first coefficient $a_{n/2}$. Taking t to be real, we obtain

$$\int_{\sigma(t)} \sum_j (-1)^j A_j x_j dx_1 \wedge \ldots \widehat{dx_j} \ldots dx_n =$$

$$= \sum_j A_j \int_0^t \left(\int_{\sigma(s)} dx_1 \wedge \ldots dx_n/df \right) ds = \sum_j A_j \operatorname{Vol}(D^n) t^{n/2},$$

where $\operatorname{Vol}(D^n)$ is the volume of the n-dimensional ball of radius 1.

Chapter 11
Complex oscillatory integrals

In the study of the asymptotic behaviour of functions one frequently has to study the asymptotic behaviour of integrals of the form

$$\int_\Gamma e^{\tau f(x)}\phi(x)dx_1 \wedge \ldots \wedge dx_n$$

for large values of the parameter τ. Here f, ϕ are holomorphic functions on $\mathbb{C}^n$, Γ is a real n-dimensional chain, lying in $\mathbb{C}^n$, and τ is a large real parameter. The function f is called the *phase*, the function ϕ is called the *amplitude*. Such integrals are called *integrals of the saddle-point method*. For examples of problems in which the need to study these arises see, for example, the book of M. B. Fedoryuk, "The saddle-point method" [110].

The saddle-point method, by which such integrals are studied, consists of the following. By Stokes' formula the integral is not changed if the chain is deformed without changing its boundary. Therefore the chain is deformed, without changing the boundary, so as to decrease on it the magnitude of the real part of the phase (for it is this that for large positive values of the parameter determines the magnitude of the integral). We can decrease the real part of the phase, moving the interior of the chain along a trajectory of the gradient of the phase. This process can be continued until the chain encounters a critical point of the phase, at which the gradient equals zero. Further deformation results in a decrease of the real part of the phase on parts of the chain not encountering the critical point, but does not change the maximum of the real part of the phase on the chain. Now the problem of studying the asymptotics of the integral leads to the local problem of estimating the integral in a neighbourhood of a critical point of the phase on which the chain is encountered and to estimate the integral in a neighbourhood of the boundary of the chain if the maximum of the real part of the phase is attained on the boundary chain.

In this way the saddle-point method consists of two parts: a topological part, consisting of deforming the chain (called a *saddle-point contour*) as indicated below, and an analytic part, consisting of an estimate of the integral along the saddle-point contour in a neighbourhood of the critical point of the phase, and also of an estimate of the integral along a neighbourhood of the boundary of the saddle-point contour (if the maximum of the real part of the phase is attained on the boundary of the chain). In applications as a rule the maximum of the real part of the phase is not attained on the boundary.

This chapter is devoted to the analytic part of the saddle-point method: we shall study the asymptotic behaviour of the integral of the saddle-point method over a chain lying in a small neighbourhood of an isolated critical point of the phase, and arranged so that on the boundary of the chain the real part of the phase is less than the real part of the critical value of the phase. We shall show that in this case the integral can be expanded in an asymptotic series of the form

$$e^{\tau f(x_0)} \sum a_{k,\alpha} \tau^\alpha (\ln \tau)^k.$$

We shall show that the power of the parameter and the power of the logarithm of the parameter in this series are connected with the (classical) monodromy operator in the $(n-1)$st homology of the Milnor fibration of the critical point of the phase. Namely, the power of the parameter is the logarithm of the eigenvalue of the monodromy operator divided by $2\pi i$ and the power of the logarithm of the parameter is less than the maximum dimension of the Jordan blocks of the monodromy operator associated with the corresponding eigenvalue.

This phenomenon has the following explanation. The boundary of the chain of the integral of the saddle-point method determines in a natural way a continuous family of integral vanishing homologies in the fibres of the Milnor fibration (see § 11.1.2). Let us consider a function on the base of the Milnor fibration, given by integrals over the class of the family of the form

$$\phi dx_1 \wedge \ldots \wedge dx_n/df,$$

where f, ϕ are the phase and the amplitude of the integral of the saddle-point method. It turns out that the integral of the saddle-point method is (correctly understood) the Laplace transform of this function. The function itself in a neighbourhood of the critical value of the phase can be decomposed into a series of fractional powers of the argument and integral powers of the logarithm of the argument (Theorem 10.8). The connection between these degrees and the monodromy operator was indicated in Theorem 10.8. We shall prove that the term-by-term Laplace transform of this series leads to an asymptotic series for the integral of the saddle-point method.

This is the first theme of this chapter.

In the second part of the chapter we shall consider an oscillatory integral on $\mathbb{R}^n$ with analytic phase, supposing, moreover, that the support of the amplitude is concentrated in a sufficiently small neighbourhood of a finite-multiplicity critical point of the phase. The phase of the function can be complexified and considered as a holomorphic function on $\mathbb{C}^n$, real on the real part of the complex space. Now we can compare the asymptotic behaviour of an oscillatory integral on $\mathbb{R}^n$ with the asymptotic behaviour of integrals of the saddle-point method with the indicated complexification of the phase. We shall prove that in a neighbourhood of a critical point of the phase there exists a saddle-point contour (called *real*), the integral over which has the same asymptotic expansion as the oscillatory integral over $\mathbb{R}^n$. This result has two corollaries. As the first corollary we obtain a new proof of Theorem 6.3 on the asymptotic expansions of oscillatory integrals. As the second corollary we obtain a geometrical meaning of the power of the parameter and the power of the logarithm of the parameter in the asymptotic expansion of the oscillatory integral: the power is expressed in terms of the Jordan structure of the classical monodromy operator of the complexification of the critical point of the phase.

This is the second theme of this chapter.

In the third part of the chapter we shall consider the integral with phase $f(x)+g(y)$ and express its asymptotics in terms of the integrals with phases f and g.

The fundamental results of the chapter (Theorems 11.1, 11.3, 11.4) are due to B. Malgrange [239].

11.1   Integrals of the saddle-point method

Let us consider the function $f : (\mathbb{C}^n, 0) \rightarrow (\mathbb{C}, 0)$, holomorphic in a neighbourhood of its own critical point at the origin. Let us consider an n-dimensional chain Γ, lying in a small neighbourhood of the critical point. Let us suppose that on the boundary of the chain the real part of the values of the function is negative. A chain with such a property we shall call *admissible*. Let us consider the holomorphic differential n-form ω, given in a neighbourhood of the critical point. We shall study the asymptotic behaviour of the integral

$$\int_{\Gamma} e^{\tau f(x)} \omega$$

for large positive values of the parameter τ. The integral is called a *complex oscillatory integral* or an *integral of the saddle-point method*.

We shall say that two chains Γ and Γ' are *equivalent* if there exists an $(n+1)$-dimensional chain V, lying in a small neighbourhood of the critical point, with the property: on the n-dimensional chain

$$\Gamma - \Gamma' + \partial V$$

(after cancellation) the real part of the values of the function is negative.

Lemma 11.1. Let Γ, Γ' be two equivalent admissible chains. Then as $\tau \to +\infty$

$$\int_\Gamma e^{\tau f}\omega - \int_{\Gamma'} e^{\tau f}\omega = o(\tau^{-N})$$

for every natural number N.

Proof. We have

$$\Gamma = \Gamma' + \partial V + \Gamma'',$$

where $\operatorname{Re} f|_{\Gamma''} < 0$. Consequently

$$\int_\Gamma - \int_{\Gamma'} = \int_{\Gamma''},$$

but the last integral is exponentially small.

We shall study the asymptotic behaviour of complex oscillatory integrals modulo terms small in comparison with τ^{-N} for every natural number N.

Example. Let $f(x) = x^3$, $\omega = dx$. An arbitrary admissible contour is equivalent to a linear combination of the two contours Γ_1, Γ_2, depicted in figure 74. The integrals over Γ_1 and Γ_2 are conjugate.

$$\int_{\Gamma_1} e^{\tau x^3}dx = (1 - e^{+2\pi i/3}) \int_0^l e^{-\tau x^3}dx = (\dots)\tau^{-1/3} \int_0^\infty e^{-x^3}dx + o(\tau^{-N}).$$

Let us make more precise the concepts of admissible and equivalent contours.

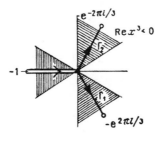

Fig. 74.

11.1.1 Precise definition

Let us suppose that the function $f : (\mathbb{C}^n, 0) \to (\mathbb{C}, 0)$ is holomorphic at the origin and has at the origin a finite-multiplicity critical point. Let us consider a pair of positive numbers ϱ, η. Let us denote by S the disk $\{t \in \mathbb{C}^1 \mid |t| < \eta\}$ and by S^- its left half: $\{t \in S \mid \operatorname{Re} t < 0\}$. Let us denote by X the set

$$\{x \in \mathbb{C}^n \mid f(x) \in S, \ \|x\| < \varrho\},$$

by X^- the set $X \cap f^{-1}(S^-)$, by X_t the set $X \cap f^{-1}(t)$ (see figure 75). We shall suppose that the pair of numbers ϱ, η are admissible for the critical point (see § 10.3.1).

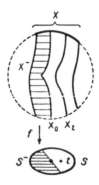

Fig. 75.

The n-dimensional chain $\Gamma \subset \mathbb{C}^n$ is said to be *admissible* if

$$\Gamma \subset X, \qquad \partial \Gamma \subset X^-.$$

Admissible chains Γ, Γ' are called *equivalent*, if there exists an $(n+1)$-dimensional chain $V \subset X$ with the property:

$$(\Gamma - \Gamma' + \partial V) \subset X^-.$$

The equivalence classes of admissible chains with the operations of addition and multiplication by scalars form a vector space, by definition coinciding with $H_n(X, X^-)$.

11.1.2 Expansion in series

To each element

$$[\Gamma] \in H_n(X, X^-)$$

we associate a family of $(n-1)$-dimensional reduced homology classes in the fibre of the Milnor fibration, lying over S^-. For this we consider the exact sequence

$$\ldots \to H_n(X) \to H_n(X, X^-) \to H_{n-1}(X^-) \to H_{n-1}(X) \to \ldots.$$

The numbers ϱ, η are admissible. Therefore X is contractible and ∂ is an isomorphism between the group $H_n(X, X^-)$ and the $(n-1)$st reduced homology group of the space X^-.

The restriction of the function f to $X' = X \setminus X_0$ is a smooth locally trivial fibration (the Milnor fibration of the critical point). In particular, $f : X^- \to S^-$ is a trivial fibration. Therefore the homology of the set X^- is isomorphic to the homology of an arbitrary fibre over S^-. In this way we have defined an isomorphism

$$\partial_t : H_n(X, X^-) \to H_{n-1}(X_t)$$

for any $t \in S^-$, where $H_{n-1}(X_t)$ is the reduced homology group. If Γ is an admissible chain, representing the element $[\Gamma] \in H_n(X, X^-)$, then taking the boundary of the chain Γ and contracting it in the set X^- to the fibre X_t, we obtain a cycle representing the element

$$\partial_t[\Gamma] \in H_{n-1}(X_t).$$

The family of classes $\partial_t[\Gamma]$ depends continuously on $t \in S^-$.

In the example on page 299 the fibre X_t of the Milnor fibration consists of the roots of the equation $x^3 = t$. To the element $[\Gamma_1] \in H_1(X, X^-)$ there corresponds a family of zero-dimensional homologies $\partial_t[\Gamma_1]$. The class $\partial_t[\Gamma_1]$ is represented by the cycle $x^0 - x^1$, where x^0 is the lower point of the fibre and x^1 is on the left (see figure 74).

Lemma 11.2. Let ω be a holomorphic differential n-form on X, and let $[\Gamma] \in H_n(X, X^-)$. Then

$$\int_{[\Gamma]} e^{\tau f} \omega \approx \int_0^{t_0} e^{-\tau t} \left(\int_{\partial_{-t}[\Gamma]} \omega/df \right) dt. \tag{1}$$

Here t_0 is a small positive number, belonging to S, on the right-hand side is an integral along a segment of the real axis, the sign $\approx$ means that the right hand side differs from the left by a term $o(\tau^{-N})$ as $\tau \to +\infty$, where N is any number.

This lemma relates the integral of the saddle-point method with the integral of a holomorphic form over a family of vanishing homology classes. The integral of the holomorphic form can be expanded in a series of powers of the parameter and powers of the logarithm of the parameter (see Chapter 10). Formula (1) allows the integral of the saddle-point method to be expanded in an analytic series.

Proof. Let us choose a chain Γ, representing $[\Gamma]$, so that $\partial\Gamma \subset X_{-t_0}$. Then for any holomorphic $(n-1)$-form ϕ on X

$$\int_\Gamma d\phi = \int_0^{-t_0} \left(\int_{\partial_t[\Gamma]} d\phi/df \right) dt. \tag{2}$$

Indeed according to Stokes' formula the integral on the left is equal to the integral of the form ϕ over $\partial\Gamma$. The cycle $\partial\Gamma$ represents $\partial_{-t_0}[\Gamma]$. Let us consider the function

$$J(t) = \int_{\partial_t[\Gamma]} \phi.$$

It is holomorphic on S^- and tends to zero as the parameter tends to zero along a radius (see Theorems 10.4, 10.7'). Consequently, its value at t_0 is equal to the integral of its derivative from 0 to $-t_0$. According to formula (3) on page 284 the derivative of the function J is given by the integral of the form $d\phi/df$. The formula (2) is proved.

For fixed τ there is on X a holomorphic $(n-1)$-form ϕ with the property:

$$d\phi = e^{\tau f}\omega$$

(by Poincaré's lemma). Then formula (2), applied to ϕ, gives the result of the lemma.

Remark. Here as in Theorem 10.8 we need accuracy in the application of Poincaré's lemma; see the remark on page 294.

Theorem 11.1 (see [239]). Let ω be a holomorphic differential n-form on X, and let $[\Gamma] \in H_n(X, X^-)$. Then as $\tau \to +\infty$ the integral

$$\int_{[\Gamma]} e^{\tau f} \omega \tag{3}$$

can be expanded in the asymptotic series

$$\sum_{\alpha, k} a_{k,\alpha} \tau^\alpha (\ln \tau)^k. \tag{4}$$

Here the parameter α runs through a finite set of arithmetic progressions, depending only on the phase and consisting of negative rational numbers. Moreover, each number α possesses the property: $\exp(-2\pi i\alpha)$ is an eigenvalue of the classical monodromy operator of the critical point of the phase. The coefficient $a_{k,\alpha}$ is equal to zero whenever the classical monodromy does not have Jordan blocks of dimension $k+1$ or more associated with the eigenvalue $\exp(-2\pi i\alpha)$.

The **proof** follows from (1). The inside integral on the right-hand side of formula (1) can be expanded on the positive real semiaxis in the series

$$\sum b_{k,\alpha} t^\alpha (\ln t)^k, \tag{5}$$

indicated in Theorem 10.8 on page 293. It is easy to show that we can obtain the asymptotic expansion of the integral (3) by calculating the asymptotic integrals of the individual terms of series (5). Using the standard formula

$$\int_0^\infty e^{-\tau t} t^\alpha (\ln t)^k \, dt = \left(\frac{d}{d\alpha}\right)^k \left(\frac{\Gamma(\alpha+1)}{\tau^{\alpha+1}}\right), \tag{6}$$

where $\Gamma(\cdot)$ is the gamma function and changing $\int_0^1$ to $\int_0^\infty$, we obtain series (4).

Remark. Series (4) determines series (5). Therefore the study of asymptotic integrals of the saddle-point method is an equivalent problem to the study of integrals of holomorphic forms over homology classes of continuous families of vanishing homologies.

11.1.3   Example

Let $f=x^{\mu+1}$. Then $x=0$ is a critical point of multiplicity μ. Let us fix chains $\Gamma_1,\ldots,\Gamma_\mu$, generating a basis in $H_1(X,X^-)$. For this let us denote by $\varepsilon_1,\ldots,\varepsilon_{\mu+1}$ the roots of the equation $x^{\mu+1}=-1$, where

$$\arg \varepsilon_j = \pi(2j-1)/(\mu+1),$$

and let us put Γ_j equal to the segment $\overrightarrow{\varepsilon_{j+1}\varepsilon_j}$, see figure 76. Let us denote by ω_l the form $x^{l-1}dx$.

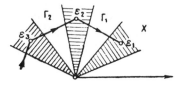

Fig. 76.

Lemma 11.3. For $1\leqslant l,j\leqslant\mu$

$$\int\limits_{[\Gamma_j]} e^{\tau f}\omega_l = a\tau^\alpha,$$

where

$$a=(\varepsilon_{j+1}^l-\varepsilon_j^l)\,\Gamma(l/(\mu+1))/(\mu+1),$$

$$\alpha=-l/(\mu+1),$$

and $\Gamma(\cdot)$ is the gamma-function.

Remark. An integral with phase $x^{\mu+1}$ and arbitrary form ω reduces by integration by parts to a linear combination of integrals with the same phase and forms ω_l, $l=1,\ldots,\mu$. The coefficients of the linear combination are formal series in $1/\tau$.

Proof of the lemma. For negative t let us denote by $\varepsilon_1(t),\ldots,\varepsilon_{\mu+1}(t)$ the roots of the equation $x^{\mu+1}=t$, where

$$\arg \varepsilon_j(t)=\pi(2j-1)/(\mu+1).$$

For such t we have that $\partial_t[\Gamma_j]$ is represented by the 0-cycle $\varepsilon_j(t)-\varepsilon_{j+1}(t)$. The 0-form ω_l/df is equal to the function $x^{l-\mu-1}/(\mu+1)$. Consequently

$$\int_{\partial_t[\Gamma_j]} \omega_l/df=(\varepsilon_{j+1}^l-\varepsilon_j^l)|t|^{l/(\mu+1)-1}/(\mu+1). \tag{7}$$

Then Lemma 11.3 follows from Lemma 11.2 and formula (6).

In conclusion let us mention an important property of the forms $\omega_1,\ldots,\omega_\mu$ of the example.

Lemma 11.4. The forms

$$\omega_1/df,\ldots,\omega_\mu/df$$

generate a basis in the vanishing cohomology in each fibre of the Milnor fibration. Furthermore

$$\det{}^2\left(\int_{\partial_t[\Gamma_j]} \omega_l/df\right)=C_\mu t^{-\mu},$$

where $1\leqslant j$, $l\leqslant\mu$, and C_μ is a non-zero constant.

Proof. In view of formula (7) it is sufficient to prove that there does not exist a non-zero polynomial

$$P=\alpha_1 x+\ldots+\alpha_\mu x^\mu$$

taking the same value at the points $\varepsilon_1,\ldots,\varepsilon_{\mu+1}$. It is clear that

$$P(\varepsilon_1)=\ldots=P(\varepsilon_{\mu+1})$$
$$=(P(\varepsilon_1)+\ldots+P(\varepsilon_{\mu+1}))/(\mu+1)=0.$$

Consequently, $P\equiv 0$.

Corollary.

$$\det \left(\int_{[\Gamma_j]} e^{\tau f} \omega_l \right) = B_\mu \tau^{-\mu/2},$$

where B_μ is a non-zero constant.

Remark. We shall prove below, with the help of Lemma 11.4 and its corollary, that for any critical point of multiplicity μ there exist forms

$$\omega_1, \ldots, \omega_\mu$$

which after division by the differential of the phase generate a basis in the vanishing cohomology in each fibre of the Milnor fibration (see Theorem 12.1).

11.2 An oscillatory integral is a special case of an integral of the saddle-point method

Let us consider a function $f:(\mathbb{C}^n, 0) \to (\mathbb{C}, 0)$, holomorphic at the origin and having a critical point at the origin. Let us suppose that the values of the function on the real subspace $\mathbb{R}^n \subset \mathbb{C}^n$ are real. Such a function can be both the phase of an oscillatory integral on $\mathbb{R}^n$ and also the phase of an integral of the saddle-point method. We shall prove that there exists an admissible n-chain Γ, lying in a neighbourhood of the origin, for which

$$\int_{\mathbb{R}^n} e^{i\tau f} \varphi \, dx \approx \int_\Gamma e^{i\tau f} \varphi \, dx \tag{8}$$

for all φ with support in a sufficiently small neighbourhood of the origin. In this way an oscillatory integral is a particular case of an integral of the saddle-point method.

Let us give three points of clarification. First, the equality is satisfied modulo terms decreasing faster than any power of the parameter τ as $\tau \to +\infty$. Second, an admissible chain for an integral with phase $i\tau f$ (instead of τf as in the previous section) is a chain on the boundary of which the imaginary part of the function f is positive. Third, the amplitude φ must be defined not only on $\mathbb{R}^n$, but also on Γ; therefore the equality (8) will be proved for amplitudes of the form $\psi\theta$, where ψ is a holomorphic function and θ is a smooth bounded function with support concentrated in a sufficiently small neighbourhood of the origin and identically equal to 1 in a still smaller neighbourhood of the origin.

According to Theorem 11.1, the asymptotics of the right-hand integral can be expressed in terms of the Jordan structure of the classical monodromy operator of the critical point. Consequently, the asymptotics of the left-hand integral are connected in the same way with the monodromy operator. Moreover it is not important that the equality (8) has been proved only for analytic amplitudes φ (see Lemma 11.5 below).

11.2.1   Change of definition

In connection with the presence of the number i in the phase of integral (8) we shall change (for this section only) the definition of admissible chain. Let us suppose that f has a finite-multiplicity critical point at the origin. Let us use the notation from §11.1.1 and denote by S^{+i} the semicircle $\{t \in S | \operatorname{Im} t > 0\}$. Let us denote by X^{+i} the set

$$X \cap f^{-1}(S^{+i}).$$

Let us call the n-chain $\Gamma \subset X$ *admissible* if $\partial\Gamma \subset X^{+i}$. Let us call two admissible chains *equivalent* if they define the same class in $H_n(X, X^{+i})$. The integral

$$\int_{\Gamma} e^{i\tau f} \omega$$

of the holomorphic n-form ω along the admissible chain Γ we shall call an *integral of the saddle-point method*. Integrals along equivalent chains have equal asymptotic expansions as $\tau \to +\infty$ (Lemma 11.1). Let us reformulate Theorem 11.1.

Theorem 11.2. Let ω be a holomorphic differential n-form on X, and let $[\Gamma] \in H_n(X, X^{+i})$. Then as $\tau \to +\infty$ the integral

$$\int_{[\Gamma]} e^{i\tau f} \omega$$

can be expanded in the asymptotic series

$$\sum a_{k,\alpha} \tau^{\alpha} (\ln \tau)^k,$$

possessing the properties indicated in Theorem 11.1.

11.2.2.

Theorem 11.3 (see [239]). Let us suppose that the value of the function f is real on the real subspace. Then there exists a class

$$[\Gamma_{\mathbb{R}}] \in H_n(X, X^{+i})$$

(called real) which possesses the following property. Let $\theta : \mathbb{C}^n \to \mathbb{R}$ be an infinitely differentiable function with support in X and identically equal to 1 in a neighbourhood of the origin. Then for any holomorphic function $\varphi : \mathbb{C}^n \to \mathbb{C}$

$$\int_{\mathbb{R}^n} e^{i\tau f} \varphi \theta \, dx_1 \wedge \ldots \wedge dx_n \approx \int_{[\Gamma_{\mathbb{R}}]} e^{i\tau f} \varphi \, dx_1 \wedge \ldots \wedge dx_n$$

modulo terms which for any N are of magnitude $o(\tau^{-N})$ as $\tau \to +\infty$.

Corollary of Theorems 11.2, 11.3 (compare with Theorem 6.3 on asymptotic expansions). The oscillatory integral

$$\int_{\mathbb{R}^n} e^{i\tau f} \psi \, dx_1 \wedge \ldots \wedge dx_n$$

can be expanded in the asymptotic series

$$\sum a_{k,\alpha} \tau^\alpha (\ln \tau)^k$$

as $\tau \to +\infty$, if the amplitude has the form indicated in Theorem 11.3. In this series the parameter α runs through a finite set of arithmetic progressions depending only on the phase and consisting of negative rational numbers. Namely each number α possesses the property: $\exp(-2\pi i\alpha)$ is an eigenvalue of the classical monodromy operator of the critical point of the phase, considered as a holomorphic function. The coefficient $a_{k,\alpha}$ is equal to zero whenever there are, in the classical monodromy, no Jordan blocks of dimension $k+1$ or more associated with the eigenvalue $\exp(-2\pi i\alpha)$.

Remark 1. With the help of Lemma 11.5 (see below) the indicated condition on the amplitude can be replaced by the condition: the support of the amplitude must lie in a sufficiently small neighbourhood of the origin.

Remark 2. It would be interesting to describe all the real classes $[\Gamma_{\mathbb{R}}]$ for different real forms of one and the same holomorphic function f.

In § 9.2 we gave an example of two complex equivalent but real nonequivalent germs of analytic functions which had different singular indexes. Consequently the real classes corresponding to these real forms are arranged differently relative to the linear functions given by integrals of the saddle-point method on $H_n(X, X^{+i})$.

Proof of Theorem 11.3. We shall construct a chain $\Gamma_{\mathbb{R}}$, representing the class $[\Gamma_{\mathbb{R}}]$ and then we shall prove the theorem for it. The chain $\Gamma_{\mathbb{R}}$ is that part of the real subspace $\mathbb{R}^n$ lying in X with edges turned up into X^{+i}. The edges are turned up along the trajectories of the vector field $i \cdot \operatorname{grad} f$. Before we make the construction more precise let us give an example. Let $f = x^3$. Then the chain $\Gamma_{\mathbb{R}}$ is depicted in figure 77a.

Let us choose a sufficiently small $r > 0$ and consider the smooth vector field ξ on X, equal to zero for $|x| < r/4$ and possessing the properties

$$\xi(\operatorname{Im} f) \geqslant 0,$$

$$\xi(\operatorname{Im} f) > 0 \quad \text{for} \quad x \geqslant r/2,$$

see figure 77b.

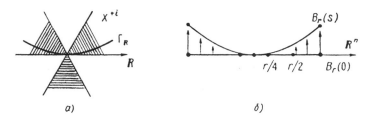

$a)$ $\qquad\qquad\qquad\qquad\qquad$ $\delta)$

Fig. 77.

Let B_r be the intersection of the ball $|x| \leqslant r$ and $\mathbb{R}^n$. For sufficiently small $s > 0$ the diffeomorphism $\exp(s\xi)$ defines in X a disk

$$B_r(s) = \exp(s\xi)(B_r),$$

the boundary of which lies in X^{+i}. Consequently, $B_r(s)$ defines a class in $H_n(X, X^{+i})$, which clearly does not depend on s. We shall show that for this class Theorem 11.3 is true.

We can assume that the support of the function θ lies in the ball $|x| < 3r/4$ and that $\theta \equiv 1$ on the ball $|x| \leqslant r/2$ (changing the function θ away from the origin does not change the asymptotic expansion of the oscillatory integral).

Then on the one hand the expressions under the integral signs in

$$\int_{B_r(s)} e^{i\tau f} \varphi\theta\, dx \quad \text{and} \quad \int_{B_r(s)} e^{i\tau f} \varphi\, dx$$

differ only on a compact set in X^{+i} and so the difference between these integrals is exponentially small.

On the other hand

$$\int_{B_r(s)} e^{i\tau f} \theta\varphi\, dx = \int_{B_r} (\exp(s\xi))^*\, [e^{i\tau f} \theta\varphi\, dx].$$

The expression under the integral sign on the right-hand side can be written in the form

$$e^{i\tau f_s} (\theta\varphi)_s dx_1 \wedge \ldots \wedge dx_n.$$

Then for sufficiently small $s > 0$

(i) $\operatorname{Im} f_s \geqslant 0$ on B_r.
(ii) f_s does not have critical points on $B_r \backslash 0$.
(iii) $f_s = f$, $(\theta\phi)_s = \theta\phi$ on $B_{r/4}$.
It follows from (i)–(iii) that $\int_{B_r} e^{i\tau f} \theta\varphi\, dx$ and $\int_{B_r} e^{i\tau f_s} (\theta\,\varphi)_s dx$ differ by $o(\tau^{-N})$ for any N as $\tau \to +\infty$. The theorem is proved.

We shall prove a lemma expressing the asymptotics of an oscillatory integral with smooth amplitude in terms of asymptotic integrals with amplitude of the form indicated in Theorem 11.3.

Lemma 11.5. Let $\varphi_1, \ldots, \varphi_\mu$ be monomials generating a basis over $\mathbb{R}$ in the local algebra

$$\mathbb{R}\{x\}/(\partial f/\partial x),$$

and let the function θ be the same as in Theorem 11.3. Then for any smooth amplitude φ with support lying in a sufficiently small neighbourhood of the origin we obtain, as $\tau \to +\infty$, the equality of the asymptotic series

$$\int_{\mathbb{R}^n} e^{i\tau f} \varphi \, dx \approx \sum_{k=1}^{\mu} c_k(\tau^{-1}) \int_{\mathbb{R}^n} e^{i\tau f} \varphi_k \theta \, dx,$$

where c_k is a suitable formal series in $1/\tau$.

Proof. We can represent φ in the form

$$\varphi = \sum a^k \varphi_k \theta + \sum h^k \partial f / \partial x_k,$$

where a^k are numbers and h^k are smooth functions with compact support. Then

$$\int e^{i\tau f} \varphi \, dx = \sum a^k \int e^{i\tau f} \varphi_k \theta \, dx + \frac{1}{i\tau} \int e^{i\tau f} \sum (-1)^{k+1} \partial h^k / \partial x_k \, dx.$$

The numbers $a^1, \ldots, a^\mu$ are the constant terms of the series $c_1(\tau^{-1}), \ldots, c_\mu(\tau^{-1})$. Transforming the right-hand integral in a way analogous to the way that the initial integral was transformed we obtain the coefficients of the formal series, standing for τ^{-1} etc.

Remark. In [286] it was proved that the series $\{c_k(\tau^{-1}) = \sum c_k^j \tau^{-j}\}$ after dividing the jth coefficient by $j!$ converges for sufficiently large τ, if the amplitude φ has the form indicated in Theorem 11.3.

11.2.3 The singular index of a simple critical point and the Coxeter number

V. I. Arnold noticed the following phenomenon (see [11, 12]): the oscillation indices of simple critical points of smooth functions of three variables are given by the formula

$$\beta = -1 - 1/N$$

where N is the Coxeter number of the corresponding group generated by reflections. Theorem 11.3 explains the presence in these formulae of the number N.

Assertion. An oscillatory integral with phase depending on three variables and having a simple critical point can be expanded in an asymptotic series of the form

$$\sum a_\alpha \tau^\alpha,$$

where α is a rational number with denominator equal to the Coxeter number of the corresponding group generated by reflections.

Indeed, in this case, the monodromy operator is the Coxeter transformation of the corresponding group (see Theorem 3.14). Therefore $M^N = \mathrm{Id}$. According to Theorem 11.3 the index α, multiplied by N, is an integer.

11.3   Complex oscillatory integrals with phase $f(x) + g(y)$

The asymptotics of integrals with phase $f + g$ can be expressed in terms of the asymptotics of the integrals with phases f and g, since the exponent of a sum is equal to the product of the exponents of the terms. In particular the asymptotics of integrals with phase

$$x_1^{\mu+1} + \ldots + x_n^{\mu+1}$$

can be expressed in terms of the asymptotics of integrals with phase

$$x_1^{\mu+1}$$

(which were calculated in § 11.1.3). From these expressions we can extract non-trivial information about the asymptotics of integrals with arbitrary phase since an arbitrary isolated critical point is found in a versal deformation of the critical point $x_1^{\mu+1} + \ldots + x_n^{\mu+1}$ (for sufficiently large μ). In particular on this path we can prove the following important theorem: if we are given a critical point of a function f of multiplicity μ, $\{\omega_1, \ldots, \omega_\mu\}$ is a sufficiently general set of holomorphic n-forms and $\{\delta_1(t), \ldots, \delta_\mu(t)\}$ is a basis, depending continuously on t, of the integral homology, vanishing at the critical point, then the function

$$\det^2 \left(\int_{\delta_j(t)} \omega_l / df \right)$$

is not identically equal to zero in a neighbourhood of the point $0 \in \mathbb{C}$, and furthermore the order of its zero at the point 0 is equal to $\mu(n-2)$ (see Theorem 12.1, and compare with Lemma 11.4).

11.3.1 Fubini's theorem for oscillatory integrals

Let $f: (\mathbb{C}^n, 0) \to (\mathbb{C}, 0)$, $g: (\mathbb{C}^l, 0) \to (\mathbb{C}, 0)$ be functions, holomorphic at the origin and having at the origin a critical point of finite multiplicity. Let us consider the function

$$f + g: (\mathbb{C}^{n+l}, 0) \to (\mathbb{C}, 0)$$

The function $f + g$ has at the origin a critical point of (finite) multiplicity equal to the product of the multiplicities of the critical points of the functions f and g. Let ω be a holomorphic n-form given in a neighbourhood of the origin in $\mathbb{C}^n$, and let η be a holomorphic l-form given in a neighbourhood of the origin in $\mathbb{C}^l$. We shall compare the asymptotics of the three integrals

$$\int e^{\tau f} \omega, \quad \int e^{\tau f} \eta, \quad \int e^{\tau(f+g)} \omega \wedge \eta,$$

choosing in a coherent way admissible chains of integration.

Let $\Gamma_1 \subset \mathbb{C}^n$, $\Gamma_2 \subset \mathbb{C}^l$ be admissible chains, respectively, for the critical points of the functions f and g. Let us suppose that

$$\mathrm{Re}\, f|_{\partial \Gamma_1} = \mathrm{Re}\, g|_{\partial \Gamma_2} = -a,$$
$$a/2 > \mathrm{Re}\, f|_{\Gamma_1}, \ \mathrm{Re}\, g|_{\Gamma_2} \geqslant -a$$

for a sufficiently small number $a > 0$ (satisfying this condition can be achieved by changing the chain to an equivalent one). Then the chain

$$\Gamma_1 \times \Gamma_2 \subset \mathbb{C}^{n+l}$$

is admissible for the critical point of the function $f + g$. It is easy to see that this construction gives a linear map from the tensor product of the groups of equivalence classes of chains, admissible for the critical points of the functions f and g, to the group of equivalence classes of chains, admissible for the critical point of the function $f + g$. It can be shown (see [322], and also Theorem 2.9), that this map is an isomorphism.

Remark. In [322] is given a topological proof of this result. We can obtain another proof using theorems about determinants, formulated at the beginning of § 11.3 (see § 13.3.5, and also Corollary 1 of Lemma 11.6).

Theorem 11.4 (see [239]).

$$\int_{\Gamma_1 \times \Gamma_2} e^{\tau(f+g)} \omega \wedge \eta = \int_{\Gamma_1} e^{\tau f} \omega \int_{\Gamma_2} e^{\tau g} \eta.$$

Theorem 11.4 is a direct corollary of Fubini's theorem.

Corollary. The asymptotic series of the left-hand integral is equal to the product of the asymptotic series of the integrals on the right-hand side.

11.3.2 Integrals with phase $f = x_1^{\mu+1} + \ldots + x_n^{\mu+1}$ (an example of the application of Theorem 11.4)

The multiplicity of the critical point of the function $x_1^{\mu+1} + \ldots + x_n^{\mu+1}$ is equal to μ^n. Let us construct μ^n n-chains $\Gamma_{j_1 \ldots j_n}$ $(1 \leqslant j_1, \ldots, j_n \leqslant \mu)$, admissible for this critical point.

Each of the chains $\Gamma_1, \ldots, \Gamma_\mu \subset \mathbb{C}$, indicated in § 11.1.3 and admissible for the critical point $x^{\mu+1}$, we change into an equivalent one so that

$$x^{\mu+1}|_{\partial \Gamma_j} = -1, \quad 1/n > \operatorname{Re} x^{\mu+1}|_{\Gamma_j} \geqslant -1.$$

Let us put the chain $\Gamma_{j_1, \ldots, j_n} \subset \mathbb{C}^n$ equal to

$$\Gamma_{j_1} \times \Gamma_{j_2} \times \ldots \times \Gamma_{j_n}.$$

It is easy to see that the chain $\Gamma_{j_1, \ldots, j_n}$ is admissible for the critical point $x_1^{\mu+1} + \ldots + x_n^{\mu+1}$.

Let us denote by ω_j, where $j = (j_1, \ldots, j_n)$, the form

$$x_1^{j_1-1} \ldots x_n^{j_n-1} dx_1 \wedge \ldots \wedge dx_n.$$

Let us denote by J the set of indices $j = (j_1, \ldots, j_n)$, for which $1 \leqslant j_1, \ldots, j_n \leqslant \mu$.

Lemma 11.6.

1. For any $j, l \in J$

$$\int_{[\Gamma_j]} e^{\tau f} \omega_l = \tau^{r_1}(\mu+1)^{-n} \prod_{k=1}^{\mu} (\varepsilon_{j_k+1}^{l_k} - \varepsilon_{j_k}^{l_k}) \Gamma \left(\frac{l_k}{\mu+1} \right),$$

where $r_l = -(l_1 + \ldots + l_n)/(\mu+1)$; $\varepsilon_1, \ldots, \varepsilon_{\mu+1}$ are the roots of the equation $x^{\mu+1} = -1$, indicated in § 11.1.3 and $\Gamma(\cdot)$ is the gamma function.

2. For any $j, l \in J$,

$$\det\left(\int_{[\Gamma_j]} e^{\tau f} \omega_l\right) = (B_\mu)^{n\mu^{n-1}} \tau^{-n\mu^n/2},$$

where B_μ is the non-zero constant defined in the corollary to Lemma 11.4.

Lemma 11.6 is clearly a corollary of Theorem 11.4 and Lemmas 11.3 and 11.4.

Remark. An integral with phase $x_1^{\mu+1} + \ldots + x_n^{\mu+1}$ and arbitrary form ω reduces by integration by parts to a linear combination of integrals with the same phase and forms ω_l, $l \in J$ (compare with Lemma 11.5).

Corollary 1. The chains Γ_j, $j \in J$, generate a basis in the group of equivalence classes of chains, admissible for the critical point of the function

$$x_1^{\mu+1} + \ldots + x_n^{\mu+1}.$$

Indeed these chains are linearly independent (Section 2 of Lemma 11.6), and their number equals the multiplicity of the critical point.

Corollary 2. We have

$$\det{}^2\left(\int_{\partial_t[\Gamma_j]} \omega_l/df\right) = C_\mu t^{\mu^n(n-2)},$$

where C_μ is a non-zero constant.

Corollary 2 obviously follows from Lemma 11.6 and formula (6) on page 303.

Remark 1. It is easy to see that Corollary 2 for the critical point

$$x_1^{\mu+1} + \ldots + x_n^{\mu+1}$$

is equivalent to the assertion of the theorem on determinants formulated at the beginning of § 11.3.

Remark 2. Lemma 11.6 and its corollary can easily be generalised to the case of a critical point of the function $x_1^{a_1} + \ldots + x_n^{a_n}$.

Chapter 12

Integrals and differential equations

In this chapter we shall prove that many-valued functions, given as integrals of a holomorphic differential form over classes of continuous families of homologies, vanishing at the critical point of a holomorphic function, are all solutions of an ordinary homogeneous linear differential equation, the order of which is not greater than the multiplicity of the critical point. The analysis of this phenomenon leads to the concept of the Gauss-Manin connection in the fibration of vanishing cohomologies associated with the Milnor fibration of the critical point.

In the first section of the chapter we shall prove a theorem about determinants, from which, in particular, follows the existence of holomorphic differential n-forms in $\mathbb{C}^n$, which after division by the differential of our function generate a basis of the cohomology of each fibre of the Milnor fibration of the critical point of this function. In the second section we shall prove the result on differential equations which was formulated above. In the third section we shall discuss the concept of the Gauss-Manin connection.

We note the introduction in § 12.2 of the concept of the Picard-Fuchs singularity of a finite-multiplicity critical point of a holomorphic function, and also the introduction in § 12.3 of the concepts of the (co)homological Milnor fibration, covariantly constant section of the (co)homological fibration and geometrical section of the cohomological fibration.

12.1 Theorem about determinants

In this section we shall prove Theorem 12.1 which is formulated below. It is one of the principal results about integrals.

Let us consider the germ $f : (\mathbb{C}^n, 0) \to (\mathbb{C}, 0)$ of a holomorphic function with a critical point of multiplicity μ. Let us consider a specialisation $f : X \to S$ of the germ and the Milnor fibration corresponding to this specialisation (see page 287). Let us choose in the fibres of the Milnor fibration bases of the integral $(n-1)$st homology, depending continuously on the point of the base. Let us denote the basis in the fibre over the point t by

$$\delta_1(t), \ldots, \delta_\mu(t).$$

(Note that the basis is a many-valued function of the point of the base.) Let us consider an arbitrary set of μ holomorphic differential n-forms $\omega_1, \ldots, \omega_\mu$, given in a neighbourhood of the origin in $\mathbb{C}^n$. Let us suppose that the space X is contained in a neighbourhood in which the forms are defined. We shall study the properties of the function

$$\det{}^2 : t \mapsto \det{}^2 \left(\int_{\delta_j(t)} \omega_l/df \right), \quad j, l = 1, \ldots, \mu.$$

Theorem 12.1 (see [364]).

1. The function $\det^2$ is single-valued in a neighbourhood of the point $t = 0$.

2. The function $\det^2$ has at $t = 0$ a zero of order not less than $\mu(n-2)$ (in particular, for $n > 1$ the function $\det^2$ is holomorphic at $t = 0$).

3. If the set of forms is sufficiently general (the finite jets of the forms at the point $0 \in \mathbb{C}^n$ do not satisfy certain complex analytic relation), then the order of the zero at $t = 0$ of the function $\det^2$ is equal to $\mu(n-2)$.

Definition. The set of forms $\omega_1, \ldots, \omega_\mu$ is called a *trivialisation* if for some family of bases $\delta_1, \ldots, \delta_\mu$ (and so also for any) the function $\det^2$ is not identically equal to zero.

Definition. The set of forms

$$\omega_1, \ldots, \omega_\mu$$

is called a *basis trivialisation* if for some family of bases $\delta_1, \ldots, \delta_\mu$ (and so also for all) the function $\det^2$ has for $t = 0$ a zero of order $\mu(n-2)$.

Corollary of Theorem 12.1. For any finite-multiplicity critical point of a holomorphic function there exists a trivialisation and furthermore there exists a basis trivialisation.

Remark 1. The first proof of the existence of a trivialisation was given in [55]. It is based on theorems (A) and (B) of H. Cartan.

Remark 2. In § 12.3 we shall define the cohomological Milnor fibration of a critical point. This is a vector bundle, the base of which is the same as the base of the usual Milnor fibration. Its fibres are the vector spaces of cohomologies of the fibres of the usual Milnor fibration. The set of forms, described in the definition of trivialisation, gives a trivialisation of the cohomological Milnor fibration over a neighbourhood of the point $t=0$.

Remark 3. The concept of trivialisation does not depend on the choice of specialisation of the germ (see § 10.3.1, and Remark 3 in § 10.3.2).

Remark 4. In Corollary 2 of Lemma 11.6 we produced a set of forms which were a basis trivialisation for the critical point of the function $x_1^N + \ldots + x_n^N$. Theorem 12.1 is derived from the existence of a basis trivialisation for the indicated critical point. For the proof of the theorem see §§ 12.1.3, 12.1.5. In §§ 12.1.1, 12.1.2 we shall prove an auxiliary results about the function $\det^2$.

Before passing on to the proof of Theorem 12.1, let us formulate the properties of trivialisations.

I. If $\omega_1, \ldots, \omega_\mu$ is a trivialisation then the Gelfand-Leray forms

$$\omega_1/df, \ldots, \omega_\mu/df$$

form a basis of the $(n-1)$st cohomology with complex coefficients in all the fibres of the Milnor fibration of the critical point 0 of the function f, lying over a sufficiently small neighbourhood of the point $t=0$.

II. Let $\omega, \omega_1, \ldots, \omega_\mu$ be holomorphic differential n-forms given in a neighbourhood of the origin in $\mathbb{C}^n$. If $\omega_1, \ldots, \omega_\mu$ is a trivialisation then on a sufficiently small punctured neighbourhood of the point $t=0$ there exists also unique holomorphic functions $p_1, \ldots, p_\mu$ with the property

$$\int_{\delta(t)} \omega/df \equiv \sum_j p_j(t) \int_{\delta(t)} \omega_j/df$$

for any continuous family δ of integral $(n-1)$st homologies in the fibres of the Milnor fibration of the critical point 0 of the function f. If, furthermore, $\omega_1, \ldots, \omega_\mu$ is a basis trivialisation, then the functions $p_1, \ldots, p_\mu$ can be holomorphically continued to the point $t=0$.

Property I is obvious. Let us prove Property II. Taking as the families δ the families $\delta_1, \ldots, \delta_\mu$ generating bases of the homologies of the fibres of the Milnor fibration, we obtain a system of linear equations in $p_1, \ldots, p_\mu$. Its determinant is non-zero by the definition of trivialisation. Solving the system of equations by Cramer's rule, we obtain the functions we are looking for in the form of quotients of pairs of determinants. Each quotient is holomorphic at $t = 0$ if $\omega_1, \ldots, \omega_\mu$ is a basis trivialisation.

12.1.1   Elementary properties of the function $\det^2$

It is convenient to consider the function $\det^2$ not on the base of the Milnor fibration of the critical point but on the base of the Milnor fibration of a versal deformation of the critical point.

Let

$$F : (\mathbb{C}^n \times \mathbb{C}^k, 0 \times 0) \to (\mathbb{C}, 0)$$

be a versal deformation of the germ f. Let us consider a specialisation $G : X \to S$ of an unfolding of the deformation and the corresponding Milnor fibration $G : X' \to S'$ (for the notation and definitions see § 10.3.1 on page 286, see also figure 73 on page 286). Let us choose in the fibres of the Milnor fibration bases of the integral $(n-1)$st homologies, continuously depending on the points of the base. If $s \in S'$, let us denote by $\delta_1(s), \ldots, \delta_\mu(s)$ a basis in the fibre over s. Let $\omega_1, \ldots, \omega_\mu$ be holomorphic differential n-forms on X. Let $F : X \to \mathbb{C}$ be a holomorphic function, representing the germ F. The forms $\omega_1, \ldots, \omega_\mu$ on each fibre of the Milnor fibration determine holomorphic differential $(n-1)$-forms

$$\omega_1/d_x F, \ldots, \omega_\mu/d_x F$$

(the forms $\omega_1, \ldots, \omega_\mu$ must be restricted to a subspace of the form $\mathbb{C}^n \times y$ and divided by $d_x F$). Let us consider on the base S' the function

$$\det^2 : s \mapsto \det^2 \left(\int_{\delta_j(s)} \omega_l/d_x F \right), \quad j, l = 1, \ldots, \mu.$$

Lemma 12.1. The function $\det^2$ is single-valued and holomorphic.

Proof. The elements of the matrix are many-valued holomorphic functions on the base (Theorem 10.5). By analytic continuation along a closed path the value of the function $\det^2$ is multiplied by the square of the determinant of the monodromy operator in the homologies. The monodromy operator is non-degenerate and integral as is its inverse. Therefore the square of its determinant is equal to 1. Consequently the function $\det^2$ is single-valued.

Remark. If n is even, then the function det is already single-valued and holomorphic. It is sufficient to verify the singlevaluedness by analytic continuation along a small path round a non-singular point of the discriminant in the base of the versal deformation. The determinant of the corresponding monodromy transformation is equal to the determinant of the Picard-Lefschetz transformation, which for even n is equal to 1.

According to the definitions in § 10.3.1, the base S' of the Milnor fibration is the complement of a hypersurface in the product of balls

$$S = B_\eta^1 \times B_\varrho^k.$$

The difference $\Sigma = S \setminus S'$ is called the discriminant.

Theorem 12.2 (see [364]). The function $\det^2$ is meromorphic at the origin of the base S and, furthermore, can be represented in the form gh^{n-2} where g and h are functions on S, holomorphic at the origin, and h being that function, the zeros of which define the discriminant (without multiplicities).

Proof. On S we are given coordinates: $y \in B_\varrho^k$ is the parameter of the deformation, $u \in B_\eta^1$ is the value of the function F. For fixed y the line of values intersects the discriminant in μ points (counting multiplicities). For general values of the coordinates y the line of values intersects the discriminant in μ different non-singular points. It is sufficient to prove that near these non-singular points the discriminant function $\det^2/h^{n-2}$ is holomorphic. Indeed then by the theorem on removing singularities the function $\det^2/h^{n-2}$ is holomorphic everywhere on S. Holomorphicity in a neighbourhood of a non-singular point of the discriminant follows from explicit calculation of the integrals along cycles, vanishing at non-degenerate critical points (see the example in § 10.3.4 on page 294).

And so let $s_0 = (u_0, y_0)$ be a non-singular point of the discriminant, in a neighbourhood of which the discriminant is the graph of the function $u = u(y)$. The function $F(\cdot, y)$ has one critical point in $X \cap (\mathbb{C}^n \times y)$ with critical value $u(y)$, and this critical point is non-degenerate. Let us consider the monodromy operator M corresponding to going round the point $(u(y), y)$ along a small circle lying in the line of values. The operator M is the Picard-Lefschetz operator of "reflection" in the homology class of a cycle, vanishing as $s \to s_0$. Using the monodromy operator, we shall change all the bases

$$\delta_1(s), \ldots, \delta_\mu(s)$$

simultaneously for all s in a small neighbourhood of the point s_0 by the same linear transformation. (This causes $\det^2$ to be multiplied by the square of the determinant of the transformation). If n is odd then M has a $(\mu - 1)$-dimensional subspace of invariant vectors and a one-dimensional subspace of anti-invariant vectors, generated by a class of a cycle, vanishing as $s \to s_0$. Let us use as a new basis

$$\delta_1(s), \ldots, \delta_\mu(s),$$

where $\delta_1(s)$ is the class of a cycle, vanishing as $s \to s_0$, and $\delta_2(s), \ldots, \delta_\mu(s)$ are any classes forming a basis of the invariant classes. If n is even then all the eigenvalues of the operator M are equal to 1, and the subspace of invariant vectors is $(\mu - 1)$-dimensional. Let us use as a new basis

$$\delta_1(s), \ldots, \delta_\mu(s),$$

where $\delta_1(s)$ is the class of a cycle, vanishing as $s \to s_0$, $\delta_2(s)$ is any homology class the intersection number of which with $\delta_1(s)$ is equal to 1, and

$$\delta_3(s), \ldots, \delta_\mu(s)$$

are any classes making, together with $\delta_1(s)$, a basis of the invariant classes. Then Theorem 12.2 follows from Lemma 12.2.

Lemma 12.2. Let ω be a holomorphic differential n-form, given in a neighbourhood of the origin in $\mathbb{C}^n \times \mathbb{C}^k$ which contains the space X. If n is odd, then in a neighbourhood of the point $s_0 \in \Sigma$ the following expansions are valid:

$$\int_{\delta_1(u, y)} \omega / d_x F = (u - u(y))^{n/2 - 1} P_1(u, y), \tag{1}$$

$$\int_{\delta_j(u, y)} \omega / d_x F = P_j(u, y), \quad j = 2, \ldots, \mu, \tag{2}$$

where $P_1, \ldots, P_\mu$ are functions, holomorphic at the point s_0. If n is even, then in a neighbourhood of the point $s_0 \in \Sigma$ expansion (1) is valid as is expansion (2) for $j = 3, \ldots, \mu$, and also expansion

$$\int_{\delta_2(u, y)} \omega / d_x F = \pm (u - u(y))^{n/2 - 1} \, \frac{\ln(u - u(y))}{2\pi i} \, P_1(u, y) + P_2(u, y), \qquad (3)$$

where $P_1, \ldots, P_\mu$ are functions, holomorphic at the point s_0.

Proof of the lemma. The restriction of the form ω to a subspace of the form $\mathbb{C}^n \times y$ is closed. By Poincaré's lemma there exists in a neighbourhood of the origin in $\mathbb{C}^n \times \mathbb{C}^k$ a holomorphic $(n-1)$-form ψ, the restriction of the differential of which to any subspace of the form $\mathbb{C}^n \times y$ coincides with the restriction of the form ω. According to formula (3) on page 284.

$$\int_{\delta_j(u, y)} \omega / d_x F = \frac{\partial}{\partial u} \int_{\delta_j(u, y)} \psi. \qquad (4)$$

If the cycle δ_j is invariant under M, then in a neighbourhood of the point s_0 the integral of the form ψ over $\delta_j(u, y)$ is single-valued and depends holomorphically on $(u, y) \in S \setminus \Sigma$. According to Theorem 10.7 this integral has a finite limit as $u \to u(y)$ along the line $y = \text{const}$. Consequently, in a neighbourhood of the point s_0 it can be extended holomorphically to Σ. Consequently its derivative (4) can also be extended holomorphically to Σ. Lemma 12.2 is proved for $j = 3, 4, \ldots, \mu$ and $j = 2$ for n odd.

In a neighbourhood of a non-degenerate critical point, with critical value $u(y)$, the function $F(\cdot, y)$ can be reduced to the form

$$z_1^2 + \ldots + z_n^2 + u(y)$$

by the holomorphic change of the variables of the form $x = x(z, y)$. In terms of these coordinates, using the calculations of the example in §10.3.4 on page 294, we obtain

$$\int_{\delta_1(u, y)} \psi = (u - u(y))^{n/2} \, Q(u, y), \qquad (5)$$

where Q is a holomorphic function. This in conjunction with (4) proves the lemma for $j=1$. It follows from the Picard-Lefschetz formula for n even that

$$\int_{\delta_2(u,y)} \psi = \pm \frac{\ln(u - u(y))}{2\pi i} \int_{\delta_1(u,y)} \psi + P(u,y), \qquad (6)$$

where P is a single-valued function in a neighbourhood of the point s_0 (the $\pm$ sign is chosen according to the parity of the number $n/2$). In an analogous way to the above, we can holomorphically extend P to Σ. Consequently the lemma is true for $j=2$ and n even. Lemma 12.2 and Theorem 12.2 are proved.

Corollary 1 of Theorem 12.2. Let

$$\omega_1, \ldots, \omega_\mu$$

be holomorphic differential n-forms on X. Let us suppose that the restriction of the function

$$\det^2 : s \mapsto \det^2 \left(\int_{\delta_j(s)} \omega_l/d_x F \right), \qquad j, l = 1, \ldots, \mu,$$

to the line $y=0$ has at the origin a zero of order $\mu(n-2)$. Then the function $\det^2$ can be represented in the form $\det^2 = gh^{n-2}$ where g is a holomorphic function on S, different from zero at the point $s=0$, and h is a holomorphic function on S, the zeros of which give the discriminant (without multiplicities). Furthermore, in this case the forms

$$\omega_1/d_x F, \ldots, \omega_\mu/d_x F$$

generate a basis of the $(n-1)$st cohomology in all the fibres of the Milnor fibration of a versal deformation, lying over the points of the base sufficiently near to the origin $s=0$.

Indeed the multiplicity of the intersection of the line $y=0$ and the discriminant at the point $s=0$ is equal to μ. Therefore the restriction of an arbitrary function, giving the discriminant without multiplicities on the line $y=0$ has at the origin a zero of order μ. According to Theorem 12.2 the quotient of the function $\det^2$ and the $(n-2)$nd power of the function giving the discriminant is invertible at the origin on S. The second assertion follows obviously.

Remark. The existence of a set of forms $\omega_1, \ldots, \omega_\mu$, for which the restriction of the function $\det^2$ to the line $y = 0$ has at the origin a zero of order $\mu(n-2)$, will be proved in § 12.1.5.

Corollary 2 of Theorem 12.2. Let

$$\omega_1, \ldots, \omega_\mu, \omega$$

be holomorphic differential n-forms on X. Let us suppose that the restriction of the function

$$\det{}^2 : s \mapsto \det{}^2 \left(\int_{\delta_j(s)} \omega_l / d_x F \right), \quad j, l = 1, \ldots, \mu,$$

to the line $y = 0$ has at the origin a zero of order $\mu(n-2)$. Then in a neighbourhood of the origin on S there exist unique holomorphic functions $p_1, \ldots, p_m$, possessing the property:

$$\int_{\delta(s)} \omega / d_x F = \sum_j p_j(s) \int_{\delta(s)} \omega_j / d_x F$$

for any continuous family δ of integral $(n-1)$st homologies in the fibres of the Milnor fibration of a versal deformation.

Proof. See Corollary 2 of Theorem 12.1.

12.1.2 Each coefficient of the expansion in series of the integral of a form along the class of a family of vanishing homologies depends only on a finite jet of the form and depends on this jet holomorphically

We shall state this result more precisely in the following lemma.

Let $f : (\mathbb{C}^n, 0) \to (\mathbb{C}, 0)$ be a function, holomorphic at the origin, and having at the origin a critical point of finite multiplicity. Let

$$\omega = g \, dx_1 \wedge \ldots \wedge dx_n$$

be a differential n-form, holomorphic at the origin. Let

$$\delta(t) \in H_{n-1}(X_t, \mathbb{Z})$$

be a class of homologies, continuously depending on a paramter, of the fibre X_t of the Milnor fibration of the critical point 0 of the function f. According to Theorem 10.8 there is in each sector of a small neighbourhood of the point $t=0$ an expansion in series

$$\int_{\delta(t)} \omega/df = \sum_{k,\alpha} \alpha_{k,\alpha} t^\alpha (\ln t)^k, \tag{7}$$

where the numbers α depend only on f and do not depend on ω, δ.

Lemma 12.3. For any k and α the coefficient $a_{k,\alpha}$ depends only on a finite jet of the function g at the point $0 \in \mathbb{C}^n$ and it depends on this jet holomorphically.

Corollary. Each coefficient of the Laurent series at the point $t=0$ of the function $\det^2$, defined in Theorem 12.1, depends only on finite jets of the forms

$$\omega_1, \ldots, \omega_\mu$$

at the point $0 \in \mathbb{C}^n$ and it depends on these jets holomorphically.

Proof of the lemma. Let us denote by J, the ideal in $\mathbb{C}\langle\langle x_1, \ldots, x_n\rangle\rangle$, generated by the functions

$$\partial f/\partial x_1, \ldots, \partial f/\partial x_n.$$

Let N be a natural number. For the proof of the first assertion of the lemma it is sufficient to prove that for $g \in (J_f)^{2N}$ all the powers α in the series (7) are greater than $N-1$. Here $(J_f)^{2N}$ is the ideal generated by $2N$-fold products of elements in J_f. Indeed the ideal J_f contains some power of the maximal ideal (in view of the finite-multiplicity of the critical point). Therefore the ideal $(J_f)^{2N}$ also contains some power of the maximal ideal. For any function g from this power of the maximal ideal all the α in series (7) are greater than $N-1$.

We shall prove that for $g \in (J_f)^{2N}$ in series (7) all α are greater than $N-1$. The proof is by induction. For $N=0$ the result is true (Theorem 10.8). Let

$$g = \sum h_j \partial f/\partial x_j,$$

where

$$h_j \in (J_f)^{2N-1}.$$

Then according to formula (3) on page 284

$$\frac{d}{dt} \int_{\delta(t)} \omega/df = \int_{\delta(t)} \sum (-1)^{j+1} \partial h_j/\partial x_j \, dx_1 \wedge \ldots \wedge dx_n/df,$$

where

$$\partial h_j/\partial x_j \in (J_f)^{2N-2}.$$

By the induction hypothesis the expansion in series of the integral on the right-hand side begins with a power greater than $N-2$. Consequently, the series of the integral of the form ω/df begins with a power greater than $N-1$.

The second part of the lemma easily follows from the obvious result: if we are given a meromorphic function, holomorphically depending on parameters, then the coefficients of its Laurent series holomorphically depend on parameters (compare with the proof of Theorem 10.2).

12.1.3   Beginning of the proof of Theorem 12.1

The first and second assertions of the theorem are corollaries of Lemma 12.1 and Theorem 12.2. According to the corollary of Lemma 12.3, it is sufficient for the proof of the third assertion of Theorem 12.1 to prove the existence of at least one set of forms for which the function $\det^2$ has at the origin a zero of order *no more than* $\mu(n-2)$. The proof of existence is derived from Corollary 2 of Lemma 11.6; see § 12.1.5. First we shall prove an auxiliary result.

12.1.4   A critical point of multiplicity μ is to be found in a versal deformation of the critical point of the function $x_1^N + \ldots + x_n^N$ for $N \geq \mu + 2$

The result is stated more precisely in the following lemma.

Lemma 12.4. Let the function $f : (\mathbb{C}^n, 0) \to (\mathbb{C}, 0)$ be holomorphic at the origin and have at the origin a critical point of multiplicity μ. Then for any $N \geq \mu + 2$ there exists a polynominal

$$P(x_1, \ldots, x_n, \delta, \varepsilon_1, \ldots, \varepsilon_n) = Q(x_1, \ldots, x_n, \delta) + \sum_{j=1}^{n} (1 + \varepsilon_j) x_j^N,$$

possessing the properties:

1. For fixed $\delta \neq 0$, $\varepsilon_1, \ldots, \varepsilon_n$ the function $P : \mathbb{C}^n \to \mathbb{C}$ and the function f are equivalent in a neighbourhood of the point 0.

2. $\quad Q(x_1, \ldots, x_n, 0) \equiv 0$.

3. There exist numbers $\delta, \varepsilon_1, \ldots, \varepsilon_n$, as small in modulus as we like, for which the hypersurface

$$\{x \in \mathbb{C}^n | P(x, \delta, \varepsilon) = 0\}$$

is non-singular away from the origin.

Corollary. Let us consider a versal deformation F of the germ of the function $x_1^N + \ldots + x_n^N$ at the origin. Let us denote by Λ the germ of the set of all values of the parameters of the deformation for which F has a unique critical point, with critical value zero, which is equivalent to the critical point 0 of the function f. Then for $N \geq \mu + 2$, Λ is not empty.

The corollary is true since the deformation indicated in Lemma 12.4 can be induced from a versal deformation.

Proof of the lemma. Let us take as Q the polynominal

$$f_{N+1}(\delta x_1, \ldots, \delta x_n),$$

where $f_{N+1}(x_1, \ldots, x_n)$ is the Taylor polynominal of degree $N + 1$ of the function f at the origin. By the theorem on finite determinacy (see Volume 1 Chapter 6) the function in a neighbourhood of the critical point of multiplicity μ is equivalent to its own Taylor polynomial of degree $\mu + 1$. Therefore Section 1 of the lemma is true. Section 2 is obvious. Section 3 is proved in an analogous way to Lemma 6.1.

12.1.5   There exists at least one set of forms for which the function $\det^2$, defined in Theorem 12.1, has at the origin a zero of order no greater than $\mu(n-2)$

Such a set was produced for a critical point of the form $x_1^N + \ldots + x_n^N$ in § 11.3.3 (see Corollary 2 of Lemma 11.6). An arbitrary critical point of finite multiplicity is to be found in a versal deformation of a critical point of the form $x_1^N + \ldots + x_n^N$. Therefore the fibre of the Milnor fibration of an arbitrary critical point is included in the fibre of the Milnor fibration of a versal deformation of a sum of powers. This inclusion induces a monomorphic inclusion of vanishing homologies. The forms of the set produced for the sum of powers can be integrated over the classes of homologies vanishing at the critical point of finite multiplicity we are studying. We shall prove that the set of forms we are seeking for the initial critical point can be chosen from the set produced for the sum of powers. Now let us give some more details.

Let $N \geqslant \mu + 2$. Let us consider a versal deformation

$$F : (\mathbb{C}^n \times \mathbb{C}^k, 0 \times 0) \to (\mathbb{C}, 0)$$

of the germ at the origin of the function $x_1^N + \ldots + x_n^N$. Let us consider a specialisation $G : X \to S$ of the unfolding of the versal deformation and the corresponding Milnor fibration $G : X' \to S'$.

Let us denote by J the set of multiindices $j = (j_1, \ldots, j_n)$, where $1 \leqslant j_1, \ldots, j_n \leqslant N$. The number of these multiindices is equal to the Milnor number μ' of the critical point of the function $x_1^N + \ldots + x_n^N : \mu' = (N-1)^n$. Let us consider on X holomorphic differential n-forms w_j, $j \in J$, where

$$\omega_j = x_1^{j_1 - 1} \ldots x_n^{j_n - 1} \, dx_1 \wedge \ldots \wedge dx_n.$$

Let us suppose that in the fibre of the Milnor fibration $G : X' \to S'$ we are given a basis $\delta_1(s), \ldots, \delta_{\mu'}(s)$, continuously depending on the point of the base, of $(n-1)$st integer homologies, where $s \in S'$. Let us consider on S' the function

$$\det^2 : s \mapsto \det^2 \left(\int_{\delta_l(s)} \omega_j / d_x F \right).$$

According to Theorem 12.2 this function is single-valued and holomorphic and is a meromorphic function on $S = S' \cup \Sigma$. According to Corollary 2 of Lemma 11.6 and Corollary 1 of Theorem 12.2 the function $\det^2$ can be represented in the form $\det^2 = gh^{n-2}$, where g is a holomorphic function on S, different from zero at the point $s = 0$, and h is a holomorphic function on S the zeros of which give the discriminant (without multiplicities).

The space S (the base of the specialisation) is the product of balls $B_\eta^1 \times B_\sigma^k$ (the coordinates in B_η^1, B_σ^k are denoted, respectively, by u, y). Let us denote by Λ the set of those $y \in B_\sigma^k$ for which the function $F(\cdot, y)$ has on $X \cap (\mathbb{C}^n \times y)$ a unique critical point with critical value zero which is equivalent to the critical point of the function f (F is the representative on X of the germ F). According to Lemma 12.4, Λ is not empty.

Let $y_0 \in \Lambda$. Let us denote by (x^0, y_0) the critical point of the function $F(\cdot, y_0)$ with zero critical value. For fixed $\varrho' > 0$ let us denote by $Y_{(u, y_0)}$ the subset of all points of the fibre $X_{(u, y_0)}$ of the specialisation, of distance less than ϱ' from (x^0, y_0); see figure 78 and compare with figure 73 on page 286.

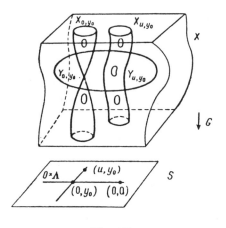

Fig. 78.

If the number ϱ' is sufficiently small, and the number $\eta' > 0$ is sufficiently small in comparison with ϱ', then the union $\cup_u Y_{(u, y_0)}$, where $u \neq 0$, $|u| < \eta'$, together with the natural projection $Y_{(u, y_0)} \mapsto (u, y_0)$ forms the Milnor fibration of the critical point (x^0, y_0) of the function

$$G(\cdot, y_0) : \mathbb{C}^n \times y_0 \to \mathbb{C} \times y_0 .$$

The fibre $Y_{(u, y_0)}$ of this fibration is homotopy equivalent to a bouquet of μ $(n-1)$-dimensional spheres. The inclusion

$$Y_{(u, y_0)} \hookrightarrow X_{(u, y_0)}$$

induces a monomorphism of $(n-1)$st homologies in accordance with Theorem 2.1.

Over the line $y = y_0$ in a neighbourhood of the point $(0, y_0)$ we make simultaneous, linear over $\mathbb{R}$, changes of all the bases $\delta_1(s), \ldots, \delta_\mu(s)$ so that

1) the first μ elements of the bases of the new family vanish as

$$(u, y_0) \mapsto (0, y_0)$$

(that is they are bases of the $(n-1)$st homology of the fibres $Y_{(u, y_0)}$ of the Milnor fibration of the critical point (x^0, y_0) of the function $G(\cdot, y_0)$);

2) the last $\mu' - \mu$ elements of the bases of the new family belong to the root subspace associated with the eigenvalue 1 of the monodromy operator, generated by going round $(0, y_0)$ on the line $y = y_0$.

Such a basis exists, since the subspace

$$H_{n-1}(Y_{(u, y_0)}) \subset H_{n-1}(X_{(u, y_0)})$$

is invariant relative to the indicated monodromy operator, and the action of this monodromy operator on the quotient space is trivial (indeed,

$$H_{n-1}(X_{(u, y_0)})/H_{n-1}(Y_{(u, y_0)}) \simeq H_{n-1}(X_{(0, y_0)})).$$

Let us consider the restriction of the function $\det^2$ to the line $y = y_0$. After changing basis the function $\det^2$ is multiplied by a number, equal to the determinant of the change of basis. The restriction of the function $\det^2$ has for $u = 0$ a zero of order $\mu(n-2)$. Indeed, the critical point (x^0, y_0) of the function $F(\cdot, y_0)$ branches under small deformations into μ non-degenerate critical points, and at each non-singular point of the discriminant the function $\det^2$ has a zero of order $n - 2$.

Our problem consists of proving the existence in the matrix

$$\left(\int_{\delta_l(u, y_0)} \omega_j/d_x F \right), \quad l, j \in J,$$

of a minor of dimension μ, occurring in the first μ rows and having for $u = 0$ a zero of order not greater than $\mu(n-2)/2$ (in the first μ rows occur the integrals over the first μ elements of the basis of homology).

The existence of such a minor follows from two remarks:

1) $\det = \sum_I \varDelta^I \bar{\varDelta}^I,$

where $I \subset J$ is a subset of μ elements, $\varDelta^I$ is the minor of dimension μ, occurring in the first μ rows and in the columns with numbers in I, and $\bar{\Delta}^I$ is the algebraic complement of the minor Δ^I.

2) For any I the minor $\bar{\Delta}^I$ can be expanded in a series of powers of the parameter u and powers of the logarithm of the parameter u, all powers of the parameter in this series being non-negative.

The first remark is the theorem on the expansion of a determinant in terms of the minors of the first μ rows. The second remark follows from Theorem 10.6 and formula (3) on page 284 in view of the coice of the last $\mu' - \mu$ elements of the basis of the homologies.

Theorem 12.1 is proved.

12.2 The integral is the solution of an ordinary homogeneous linear differential equation with a regular singular point

12.2.1 Integrals of a single form

Theorem 12.3. Let the function $f : (\mathbb{C}^n, 0) \to (\mathbb{C}, 0)$ be holomorphic in a neighbourhood of the origin and have at the origin a critical point of finite multiplicity. Let us be given a holomorphic differential n-form ω in a neighbourhood of the origin in $\mathbb{C}^n$. Then in a sufficiently small punctured neighbourhood of the point $0 \in \mathbb{C}$ there exist unique holomorphic functions $p_1, \ldots, p_l$, for which the ordinary differential equation

$$I^{(l)} + p_1 I^{(l-1)} + \ldots + p_l I = 0 \tag{8}$$

has the property: all its solutions are linear combinations of many-valued holomorphic functions of the form

$$I(t) = \int_{\delta(t)} \omega/df, \tag{9}$$

where δ is an arbitrary continuous family of integral $(n-1)$st homologies in the fibres of the Milnor fibration of the critical point 0 of the function f.

Definition. Equation (8) is called the *Picard-Fuchs equation* of the form ω.

Example. Let f be a quasihomogeneous polynomial of type $(\alpha_1, \ldots, \alpha_n)$ of weight 1. Let ω be a quasihomogeneous polynomial differential n-form of type

$(\alpha_1, \ldots, \alpha_n)$ of weight r. This means that relative to the stretch function

$$g_\lambda : (x_1, \ldots, x_n) \mapsto (\lambda^{\alpha_1} x_1, \ldots, \lambda^{\alpha_n} x_n), \qquad \lambda \in \mathbb{C},$$

the polynomial and the form have the properties:

$$f \circ g_\lambda = \lambda f, \qquad g_\lambda^* \omega = \lambda^r \omega.$$

Let us suppose that f has an isolated critical point at the origin and that there exists at least one class of homologies, vanishing at the critical point of the polynomial f, over which the integral of the form ω is different from zero. Then

$$dI/dt = (r-1)I/t$$

is the Picard-Fuchs equation of the form ω. In particular, for $f = x_1^2 + \ldots + x_n^2$

$$dI/dt = (n/2 - 1)I/t$$

is the Picard-Fuchs equation of the form $dx_1 \wedge \ldots \wedge dx_n$. Indeed

$$g_\lambda^* (\omega/df) = \lambda^{r-1} \omega/df.$$

The stretch function g_λ induces an isomorphism between the homologies of the fibres of the Milnor fibration. Therefore all integrals of the form ω/df have the form $\mathrm{const} \cdot t^{r-1}$.

Remark. We can show (using Corollary 1 of the theorem about determinants) that the order of a Picard-Fuchs equation of a sufficiently general form is equal to the multiplicity of the critical point.

Proof of the theorem. Let $\delta_1(t), \ldots, \delta_\mu(t)$ be a basis, depending continuously on t, of the integral $(n-1)$st homologies in the fibre of the Milnor fibration of the critical point of the function f. Let us consider the many-valued vector function

$$I(t) = \left(\int_{\delta_1(t)} \omega/df, \ldots, \int_{\delta_\mu(t)} \omega/df \right).$$

For any natural number k let us denote by $L_k(t) \subset \mathbb{C}^\mu$ the subspace generated by the vectors

$$I(t), I^{(1)}(t), \ldots, I^{(k)}(t).$$

The subspace $L_k(t)$ depends on the choice of the argument t. However, its dimension does not depend on the choice of the argument t (under change of argument all the vectors are multiplied by the monodromy operator). For all t with sufficiently small modulus the dimensions of the subspaces $L_k(t)$ are equal. Indeed, the dimension of the subspace is the dimension of the maximal non-zero minor of the matrix, consisting of the coordinates of vectors generating the subspace. According to Theorem 10.8, the minors of the matrix consisting of the coordinates of the vectors $I, \ldots, I^{(k)}$ can be expanded in series in a neighbourhood of the point $t = 0$. Therefore, if a minor is not identically equal to zero, then it is not mapped to zero in a sufficiently small punctured neighbourhood of the point $t = 0$. Let, then, l be the smallest natural number for which the vector $I^{(l)}(t)$ is a linear combination of the vectors

$$I(t), I^{(1)}(t), \ldots, I^{(l-1)}(t)$$

(for all t with sufficiently small modulus). We have

$$I^{(l)}(t) + p_1(t) I^{(l-1)}(t) + \ldots + p_l(t) I(t) = 0, \tag{10}$$

where $p_1, \ldots, p_l$ are holomorphic functions in a punctured neighbourhood of zero. These functions are single-valued, since under change of argument t the function I and all its derivatives are multiplied by the monodromy operator. By construction each coordinate of the vector function I is a solution of equation (10), and the linear combinations of the coordinates generate the l-dimensional solution space. The theorem is proved.

Corollary. The order of the differential equation (8) is not greater than the multiplicity of the critical point 0 of the function f.

Definition. Let

$$I^{(l)} + p_1 I^{(l-1)} + \ldots + p_l I = 0$$

be a differential equation, the coefficients of which are defined in a punctured neighbourhood of the point $t = 0$ and are holomorphic. Then $t = 0$ is called a *singular* point of the equation if at least one of the coefficients cannot be holomorphically continued to the point $t=0$. The point $t=0$ is called *regular singular* point, if the point $t=0$ is singular and the coefficients of the equation can be represented in the form

$$p_j = P_j/t^j, \quad j = 1, \ldots, l,$$

where the functions $P_1, \ldots, P_l$ are holomorphic at $t=0$.

Example. Let f and ω be the same as in the previous example. The Picard-Fuchs equation of the form ω has for $t=0$ a non-singular point if $r=1$ and a regular singular point if $r \neq 1$.

Theorem 12.4 (see [79]). A necessary and sufficient condition for the singular point $t=0$ to be a regular singular point of the differential equation

$$I^{(l)} + p_1 I^{(l-1)} + \ldots + p_l I = 0$$

is that an arbitrary solution of the equation in an arbitrary sector $a < \arg t < b$ grows no faster, as $t \to 0$, than a suitable power of the parameter $I(t) = o(t^{-N})$ for some N.

Corollary. For the Picard-Fuchs equation of a holomorphic differential form (see (8)) the point $t=0$ is either a non-singular or a regular singular point.
    Indeed, see Theorem 10.7.

Idea of the proof of sufficiency. We use induction on l. For $l=1$ the theorem is true, since

$$I = \text{const } e^{-\int p_1 dt}.$$

Let us suppose that we have proved sufficiency for $l=m$ and let us prove sufficiency for $l=m+1$. First, the initial equation has a solution of the form

$$I_0(t) = t^\alpha \phi(t),$$

where $\alpha \in \mathbb{C}$, and ϕ is a function, holomorphic at $t=0$. Indeed if

$$I_1, \ldots, I_{m+1}$$

is a fundamental system of solutions, then we can take as I_0 the eigenvector $\sum c_j I_j$ of the monodromy transformation of the solutions (α and the eigenvalue λ are connected by the relation $\lambda = e^{2\pi i \alpha}$). Second, for any j,

$$t^j I_0^{(j)} / I_0$$

is a holomorphic function. Third, the change

$$I = I_0 \int J \, dt$$

reduces the equation to a homogeneous linear equation of order m. The coefficients of the new equation can be explicitly calculated in terms of the coefficients of the original equation and expressions of the form $I_0^{(j)} / I_0$. By the induction hypothesis the new equation has a regular singular point. Returning to the original equation and using explicit formulae for the coefficients, we obtain the regularity of its critical point.

Idea of the proof of necessity. The change of variables

$$y_1 = I, \; y_2 = t I^{(1)}, \ldots, y_l = t^{l-1} I^{(l-1)}$$

reduces the equation to the system of homogeneous linear first order equations:

$$dy/dt = Ay,$$

in which the matrix A has for $t = 0$ a pole of first order. Then necessity follows from Theorem 12.6 which is discussed below.

The Picard-Fuchs equation describes integrals of a single form. Let us consider a system of differential equations, describing the integrals of forms generating a basis of the vanishing cohomology. Such a system contains information, not only about the forms, but about the critical point of the function.

12.2.2 Integrals of forms generating a basis in the vanishing cohomology

Let the function $f : (\mathbb{C}^n, 0) \to (\mathbb{C}, 0)$ be holomorphic at the point $0 \in \mathbb{C}^n$ and have there a critical point of multiplicity μ. Let $\omega_1, \ldots, \omega_\mu$ be holomorphic differential n-forms, given in a neighbourhood of the point $0 \in \mathbb{C}^n$ and constituting a trivialisation for the critical point 0 of the function f (see page 317).

Theorem 12.5. In a sufficiently small punctured neighbourhood of the point $0 \in \mathbb{C}$ there exists a unique $\mu \times \mu$ matrix A of holomorphic functions for which the system of ordinary differential equations

$$\frac{dI^j}{dt} = \sum_{k=1}^{\mu} A_k^j I^k, \quad j = 1, \ldots, \mu, \tag{11}$$

possesses the property: all the solutions of the system are linear combinations of vector functions of the form

$$I(t) = \left(\int_{\delta(t)} \omega_1/df, \ldots, \int_{\delta(t)} \omega_\mu/df \right), \tag{12}$$

where δ is an arbitrary continuous family of integral $(n-1)$st homologies in the fibres of the Milnor fibration of the critical point 0 of the function f.

Definition. The system (11) is called the *Picard-Fuchs equation of the trivialisation*

$$\omega_1, \ldots, \omega_\mu.$$

The **proof** of the theorem is analogous to the proof of Theorem 12.3.

Remark. The coefficients of the matrix A are meromorphic at the point $t = 0$ in view of Theorem 10.8.

Example of the Picard-Fuchs equations of a trivialisation. Let

$$f = x_1^{\mu+1} + \ldots + x_n^{\mu+1}.$$

Let $\omega_j, j \in J$ be the set of forms, defined before Lemma 11.6 on page 314. These forms are a trivialisation for the critical point of the function f (Corollary 2 of Lemma 11.6). The Picard-Fuchs equations of this trivialisation are the system

$$dI^j/dt = ((j_1 + \ldots + j_n)/(\mu+1) - 1)I^j/t, \quad j \in J.$$

12.2.3 The lattice in the solution space

There can be chosen a μ-dimensional integer lattice in the linear space of solutions of the Picard-Fuchs equations of a trivialisation: a solution belongs to this lattice if it has the form (12), that is if its coordinates are integrals over a class of the continuous family of integral homologies. This lattice possesses two properties:

1. A basis over $\mathbb{Z}$ of the lattice is a basis over $\mathbb{C}$ of the μ-dimensional complex space of all solutions of the system.

2. Let us consider the monodromy operator of the solutions, that is the linear operator in the solution space, generated by analytic continuations of solutions along a path going once anticlockwise round the point $t=0$. Then both the monodromy operator itself and its inverse preserve the lattice.

Let us denote by V the space of solutions, by $V_{\mathbb{Z}}$ the lattice, and by M the monodromy operator. Then properties 1 and 2 can be written:

$$V = V_{\mathbb{Z}} \otimes_{\mathbb{Z}} \mathbb{C}, \quad M(V_{\mathbb{Z}}) = M^{-1}(V_{\mathbb{Z}}) = V_{\mathbb{Z}}.$$

Property 1 corresponds to the assertion: the natural image of the group $H_{n-1}(X_t, \mathbb{Z})$ in $H_{n-1}(X_t, \mathbb{C})$ forms a μ-dimensional integer lattice, a basis for which over $\mathbb{Z}$ is a basis over $\mathbb{C}$ in $H_{n-1}(X_t, \mathbb{C})$ (here X_t is the fibre of the Milnor fibration of the critical point 0 of the function f). Property 2 corresponds to the assertion: the monodromy operator, acting on $H_{n-1}(X_t, \mathbb{C})$, preserves the image of the group $H_{n-1}(X_t, \mathbb{Z})$.

Let us define the real subspace in the space of solutions as the real vector space generated by the lattice:

$$V_{\mathbb{R}} = V_{\mathbb{Z}} \otimes_{\mathbb{Z}} \mathbb{R}.$$

The space V of all solutions of the system is the complexification of the real subspace:

$$V = V_{\mathbb{R}} \oplus i V_{\mathbb{R}}, \ i^2 = -1.$$

This decomposition defines in V an operation of complex conjugation.

Remark. The relationship between the asymptotics of solutions and the operation of complex conjugation leads to the concept of the mixed Hodge structure of a critical point of a holomorphic function (see Chapters 13, 14 below).

A structure, analogous to a lattice, exists in the space of solutions of the Picard-Fuchs equation of a holomorphic differential n-form (see (8)). That is, we can choose in the linear space of its solutions a module over $\mathbb{Z}$ with μ generators: a solution belongs to the module if it has the form (9). In this case also the monodromy operator and its inverse preserve the module.

12.2.4   The change in the system of equations under a change of trivialisation

Let us make explicit what happens to the system of equations (11) under change of trivialisation. Let $\omega_1', \ldots, \omega_\mu'$ be the new trivialisation. Put

$$I_l'^j(t) = \int_{\delta_l(t)} \omega_j'/df, \quad j, l = 1, \ldots, \mu.$$

By the definition of trivialisation there exists for any t from a sufficiently small punctured neighbourhood of the point $t=0$ a unique invertible $\mu \times \mu$ matrix $Q(t)$ for which

$$I^j(t) = \sum_r Q_r^j(t) I''(t), \quad j = 1, \ldots, \mu.$$

The coefficients of the matrix Q (just as the coefficients of the matrix A above) are single-valued holomorphic functions, defined in a punctured neighbourhood of the point $t=0$. The coefficients of the matrix Q are meromorphic at $t=0$ by virtue of Theorem 10.8. The new system of equations has the form

$$dI'/dt = (-Q^{-1}dQ/dt + Q^{-1}AQ)I'.$$

The solutions of the old system are connected with the solutions of the new system by the meromorphic transformation

$$I = QI'. \tag{13}$$

This transformation carries the lattice in the space of solutions of the new system into the lattice in the space of solutions of the old system. The monodromy operator in the space of solutions commutes with this transformation.

Lemma 12.5. If the forms

$$\omega_1', \ldots, \omega_\mu'$$

make up a basis trivialisation (see page 317), then the coefficients of the matrix Q are holomorphic for $t=0$.

Proof. See Property II of trivialisations on page 317.

Corollary. If both the forms $\omega_1,\ldots,\omega_\mu$ and the forms $\omega_1',\ldots,\omega_\mu'$ are basis trivialisations, then the coefficients of the matrix Q are holomorphic at $t=0$ and the matrix $Q(0)$ is invertible.

Therefore the Picard-Fuchs equations of two basis trivialisations are mapped onto each other by a holomorphic invertible transformation (13) of the unknown functions. This transformation preserves the lattice in the space of solutions of these equations.

We shall prove that, conversely, any holomorphic change of unknown functions in the Picard-Fuchs equations of a basis trivialisation, which is invertible for $t=0$, can be induced by a transition to a new basis trivialisation.

Lemma 12.6. Let the forms $\omega_1',\ldots,\omega_\mu'$ be a basis trivialisation (see page 317). Q be an invertible $\mu\times\mu$ matrix of holomorphic functions, given on a neighbourhood of the point $0\in\mathbb{C}$. Then there exist forms $\omega_1,\ldots,\omega_\mu$, which are a basis trivialisation and for which the transition (13) from the Picard-Fuchs equations of the trivialisation $\omega_1,\ldots,\omega_\mu$ to the Picard-Fuchs equations of the trivialisation

$$\omega_1',\ldots,\omega_\mu'$$

is given by the matrix Q.

Proof. As the required forms we can take the forms

$$\omega_j=\sum_r Q(f)_r^j\omega_r', \quad j=1,\ldots,\mu,$$

where $Q(f)$ is a matrix function in a neighbourhood of the origin in $\mathbb{C}^n$, induced from the matrix function Q by the map $f:(\mathbb{C}^n,0)\to(\mathbb{C},0)$.

12.2.5 The definition of the Picard-Fuchs singularity
of a finite-multiplicity critical point

Let A be a $\mu \times \mu$ matrix of holomorphic functions, given in a punctured neighbourhood of the point $0 \in \mathbb{C}$. Let us consider the system of differential equations

$$dI^j/dt = \sum_{k=1}^{\mu} A_k^j I^k, \quad j = 1, \ldots, \mu. \tag{14}$$

Let us suppose that in the μ-dimensional space of solutions of this system there is given a μ-dimensional integer lattice, a basis of which is a basis over $\mathbb{C}$ of the whole solution space (that is, if V is the solution space and $V_{\mathbb{Z}} \subset V$ is the lattice then $V = V_{\mathbb{Z}} \otimes_{\mathbb{Z}} \mathbb{C}$). Let us consider the monodromy operator M of the solutions (that is the linear operator in the solution space generated by analytic continuation of solutions along a path going once clockwise round the point $t = 0$). Let us suppose that the monodromy operator and its inverse preserve the lattice (that is $M(V_{\mathbb{Z}}) = M^{-1}(V_{\mathbb{Z}}) = V_{\mathbb{Z}}$). In this case the system (14) is said to be *rigged* and the lattice in the space of solutions is called a *rigged* system.

Examples of rigged systems satisfy the Picard-Fuchs equations of a trivialisation (see (11)), and the rigging of these systems give solutions of the form (12).

Remark. A system of differential equations can be rigged if and only if in the space of its solutions there exists a basis, with respect to which the monodromy operator and its inverse are integral.

Let us consider two rigged systems of μ equations. Let us denote their matrices, solution spaces and riggings, respectively, by

$$A, \, V, \, V_{\mathbb{Z}}, \, A', \, V', \, V_{\mathbb{Z}}'.$$

These systems are said to be *equivalent* if in a neighbourhood of the point $0 \in \mathbb{C}$ there exists an invertible $\mu \times \mu$ matrix Q of holomorphic functions for which the change of variables

$$I^j(t) = \sum_r Q_r^j(t) I''(t), \quad j = 1, \ldots, \mu,$$

maps the first system into the second, moreover preserving the rigging, that is

$$A' = -Q^{-1}dQ/dt + Q^{-1}AQ, \quad V_{\bar{z}} = QV'_{\bar{z}}.$$

An equivalence class of systems is called a *singularity* of rigged systems of differential equations.

According to Lemmas 12.5 and 12.6 the Picard-Fuchs equations of basis trivialisations constitute exactly one equivalence class. This equivalence class is called the *Picard-Fuchs singularity of the critical point* 0 of the holomorphic function f.

Problem. Describe the Picard-Fuchs singularity of a critical point.

For example, indicate the normal form of the singularity. Show which singularities of rigged systems can be Picard-Fuchs singularities of critical points.

Reasonable questions about Picard-Fuchs singularities of critical points have, very likely, reasonable answers. For example, a Picard-Fuchs singularity has a diagonal representation (that is a representation of the form

$$dI^j/dt = a_j(t)I^j, \quad j = 1, \ldots, \mu)$$

if and only if the critical point is holomorphically equivalent to the critical point of a quasihomogeneous function. (Proof: The Picard-Fuchs singularity of the critical point of a quasihomogeneous function has a diagonal representation according to [341] or [239, Example (6.7)], and the converse assertion can easily be derived from [306] (compare with [363])).

Conjecture. The Picard-Fuchs singularities of finite-multiplicity critical points are different for holomorphically inequivalent critical points, at least for nearby ones.

This conjecture is analogous to Torelli's theorem in algebraic geometry (see [141], [9]).

As a circumstance stimulating the study of the problem and supporting the conjecture we note that the Picard-Fuchs singularity of a critical point determines the mixed Hodge structure of the critical point. See § 13.2 below for further details.

12.2.6   The Picard-Fuchs equations of a trivialisation have a regular singular point

We recall a classical theorem from the theory of ordinary linear differential equations.

Theorem 12.6 (see [79]). Let A be a $\mu \times \mu$ matrix of holomorphic functions in a punctured neighbourhood of the point $0 \in \mathbb{C}$. Let us consider the system of μ homogeneous linear differential equations

$$dI/dt = AI, \; I(t) \in \mathbb{C}^{\mu}. \tag{15}$$

Then the following properties of the system of equations are equivalent.

1. Let

$$I = (I^1, \ldots, I^{\mu})$$

be an arbitrary solution of the system of equations. Then in an arbitrary sector $a < \arg t < b$ each coordinate of the solution grows, as $t \to 0$, no faster than a power of the form:

$$I_j(t) = o(t^{-N})$$

for some N.

2. There exists an invertible $\mu \times \mu$ matrix Q of holomorphic functions in a punctured neighbourhood of the point $t = 0$, meromorphic at the point $t = 0$, for which the substitution $I = QI'$ transforms the original system of equations into a system of equations with a simple pole at the point $t = 0$, that is the matrix of functions

$$Q^{-1}AQ - Q^{-1}dQ/dt$$

has at the point $t = 0$ a pole of order no higher than the first.

Definition. The point $t = 0$ is called a *singular* point of the system of equations (15), if at least one of the coordinates of the matrix A cannot be continued holomorphically to the point $t = 0$. A singular point of the system of equations (15) is called *regular* if the system satisfies the properties 1 and 2 indicated in Theorem 12.6.

The Picard-Fuchs equation of a trivialisation has at the point $t=0$ a regular singular or a non-singular point in accordance with Theorem 10.7.

Remark. The point $t=0$ is a non-singular point of the Picard-Fuchs equation of the trivialisation if and only if the critical point 0 of the function f is non-degenerate, the number of variables of the function is equal to two and the trivialisation is a basis trivialisation. (The proof follows easily from Theorem 10.8 and Theorem 4.6.).

Idea of the proof of Theorem 12.6.
$(1) \Rightarrow (2)$. Let M be the monodromy transformation of the solutions, generated by going anticlockwise round $t=0$. Let $\ln M$ be one of the possible values of the logarithm of the matrix M. Let B be the matrix of a fundamental system of solutions of the equation (15). Then

$$P = B \exp(-\ln t \, \ln M / 2\pi i)$$

is a matrix of meromorphic functions. The change $I = PI'$ reduces the equation to the form

$$dI'/dt = -(\ln M / 2\pi i t) I'. \tag{16}$$

Remark 1. The matrix

$$\exp(-\ln t \, \ln M / 2\pi i)$$

is the matrix of a fundamental system of solutions of the equation (16). Therefore the solutions of the equation (15) can be expanded in a series of the form

$$\sum a_{k,\alpha} t^{\alpha} (\ln t)^k$$

(in the presence of the property indicated in section 1 of the theorem); compare with the proof of Theorem 10.2.

Remark 2. An arbitrary system of equations (15) can be reduced by change of variables $I = PI'$ to the form (16). However P will not always be a meromorphic matrix.

$(2) \Rightarrow (1)$. Since A has a pole of first order,

$$d\|B\|/dr \leqslant \text{const}\,\|B\|/r$$

where $r = |t|$. Therefore

$$\|B\| < B_0 r^{-\text{const}}.$$

For more detail see [111].

12.3   The Gauss-Manin connection

In this section we shall discuss a geometrical interpretation of a system of differential equations which are satisfied by the integrals of holomorphic forms over the classes of continuous families of integral vanishing homologies.

With an arbitrary locally trivial fibration π there is associated the complex vector bundle of kth (co)homologies of its fibres. In this vector bundle there is defined a natural operation of translation of the fibres over curves in the base (since the (co)homologies of nearby fibres of the initial fibration are canonically isomorphic). The operation of translation is called the *Gauss-Manin connection* in the (co)homology fibration.

A *many-valued covariantly constant section* of the (co)homology fibration is a many-valued section (that is a section of the (co)homology fibration, lifted to the universal cover of the base), the values of which are invariant under the operation of translation. Covariantly constant sections make up the first important class of sections of the (co)homology fibrations.

Let us suppose now that the fibration π is complex analytic. Let us consider on the total space of the fibration a holomorphic differential k-form, the restriction of which to the fibre is closed. The form determines a (single-valued) section of the fibration of kth cohomologies: a point of the base is mapped to the cohomology class of the restriction of the form to the fibre lying over the point. The sections obtained in such a geometrical way form the second important class of sections of the cohomology fibration (the class of *geometric sections*, for a precise definition see § 12.3.2). The fibrations of the kth homologies and kth cohomologies are in a natural way conjugate. Having a covariantly constant section of the fibration of kth homologies and a geometric section of the fibration of kth cohomologies and calculating pointwise the value of one section on the values of the other, we obtain a many-valued function on the base. The values of this function are none other than the integrals of the form giving the geometric section over a cycle representing in the fibres the homology which is the values of the covariantly constant section. Such functions on the base of the Milnor fibration were considered in Chapter 10.

Let us suppose that there is given a basis of geometric sections and it is known how to translate cohomologies with respect to coordinates given by this basis. Let us suppose that we want to find the coordinates of the covariantly constant sections. Then to find them it is necessary to solve a system of differential equations of first order. This system of differential equations for the case of the Milnor fibration of a critical point is conjugate to the Picard-Fuchs equation of the trivialisation given by a set of forms defining the bases of the geometric sections (see § 12.2). Now let us discuss these facts in more detail.

12.3.1 The (co)homological fibration associated with a locally trivial fibration

Let $\pi : X \to B$ be a locally trivial fibration (not necessarily a vector bundle). For any $k \geqslant 0$ we define complex vector bundles of k-dimensional homology and k-dimensional cohomology of the fibres of the fibration π. Put

$$H_k = \cup_{b \in B} H_k(X_b, \mathbb{C}).$$

Denote by π_* the natural projection $H_k \to B$.

Let $U \subset B$ be a contractible open subset. Then $\pi^{-1}(U)$ is homeomorphic to a direct product of the fibre and the set U. The natural inclusion of an arbitrary fibre over U into $\pi^{-1}(U)$ induces an isomorphism of (co)homologies. In this way we have defined a trivialisation of the projection π_* over open contractible subsets of the base B. The constructed trivialisations define over

$$\pi_* : H_k \to B$$

the structure of a complex vector bundle. This bundle is called the *bundle of k-dimensional homology associated with π*.

The *bundle of k-dimensional cohomology* associated with the bundle π is defined analogously. The bundles of k-dimensional homology and k-dimensional cohomology are, in a natural way, conjugate (since k-dimensional homology and k-dimensional cohomology are conjugate).

Remark. The transition functions of the constructed trivialisations are locally constant.

In the (co)homology bundle there is defined a natural operation of parallel translation of fibres over curves in the base. (If a curve is given in the base then there is induced by it from π a fibration over a segment. This fibration is trivial.

Therefore the (co)homology of the fibres over the initial and final points of the curve are canonically isomorphic.) The operation of parallel translation possesses the following properties:

1. The map of the fibre over the initial point of the path into the fibre over the terminal point of the path is a linear isomorphism.

2. The map does not depend on the choice of curve in the homotopy class of curves with fixed ends.

The operation of translation of (co)homologies is called the *Gauss-Manin connection* in the (co)homological fibration. The operation of translation in the homological fibration and in the cohomological fibration are coordinated – translation commutes with dualization.

A section of the (co)homological fibration over an open subset of the base is said to be *covariantly constant* if its values are invariant relative to parallel translation along any curves lying in this open set.

If the base *B* is locally simply connected (a manifold, for example), then any vector of an arbitrary fibre of the (co)homological fibration can be extended, in a unique way, to a covariantly constant section over a sufficiently small neighbourhood of its projection into the base (for this the vector must be parallel translated into nearby fibres). Further extension of this covariantly constant section leads to a many-valued covariantly constant section over the entire base, that is to a covariantly constant section of the (co)homological fibration, lifted to the universal cover of the base.

In each fibre of the (co)homological fibration there are additional structures: a real subspace and in it an integral lattice. The real subspace is the natural image in the (co)homologies of the fibres of the fibration π with complex coefficients of the (co)homologies with real coefficients; the integral lattice in the real subspace is the natural image of the (co)homology groups of the fibres with integer coefficients. A fibre of the (co)homological fibration is the complexification of its real subspace, the lattice generates the real subspace (under addition and real scalar multiplication). The real subspace and the integral lattice are invariant relative to parallel translation.

Let us suppose that the base of the fibration π is a holomorphic manifold. Then the (co)homological fibration, associated with π possesses the canonical structure of a holomorphic vector bundle. Indeed, we define the holomorphic sections of the (co)homological fibration as sections having holomorphic coordinates relative to an arbitrary frame of the (co)homological fibration, formed from covariantly constant sections. This definition is well-defined, since the transition functions between frames, arising from covariantly constant sections, are locally constant.

Remark. In connection with the Gauss-Manin connection see [241, 242, 287].

12.3.2 The (co)homological Milnor fibration of a deformation

Let $f : (\mathbb{C}^n, 0) \to (\mathbb{C}, 0)$ be the germ of a holomorphic function having a critical point of finite multiplicity μ. Let

$$F : (\mathbb{C}^n \times \mathbb{C}^k, 0 \times 0) \to (\mathbb{C}, 0)$$

be a deformation of the germ. Let $G : X \to S$ be a specialisation of the deformation and let $G : X' \to S'$ be the corresponding Milnor fibration (see page 287). The bundle of $(n-1)$st (co)homologies associated with the Milnor fibration is called the *(co)homological Milnor fibration* of the deformation (more precisely of the specialisation of the deformation). The (co)homological Milnor fibration possesses the canonical structure of a holomorphic vector bundle. In the (co)homological Milnor fibration there is given a Gauss-Manin connection.

Remark. The fibre of the Milnor fibration has the homotopy type of a bouquet of $(n-1)$-dimensional spheres. Therefore the associated bundle of kth (co)homologies is interesting only for $k = n-1$.

Let us define the class of geometric sections of the cohomological Milnor fibration. Let ω be a holomorphic differential n-form on X. For any point $b \in S'$ the restriction of the Gelfand-Leray form to the fibre X_b of the Milnor fibration defines the cohomology class

$$[\omega/d_x F|_{X_b}] \in H^{n-1}(X_b, \mathbb{C}).$$

The section

$$b \mapsto [\omega/d_x F|_{X_b}]$$

of the cohomological Milnor fibration is called the *geometric section* of the form ω and is denoted by $s[\omega]$.

Let δ be a many-valued covariantly constant section of the homological Milnor fibration. Let us consider on S' the many-valued function $\langle s[\omega], \delta \rangle$. According to our definition

$$\langle s[\omega], \delta \rangle = \int_\delta \omega/d_x F.$$

Functions of this sort were considered in Chapter 10.

The function $\langle s[\omega], \delta \rangle$ is holomorphic according to Theorem 10.5. Therefore a geometric section of a holomorphic differential n-form is a holomorphic section of the cohomological Milnor fibration.

The collection of geometric sections is a module over the ring of holomorphic functions on the base S of a specialisation of the deformation: if $s[\omega]$ is a geometric section, and g is a holomorphic function on S, then the section $gs[\omega]$ is a geometric section of the form $s[(g \circ G) \cdot \omega]$.

According to Theorems 12.1 and 12.2 there exists a set of μ geometric sections, the values of which generate bases in all the fibres of the cohomological Milnor fibration, lying over the points of the base which are sufficiently near to the origin $0 \in S$.

Remark. The module of germs of geometric sections at the origin of the base S of a versal deformation of a critical point of the germ f is a free module of rank μ over the ring of germs of holomorphic functions at the origin of the base S. It is not hard to show, using the corollaries of Theorem 12.2 and the results of § 12.1.5 (see below the case of a trivial deformation of a germ). A basis of the module is given by geometric sections of forms for which the restriction of the function $\det^2$ to the axis of values, passing through the origin, has at the origin a zero of minimal multiplicity, see the corollary of Theorem 12.2.

12.3.3   The Gauss-Manin connection and the Picard-Fuchs equations of a trivialisation

Let us restrict ourselves further to the case of a trivial deformation of a germ f. Let us consider a specialisation $f: X \to S$ of the germ and the corresponding Milnor fibration $f: X' \to S'$. Let $s_1, \ldots, s_\mu$ be a basis over S' of the holomorphic sections of the (co)homological Milnor fibration.

Lemma 12.7. On S' there exists a unique matrix B of holomorphic functions, for which the system of differential equations

$$dI^j/dt = \sum B_k^j I^k, \quad j = 1, \ldots, \mu, \tag{17}$$

have the property: the section $\sum I^j s_j$ is covariantly constant in the Gauss-Manin connection if and only if

$$(I^1, \ldots, I^\mu)$$

is a solution of the system.

The system of differential equations (17) is called the equations of covariantly constant sections with respect to the frame $s_1, \ldots, s_\mu$.

Proof. The set of many-valued covariantly constant sections of the (co)homological Milnor fibration forms a μ-dimensional complex vector space. Expanding the covariantly constant sections in terms of the basis $s_1, \ldots, s_\mu$, we obtain a μ-dimensional complex vector space of holomorphic vector functions

$$(I^1, \ldots, I^\mu),$$

invariant relative to analytic continuation of the vector functions around the point 0. The existence and uniqueness of a system of differential equations, such that the vector functions of such a space serve as their solutions, is proved analogously to Theorems 12.3 and 12.5.

Let $\omega_1, \ldots, \omega_\mu$ be holomorphic differential n-forms on X, the geometric sections of which

$$s_j = s[\omega_j], \quad j = 1, \ldots, \mu$$

form a basis of the sections of the cohomological Milnor fibration. The set of such forms was called in §12.1 a trivialisation (their geometric sections give a trivialisation of the cohomological Milnor fibration).

With such set of forms there are connected two systems of differential equations. On the one hand, there are the equations (17) of covariantly constant sections with respect to the frame $s_1, \ldots, s_\mu$, on the other hand there are the equations (11) of the Picard-Fuchs trivialisation.

Lemma 12.8. The matrix B of the equations of the covariantly constant sections and the matrix A of the equations of the Picard-Fuchs trivialisation are connected by the relationship

$$A + B^* = 0.$$

Corollary. Let $s^1, \ldots, s^\mu$ be the frame of the homological Milnor fibration conjugate to the frame $s_1, \ldots, s_\mu$. Then the map

$$(I^1, \ldots, I^\mu) \mapsto \sum I^j s^j$$

gives an isomorphism of the space of solutions of the Picard-Fuchs equations
(11) and the space of covariantly constant sections of the homological Milnor
fibration. Under this isomorphism the lattice in the space of solutions of the
Picard-Fuchs equations, given in § 12.2.3, maps to the lattice of covariantly
constant sections of the homological Milnor fibration formed by sections, the
values of which belong to the natural image of the homology with integer
coefficients.

The **proof** of the lemma follows easily from the definitions.

Lemma 12.8 and its corollary give a geometrical interpretation of the
equations of the Picard-Fuchs trivialisation.
Let us consider a basis trivialisation

$$\omega_1, \ldots, \omega_\mu$$

(see page 317). According to Property II of trivialisations (page 318) there exist
for any geometric section $s[\omega]$ holomorphic functions

$$p_1, \ldots, p_\mu,$$

given in a neighbourhood of the point $0 \in S$, for which

$$s[\omega] = \sum_j p_j s[\omega_j].$$

Consequently, the geometric sections of a basis trivialisation generate a basis
of the module of germs of geometric sections at the point $0 \in S$. This property singles out basis trivialisations among all trivialisations (Lemmas 12.5
and 12.6).
In this way, the Picard-Fuchs singularity of a critical point (see § 12.2.5) is the
class of equations of covariantly constant sections of the homological Milnor
fibration with respect to frames dual to frames of geometric sections generating a
basis of the module of germs of geometric sections at the point $0 \in S$.

Chapter 13

The coefficients of series expansions of integrals, the weight and Hodge filtrations and the spectrum of a critical point

Let us consider the integral of a holomorphic differential form over a homology class of a continuous family of integral homologies of fibres of the Milnor fibration of a critical point. The function given by the integral can be expanded in a series of powers of the parameter and powers of the logarithm of the parameter of the family (Chapter 10). Each coefficient of the series depends linearly both on the form and on the continuous family of integral homologies. If the form is fixed but the continuous family varies, then each coefficient of the series is a linear function of the continuous families. Linear combinations, over $\mathbb{C}$, of continuous families of integral homology classes form the space of covariantly constant (with respect to the Gauss-Manin connection) sections of the homological Milnor fibration. Therefore (if the form is fixed) each coefficient of the series is a linear function on the space of covariantly constant sections of the homological Milnor fibration, that is it is a covariantly constant section of the cohomological Milnor fibration.

This can be seen in a different way as follows. A holomorphic form corresponds to a (single-valued) geometric section of the cohomological Milnor fibration. Let us choose in the cohomological fibration a covariantly constant frame and let us decompose the geometric section in terms of this frame. The coefficients of the decomposition could be many-valued functions. It is not hard to convince oneself that the coefficients of the decomposition have the form

$$\sum b_{k,\alpha} t^{\alpha} (\ln t)^k.$$

Regrouping the terms, we obtain an expansion of the geometric section in a series of powers and logarithms of the parameter with coefficients which are covariantly constant sections. In order to define the integral of a form over homology classes of covariantly constant families of homologies, it is sufficient to determine the values of the coefficients of the series on a given covariantly constant section of the homological Milnor fibration.

And so a geometric section can be expanded in a series

$$s[\omega] = \sum_{k,\alpha} t^\alpha (\ln t)^k A^\omega_{k,\alpha}/k\,!, \tag{1}$$

where ω is a form, and $A^\omega_{k,\alpha}$ is a covariantly constant section. The coefficients $A^\omega_{k,\alpha}$ completely determine the cohomological properties of the form ω/df, given on the fibres of the Milnor fibration. Furthermore, the collection of coefficients of all the forms contains information about the critical point.

In § 13.1 we shall prove the fundamental properties of the coefficients:

1. $A^\omega_{k,\alpha}(t) = (-\ln M_u/2\pi i)^k A^\omega_{0,\alpha}(t),$

where M_u is the unipotent part of the classical monodromy operator M in cohomology.

2. The section $A^\omega_{k,\alpha}$ belongs to the root subspace of the operator M corresponding to the eigenvalue $\exp(-2\pi i\alpha)$.

3. Each section $A^\omega_{k,\alpha}$ depends holomorphically on the form ω (more precisely on its finite jet at the original critical point).

We shall define further the *principal part* of a form (of a geometric section) as the sum of terms of the series (1) with fixed α, which is the smallest of those occurring in the series. This smallest α is called the *order* of the form. We shall formulate a results on the calculation of the order of a form.

Section § 13.2 is fundamental. In that section we construct from the principal parts of all the forms a decreasing filtration* in the fibres of the cohomological Milnor fibration. This is called the *Hodge filtration*. We construct further, using the Jordan structure of the monodromy operator, an increasing filtration in the fibres of the cohomological Milnor fibration. It is called the *weight filtration*. The subspaces of the filtrations in the different fibres are coherent: they form a holomorphic subfibrations of the cohomological Milnor fibration. In § 13.2 we shall formulate a theorem about the mixed Hodge structure in the fibres of the cohomological Milnor fibration of an isolated critical point. Chapter 15 will be devoted to a discussion of this theorem.

In § 13.3 we shall define numerical characteristics of the weight and Hodge filtrations – the *spectral pairs* of a critical point. The set of spectral pairs is an unordered set of μ pairs of numbers, where μ is the multiplicity of the critical point. The first of the numbers in the pair is a rational number, the second is an integer. The unordered set of first numbers of the pairs is called the *spectrum of the critical point*. The spectral numbers are the logarithms of the eigenvalues of

---

* A *filtration* of a linear space is an ordered sequence of linear subspaces.

the monodromy operator divided by $2\pi i$. The choice of branch of the logarithm is made with the help of the principal part of the holomorphic forms. The second numbers in the spectral pairs are the Jordan levels, renormalised in the right way, of elements of the Jordan basis. The spectrum of a critical point possesses remarkable properties:

1. The sum of all the spectral numbers is equal to

$$\mu(n/2 - 1),$$

where μ is the multiplicity of the critical point and n is the number of variables.

2. The spectrum is symmetric relative to the point $n/2 - 1$.

3. The spectrum of a non-degenerate critical point consists of one number $n/2 - 1$. If the spectrum is concentrated at the point $n/2 - 1$, that is, if it consists of several numbers $n/2 - 1$, then the critical point is non-degenerate.

4. The spectrum does not change under deformations of the critical point which do not change its multiplicity.

5. If $\{\alpha_i\}$, $i = 1, \ldots, \mu$, is the spectrum of the critical point of the germ $f : (\mathbb{C}^n, 0) \to (\mathbb{C}, 0)$, and $\{\beta_j\}$, $j = 1, \ldots, \eta$, is the spectrum of the critical point of the germ $g : (\mathbb{C}^l, 0) \to (\mathbb{C}, 0)$, then

$$\{\alpha_i + \beta_j + 1\}, \quad i = 1, \ldots, \mu, \quad j = 1, \ldots, \eta,$$

is the spectrum of the critical point of the germ

$$f + g : (\mathbb{C}^n \times \mathbb{C}^l, 0 \times 0) \to (\mathbb{C}, 0).$$

These, and other properties of the spectrum also, are discussed in §13.3 and in Chapter 14.

13.1 Coefficients of expansions in series

13.1.1 Coefficients and the monodromy operator

Let us consider the germ

$$f : (\mathbb{C}^n, 0) \to (\mathbb{C}, 0)$$

of a holomorphic function, with a critical point of finite multiplicity. Let us consider a specialisation $f : X \to S$ of it and the corresponding Milnor fibration $f : X' \to S'$ (see page 287). If ω is a holomorphic differential n-form on X and δ is a covariantly constant section of the homological Milnor fibration, then

$$\int_{\delta(t)} \omega/df = \sum_{k,\alpha} a_{k,\alpha} t^{\alpha} (\ln t)^k / k!, \tag{2}$$

according to Theorem 10.8. In this formula the coefficients $1/k!$ are inserted in order to simplify subsequent formulae. In the series (2) the numbers k and α do not depend on the forms or the sections but are determined only by the monodromy operator of the Milnor fibration. For a fixed form and fixed k and α the coefficients $a_{k,\alpha}$ determine the covariantly constant many-valued section $A_{k,\alpha}^{\omega}$ of the cohomological Milnor fibration by the formula

$$\langle A_{k,\alpha}^{\omega}(t), \delta(t) \rangle = a_{k,\alpha}(\omega, \delta).$$

By definition

$$s[\omega](t) = \sum_{k,\alpha} t^{\alpha} (\ln t)^k A_{k,\alpha}^{\omega}(t)/k!, \tag{3}$$

where $s[\omega]$ is the geometric section corresponding to the form ω. The series (3) converges in each sector

$$a < \arg t < b$$

if the modulus of the parameter t is sufficiently small.

Lemma 13.1.

1. $A_{k,\alpha}^{\omega} = (-\ln(M_u)/2\pi i)^k A_{0,\alpha}^{\omega},$ \hfill (4)

where M_u is the unipotent part of the operator in the $(n-1)$st cohomology.

2. The section $A_{k,\alpha}^{\omega}$ belongs to the root subspace of the eigenvalue $\exp(-2\pi i\alpha)$ of the monodromy operator in the $(n-1)$st cohomologies.

Remark. The monodromy operator M in cohomologies (or in the covariantly constant sections of the cohomological Milnor fibration) is the monodromy operator in the Gauss-Manin connection generated by going anticlockwise round the point $t=0$. If M_{hom} is the monodromy operator in the $(n-1)$st homologies, defined in an analogous fashion, then

$$M^* = M_{\text{hom}}^{-1}$$

where * denotes the dual operator.

The **proof** of the lemma follows from the following remark. Analytic continuation round the point $t = 0$ of the left-hand side of formula (2) leads to the integral of the form ω/df over $M_{\text{hom}}\delta(t)$. Analytic continuation of the right-hand side round the point $t = 0$ consists of substituting the expression $te^{2\pi i}$ in place of t. By equating the two methods of doing the analytic continuation, we obtain the lemma. For further details see [364].

Corollary of Lemma 13.1.

1. The geometric section $s[\omega]$ is determined by the sections $A_{k,\alpha}^{\omega}$ with $k=0$.
2. If, for any α, $A_{0,\alpha}^{\omega}=0$, then for all k, $A_{k,\alpha}^{\omega}=0$.
3. For any α

$$t^{\alpha}(A_{0,\alpha}^{\omega}(t) + \ldots + (\ln t)^{n-1}A_{n-1,\alpha}^{\omega}(t)/(n-1)!) =$$
$$= \exp(\ln t(\alpha\,\text{Id} - \ln(M_u)/2\pi i))A_{0,\alpha}^{\omega}(t).$$

(Note that the dimensions of the Jordan blocks of the monodromy operator are not greater than n, therefore the coefficients $A_{k,\alpha}^{\omega}$ with $k \geqslant n$ are equal to zero).

4. The section given by the previous formula is a holomorphic single-valued section of the cohomological Milnor fibration.

Lemma 13.2. Each section $A_{k,\alpha}^{\omega}$ depends only on a finite jet of the form ω at the point $0 \in \mathbb{C}^n$ and, furthermore, it depends on this jet holomorphically.

Lemma 13.2 is a corollary of Lemma 12.3.

13.1.2 Elementary sections

Let A be a covariantly constant section of the cohomological Milnor fibration, belonging to the root subspace of the eigenvalue λ of the monodromy operator in cohomology. Let α be a rational number with the property: $\exp(-2\pi i\alpha) = \lambda$. Let us define the section $s[A, \alpha]$ of the cohomological fibration by the formula

$$s[A, \alpha] = \exp\left[\ln t(\alpha \mathrm{Id} - \ln(M_u)/2\pi i)\right] A.$$

The section $s[A, \alpha]$ is called an *elementary section* of order α generated by the section A.

According to Lemma 13.2 a geometric section is a sum of elementary sections:

$$s[\omega] = \sum_\alpha s[A^\omega_{0,\alpha}, \alpha].$$

Lemma 13.3. 1. The section $s[A, \alpha]$ is a single-valued holomorphic section of the cohomological fibration.

2. If the covariantly constant sections

$$A_1, \ldots, A_l$$

belong to the root subspace of the eigenvalue λ of the operator M and are linearly independent over $\mathbb{C}$, then the values of the sections

$$s[A_1, \alpha], \ldots, s[A_l, \alpha]$$

are linearly independent at each point of the base of the cohomological fibration.

3. We have

$$t V_{\partial/\partial t} s[A, \alpha] = \alpha s[A, \alpha] + s[-\ln(M_u) \cdot A/2\pi i, \alpha],$$

where $V_{\partial/\partial t}$ is differentiation with respect to the Gauss-Manin connection, that is differentiation of the coordinates of a section with respect to a covariantly constant frame.

Proof. Sections 1 and 3 are obvious. Section 2 follows from the linear independence of covariantly constant sections and the non-degeneracy of the linear transformation $\exp[\cdot]$.

13.1.3   The order and the principal part of a form

Let ω be a holomorphic differential n-form on X (the domain of the specialisation of the germ f). The *order* of the form (or of the geometric section determined by the form) is the smallest number α for which the coefficient $A_{0,\alpha}^\omega$ is different from zero. The order will be denoted by $\alpha(\omega)$.

Remember that, according to Lemma 13.1, if $A_{0,\alpha}^\omega = 0$ then $A_{k,\alpha}^\omega = 0$ for all k.

The *principal part* of a form (or the geometric section determined by the form) is the single-valued holomorphic section $s_{\max}[\omega]$ of the cohomological Milnor fibration of the critical point of the germ f given by the formula

$$s_{\max}[\omega] = t^{\alpha(\omega)} (A_{0,\alpha(\omega)}^\omega + \ldots + (\ln t)^{n-1} A_{n-1,\alpha(\omega)}^\omega / (n-1)!).$$

Remark 1. Let η be a holomorphic differential $(n-1)$-form on X. Its restriction to the fibre of the Milnor fibration determines a geometric section $s[\eta]$. By construction

$$s[\eta] = s[df \wedge \eta].$$

This equation determines the order and the principal part of the form η and its geometric section.

Remark 2. If the form defines the zero geometric section, then we put its order equal to $+\infty$, and its principal part equal to the zero section of the cohomological fibration.

Example. Let f be a quasihomogeneous polynomial and let ω be a quasihomogeneous n-form. Then the integral of its Gelfand-Leray form on the homology classes of a covariantly constant section has the form

$$\text{const} \cdot t^{\alpha-1},$$

where α is the ratio of the degrees of the quasihomogeneous form and the polynomial. Therefore the principal part of the form is equal to its geometric section, that is

$$s[\omega](t) = s_{\max}[\omega](t) = t^{\alpha-1} A_{0,\alpha-1}^\omega(t).$$

Let us formulate a useful property of orders and principal parts.

Lemma 13.4 (see [364]). Let $\omega_1, \ldots, \omega_\mu$ be holomorphic differential n-forms on X. Let $\delta_1, \ldots, \delta_\mu$ be a basis of covariantly constant sections of the homological

Milnor fibration of the critical point of the germ f. Let us denote by *ord* the order of zero at $t=0$ of the function

$$\det^2 : t \mapsto \det^2 \left(\int_{\delta_l(t)} \omega_j/df \right), \quad j, l=1,\ldots,\mu,$$

then

$$\mathrm{ord}/2 \geqslant \alpha(\omega_1)+ \ldots + \alpha(\omega_\mu),$$

equality holding if and only if the principal parts of the forms

$$\omega_1, \ldots, \omega_\mu$$

are linearly independent sections of the cohomological Milnor fibration.
   The **proof** is obvious.

13.1.4   The order of a form and the index of the principal part of the asymptotics of an oscillatory integral

Let us consider a complex oscillatory integral with phase f over an admissible chain concentrated in a neighbourhood of the critical point of the germ f, that is the integral

$$\int_{[\Gamma]} e^{\tau f} \omega,$$

where $[\Gamma] \in H_n(X, X^-)$, and ω is a holomorphic differential n-form on X (see § 11.1). According to Theorem 11.1 as $\tau \to +\infty$ the integral can be expanded in the series

$$\sum a_{k,\alpha} \tau^\alpha (\ln \tau)^k.$$

Let us denote by $\beta(\omega, [\Gamma])$ the largest α occurring in this series, that is $\beta(\omega, [\Gamma])$ is the index of the principal term of the asymptotic series.

Lemma 13.5. For any chain $[\Gamma] \in H_n(X, X^-)$ we have the inequality

$$\beta(\omega, [\Gamma]) \leqslant -\alpha(\omega) - 1,$$

where $\alpha(\omega)$ is the order of the form ω. Furthermore there exists $[\Gamma]$ for which

$$\beta(\omega, [\Gamma]) = -\alpha(\omega) - 1.$$

Proof. See formulae (4)–(6) on page 303.

13.1.5 The complex oscillation index

Let us denote by α_{min} the smallest number among the orders of holomorphic n-forms on X. The number

$$-(1 + \alpha_{min})$$

is called the *complex oscillation index* of the critical point of the germ f.

According to Lemma 13.5 the complex oscillation index is the largest possible value of the index of the principal term of the asymptotics of the complex oscillatory integral with phase f over an admissible chain concentrated in a neighbourhood of a critical point of the germ f.

Example. For the critical points of the functions $x^{\mu+1}$, and $x_1^2 + \ldots + x_n^2$ the complex oscillation index is equal, respectively, to $-1/(\mu+1)$, $-n/2$.

Lemma 13.6. Let us suppose that the germ f, restricted to the real subspace $\mathbb{R}^n \subset \mathbb{C}^n$, takes only real values. Then the oscillation index of the germ $f|\mathbb{R}^n$, defined in Chapter 6 is no greater than the complex oscillation index of the germ f.

Proof. The complex oscillation index is equal to the largest possible value of the order of the complex oscillatory integral with phase f for all possible $[\Gamma]$, ω. The real oscillation index is equal to the largest possible value of the order of the complex oscillatory integral with phase f on the real contour $[\Gamma_\mathbb{R}]$ for all possible ω (see Theorem 11.3).

As example 1 of Chapter 9 shows the complex oscillation index can be strictly greater than the real oscillation index.

Definition. The *complex singular index* of the critical point of a germ f is the number

$$n/2 - (1 + \alpha_{min}),$$

equal to the complex oscillation index increased by $n/2$.

Example. For the critical points of the functions $x^{\mu+1}$, and $x_1^2 + \ldots + x_n^2$ the complex singular index is equal, respectively, to $1/2 - 1/(\mu+1)$, 0.

The complex singular indices are equal for stably equivalent points, see § 13.3.5. The complex singular index is non-negative, see § 13.3.3.

13.1.6 The order of a form and the resolution of the critical point of the germ f

According to Theorem 7.3 the oscillation index of a critical point of a real analytic function is equal to the weight of the resolution of the singularity. We shall formulate a complex analytic analogue of this result.

Let $\pi : Y \to X$ be a resolution of the singularity of the critical point 0 of the function $f : X \to S$ (where $f : X \to S$ is a specialisation of the germ f). This means that

1. Y is a non-singular complex analytic n-dimensional manifold, π is a proper holomorphic map, inducing a biholomorphism between $Y \backslash \pi^{-1}(0)$ and $X \backslash 0$.

2. the fibre $(f \circ \pi)^{-1}(0)$ is a union of smooth divisors on Y, intersecting normally (see Sections 1–4 of the definition of resolution on page 195–196).

A resolution of the singularity exists according to Hironaka's theorem [158].

Let us decompose the preimage of the critical point of the function f into the union of non-singular irreducible $(n-1)$-dimensional complex analytic sets

$$E_1, \ldots, E_N.$$

To each of these irreducible components we can associate two well-defined non-negative integers: the multiplicities of zero on this component of, respectively, the function $f \circ \pi$ and the Jacobian of the map π. Let us denote these numbers by, respectively, k and m. The number

$$-(m+1)/k$$

is called the *weight of the component* (compare with § 7.3.1). The maximum of the weights of the components

$$E_1, \ldots, E_N$$

is called the *weight of the resolution of the singularity.*

Let ω be a holomorphic differential n-form on X. To the form $\pi^*\omega$ on Y and an arbitrary component E_i, $i = 1, \ldots, N$, we can associate two well-defined non-negative integers: the multiplicities of zero on this component of, respectively, the function f and the form $\pi^*\omega$. Let us denote these numbers, respectively, by k and m. The number

$$-(m+1)/k$$

is called the *weight of the component relative to ω.* The maximum of the weights of the components

$$E_1, \ldots, E_N$$

is called the *weight relative to ω of the resolution of the singularity.*

It is easy to see that the weight of the resolution of the singularity is equal to the weight of the resolution of the singularity relative to the form

$$dx_1 \wedge \ldots \wedge dx_n.$$

Theorem 13.1 (see [364]).

1. The complex oscillation index of the critical point of a germ f is not greater than the weight of the resolution of the singularity of the critical point. The complex oscillation index is equal to the weight of the resolution if the weight of the resolution is not less than -1.

2. Let ω be a holomorphic differential n-form on X. Let $\alpha(\omega)$ be its order. Then the number

$$-(\alpha(\omega)+1)$$

is not greater than the weight relative to ω of the resolution of the singularity of the critical point of the germ f. Furthermore $-(\alpha(\omega)+1)$ is equal to the weight relative to ω, if the weight relative to ω is not less than -1.

Remark. This theorem is analogous to Theorem 7.5 and is a more precise version of Theorem 10.7'. The results of the theorem concerning inequalities were proved also in [387].

For the proofs of the inequalities in the results of the theorem we need to lift an arbitrary form ω to Y and estimate in terms of the numbers k and m the integral of its Gelfand-Leray form in a neighbourhood of an arbitrary point of the preimage of the critical point of the germ f. For the proofs of the equalities in the results we must produce, for an arbitrary form ω, relative to which the weight of the resolution is not less than -1, a continuous family of vanishing homologies, the integrals over which of the form ω/df have the required order. For this we define in the right way the limit of the form $\pi^*(\omega/df)$ on each non-compact divisor

$$E_i \backslash (\cup_{j \neq i} E_j) \cap E_i, \quad i = 1, \ldots, N,$$

more precisely on a suitable $k(E_i)$-fold cover of this non-compact divisor. The limiting form is a holomorphic $(n-1)$-form on covers, meromorphic near pairwise intersections of divisors. If the weight relative to the form ω is not less than -1, then on one of the coverings the poles of the limiting form have no higher than the first order. According to Deligne's theorem (see [92, 142]) such a form generates a non-zero cohomology class. On this particular cover we choose an $(n-1)$-dimensional cycle, over which the integral of the limiting form is different from zero. Movement of this cycle off the cover of the divisor in the level hypersurface of the function $f \circ \pi$ determines the required family of vanishing homologies. See [364], Chapter 4.

13.1.7 The order of a form and Newton polyhedra

Let us reformulate the result of Theorem 13.1 in terms of the Newton polyhedra of the critical point of the germ f and the form ω.

Let ω be a holomorphic differential n-form on X. Let us define a rational number, called the remoteness of the polyhedra of the germ f and the form ω. Let us write ω in the form

$$g dx_1 \wedge \ldots \wedge dx_n,$$

and let us consider the Newton polyhedron

$$\Gamma(x_1 \ldots x_n g)$$

of the Taylor series of the function g, calculated at the critical point of the germ f, and multiplied by the product of all the variables. Let us consider a second polyhedron – the Newton polyhedron of the Taylor series of the critical point of the germ f. The *remoteness of the polyhedra of the germ f and the form ω is minus* the reciprocal of the coefficient of inscription of the first polyhedron in the second (see § 8.3.1, page 254).

Theorem 13.2 (see [364] and also [106, 387]). Let us suppose that the Taylor series of the critical point of the germ f has a $\mathbb{C}$-nondegenerate principal part.

1. The complex oscillation index of the critical point of the germ f is not greater than the remoteness of the Newton polyhedron of the critical point. The complex oscillation index is equal to the remoteness if the Newton polyhedron is remote (see the definitions in Chapter 6).

2. Let ω be a holomorphic differential n-form on X. Let $\alpha(\omega)$ be its order. Then the number

$$-(\alpha(\omega)+1)$$

is not greater than the remoteness of the polyhedron of the germ f and the form ω. Furthermore the number

$$-(\alpha(\omega)+1)$$

is equal to the remoteness of the polyhedra of the germ and the form if the remoteness is not less than -1.

Theorem 13.2 is derived from Theorem 13.1 in the same way as Theorems 8.3 and 8.4 were derived from Theorem 7.5.

Example. Let

$$f = x_1^6 + x_1^3 x_2^2 + x_2^6,$$

and

$$\omega = x_1 x_2 dx_1 \wedge dx_2.$$

Then the complex oscillation index of the germ f is equal to $-7/18$ and the order of the form ω is equal to $-7/9$.

We now state the assertion from [380], cf. [314], which strengthens Theorem 13.2: The number $-(\alpha(\omega)+1)$ is equal to the remoteness of the polyhedron of the germ and the form if the remoteness cannot be increased by adding to ω a form of the type $df \wedge d\eta$.

There exist complex analytic analogues of Theorems 6.5 and 8.5, for which see [364].

**13.2   The Hodge and weight filtrations in the fibres
        of the cohomological Milnor fibration of a critical point**

13.2.1   The Hodge filtration

Among the holomorphic sections of the cohomological Milnor fibration we can
single out a class of geometric sections. This consists of the sections generated by
the Gelfand-Leray forms of holomorphic n-forms. There are many geometric
sections (from them we can obtain a basis of the holomorphic sections of the
cohomological fibration). However, geometric sections behave in a very special
way in relation to covariantly constant sections as the point of the base tends to
the point $t=0$. A characteristic of the asymptotic properties of geometric
sections is the concept of the Hodge filtration of the cohomological Milnor
fibration, defined below. The Hodge filtration is a decreasing sequence of
analytic subfibrations of the cohomological Milnor fibration. The cohomology
class of a fibre of the Milnor fibration (or, which is the same thing, a vector of the
fibre of the cohomological Milnor fibration) belongs to the subfibration with
number k if it is the value of the principal part of a geometric section, the order of
which is not greater than $n-k-1$.

Now let us give a more precise definition. Let us define a sequence of analytic
subfibrations

$$F^k(f^*): F^k \to S', \quad k \in \mathbb{Z},$$

of the cohomological Milnor fibration of the critical point of the germ f. We
shall call this sequence the *Hodge filtration* of the cohomological Milnor
fibration $f^*: H^{n-1} \to S'$ of the critical point of the germ f. Let us define
subfibrations giving their sections. Let the fibres

$$F_t^k \ (t \in S')$$

be the linear subspaces generated by the values of the principal parts of geometric
sections or order not greater than $n-k-1$. If a section of such an order does not
exist then we put $F_t^k = \{0\}$. In other words F_t^k are subspaces spanned by vectors of
the form

$$s_{\max}[\omega](t),$$

where ω is a holomorphic differential n-form on X, of order not greater than
$n-k-1$. Let us list some obvious corollaries of the definition.

Lemma 13.7. 1. For any $k \in \mathbb{Z}$, $t \in S'$, the subspace

$$F_t^k \subset H^{n-1}(X_t, \mathbb{C})$$

is the direct sum of its intersections with the root subspaces of the monodromy operator, that is the Hodge filtration is invariant relative to the semisimple part of the monodromy operator.

2. For any $k \in \mathbb{Z}$

$$F^k(f^*): F^k \to S'$$

is an analytic subfibration of the cohomological Milnor fibration.

3. The subfibrations of the Hodge filtration form a decreasing filtration, that is for any k

$$F^{k+1} \subset F^k$$

4. If $k \geqslant n$, then $F^k(f^*)$ is a subfibration of rank 0.

5. There exists a k for which F^k is the same as H^{n-1}.

Proof. Section 1 is a corollary of Section 2 of Lemma 13.1.

Let us prove Section 2. If the form is multiplied by f then the geometric sections of the form are multiplied by t. Therefore the subspace F_t^k is generated by the principal parts of the geometric sections, the weights of which lie within the limits

$$n - k - 2 < \alpha(\omega) \leqslant n - k - 1.$$

Now Section 2 is a corollary of Section 2 of Lemma 13.3.

Section 3 is obvious. Section 4 is a corollary of Theorem 10.8.

For the proof of Section 5 we need to produce a set of forms for which the principal parts form a basis of the sections of the cohomological Milnor fibration. According to the theorem on determinants, there exist on X holomorphic n-forms

$$\omega_1, \ldots, \omega_\mu,$$

for which the geometric sections form a basis of the sections of the cohomological fibration. If the principal parts of these forms form a basis of the sections then Section 5 is proved. If they do not form a basis then we must modify the forms.

For this we must take a suitable non-degenerate $\mu \times \mu$ matrix Q of holomorphic functions on S, induce by the map f a matrix $Q(f)$ of holomorphic functions on X and consider the new forms

$$\omega_i' = \sum_j Q(f)_i^j \omega_j, \quad j = 1, \ldots, \mu.$$

The existence of a matrix for which the principal parts of the new forms form a basis of the sections easily follows from the choice of the forms

$$\omega_1, \ldots, \omega_\mu.$$

For further details see [364].

Remark. From the formulation of the theorem below on mixed Hodge structures (see also [364]) it follows that F^0 is the same as H^{n-1}.

Example 1. Let

$$f = x^{\mu+1}.$$

The cohomological Milnor fibration is a fibration of rank μ of the reduced zeroth cohomologies of the fibres of the Milnor fibration (each of which consists of $(\mu+1)$ points). According to Lemma 11.3

$$\{0\} = F^1 \subset F^0 = H^0.$$

Example 2. Let

$$f = x_1^2 + \ldots + x_n^2.$$

The cohomological Milnor fibration is one-dimensional. According to the example in § 10.3.4,

$$\{0\} = F^{[n/2]+1} \subset F^{[n/2]} = H^{n-1},$$

and the subfibration $F^{[n/2]}(f^*)$ is generated by the principal part of the form

$$dx_1 \wedge \ldots \wedge dx_n.$$

13.2.2   The weight filtration of a linear operator
(see [140, 142, 320])

With a linear nilpotent operator (that is with a linear operator the eigenvalues of which equal zero) we can associate an increasing filtration in the space where the operator acts. We shall list below three equivalent definitions of it. The first of these is the least invariant and the most intelligible. The last definition is the most usual and probably the least intelligible.

Let H be a finite-dimensional vector space, let $N: H \to H$ be a nilpotent linear operator, and let k be an integer. The sequence of subspaces

$$\{0\} \subset \ldots \subset W_l \subset W_{l+1} \subset \ldots \subset H \tag{5}$$

defined below is called the *weight filtration of the operator N with central index k*.

Definition 1. Let us consider an arbitrary Jordan basis of the operator N. Each subspace of the weight filtration is generated by a set, defined below, of vectors of the Jordan basis. Let us divide the elements of the basis into groups, putting into one group vectors of one Jordan block. We shall depict one group:

$$\square \to \square \to \ldots \to \square,$$

where the boxes are the vectors of the group, and the arrows are the action of the operator, the last vector mapping to zero. Now we shall depict all the groups one above the other, arranging them symmetrically relative to a vertical axis and reserving for the arrows the space of one box, see figure 79. To each vector of the

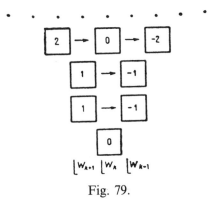

Fig. 79.

basis we assign an integer – the signed distance to the axis of symmetry (in figure 79 the numbers are written in the boxes). Let $W_{k+l} \subset H$ be the subspace generated by vectors of the basis with number no more than l (see figure 79).

Definition 2. Let $h \in H$. Put

$$l_+(h) = \min \{l | h \in \ker N^l\},$$
$$l_-(h) = \max \{l | h \in \operatorname{Im} N^l\}.$$

Let us define the subspace W_{k+l} by the property

$$h \in W_{k+l} \Leftrightarrow l_+(h) - l_-(h) \leqslant l+1.$$

Definition 3 (in the form of a lemma, see [320]). There exists a unique filtration (5), possessing the following properties (i), (ii).

(i) $N(W_l) \subset W_{l-2}, \quad l \in \mathbb{Z}.$

Put

$$gr_l W = W_l / W_{l-1}.$$

According to (i) N induces a map $gr_l W \to gr_{l-2} W$.

(ii) $N^l : gr_{k+l} W \to gr_{k-l} W$

is an isomorphism, for $l \in \mathbb{Z}$.

Lemma 13.8 (see [320]). Definitions 1–3 are equivalent.

13.2.3 The weight filtration of the cohomological Milnor fibration

Let us define a sequence of analytic subfibrations

$$W_l(f^*) : W_l \to S', \quad l \in \mathbb{Z}$$

of the cohomological Milnor fibration of the critical point of the germ f. We shall call this sequence the *weight filtration* of the cohomological Milnor fibration of the critical point of the germ f.

First we shall define the intersections of the fibres of the subfibrations

$$\{W_l(f^*)\}$$

with the root subspaces of the monodromy operator, and then we shall put the fibres of the subfibration equal to the direct sum of their intersections with the root subspaces.

Let us denote by $H_\lambda^{n-1}(X_t, \mathbb{C})$ the root subspace of the eigenvalue λ of the monodromy operator in $H^{n-1}(X_t, \mathbb{C})$. Let us denote by N the logarithm of the unipotent part of the monodromy operator. Now N is a nilpotent operator, commuting with the monodromy operator and so preserving its root subspaces.

Suppose $\lambda \neq 1$. Then as the weight filtration in

$$H_\lambda^{n-1}(X_t, \mathbb{C})$$

we take the weight filtration of the operator N with central index $n-1$. Suppose $\lambda = 1$. Then as the weight filtration in

$$H_1^{n-1}(X_t, \mathbb{C})$$

we take the weight filtration of the operator N with central index n. Put the fibre over t of the fibration $W_l(f^*)$ equal to

$$W_{l,t} = \oplus_\lambda W_l(H_\lambda^{n-1}(X_t, \mathbb{C})).$$

Let us list the obvious corollaries of the definition.

Lemma 13.9. 1. The weight filtration is an increasing sequence of analytic subfibrations of the cohomological Milnor fibration.

2. The weight filtration is invariant relative to the Gauss-Manin connection (that is the covariant derivatives of sections of each subfibration belong to the same subfibration).

3. The weight filtration is invariant relative to the action of the semisimple part of the monodromy operator.

4. For any $l \in \mathbb{Z}$ and for any point $t \in S'$

$$N(W_{l,t}) \subset W_{l-2,t}.$$

Furthermore, if $\lambda \neq 1$, then

$$N^l : gr_{n-1+l}W(H_\lambda^{n-1})_t \rightarrow gr_{n-1-l}W(H_\lambda^{n-1})_t$$

is an isomorphism, and if $\lambda = 1$, then

$$N^l : gr_{n+l}W(H_1^{n-1})_t \rightarrow gr_{n-1-l}W(H_1^{n-1})_t$$

is an isomorphism.

The **proof** is obvious.

Let us formulate one more property of the weight filtration. Remember that in the fibres of the cohomological Milnor fibration there is an operation, invariant relative to the Gauss-Manin connection, of complex conjugation: the space

$$H^{n-1}(X_t, \mathbb{C})$$

is the complexification of the natural image in $H^{n-1}(X_t, \mathbb{C})$ of the space $H^{n-1}(X_t, \mathbb{R})$. Furthermore there is in the fibres an integral structure which is invariant relative to the Gauss-Manin connection: the lattice in $H^{n-1}(X_t, \mathbb{C})$ being the natural image of the group $H^{n-1}(X_t, \mathbb{Z})$.

Lemma 13.10. For any $l \in \mathbb{Z}$ the fibres of the subfibration $W_l(f^*)$ are invariant relative to conjugation. Furthermore the fibre of the subfibration $W_l(f^*)$ can be given in terms of coordinates with respect to a basis of the integer lattice by equations with integer coefficients.

The proof follows easily from the definition of a weighted filtration.

Example 1. Let $f = x^{\mu+1}$. The eigenvalues of the monodromy operator are

$$\exp(2\pi i k/(\mu+1)), \quad k = 1, \ldots, \mu.$$

Therefore $\{0\} = W_{-1} \subset W_0 = H^0$.

Example 2. Let

$$f = x_1^2 + \ldots + x_n^2.$$

Then $(-1)^n$ is the unique eigenvalue of the monodromy operator. Therefore

$$\{0\} = W_{2[n/2]-1} \subset W_{2[n/2]} = H^{n-1}.$$

Example 3. Let us suppose that the monodromy operator in the cohomology, vanishing at the critical point of a holomorphic function in n variables, has finite order (for example, for the critical point of a (semi)quasihomogeneous function). Then

$$\{0\} = W_{n-2} \subset W_{n-1} \subset W_n = H^{n-1},$$

where $W_{n-1} = \bigoplus_{\lambda \neq 1} H^{n-1}_\lambda$.

Remark 1. The dependence, indicated in the definition of the weight filtration, of the central index of the filtration on the eigenvalue of the monodromy operator is motivated by a theorem, formulated below, on mixed Hodge structures. The central index is chosen so that the weight filtration, together with the Hodge filtration, makes up a mixed Hodge structure.

Remark 2. According to Theorem 3.12 the dimension of the Jordan blocks of the monodromy operator of a critical point of a function of n variables is not greater than n. Therefore a priori

$$\{0\} = W_{-1}(H^{n-1}_{\lambda \neq 1}) \subset W_0(H^{n-1}_{\lambda \neq 1}) \subset \ldots \subset W_{2n-2}(H^{n-1}_{\lambda \neq 1}) = H^{n-1}_{\lambda \neq 1},$$
$$\{0\} = W_0(H^{n-1}_1) \subset W_1(H^{n-1}_1) \subset \ldots \subset W_{2n-1}(H^{n-1}_1) = H^{n-1}_1.$$

However, according to the theorem formulated below on mixed Hodge structures the dimension of the Jordan blocks associated with the eigenvalue 1 is not greater than $n-1$. Therefore

$$W_1(H^{n-1}_1) = \{0\}, \qquad W_{2n-2}(H^{n-1}_1) = H^{n-1}_1, \qquad W_{2n-2} = H^{n-1}.$$

13.2.4 The reciprocal arrangement of the weight and Hodge filtrations (elementary properties)

Lemma 13.11. For any $k, l \in \mathbb{Z}$ the intersection of the spaces F^k, W_l together with the projection onto the base forms a holomorphic subfibration

$$F^k \cap W_l(f^*): F^k \cap W_l \to S'$$

of the cohomological Milnor fibration of the critical point of the germ f.

Proof. It is sufficient to prove that all the fibres of the projection

$$F^k \cap W_l \to S'$$

have equal dimensions. Since the subfibration $F^k(f^*)$ is generated by the principal parts of forms with given orders, it is sufficient to prove that the values of the principal part of an arbitrary form belong or do not belong to W_l simultaneously for all points of the base (see Section 2 of Lemma 13.3). Let us prove this. According to formula (4) on page 354 the value of the principal part $s_{\max}[\omega](t)$ at the point $t \in S'$ belongs to W_l if and only if the vector $A^{\omega}_{0,\alpha(\omega)}(t)$ belongs to W_l. This vector belongs to W_l if and only if the section $A^{\omega}_{0,\alpha(\omega)}$ is a section of the subfibration $W_l(f^*)$, which is what we were trying to prove.

For any l let us denote by

$$gr_l W(f^*) : gr_l W \to S'$$

the quotient fibration of the subfibrations $W_l(f^*)$, $W_{l-1}(f^*)$; its fibre is the quotient space

$$gr_l W_t = W_{l,t}/W_{l-1,t},$$

where $t \in S'$. The fibration $gr_l W(f^*)$ possesses an induced Gauss-Manin connection in view of Section 2 of Lemma 13.9. In the fibres of the fibration

$$gr_l W(f^*)$$

there are given a real and an integral structure (in view of Lemma 13.10) and an action of the semisimple part of the monodromy operator (in view of Section 3 of Lemma 13.9).

Remark. The action of the monodromy operator in the fibres of the fibration

$$gr_l W(f^*)$$

is the same as the action of its semisimple part, see Sections 3 and 4 of Lemma 13.9.

Corollary of Lemma 13.11. The projection into $gr_l W(f^*)$ of the fibration

$F^k \cap W_l(f^*)$ defines a subfibration

$$F^k gr_l W(f^*): F^k gr_l W \to S';$$

its fibre is the quotient space

$$F^k gr_l W_t = (F_t^k \cap W_{l,t} + W_{l-1,t})/W_{l-1,t},$$

where $t \in S'$.

The subfibration

$$F^k gr_l W(f^*) \subset gr_l W(f^*)$$

is invariant relative to the action of the semisimple part of the monodromy operator (Lemmas 13.7 and 13.9) and invariant relative to the Gauss-Manin connection (Section 3 of Lemma 13.3).

Let us consider the operator N – the logarithm of the unipotent part of the monodromy operator. The operator N defines a morphism of fibrations

$$N: gr_l W \to gr_{l-2} W$$

(that is a linear map on the fibres, commuting with the projection onto the base and preserving the class of holomorphic sections).

Lemma 13.12. For any $k, l \in \mathbb{Z}$

$$N(F^k gr_l W) \subset F^{k-1} gr_{l-2} W.$$

Proof. It is sufficient to prove that if a holomorphic differential n-form ω possesses the properties:
(i) it has order no greater than $n - k - 1$;
(ii) its principal part is a section of the subfibration

$$W_l(f^*);$$

then the projection of the section $N s_{\max}[\omega]$ into the quotient fibration $gr_{l-2} W(f^*)$ is a section of the subfibration

$$F^{k-1} gr_{l-2} W(f^*).$$

Note that according to Lemma 13.1

$$Ns_{\max}[\omega] = -2\pi it^{\alpha}(A^{\omega}_{1,\alpha(\omega)} + \ldots + (\ln t)^{n-1} A^{\omega}_{n-1,\alpha(\omega)}/(n-2)!). \tag{6}$$

Therefore it is sufficient to produce two forms and a linear combination of them, for which the principal part possesses the properties:
(iii) it is a section of the subfibration $W_{l-2}(f^*)$;
(iv) its projection into $gr_{l-2}W(f^*)$ is proportional to the projection of the section $Ns_{\max}[\omega]$;
(v) its projection into $gr_{l-2}W(f^*)$ is a section of the subfibration

$$F^{k-1}gr_{l-2}W(f^*).$$

The first form is $f\omega$,

$$s_{\max}[f\omega] = t^{\alpha(\omega)+1}(A^{\omega}_{0,\alpha(\omega)} + \ldots + (\ln t)^{n-1} A^{\omega}_{n-1,\alpha(\omega)}/(n-1)!).$$

The second form is the arbitrary form

$$\psi = df \wedge \eta,$$

where $d\eta = \omega$.
According to formula (3) on page 284

$$s_{\max}[\psi] = \int_0^t s_{\max}[\omega](u)\,du = t^{\alpha(\omega)+1}(A^{\omega}_{0,\alpha(\omega)}/(\alpha(\omega)+1) -$$
$$- A^{\omega}_{1,\alpha(\omega)}/(\alpha(\omega)+1)^2 + \ldots), \tag{7}$$

where we have ommitted terms in which there occur either $A^{\omega}_{k,\alpha(\omega)}$, with $k \geqslant 2$, or $(\ln t)^k$ with $k \geqslant 1$. A linear combination with the properties (iii)–(v) is the form

$$f\omega - (\alpha(\omega)+1)\psi.$$

Indeed,

$$s_{\max}[f\omega - (\alpha(\omega)+1)\psi] = t^{\alpha(\omega)+1}(A^{\omega}_{1,\alpha(\omega)}/(\alpha(\omega)+1) + \ldots), \tag{8}$$

where we have left out terms of the same sort. According to Lemma 13.1 we have properties (iii) and (iv). Since the order of the linear combination is equal to

$$\alpha(\omega)+1,$$

we have property (v). The lemma is proved.

Remark. As formulae (6) and (8) show, the operation of transition from $s_{\max}[\omega]$ to $s_{\max}[f\omega-(\alpha(\omega)+1)\psi]$ is very like the application to $s_{\max}[\omega]$ of the logarithm of the unipotent part of the monodromy operator. In addition, the forms $f\omega$ and $f\omega-(\alpha(\omega)+1)\psi$ generate one and the same element in

$$\Omega^n(X)/df \wedge \Omega^{n-1}(X),$$

where $\Omega^p(X)$ is the holomorphic p-forms on X. Therefore the formulae (6) and (8) show an analogy between the action in cohomologies of the logarithm of the unipotent part of the monodromy and the action in

$$\Omega^n(X)/df \wedge \Omega^{n-1}(X)$$

of the operation of multiplication by f. For further details see § 14.3.5 and [363].

Let us now formulate a theorem about the mixed Hodge structure. Chapter 14 will be devoted to a discussion of the results of this theorem and also its corollaries. For a detailed definition of the mixed Hodge structure see § 14.1.

Theorem 13.3. (About the mixed Hodge structure, see [361, 362, 364]). The weight and Hodge filtrations form a mixed Hodge structure in the fibres of the cohomological Milnor fibration of a critical point, that is for any $k, l \in \mathbb{Z}, t \in S'$

$$gr_l W_t = F^k gr_l W_t \oplus \overline{F^{l-k+1} gr_l W_t}, \tag{9}$$

where $\oplus$ is the direct sum and the bar denotes conjugation.

Remark. The mixed Hodge structure in vanishing cohomologies was defined by Steenbrink in [343]. The weight filtration, defined in § 13.2.3, is the same as the weight filtration of Steenbrink. The Hodge filtration, defined in § 13.2.1, differs, generally speaking, from the Hodge filtration of Steenbrink. The Hodge filtration of §13.2.1 and the Hodge filtration of Steenbrink coincide on the quotient fibration of the weight filtration. The Hodge filtration of §13.2.1 and the Hodge filtration of Steenbrink can be expressed simply in terms of each other: the Hodge subfibration $F^k(f^*)$ of §13.2.1 is generated by the principal parts of forms of orders belonging to the half-open interval

$$(n-k-2, \, n-k-i];$$

if $s_{max}[\omega]$ is one such principal part, then

$$(\nabla_{\partial/\partial t})^{n-k-1} s_{max}[\omega]$$

is a section of the Hodge subfibration $F_{St}^k(f^*)$ of Steenbrink. The definition of the Hodge filtration of Steenbrink uses the resolution of the singularity of the critical point of the germ f and does not use the asymptotics of integrals of holomorphic forms. For further details see [343, 362, 364, 365].

For the **proof** of the theorem see [362, 364] (the scheme of the proof is laid out in [362], for the missing details and for a different proof for the case $n=2$ see [365]). The proof is derived from deep and non-trivial theory of deformations of Hodge structures of the cohomology of compact non-singular Kähler manifolds. This theory was hammered out by Griffiths, Schmidt and Deligne (see [140, 142, 320, 92]). In the cohomologies of compact non-singular Kähler manifolds there exists a natural filtration – a Hodge structure (see [71, 405]). Each manifold corresponds to a point in the classification space of all Hodge structures. If the manifold depends holomorphically on parameters, then there is a holomorphic mapping of the parameter space into the classification space of Hodge structures (see [140, 142, 320]). This mapping is called the *period map*. The period map possesses very special properties, connected with the negative curvature of the classification space of Hodge structures. If the parameter space of a family of manifolds is a punctured disc, then by studying the asymptotics of the period map as the parameter of the family tends to a distinguished point of the disc we can obtain information about how the non-singular manifolds of the family become degenerate at it (see [142, 320]). In the local situation, indicated in Theorem 13.3, this theory is applied in the following way. As a representative of the germ f we take a polynomial P for which the point 0 is a unique critical point with zero as the critical value. We consider the compactification Y_t in $\mathbb{C}P^n$ of the level hypersurface of the polynomial (t is the level value). The polynomial can be chosen so that for small $t \neq 0$ the hypersurface Y_t is non-singular, but the hypersurface Y_0 has a unique singular point at 0. The fibre X_t of the Milnor fibration of the critical point 0 of the polynomial P is a part of the hypersurface Y_t. We can choose the polynomial so that the inclusion $X_t \hookrightarrow Y_t$ induces an epimorphism

$$H^{n-1}(Y_t) \to H^{n-1}(X_t)$$

(for this we need the degree of the polynomial to be sufficiently high, see [317]). There is a natural map (defined modulo homotopy)

$$p: Y_t \to Y_0,$$

for which the fibre X_t of the Milnor fibration is mapped to the singular point, see figure 80. According to the exact sequence of the pair $X_t \subset Y_t$ there is an isomorphism

$$\pi: H^{n-1}(Y_t)/p^*H^{n-1}(Y_0) \to H^{n-1}(X_t).$$

From the theory of deformations of Hodge structures it follows that on

$$H^{n-1}(Y_t)/p^*H^{n-1}(Y_0)$$

there are two natural filtrations, called the weight and the Hodge, which possess the properties indicated in formula (9) (see [320, 78, 317]). It can be proved that the isomorphism π maps these filtrations into our weighted and Hodge filtrations (more precisely the Hodge filtration maps into the Hodge filtration of Steenbrink, see [362]). This proves Theorem 13.3.

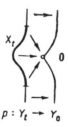

$$p: Y_t \to Y_0$$

Fig. 80.

Remark. Let us clarify why in the definition of the weight filtration in §13.2 the central index depends on the eigenvalue of the monodromy operator.
   On the space

$$H^{n-1}(Y_t)$$

there acts the monodromy operator induced by the parameter t going round the point $t=0$. For obvious reasons $p^*H^{n-1}(Y_0)$ is contained in the subspace of eigenvectors associated with the eigenvalue 1. It can be proved that $p^*H^{n-1}(Y_0)$ coincides with this subspace. The proof can be derived from a theorem on invariant cycles (see [78, 342]). It is not hard also to derive this result directly from the fundamental theorem of Schmid in [320] with the help of Theorem 12.1 on determinants (compare with the proof of Lemma 2 in [362]).

We mentioned earlier that the weight filtration on

$$H^{n-1}(Y_t)/p^*H^{n-1}(Y_0)$$

is induced by the defined filtration on $H^{n-1}(Y_t)$ (see [317]). This filtration on $H^{n-1}(Y_t)$ is the weight filtration of the logarithm of the unipotent part of the monodromy operator with central index $n-1$ (it is defined on all the root subspaces in the same way). Since the kernel of the epimorphism

$$H^{n-1}(Y_t) \to H^{n-1}(X_t)$$

is a subspace of invariant vectors, the projection into $H^{n-1}(X_t)$ of the indicated weight filtration on $H^{n-1}(Y_t)$ coincides with the weighted filtration from § 13.2.3.

13.2.5   First corollaries

For any $k, l \in \mathbb{Z}$, $t \in S'$, the Hodge subspace

$$F^k gr_l W_t$$

can be decomposed into the direct sum of its intersections with the root subspaces of the action of the monodromy operator on $gr_l W_t$:

$$F^k gr_l W_t = \oplus_\lambda F^k gr_l W_{t,\lambda},$$

where λ are the eigenvalues of the monodromy operator. The monodromy operator preserves the integral structure of the space $gr_l W_t$, since conjugation interchanges the root subspaces corresponding to the eigenvalues λ and $\bar{\lambda}$. In this way from (9) we obtain Corollary 1.

Corollary 1. For any eigenvalue λ

$$gr_l W_{t,\lambda} = F^k gr_l W_{t,\lambda} \oplus \overline{F^{l-k+1} gr_l W_{t,\bar{\lambda}}}. \tag{10}$$

Corollary 2 (see [343]). The dimension of an arbitrary Jordan block of the monodromy operator is not greater than n. In addition, the dimension is not greater than $n-1$ if the eigenvalue of the block is equal to 1.

Example. If $n=1$ then all the eigenvalues of the monodromy operator are not equal to 1, and in the space of vanishing cohomologies there exists a basis of eigenvectors of the monodromy operator (see the example on page 370).

Proof of Corollary 2. According to Lemma 13.7 for $k \geqslant n$ the space

$$F^k gr_l W_t$$

is zero-dimensional. According to formula (10) the space $gr_l W_{t,\lambda}$ is zero-dimensional for $l \geqslant 2n-1$, which is what we were trying to prove (see § 13.2.3).

Corollary 3. For any k, $l \in \mathbb{Z}$, $t \in S'$, the operator N induces isomorphisms

$$N^l : F^k gr_{n-1+l} W_{t,\lambda} \overset{\sim}{\to} F^{k-l} gr_{n-1+l} W_{t,\lambda}, \tag{11}$$

if $\lambda \neq 1$, and

$$N^l : F^k gr_{n+l} W_{t,\lambda=1} \overset{\sim}{\to} F^{k-l} gr_{n-l} W_{t,\lambda=1}, \tag{12}$$

if $\lambda = 1$.

Proof of Corollary 3. According to Lemma 13.12 the image of the left-hand side is contained in the right-hand side. According to Section 4 of Lemma 13.9 and formula (10) the image of the left-hand side cannot be smaller than the right-hand side.

13.3   The spectrum of a critical point

13.3.1   The spectral pairs of a critical point

We shall define an unordered set of μ pairs of numbers which characterise the reciprocal arrangement of the weight and Hodge filtrations.

For any k, $l \in \mathbb{Z}$ let us consider the quotient fibration

$$gr^k F \, gr_l \, W(f^*) = F^k gr_l \, W(f^*)/F^{k+1} gr_l \, W(f^*).$$

According to Lemma 13.7 there is induced on this fibration an action of the semisimple part of the monodromy operator. Let us fix a fibre of this fibration and put in correspondence with each eigenvalue λ of the action in the fibre of the semisimple part the pair

$$
\begin{aligned}
&(n-1-l_k(\lambda), l), \quad \text{if} \quad \lambda \neq 1, \\
&(n-1-l_k(\lambda), l-1), \quad \text{if} \quad \lambda = 1;
\end{aligned}
\tag{13}
$$

here the number $l_k(\lambda)$ is determined by the conditions:

$$\exp(2\pi i l_k(\lambda)) = \lambda, \quad k \leqslant \operatorname{Re} l_k(\lambda) < k+1.$$

The unordered set of μ pairs of numbers, constructed in this way (for all k, l, λ) is called the *set of spectral pairs of the critical point of the germ* f. The unordered set of the μ first elements of the pairs is called the *set of spectral numbers (or spectrum) of the critical point*. It is clear that these sets do not depend on the choice of fibre of the fibration

$$\{gr^k Fgr_l \, W(f^*)\}.$$

Let us make clear why the eigenvalue λ is put in correspondence with the number $n-1-l_k(\lambda)$, which is one of the values of the logarithm of the number $1/\lambda$ divided by $2\pi i$. According to the definition of the Hodge filtration, the quotient fibration

$$F^k(f^*)/F^{k+1}(f^*)$$

is generated by the values of the principal parts of forms of degree not greater than $n-1-k$. Let ω be a form, the order of which belongs to

$$(n-2-k, \, n-1-k]$$

and the principal part of which generates a non-zero section in

$$F^k(f^*)/F^{k+1}(f^*).$$

According to Lemma 13.1 this non-zero section is contained in the root subspace of the eigenvalue $\lambda = \exp(-2\pi i\alpha(\omega))$ of the action of the semisimple part of the monodromy operator. Expressing the order of the form ω in terms of λ we obtain

$$n-1-l_k(\lambda) = n-1-k-l_0(\lambda) = \alpha(\omega).$$

In this way, $n-1-l_k(\lambda)$ is the smallest number α for which there exist forms of order α, the principal parts of which generate in

$$F^k(f^*)/F^{k+1}(f^*)$$

the root subspace of the eigenvalue λ of the action of the semisimple part of the monodromy operator. Moreover the number α enters into the spectrum exactly as often as the difference of the dimensions of the spaces of the principal parts of the forms of order α and of the forms of order $\alpha-1$. The second number of the spectral pair indicates on which levels of the weight filtration the new principal parts arise. For example, for $k=n-1$

$$F^{n-1}(f^*)/F^n(f^*) = F^{n-1}(f^*),$$

the numbers $n-1-l_{n-1}(\lambda)$ belong to the half-open interval $(-1,0]$. If $\omega_1, \ldots, \omega_s$ are holomorphic differential n-forms of non-positive order, the principal parts of which form a basis of the sections of the fibration $F^{n-1}(f^*)$, then the orders of these forms form the part of the spectrum belonging to $(-1,0]$.

We turn out attention to the dependence on the eigenvalue λ of the second number in the characteristic pair, see (13). In the definition of the weight filtration the index on the root subspace associated with the eigenvalue 1 was increased by 1, in order to agree with the result of Theorem 13.3. Now in (13) this index has been decreased by 1. This decrease is motivated by simple formulations of results about such numbers. For example see the symmetries (iii), (iv), indicated below, of the spectral pairs.

In conclusion we introduce one more definition of spectrum.

Lemma 13.13 (see [365, 364]). Let $\omega_1, \ldots, \omega_\mu$ be holomorphic differential n-forms on X. Let us suppose that

1. the sum of the orders of these forms is equal to $\mu(n/2-1)$.
2. the function

$$\det^2 : t \mapsto \det^2 \left(\int_{\delta_{l(t)}} \omega_j/df \right), \quad j, l = 1, \ldots, \mu,$$

where $\delta_1, \ldots, \delta_\mu$ is a basis of the covariantly constant sections of the homological Milnor fibration, has for $t=0$ a zero of order $\mu(n-2)$. Then the orders

$$\alpha(\omega_1), \ldots, \alpha(\omega_\mu)$$

of these forms are the spectrum of the critical point 0 of the germ f.

The **proof** follows easily from Lemma 13.4, Theorem 12.1 on determinants and the definitions of the Hodge filtration and the spectrum.

Corollary. The sum of the spectral numbers of the critical point of the germ f is equal to $\mu(n/2-1)$.

Note that the proof of this result does not use the theorem on mixed Hodge structures.

Remark. From the definition of spectral pairs it follows easily that the sum of all the second elements of the spectral pairs is equal to $\mu(n-1)$.

13.3.2   Hodge numbers

Analogous to the spectral pairs, we shall define the Hodge numbers

$$h^{k,m}, h_\lambda^{k,m}.$$

We shall denote by $h_\lambda^{k,m}$ the dimension of the root subspace of the eigenvalue λ of the action in the fibre of the fibration

$$gr^k F gr_{k+m} W(f^*)$$

of the semisimple part of the monodromy operator. We shall denote by $h^{k,m}$ the rank of the fibration

$$gr^k F\, gr_{k+m} W(f^*).$$

By definition,

$$h^{k,m} = \sum_{\lambda} h^{k,m}_{\lambda}.$$

The spectral pairs and the Hodge numbers $h^{k,m}_{\lambda}$ mutually determine each other.

Example 1. Let $f = x^{\mu+1}$. Then $h^{0,0} = \mu$; $h^{0,0}_{\lambda} = 1$ for $\lambda = \exp(2\pi ik/(\mu+1))$, $k = 1, \ldots, \mu$; the spectral pairs are

$$(-k/(\mu+1), 0),$$

where $k = 1, \ldots, \mu$.

Example 2. Let

$$f = x_1^2 + \ldots + x_n^2.$$

Then

$$h^{[n/2],[n/2]} = h^{[n/2],[n/2]}_{(-1)^n} = 1;$$

the spectral pair is $(n/2 - 1,\ n - 1)$.

13.3.3 Symmetries

Lemma 13.14 (see [343, 364]). For any k, m we have the symmetries

(i) $h^{k,m}_{\lambda} = h^{m,k}_{\bar{\lambda}}$,

(ii) $h^{k,m}_{\lambda} = h^{n-1-m, n-1-k}_{\lambda}$, if $\lambda \neq 1$,

(iii) $h^{k,m}_1 = h^{n-m, n-k}_1$.

For the derivation of Lemma 13.14 from Theorem 13.3 see § 14.2.

Let us formulate these symmetries in terms of the spectral pairs.

(i) The number of pairs equal to (α, l) is the same as the number of pairs equal to

$$(2n - 3 - l - \alpha, l)$$

(compare with the examples).

(ii), (iii) The number of pairs equal to (α, l) is the same as the number of pairs equal to

$$(\alpha + l - n + 1, 2n - 2 - l).$$

The relations (i)–(iii) describe the symmetries of the dimensions of the spaces of the principal parts of the forms of fixed order, which are sections of a fixed fibration of the weight filtration (more precisely they describe the symmetries of increments of dimensions).

Combining (i) with (ii) and (iii) we obtain

Corollary.

1. The spectrum of the critical point is arranged symmetrically relative to the point $n/2 - 1$. The set of spectral pairs, as a subset of $\mathbb{R}^2$, is centrally symmetric relative to the point $(n/2 - 1, n - 1)$.

2. The complex singular index (see §13.1.5) is non-negative.

Indeed the complex singular index is equal to $n/2 - (1 + \alpha_{min})$, where α_{min} is the smallest spectral number.

Note the following property of the spectrum. If the spectrum of the critical point is concentrated at the point $n/2 - 1$, that is if it consists of several numbers $n/2 - 1$, then the critical point is non-degenerate and the spectrum consists of one number $n/2 - 1$. This property is a direct corollary of Theorem 3.7 on the trace of the monodromy operator.

13.3.4 The spectrum and the Newton polyhedron

The spectrum can be expressed in terms of the geometry of the Newton polyhedron of the Taylor series of the critical point, if the principal part of the Taylor series is $\mathbb{C}$-nondegenerate, see [314, 380] and also [343, 17, 89].

Let us express the spectrum in terms of the Newton polyhedron for $n = 2$. For $n = 2$ the spectrum is symmetrical relative to the point 0 and lies in the interval

$(-1, 1)$, therefore it is sufficient to describe that part of the spectrum belonging to the interval $(-1, 0]$. According to §13.3.1, it is sufficient for a description of this part of the spectrum to indicate forms of non-negative orders, the principal parts of which generate a basis of the sections of the fibration $F^1(f^*)$, and to calculate the orders of the indicated forms.

Let us consider the Newton polyhedron of the Taylor series of the critical point of the germ $f : (\mathbb{C}^n, 0) \rightarrow (\mathbb{C}, 0)$. Let us call the monomial x^m *subdiagrammatic* if the vector

$$m + (1, \ldots, 1)$$

does not belong to the interior of the Newton polyhedron.

Example. Let

$$f = x_1^6 + x_1^3 x_2^2 + x_2^5.$$

We have depicted in figure 81 the vectors $m + (1, 1)$ for the subdiagrammatic monomials x^m.

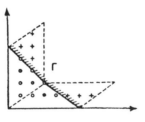

Fig. 81.

With each subdiagrammatic monomial we put in correspondence the form

$$x^m dx_1 \wedge \ldots \wedge dx_n,$$

which we shall call *subdiagrammatic*. Let us suppose that the principal part of the Taylor series of the critical point of the germ f is $\mathbb{C}$-nondegenerate. Then the order of each subdiagrammatic form is non-positive and can be calculated from the degrees of the monomials with the help of Theorem 13.2 (the order is equal to the remoteness of the polygons of the germ f and the form).

In the example the orders of the subdiagrammatic forms are equal to

$$-3/5, \ -5/12, \ -3/12, \ -1/12, \ -2/5, \ -1/5, \ -1/5, \ 0, \ 0, \ 0.$$

Theorem 13.4 (see [364]). If the principal part of the Taylor series of the critical point of the germ $f : (\mathbb{C}^n, 0) \to (\mathbb{C}, 0)$ is $\mathbb{C}$-nondegenerate then the principal parts of the subdiagrammatic forms form a basis of the sections of the fibration $F^{n-1}(f*)$.

Corollary. The orders of subdiagrammatic forms constitute the part of the spectrum of the critical point of the germ f belonging to the interval $(-1, 0]$. In particular for $n=2$ the orders of the subdiagrammatic forms completely determine the spectrum.

Remark 1. For $n=2$ it is convenient in describing the spectrum to mark, not only the indices of subdiagrammatic monomials, but also the indices of symmetric superdiagrammatic monomials. The remoteness of pairs of polyhedra of the germ f (with $\mathbb{C}$-nondegenerate principal part) and distinguished monomials make up the spectrum of the critical point of the germ f. See figure 82 where we

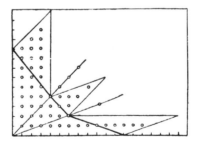

Fig. 82.

have depicted the indices, moved to $(1, 1)$, of the subdiagrammatic and distinguished superdiagrammatic monomials for

$$f = x_2^9 + x_1^4 x_2^4 + x_1^6 x_2^2 + x_1^{12}.$$

Remark 2. The corollary of Theorem 13.4 is sufficient for a description of the spectrum also for $n = 3$. The part of the spectrum belonging to the half-open interval $(-1, 0]$ is given by the corollary. The part of the spectrum belonging to the half-open interval $[1, 2)$ is determined by the symmetry of the spectrum relative to the number $1/2$. The remaining part of the spectrum belongs to the interval $(0, 1)$. Each spectral number is the logarithm of an eigenvalue of the monodromy operator divided by $2\pi i$. In order to describe the part of the spectrum in $(0, 1)$ we need to calculate on the Newton polyhedron all the eigenvalues of the monodromy operator (see Theorem 3.13 from Chapter 3, and [359]), to mark off the eigenvalues for which the corresponding spectral numbers belong to the union

$$(-1, 0] \cup [1, 2),$$

to take the branches of the logarithms of the remaining eigenvalues which after division by $2\pi i$ lie in $(0, 1)$. These logarithms, divided by $2\pi i$ form the remaining part of the spectrum.

Proof of the theorem. The principal parts of the subdiagrammatic forms are linearly independent since by Theorem 13.2 the order of a linear combination of subdiagrammatic forms is equal to the minimum of the orders of the terms. The principal parts of the subdiagrammatic forms form a basis of the sections of the fibration $F^{n-1}(f^*)$, since by Theorem 13.2 the order of a superdiagrammatic form is positive.

Let us describe the spectrum of the finite-multiplicity critical point of the germ $f : (\mathbb{C}^n, 0) \to (\mathbb{C}, 0)$ of a quasihomogeneous function. Let f have type $(\alpha_1, \ldots, \alpha_n)$ and weight 1. Let $\{x^m | m \in I\}$ be a set of monomials, projecting to a basis over $\mathbb{C}$ of the local algebra

$$\mathbb{C}\{x\}/(\partial f/\partial x).$$

For the index $m \in I$ put

$$l(m) = (m_1 + 1)\alpha_1 + \ldots + (m_n + 1)\alpha_n - 1.$$

Theorem 13.5 ([341], see also [364]). The numbers

$$l(m), \ m \in I,$$

make up the spectrum of the critical point of the germ f of the quasihomogeneous function.

According to [364] the spectrum of the critical point of the germ of a
semiquasihomogeneous function is the same as the spectrum of the critical point
of the germ of the corresponding quasihomogeneous function.

For semiquasihomogeneous germs it is easy to indicate the spectral pairs: all
the second elements of the spectral pairs are equal to $n-1$ (see the example on
page 371).

For any Hodge subfibration $F^k(f^*)$ we can indicate forms, the principal parts
of which generate a basis of the sections of this subfibration. Namely we put the
monomial x^m in correspondence with the form

$$\omega_m = x^m dx_1 \wedge \ldots \wedge dx_n.$$

Theorem 13.6 (see [341, 364]). Let $f : (\mathbb{C}^n, 0) \to (\mathbb{C}, 0)$ be the germ of a semiquasihomogeneous function. Then

  1. For any $m \in I$ the order of the form ω_m is equal to $l(m)$.
  2. For any $k \in \mathbb{Z}$ the principal parts of the forms

$$\{\omega_m | m \in I, \ l(m) \leqslant n - 1 - k\}$$

make up a basis of the sections of the Hodge fibration $F^k(f^*)$.

We shall give a table, compiled by V. V. Goryunov [131], of the spectra of
simple, uni- and bimodal critical points for $n = 3$. In this table (see page 389)
there is indicated for each point the numbers N, L_r; the spectrum $\{\alpha_r\}$ is given by
the formula

$$\alpha_r = (L_r/N) - 1.$$

In view of the symmetry relative to the number $1/2$ all the spectra, except the
spectra of the points A_μ, D_μ, $T_{p,q,t}$ are written out up to the half-way mark, that is
for $r \leqslant \mu/2$. For the notation of the critical point see Volume 1.

Class	N	(L_r)						Class	N	(L_r)				
A_μ	$\mu+1$	$\mu+1+k$	$1\le k\le\mu$					E_6	12	13	16	17		
D_μ	$2\mu-2$	$3\mu-3$	$2\mu-1+2k$					E_7	18	19	23	25		
			$0\le k\le\mu-2$					E_8	30	31	37	41	43	

Class	N	(L_r)						Class	N	(L_r)				
P_8	3		3	4	4		4	Z_{12}	22	21	25	27	29	31	33
X_9	4		4	5	5		6	Z_{13}	18	17	20	22	23	25	26
J_{10}	6	6	7	8	8		9	W_{12}	20	19	23	24	27	28	29
Q_{10}	24	23	29	31	32	35		W_{13}	16	15	18	19	21	22	23
Q_{11}	18	17	21	23	24	25		E_{12}	42	41	47	53	55	59	61
Q_{12}	15	14	17	19	20	20	22	E_{13}	30	29	33	37	39	41	43
S_{11}	16	15	19	20	21	23		E_{14}	24	23	26	29	31	32	34 35
S_{12}	13	12	15	16	17	18	19	$T_{p,q,t}$	pqt	pqt $2pqt$ $(p+k_1)qt$					
U_{12}	12	11	14	15	15	17	18			$(q+k_2)pt$ $(t+k_3)pq$					
Z_{11}	30	29	35	37	41	45				$0<k_1<p$ $0<k_2<q$ $0<k_3<t$					

$J_{3,p}$	18 $(p+9)$	17 $(p+9)$ 23 $(p+9)$ 25 $(p+9)$ 9 $(2p+17+2k)$	
		$1\le k\le(p+10)/2$	
$Z_{1,p}$	14 $(p+7)$	13 $(p+7)$ 17 $(p+7)$ 19 $(p+7)$ 21 $(p+7)$ 7 $(2p+13+2k)$	
		$1\le k\le(p+7)/2$	
$W_{1,p}$	12 $(p+6)$	11 $(p+6)$ 14 $(p+6)$ 16 $(p+6)$ 17 $(p+6)$ 6 $(2p+11+2k)$	
		$1\le k\le(p+7)/2$	
$W_{1,p}^{\#}$	12 $(p+12)$	11 $(p+12)$ 17 $(p+12)$ 12 $(p+12+k)$ $1\le k\le(p+11)/2$	
$Q_{2,p}$	12 $(p+6)$	11 $(p+6)$ 15 $(p+6)$ 16 $(p+6)$ 17 $(p+6)$ 6 $(2p+11+2k)$	
		$1\le k\le(p+6)/2$	
$S_{1,p}$	10 $(p+5)$	9 $(p+5)$ 12 $(p+5)$ 13 $(p+5)$ 14 $(p+5)$ 5 $(2p+9+2k)$	
		$1\le k\le(p+6)/2$	
$S_{1,p}^{\#}$	10 $(p+10)$	9 $(p+10)$ 13 $(p+10)$ 10 $(p+10+k)$ $1\le k\le(p+10)/2$	
$U_{1,p}$	9 $(p+9)$	8 $(p+9)$ 11 $(p+9)$ 13 $(p+9)$ 9 $(p+9+k)$ $1\le k\le(p+8)/2$	

Class	N	(L_r)									
Q_{16}	21	19	22	25	26	28	28	29	31		
Q_{17}	30	27	31	35	37	39	40	41	43		
Q_{18}	48	43	49	55	59	61	64	65	67	71	
S_{16}	17	15	18	20	21	22	23	24	25		
S_{17}	24	21	25	28	29	31	32	33	35		
U_{16}	15	13	16	18	18	19	21	21	22		
W_{17}	20	18	21	23	24	26	27	28	29		
W_{18}	28	25	29	32	33	36	37	39	40	41	
Z_{17}	24	22	25	28	29	31	32	34	35		
Z_{18}	34	31	35	39	41	43	45	47	49	51	
Z_{19}	54	49	55	61	65	67	71	73	77	79	
E_{18}	30	28	31	34	37	38	40	41	43	44	
E_{19}	42	39	43	47	51	53	55	57	59	61	
E_{20}	66	61	67	73	79	83	85	89	91	95	97

13.3.5   The spectrum of a direct sum of critical points is equal to the sum of the spectra plus 1

Let $f:(\mathbb{C}^n,0)\to(\mathbb{C},0)$, $g:(\mathbb{C}^l,0)\to(\mathbb{C},0)$ be germs of holomorphic functions at critical points of (finite) multiplicities, respectively, μ, η. Let us consider the direct sum of these germs:

$$f+g:(\mathbb{C}^n\times\mathbb{C}^l,0\times 0)\to(\mathbb{C},0).$$

The germ $f+g$ has at 0×0 a critical point of multiplicity $\mu\cdot\eta$.

Theorem 13.7 (see [215]). If $\{\alpha_i\}$ is the spectrum of the critical point of the germ f and $\{\beta_j\}$ is the spectrum of the critical point of the germ g, then $\{\alpha_i+\beta_j+1\}$ is the spectrum of the critical point of the germ $f+g$ (here $i=1,\ldots,\mu$, $j=1,\ldots,\eta$).

Corollary 1. If $\{\alpha_i\}$ is the spectrum of the critical point of the germ $f:(\mathbb{C}^n,0)\to(\mathbb{C},0)$, then $\{\alpha_i+1/2\}$ is the spectrum of the critical point of the germ

$$f+z^2:(\mathbb{C}^{n+1},0)\to(\mathbb{C},0).$$

Corollary 2. The complex singular indices are equal for stably equivalent critical points.

The **proof** of the theorem uses complex oscillatory integrals with phases

$$f,g,f+g$$

and Fubini's theorem for such integrals. We shall use the notation of § 11.3.

In § 11.3 we indicated a mapping from the tensor product of the groups of admissible chains for the critical points of the germs f and g to the group of admissible chains for the critical point of the germ $f+g$

$$H_n(X^f,X^{-,f};\mathbb{C})\otimes H_l(X^g,X^{-,g};\mathbb{C})\to H_{n+l}(X^{f+g},X^{-,f+g};\mathbb{C}).$$

Lemma 13.15. This map is an isomorphism.

Proof. Let us denote the map by h. Let

$$\delta_1, \ldots, \delta_\mu \in H_n(X^f, X^{-,f}),$$

$$\gamma_1, \ldots, \gamma_\eta \in H_l(X^g, X^{-,g}),$$

be bases. It is sufficient to produce on X^{f+g} $\mu \cdot \eta$ holomorphic differential $(n+l)$-forms

$$\psi_1, \ldots, \psi_{\mu\eta},$$

for which

$$\det \left(\int_{h(\delta_i \otimes \gamma_j)} e^{\tau(f+g)} \psi_e \right) \not\equiv 0.$$

By the theorem on determinants there exist n-forms $\omega_1, \ldots, \omega_\mu$ on X^f, for which

$$\det \left(\int_{\delta_i} e^{\tau f} \omega_e \right) \not\equiv 0$$

(see Lemma 11.2). Analogously there exist l-forms $\phi_1, \ldots, \phi_\mu$ on X^g, for which

$$\det \left(\int_{\gamma_j} e^{\tau g} \phi_e \right) \not\equiv 0.$$

We can take as the forms $\{\psi_s\}$ the forms $\{\omega_i \wedge \phi_j\}$, see Fubini's Theorem 11.4. The lemma is proved.

In § 11.1 we constructed the isomorphism

$$H_{n-1}(X_t^f) \cong (X^f, X^{-,f})$$

for each $t \in S^-$. According to Lemma 13.15 this isomorphism induces an isomorphism

$$H_{n-1}(X_t^f) \otimes H_{l-1}(X_t^g) \cong H_{n+l-1}(X_t^{f+g})$$

for each $t \in S^-$. It is easy to see that this isomorphism can be extended to an isomorphism of the homologies of the fibration

$$f_* \otimes g_* \cong (f+g)_* \tag{14}$$

(here $*$ denotes the homological Milnor fibration). For this we need to consider not the pair $(X^f, X^{-,f})$ but the pairs

$$(X^f, X \cap f^{-1}(\{t \in S | \mathrm{Re}\,(e^{i\phi}t) < 0\}));$$

analogous changes must be made to the pairs

$$(X^g, X^{-,g}), \; (X^{f+g}, X^{-,f+g}).$$

Under the indicated isomorphism of fibrations, the tensor product of co-variantly constant sections is covariantly constant, that is the Gauss-Manin connection in the fibration $(f+g)_*$ isomorphic to

$$V^f \otimes id^g + id^f \otimes V^g$$

is the tensor product of the Gauss-Manin connections of the fibrations f_*, g_*.

If ω is a holomorphic differential n-form on X^f, ϕ is a holomorphic differential l-form on X^g, then $\omega \wedge \phi$ is a holomorphic differential $(n+l)$-form in a neighbourhood of the point

$$0 \times 0 \in \mathbb{C}^n \times \mathbb{C}^l.$$

Lemma 13.15 allows us to express the geometric sections of the form $\omega \wedge \phi$ in terms of the geometric sections of the forms ω, ϕ.

Lemma 13.16 (see [364]). If

$$s[\omega] = \sum t^\alpha (\ln t)^k A^\omega_{k,\alpha}/k!,$$

$$s[\psi] = \sum t^\beta (\ln t)^s A^\varphi_{s,\beta}/s!,$$

then

$$s[\omega \wedge \varphi] = \sum \sum \frac{\partial^{k+s}}{\partial \alpha^k \partial \beta^s} [t^{\alpha+\beta+1} B(\alpha+1, \beta+1)] A^\omega_{k,\alpha} \otimes A^\varphi_{s,\beta}/(k!s!),$$

where B is the beta function.

Remark. The section

$$A^{\omega}_{k,\alpha} \otimes A^{\eta}_{s,\beta}.$$

should be considered as a section of the fibration $(f+g)_*$ in view of the isomorphism (14).

Lemma 13.16 follows from Lemma 11.2 and Fubini's theorem.

Corollary. The order of the form $\omega \wedge \eta$ is equal to the sum of the orders of the forms ω, η plus 1.

Now for the proof of Theorem 13.7 it is sufficient to take advantage of Lemma 13.13. Namely, let $\omega_1, \ldots, \omega_\mu$ be holomorphic n-forms on X^f, which together with f satisfy the conditions of Lemma 13.13. Let $\phi_1, \ldots, \phi_\eta$ be holomorphic l-forms on X^g which together with g satisfy the conditions of Lemma 13.13. It is not hard to convince oneself that the forms

$$\{\omega_i \wedge \phi_j\}$$

together with $f+g$ also satisfy the conditions of Lemma 13.13. Consequently the orders of these forms (the numbers $\alpha(\omega_i) + \alpha(\phi_j) + 1$) make up the spectrum of the critical point of the germ $f+g$.

By carrying out the reasoning accurately we can prove the following strengthening of Theorem 13.7: if $\{(\alpha_i, v_i)\}$ are the spectral pairs of the germ f, and $\{(\beta_j, u_j)\}$ are the spectral pairs of the germ g, then

$$\{\alpha_i + \beta_j + 1, v_i + u_j + 1)\}$$

are the spectral pairs of the germ $f+g$.

In [364] there are given explicit formulae, relating, with the help of the operations of tensor product and sum, the weight and Hodge filtrations of the cohomological Milnor fibration of the critical points of the germs

$$f, g, f+g.$$

Chapter 14

The mixed Hodge structure
of an isolated critical point
# of a holomorphic function

A mixed Hodge structure in a vector space is two filtrations of the space, satisfying the axioms indicated below. In the space of cohomologies, vanishing at the critical point of the holomorphic function, there is a natural mixed Hodge structure. The role of the above-mentioned filtrations is played by the weight and Hodge filtrations, introduced in Chapter 13. The weight filtration is constructed from the Jordan structure of the monodromy operator and reflects the behaviour of integrals over vanishing cycles under analytic continuation of the integrals round critical values of the parameter. The Hodge filtration is constructed by starting from a comparison of the rates of convergence to zero of integrals over vanishing cycles as the parameter of the integrals converges to the critical value. As is well-known, there are in geometry two theories which study a function in a neighbourhood of a critical point: Morse theory and Picard-Lefschetz theory. Morse theory studies the reconstruction of a level hypersurface of the function as the level tends to the critical value. Picard-Lefschetz theory studies the transformation of a level hypersurface of a function as the level goes round the critical value in the complex plane. In this sense the theory of mixed Hodge structures of critical points is a synthesis of Morse theory and Picard-Lefschetz theory. The mixed Hodge structure in the vanishing cohomology plays an outstanding role in the local theory of singularities.

In this chapter we shall discuss the interactions of the mixed Hodge structure with other characteristics of the critical point.

14.1   The definition of a mixed Hodge structure

A mixed Hodge structure is an additional structure, existing in the cohomologies of complex manifolds and induced by the complex structure of the manifold.

Example. Let us consider a complex non-singular projective curve

$$X \subset \mathbb{C}P^2$$

of genus 1, in an affine chart given by the equation

$$y^2 = P_3(x),$$

where P_3 is a polynomial of degree three without multiple roots. Let us consider its cohomology with complex coefficients. The dimensions of the spaces

$$H^0(X, \mathbb{C}), \ H^1(X, \mathbb{C}), \ H^2(X, \mathbb{C})$$

are equal, respectively, to 1, 2, 1. In each of these spaces there is given a real subspace – the image of the natural inclusion of cohomology with real coefficients. Furthermore in the real subspace there can be chosen an integer lattice – the image of the natural inclusion of cohomology with integer coefficients. The real subspace with the lattice is always present in cohomologies with complex coefficients. The following object is a manifestation of the complex structure on X. Let us consider on X the differential 1-form

$$\omega = dx/y.$$

It is easy to convince oneself that, everywhere on X, ω is a regular holomorphic 1-form. The form ω is closed and, consequently, defines a cohomology class

$$[\omega] \in H^1(X, \mathbb{C}).$$

Any other holomorphic 1-form ω' is proportional to ω (indeed, ω'/ω is a bounded holomorphic function and, consequently, is a constant). The complex structure gives an orientation of the curve. For this orientation

$$i \int_X \omega \wedge \bar{\omega} > 0.$$

In particular this means that the class $[\omega]$ is different from zero. In this way the holomorphic differential 1-forms on X generate in $H^1(X, \mathbb{C})$ a one-dimensional subspace F, and this subspace possesses the property

$$H^1(X, \mathbb{C}) = F \oplus \bar{F},$$

where the bar denotes conjugation relative to the real subspace.

Standard theorems from the theory of elliptic curves assert (see [259, 328]): the vector space H^1 together with the indicated structures (the real subspace with the integral lattice and the subspace F) determine the curve X.

Let us suppose that we have two curves X, X' of genus 1 and a holomorphic map $f: X \to X'$. Let us consider the induced map

$$f^*: H^1(X', \mathbb{C}) \to H^1(X, \mathbb{C}).$$

It is clear that under this map the real subspace and the integer lattice map, respectively, into the real subspace and the integer lattice. Furthermore, since the preimage of a holomorphic form is holomorphic,

$$f^*F \subset F.$$

This remark shows that the structure described above in the cohomologies is functorial and carries non-trivial information about holomorphic maps. For example, if there is no non-zero linear map

$$(H^1(X', \mathbb{C}), F) \to (H^1(X, \mathbb{C}), F),$$

preserving the real subspace and the integer lattice, then any holomorphic map $X \to X'$ is a map onto a point.

The construction of the subspace F in the first cohomologies of a curve of genus 1 can be generalised to a complex non-singular projective manifold X of arbitrary dimension [71, 405]. For any non-negative integer l we can distinguish in the cohomology space $H^l(X, \mathbb{C})$, besides the real subspace with the integer lattice, subspaces

$$H^{l,0}, H^{l-1,1}, \ldots, H^{0,l},$$

where $H^{k,l-k}$ is the subspace of all lth cohomology classes, represented by closed differential forms which, when written down in terms of local coordinates

$$\sum a_{i1,\ldots,i_r,j_1,\ldots,j_{l-r}} dz_{i1} \wedge \ldots \wedge dz_{i_r} \wedge d\bar{z}_{j_1} \wedge \ldots \wedge d\bar{z}_{j_{l-r}}$$

have in each term exactly k holomorphic differentials and exactly $l-k$ antiholomorphic differentials, that is in each term $r=k$. Hodge's theorem (see [71, 405]) asserts that

$$H^l(X, \mathbb{C}) = \bigoplus_k H^{k,l-k}, \qquad \overline{H^{k,l-k}} = H^{l-k,k}. \tag{1}$$

As in the case of curves, the subspaces $H^{k,l-k}$ carry valuable information about the various characteristics of the manifold (see [71, 405, 140, 92, 187]). It is clear that the subspace $H^{k,l-k}$ is preserved by holomorphic maps of manifolds: if $f: X \to X'$ is a holomorphic map, then for any k

$$f^*(H^{k,l-k}) \subset H^{k,l-k}.$$

To study the dependence of the expansion (1) on the complex structure on X and also to generalise the construction to manifolds with singularities, it turns out that the right object of study is not the sequence of subspaces

$$H^{l,0}, \ldots, H^{0,l},$$

but rather the sequence of subspaces

$$\{0\} \subset F^l \subset F^{l-1} \subset \ldots \subset F^0 = H^l(X, \mathbb{C}),$$

where

$$F^k = H^{l,0} \oplus H^{l-1,1} \oplus \ldots \oplus H^{k,l-k}.$$

P. Deligne [92] distinguished in the cohomology of a quasiprojective algebraic variety (with any singularities) two natural filtrations: the Hodge $\{F^k\}$ and the weight $\{W_l\}$, and proved that these filtrations possess a property generalising property (1) and are functorial relative to algebraic maps of varieties. These filtrations are called by Deligne mixed Hodge structure in the cohomologies. The properties of these filtrations are taken as a basis of the formal definitions cited below.

Remark. In the examples cited above the manifold was non-singular and compact. For non-singular compact manifolds the weight filtration is trivial:

$$\{0\} = W_{l-1} \subset W_l = H^l(X, \mathbb{C}).$$

The mixed Hodge structure in vanishing cohomologies was defined by J. Steenbrink in [343].

14.1.1 Hodge structures (see [142])

Let $H_{\mathbb{R}}$ be a finite-dimensional vector space over $\mathbb{R}$, containing a lattice $H_{\mathbb{Z}}$, and let $H = H_{\mathbb{R}} \otimes_{\mathbb{R}} \mathbb{C}$ be its complexification.

Definition. A *Hodge structure of weight l* on H consists of a decomposition of H into a direct sum

$$H = \bigoplus_{k+m=l} H^{k,m},$$

such that $H^{k,m} = \overline{H^{m,k}}$ (the bar denotes conjugation). The numbers

$$h^{k,m} = \dim_{\mathbb{C}} H^{k,m}$$

are called the *Hodge numbers.*

For any two Hodge structures H, H' of weight l the direct sum $H \oplus H'$ carries an obvious Hodge structure of weight l. Analogously if H and H' have possibly distinct weights l and l', then

$$H \otimes H', \ \mathrm{Hom}\,(H, H'), \ \Lambda^p H, \ H^*$$

inherit Hodge structures of weights, respectively $l + l'$, $l - l'$, pl, $-l$.

Namely, $\lambda \in \mathrm{Hom}\,(H, H')$ has type (k, m) if, for all r, s

$$\lambda(H^{r,s}) \subset (H')^{k+r,m+s}.$$

In particular, this definition conforms with $H^* = \mathrm{Hom}\,(H, \mathbb{C})$ where $\mathbb{C}$ carries the trivial structure of weight 0, $H \otimes H'$ can be considered as

$$\mathrm{Hom}\,(H^*, H'),$$

and $\otimes^p H$ induces a Hodge structure on its subspace $\Lambda^p H$.

Definition. A linear map $\phi : H \to H'$ of vector space with Hodge structures is called a morphism of type (r, r) if it is defined over $\mathbb{Q}$ relative to the lattices $H_{\mathbb{Z}}$, $H'_{\mathbb{Z}}$ and if

$$\phi(H^{k,m}) \subset (H')^{k+r,m+r}$$

for all k, m.

Remark. A map ϕ is said to be defined over $\mathbb{Q}$ relative to the lattices $H_{\mathbb{Z}}$, $H'_{\mathbb{Z}}$ if the elements of its matrix, with respect to bases consisting of vectors of the lattices, are rational.

With each Hodge structure

$$H = \bigcup_{k+m=l} H^{k,m}$$

of weight l there is connected a *Hodge filtration*

$$\{0\} \subset \ldots \subset F^{k+1} \subset F^k \subset F^{k+1} \subset \ldots \subset H,$$

where

$$F^k = \bigoplus_{i \geq k} H^{i,l-i},$$

(see figure 83).

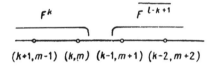

$$(k+1, m-1) \quad (k, m) \quad (k-1, m+1) \quad (k-2, m+2)$$

Fig. 83.

A Hodge filtration defines a Hodge structure:

$$H^{k,m} = F^k \cap \overline{F^m}. \tag{2}$$

Conversely, a decreasing filtration $\{F^k\}$ on H arises from some Hodge structure of weight l if and only if

$$F^k \oplus \overline{F^{l-k+1}} = H$$

for all k.

In terms of the new description, a linear map $\phi : H \to H'$, defined over $\mathbb{Q}$, is a morphism of type (r, r) if and only if it preserves the Hodge filtration with displacement of the indices by r:

$$\phi(F^k) \subset F'^{k+r}$$

for all k.

Let us consider a Hodge structure

$$H = \bigoplus_{k+m=l} H^{k,m}$$

and a bilinear form S on H. Let us suppose that the values of the form on pairs of vectors of the lattice $H_{\mathbb{Z}}$ are rational. Let us suppose also that the form is symmetric if l is even and antisymmetric if l is odd.

Definition. The Hodge structure is *polarised by the form S* if

$$S(H^{k,m}, H^{k',m'}) = 0,$$

for $(k,m) \neq (m', k')$, and

$$(\sqrt{-1})^{k-m} S(v, \bar{v}) > 0$$

for $v \in H^{k,m}$, $v \neq 0$.

An example of the polarisation of a Hodge structure is the Hodge bilinear form on the primitive cohomologies of a smooth projective variety (see [405, Chapter V]).

14.1.2 The mixed Hodge structure (see [142])

Let H, $H_{\mathbb{R}}$, $H_{\mathbb{Z}}$ be as in § 14.1.1.

Definition. A *mixed Hodge structure* on H consists of two filtrations:

$$\{0\} \subset \ldots \subset W_{l-1} \subset W_l \subset W_{l+1} \subset \ldots \subset H$$

is the *weight* filtration, defined over $\mathbb{Q}$ relative to the lattice $H_{\mathbb{Z}}$, and

$$\{0\} \subset \ldots \subset F^{k+1} \subset F^k \subset F^{k-1} \subset \ldots \subset H$$

is the *Hodge* filtration. It is required that for any l the filtration on

$$gr_l W = W_l / W_{l-1},$$

induced by the Hodge filtration makes up on $gr_l W$ a pure Hodge structure of weight l (the induced filtration is the filtration

$$F^k gr_l W = (F^k \cap W_l + W_{l-1})/W_{l-1}).$$

In other words it is required that for any $k, l \in \mathbb{Z}$

$$gr_l W = F^k gr_l W \oplus \overline{F^{l-k+1} gr_l W}.$$

Compare with the result of Theorem 13.3.

The concept of mixed Hodge structure contains the concept of Hodge structure as a particular case: if $(H, \{F^k\})$ is a Hodge structure of weight l, we can take as the weight fibration

$$\{0\} = W_{l-1} \subset W_l = H,$$

then $(H, \{F^k\}, \{W_m\})$ is a mixed Hodge structure.

Definition. A *morphism of type* (r, r) of mixed Hodge structures

$$(H, \{F^k\}, \{W_l\}), \quad (H', \{F'^k\}, \{W_l'\})$$

is a linear map

$$\phi: H \to H',$$

defined over $\mathbb{Q}$ relative to the lattices $H_{\mathbb{Z}}$, $H'_{\mathbb{Z}}$ and possessing the properties:

$$\phi(W_l) \subset W'_{l+2r}, \ \phi(F^k) \subset F'^{k+r}$$

for any l, k.

A morphism of type (r, r) induces a map

$$\phi: gr_l W \to gr_{l+2r} W',$$

which is a morphism of type (r, r) of pure Hodge structures of weights $l, l+2r$, respectively.

We can extend to the mixed Hodge structure in a natural way the operations of direct sum, tensor product and conjugation.

Example. If

$$(H, \{F^k\}, \{W_l\})$$

is a mixed Hodge structure, we can define a mixed Hodge structure on the conjugate space H^*. Put

$$H_{\mathbb{Z}}^* = \operatorname{Hom}(H_{\mathbb{Z}}, \mathbb{Z}),$$

$$(F^*)^k = \operatorname{ann} F^{1-k},$$

$$W_l^* = \operatorname{ann} W_{-l-1},$$

where ann is the annihilator.

14.2   Discussion of Theorem 13.3 about mixed Hodge structures

14.2.1   Examples

The critical point of the germ $f : (\mathbb{C}, 0) \to (\mathbb{C}, 0)$ of a holomorphic function of one variable

After a suitable change of variables,

$$f = x^{\mu+1},$$

where μ is the multiplicity of the critical point. Let $f : X \to S$ be a specialisation of the germ, $f^* : H^0 \to S'$ be the corresponding cohomological Milnor fibration (page 347). According to the examples on pages 366, 370, we have for any $t \in S'$

$$\{0\} = W_{-1,t} \subset W_{0,t} = \tilde{H}^0(X_t, \mathbb{C}),$$

$$\{0\} = F_t^1 \subset F_t^0 = \tilde{H}^0(X_t, \mathbb{C}).$$

Now $gr_0 W_t = \tilde{H}^0(X_t, \mathbb{C})$ is the unique non-trivial quotient space of the weight filtration. On $\tilde{H}^0(X_t, \mathbb{C})$ the Hodge filtration induces a pure Hodge structure of weight 0:

$$\tilde{H}^0(X_t, \mathbb{C}) = H_t^{0,0},$$

where $H_t^{0,0} = F_t^0 \cap \overline{F_t^0}$ (see formula (1)). Theorem 13.3 for f is proved.

The non-degenerate critical point of the germ $f = x_1^2 + \ldots + x_n^2$

According to the example on pages 366, 370 we have for any $t \in S'$

$$\{0\} = W_{2[n/2]-1,t} \subset W_{2[n/2],t} = H^{n-1}(X_t, \mathbb{C}),$$

$$\{0\} = F_t^{[n/2]+1} \subset F_t^{[n/2]} = H^{n-1}(X_t, \mathbb{C}).$$

Now $gr_{2[n/2]}W_t = H^{n-1}(X_t, \mathbb{C})$ is the unique non-trivial quotient space of the weight filtration. On $H^{n-1}(X_t, \mathbb{C})$ the Hodge filtration induces a pure Hodge structure of weight $2[n/2]$:

$$H^{n-1}(X_t, \mathbb{C}) = H_t^{[n/2],[n/2]},$$

where $H_t^{[n/2],[n/2]} = F_t^{[n/2]} \cap \overline{F_t^{[n/2]}}$ (see formula (2)). Theorem 13.3 for f is proved.

The finite-multiplicity critical point of a germ $f : (\mathbb{C}^2, 0) \rightarrow (\mathbb{C}, 0)$ of a holomorphic function of two variables

This is the first non-trivial case. Let us reformulate for this case the result of the theorem on mixed Hodge structures.

Let $f : X \rightarrow S$ be a specialisation of the germ f, and let $t \in S'$. According to Lemma 13.7 the Hodge filtration has the form

$$\{0\} \subset F_t^1 \subset F_t^0 \subset \ldots \subset H^1(X_t, \mathbb{C}).$$

According to Corollary 2 in § 13.2.5, the dimensions of the Jordan blocks of the monodromy operator are not greater than 2, and the root subspace of the eigenvalue 1 of the monodromy operator consists of eigenvectors. Therefore the weight filtration has the form

$$\{0\} \subset W_{0,t} \subset W_{1,t} \subset W_{2,t} = H^1(X_t, \mathbb{C}),$$

where $W_{1,t}$ is generated by all the eigenvectors of the monodromy operator with eigenvalues not equal to 1, and $W_{0,t}$ is generated by the eigenvectors of the Jordan blocks of dimension 2.

The theorem on mixed Hodge structures asserts that the Hodge filtration induces on the quotient spaces of the weight filtration pure Hodge structures, that is in the given case

$$gr_2 W_t = H_t^{1,1}, \tag{3}$$

$$gr_1 W_t = H_t^{1,0} \oplus H_t^{0,1},$$
$$gr_0 W_t = H_t^{1,-1} \oplus H_t^{0,0} \oplus H_t^{-1,1}, \tag{4}$$

where

$$H^{k,m} = F^k gr_{k+m} W \cap \overline{F^m gr_{k+m} W}.$$

According to Corollary 3 in § 13.2.5, $H_t^{1,-1}$ and $H_t^{-1,1}$ are empty and

$$gr_0 W_t = H_t^{0,0}. \tag{5}$$

Corollary. We have $F_t^0 = H^1(X_t, \mathbb{C})$. In other words an arbitrary class of vanishing cohomologies, belonging to one of the root subspaces of the monodromy operator is the value of the principal part of a holomorphic 2-form on X, the order of which is less than 1.

Remark. For germs of functions in n variables $F_t^0 = H^{n-1}(X_t, \mathbb{C})$. The proof is analogous.

The results (3)–(5) can also be reformulated in the language of principal parts of holomorphic 2-forms on X.

(i) Each eigenvector in $W_{0,t}$ is the value of the principal part of some form whose order belongs to the interval $(0, 1)$.

(ii) If the degree of the form is not greater than 0, then the value at t of the principal part of the form does not belong to $W_{0,t}$.

(iii) Each eigenvector associated with the eigenvalue 1 is the value of the principal part of some form of degree 0.

(iv) Each eigenvector from $W_{0,t}$ is the value of a coefficient $A_{1,\alpha(\omega)}^\omega$, where ω is some holomorphic 2-form on X, the order $\alpha(\omega)$ of which belongs to the interval $(-1, 0)$.

(v) Let us denote by $\mathscr{F}_t^1$ the projection into $gr_1 W_t$ of the subspace in $W_{1,t}$ generated by the values at t of the principal parts of forms of degree less than 0. Then

$$gr_1 W_t = \mathscr{F}_t^1 \oplus \overline{\mathscr{F}_t^1}.$$

From (i)–(v), in particular, it is easy to deduce that any elementary section of non-negative order is a geometric section of some 2-form. For example, if

$$A : t \mapsto A(t)$$

is a covariantly constant single-valued section of the cohomological Milnor fibration, then there exists a 2-form ω, for which

$$[\omega/df|_{X_t}] = A(t) \quad \text{for all} \quad t.$$

14.2.2 The symmetry of the Hodge numbers

Let us deduce Lemma 13.14 from Theorem 13.3. According to Theorem 13.3 we have for any l

$$gr_l W_t = \bigoplus_{k+m=l} H_t^{k,m}, \qquad H^{k,m} = \overline{H_t^{m,k}}. \tag{6}$$

The monodromy operator on $gr_l W_t$ preserves the real structure and the induced Hodge filtration. Therefore the monodromy operator preserves the decomposition (6). Let us denote by $H_\lambda^{k,m}$ the root subspace of the eigenvalue λ of the action of the monodromy on $H^{k,m}$.

By construction

$$F^k gr_l W_t = H^{k,l-k} \oplus H^{k+1,l-k-1} \oplus \dots.$$

Therefore

$$H^{k,l-k} \cong F^k gr_l W_t / F^{k+1} gr_l W_t,$$

$$\dim H^{k,l-k} = h^{k,l-k}, \qquad \dim H_\lambda^{k,l-k} = h_\lambda^{k,l-k}.$$

Now assertion (i) of Lemma 13.14 follows from (6).

According to corollary 2 in § 13.2.5 we have for any k, l

$$N^l (H_\lambda^{k,n-1+l-k}) = H_\lambda^{k-l,n-1-k},$$

if $\lambda \neq 1$, and

$$N^l (H_1^{k,n+l-k}) = H^{k-l,n-k}.$$

This establishes assertions (ii) and (iii) of Lemma 13.14.

14.2.3 Functoriality of the mixed Hodge structure in vanishing cohomologies

Let $f : (\mathbb{C}^n, 0) \to (\mathbb{C}, 0)$ be the germ of a holomorphic function with a critical point of finite multiplicity. Let $g : (\mathbb{C}^n, 0) \to (\mathbb{C}^n, 0)$ be the germ of a holomorphic map of finite multiplicity. Let us suppose that the germ

$$f \circ g : (\mathbb{C}^n, 0) \to (\mathbb{C}, 0)$$

also has a critical point of finite multiplicity. The germ g induces a map from a level hypersurface of the germ $f \circ g$ to a hypersurface of the same level of the germ f. It is not hard to convince oneself that this map gives a linear map g^* from the cohomologies, vanishing at the critical point of the germ f, into the cohomologies, vanishing at the critical point of the germ $f \circ g$, or more precisely gives a morphism from the cohomological Milnor fibration of the critical point of the germ f to the cohomological Milnor fibration of the critical point of the germ $f \circ g$. According to Theorem 13.3, in the fibres of these cohomological fibrations the weight and Hodge filtrations form a mixed Hodge structure.

Theorem 14.1 (see [364]). The map g^* is a morphism of type $(0, 0)$ of mixed Hodge structures. Namely, g^* has zero kernel and for any k, l

$$g^*(F^k(f^*)) \subset F^k((f \circ g)^*),$$

$$g^*(W_l(f^*)) \subset W_l((f \circ g)^*).$$

Proof. Let X_t^f, X_t^{fg} be the fibres of the Milnor fibrations of the germs f, $f \circ g$, respectively. Let us suppose that the specialisations of the germs and the representation g of the germ g are chosen so that

$$g(X_t^{fg}) \subset X_t^f.$$

The map

$$g_* : H_{n-1}(X_t^{fg}, \mathbb{C}) \to H_{n-1}(X_t^f, \mathbb{C})$$

is an epimorphism. Indeed if $[\sigma] \in H_{n-1}(X_t^f, \mathbb{C})$, σ is a representative of the class $[\sigma]$, and $\{g^{-1}(\sigma)\}$ is the full preimage of the cycle σ in X_t^{fg}, then

$$g_*[\{g^{-1}(\sigma)\}] = k [\sigma],$$

where k is the multiplicity of the germ g.

Let us prove the weight filtration is functorial. It is clear that the map g^* maps covariantly constant sections into covariantly constant sections. In addition the map g^* commutes with the monodromy operator. According to the second definition of the weighted filtration in § 13.2.2, this implies the second inclusion of the theorem.

Let us prove that the Hodge filtration is functorial. Let ω be a holomorphic differential n-form on X^f. It is clear that the order of the form $g^*(\omega)$ is equl to the order of ω. Furthermore the principal part of the form $g^*(\omega)$ is equal to the image of the principal part of the form relative to the monomorphism g^*. This proves the first inclusion theorem.

Remark. Other manifestations of the functoriality of the mixed Hodge structure in vanishing cohomologies are formulae relating the spectra of the critical points of the germs $f(x)$, $g(y)$, $f(x)+g(y)$. See § 13.3.5 and also [364].

14.2.4 Reformulation of the theorem on mixed Hodge structures in the language of complex oscillatory integrals

Let $f:(\mathbb{C}^n,0)\to(\mathbb{C},0)$ be the germ of a holomorphic function at a critical point of finite multiplicity. Let us consider complex oscillatory integrals with phase f on admissible chains, concentrated in a neighbourhood of the critical point of the germ f, that is integrals of the form

$$\int_{[\Gamma]} e^{\tau f}\omega,$$

where $[\Gamma]\in H_n(X,X^-)$, and ω is a holomorphic differential n-form on X (see § 11.1).

For fixed amplitude ω integration of the expression $e^{\tau f}\omega$ over admissible chains defines a linear function on admissible chains depending on the parameter τ. As $\tau\to+\infty$ such a linear function can be expanded in an asymptotic series (Theorem 11.1). The principal part of this asymptotic series is called the principal part of the amplitude ω. Below, using all the principal parts of all the amplitudes we shall construct a filtration on the space conjugate to $H_n(X,X^-;\mathbb{C})$. This filtration is called the Hodge filtration. Further, on the same space, with the help of the monodromy operator, we shall construct another filtration, called the weight filtration.

Theorem 14.2. The weight and Hodge filtrations constructed below form a mixed Hodge structure on the space conjugate to

$$H_n(X, X^-; \mathbb{C}).$$

Further, it will be easy to see that Theorem 14.2 is a reformulation of Theorem 13.3 in view of Lemma 11.2.

Remark. The Hodge filtration constructed below depends on the parameter τ. Theorem 14.2 is true for any positive value of the parameter of the Hodge filtration.

Let us denote by H^* the space conjugate to $H_{n-1}(X, X^-; \mathbb{C})$. In H^* there is a natural real subspace $H_{\mathbb{R}}^*$ and an integer lattice $H_{\mathbb{Z}}^*$ (for example, $H_{\mathbb{Z}}^*$ consists of linear functions taking integer values on the natural image in $H_{n-1}(X, X^-; \mathbb{C})$ of the group $H_{n-1}(X, X^-; \mathbb{Z})$). Let us define a Hodge filtration in H^*.

Let ω be a holomorphic n-form on X. According to Theorem 11.1

$$\int_{[\,]} e^{\tau f} \omega \approx \Sigma \tau^\alpha (\ln \tau)^k B_{\alpha,k}^\omega [\;],$$

where $B_{k,\alpha}^\omega \in H^*$. Let us call the *weight* of the form ω the largest number α for which the coefficient $B_{0,\alpha}^\omega$ is different from zero (compare with the definition of the order of a form in §13.1.3). Let us denote the weight by $\beta(\omega)$.

Remark. According to formula (6) on page 303 the sum of the order of a form and its weight is equal to -1.

The *main part* of a form ω is the expression

$$\tau^{\beta(\omega)} (B_{0,\beta(\omega)}^\omega + \ldots + (\ln \tau)^{n-1} B_{n-1,\beta(\omega)}^\omega).$$

The main part is a vector of the space H^* depending on the parameter τ.

Let us fix a positive number τ. We define the subspace $F_\tau^k \subset H^*$ by the property: F_τ^k is the linear span of the main parts of all the forms of weight not less than $k - n$ (in the principal parts the parameter τ is fixed). If a main part of such a weight does not exist then we put

$$F_\tau^k = \{0\}.$$

Let us call the filtration $\{F_\tau^k\}$, $k \in \mathbb{Z}$, the *Hodge filtration.*

Let us define the weight filtration in H^*. As shown in § 11.1,

$$H_n(X, X^-; \mathbb{C}) \cong H_{n-1}(X_t, \mathbb{C}),$$

where $t \in S^-$. Therefore $H^* \cong H^{n-1}(X_t, \mathbb{C})$. The weight filtration in $H^{n-1}(X_t, \mathbb{C})$ was defined in § 13.2. Let us take as the *weight filtration* in H^* the filtration induced from the weight filtration in $H^{n-1}(X_t, \mathbb{C})$.

14.3 Survey of results on the mixed Hodge structure

14.3.1 Mixed Hodge structure and intersection form

Let $f : (\mathbb{C}^n, 0) \rightarrow (\mathbb{C}, 0)$ be the germ of a holomorphic function at a critical point of multiplicity μ. Let us consider the intersection form S on the $(n-1)$st homology $H_{n-1}(X_t, \mathbb{R})$, vanishing at the critical point. Let us denote by μ_0 the dimension of the kernel of the form S. If n is even, then the form S is antisymmetric and μ_0 is the unique real invariant of the form S. If n is odd then the form S is symmetric and by a real linear transformation the form S can be diagonalised. Let μ_+ and μ_- be the numbers of positive and negative coefficients of the diagonalisation. The numbers μ_0, μ_+, μ_- form a complete set of real invariants of the form S.

Let us denote by $h_\lambda^{k,m}$ the Hodge numbers of the mixed Hodge structure in the cohomologies vanishing at the critical point of the germ f.

Theorem 14.3 (see [343]). Using the above notation

$$\mu_0 = \sum_{k+m \leqslant n} h_1^{k,m} - \sum_{k+m \geqslant n+2} h_1^{k,m}.$$

If n is odd, then

$$\mu_+ = \sum_{\substack{k+m=n+1 \\ m \text{ even}}} h_1^{k,m} + 2 \sum_{\substack{k+m \geqslant n+2 \\ m \text{ even}}} h_1^{k,m} + \sum_{\lambda \neq 1} \sum_{m \text{ even}} h_\lambda^{k,m},$$

$$\mu_- = \sum_{\substack{k+m=n+1 \\ m \text{ odd}}} h_1^{k,m} + 2 \sum_{\substack{k+m \geqslant n+2 \\ m \text{ odd}}} h_1^{k,m} + \sum_{\lambda \neq 1} \sum_{m \text{ odd}} h_\lambda^{k,m}.$$

Corollary 1. The form S is non-degenerate if and only if the number 1 is not an eigenvalue of the monodromy operator.

Corollary 2 (see [343]).

> If n is even then $\mu - \mu_0$ is even.
> If $n \equiv 3 \bmod 4$ then $\mu - \mu_-$ is even.
> If $n \equiv 1 \bmod 4$ then $\mu - \mu_+$ is even.

If f is the germ of a quasihomogeneous function, then the Hodge numbers $h_\lambda^{k,m}$ can be expressed in terms of the quasihomogeneous structure of the local algebra of the critical point (see Theorems 13.4, 13.5). Let us formulate Theorem 14.3 in this case.

Let $f : (\mathbb{C}^n, 0) \to (\mathbb{C}, 0)$ be the germ of a quasihomogeneous function of type $(\alpha_1, \ldots, \alpha_n)$ and weight 1. Let us suppose that 0 is a critical point of finite multiplicity of the germ f. Let $\{x^m | m \in I\}$ be a set of monomials, projecting into a basis over $\mathbb{C}$ of the local algebra $\mathbb{C}\{x\}/(\partial f/\partial x)$. For $m \in I$ put

$$l(m) = (m_1 + 1)\alpha_1 + \ldots + (m_n + 1)\alpha_n - 1.$$

Theorem 14.4 (see [341]). Using the above notation,

$$\mu_0 = \#\{m \in I | l(m) \in \mathbb{Z}\}.$$

If n is odd, then

$$\mu_+ = \#\{m \in I | l(m) \notin \mathbb{Z}, \ [l(m)] \text{ is odd}\},$$

$$\mu_- = \#\{m \in I | l(m) \notin \mathbb{Z}, \ [l(m)] \text{ is even}\}.$$

Example 1. The singularity $A_\mu : f = x_1^{\mu+1} + x_2^2 + x_3^2$,

$$\alpha = (1/(\mu+1), \tfrac{1}{2}, \tfrac{1}{2}),$$

$$I = \{(m_1, 0, 0) | m_1 = 0, \ldots, \mu - 1\},$$

$$l(m) = (m_1 + 1)/(\mu + 1).$$

For any $m \in I$ we have $l(m) \in (0, 1)$. The intersection form is negative definite.

Example 2. Let $f = x_1^{a_1} + \ldots + x_n^{a_n}$. The intersection form is non-degenerate if the numbers

$$a_1, \ldots, a_n$$

are pairwise coprime.

Let f be the germ of a quasihomogeneous function. Let us formulate a theorem relating the intersection form in this case with the operation of multiplication in the local algebra of the critical point of the germ.

Let X_t be a fibre of the Milnor fibration of the critical point of the germ f of a quasihomogeneous function. Let us consider the Poincaré homomorphism

$$\pi : H_{n-1}(X_t, \mathbb{C}) \to H^{n-1}(X_t, \mathbb{C}).$$

It is not hard to convince oneself that the image of the homomorphism is

$$\oplus_{\lambda \neq 1} H^{n-1}(X_t, \mathbb{C})_\lambda,$$

where the index λ denotes the root subspace of the eigenvalue λ of the monodromy operator. Let us define on the image of the homomorphism the form S^* by the formula

$$S^*(\cdot, \cdot) = S(\pi^{-1}(\cdot), \pi^{-1}(\cdot)).$$

For any $m \in I$ put

$$\omega_m = x^m dx_1 \wedge \ldots \wedge dx_n.$$

According to Theorem 13.6 the geometric sections of the forms ω_m, $m \in I$, make up a basis of the sections of the cohomological Milnor fibration. If $l(m) \notin \mathbb{Z}$, then the values of the geometric section of the form ω_m belong to the image of the Poincaré homomorphism. The form ω is said to be *primitive* if $l(m) \notin \mathbb{Z}$.

Let us denote by J the class of the Hessian det $(\partial^2 f / \partial x_i \partial x_j)$ in the local algebra

$$Q = \mathbb{C}\{x\}/(\partial f / \partial x).$$

A linear functional $\alpha : Q \to \mathbb{C}$ is said to be *admissible* if $\alpha(J) \neq 0$ and α is quasihomogeneous (that is equal to zero on elements of the algebra Q, the quasihomogeneous degrees of which are different from the degree of the element J).

For $m \in I$ put

$$R(m) = l(m)(l(m) - 1) \ldots (l(m) - [(n-2)/2]), \quad \text{if} \quad n \geq 2,$$

$$R(m) = 1, \quad \text{if} \quad n = 1.$$

Theorem 14.5 (see [126] and also [377]). The values of the geometric sections of two primitive forms ω_m, $\omega_{m'}$ are orthogonal relative to the form S^* if the sum of the orders of the forms (that is the number $l(m) + l(m')$) is not an integer or if this sum is less than $n - 2$. There exists a linear functional α on the local algebra Q, with the property: for any two primitive forms ω_m, $\omega_{m'}$, the sum of the orders of which is $n - 2$,

$$S^*(s[\omega_m], s[\omega'_m]) = \text{const} \cdot \alpha(x^m \cdot x^{m'}) t^{n-2} / (R(m) \cdot R(m')),$$

where $\text{const} = 1$ if n is odd, and $\text{const} = l(m) - l(m')$ if n is even; $s[\omega]$ is a geometric section of the form ω; t is a coordinate in the base of the Milnor fibration. For an explicit formula for the functional α see [377].

Example. $A_\mu : f = x_1^{\mu+1} + x_2^2 + x_3^2$. The forms

$$\omega_m = x_1^m dx_1 \wedge dx_2 \wedge dx_3, \quad m = 0, 1, \ldots, \mu - 1$$

are primitive, and

$$S^*(s[\omega_m], s[\omega_{m'}]) = \begin{cases} -4\pi^2 t(\mu+1)/(m+1)(\mu-m) & \text{of } m+m' = \mu - 1 \\ 0 & \text{otherwise}. \end{cases}$$

The relation between the local residue and the intersection form is discussed in [377].

14.3.2 The mixed Hodge structure and deformations

Let us suppose that a critical point of a holomorphic function can, by deformation of the function, be decomposed into several simpler critical points.

Problem. How is the mixed Hodge structure of the initial critical point related to those of the critical points obtained by the decomposition?

Certainly, there are "conservation laws", formulated in terms of the mixed Hodge structures, for decompositions of critical points. Numerous examples prompt the following conjecture.

Let us order the spectrum of the critical point:

$$\alpha_1 \leqslant \alpha_2 \leqslant \ldots \leqslant \alpha_\mu.$$

Conjecture (V. I. Arnold [19]). The spectrum is semicontinuous in the following sense: if a critical point P adjoins a (simpler) critical point P' (with $\mu' < \mu$), then

$$\alpha_k \leqslant \alpha_k'.$$

Remarks.

(1) Even in simple and explicitly calculated cases, such as the quasihomogeneous case or the case of the critical point of a function of two variables whose Taylor series has non-degenerate principal part, this conjecture is a non-trivial arithmetical result about integer points inside convex polyhedra.

(2) V. V. Goryunov [131] verified the conjecture for simple critical points adjoining simple, for unimodal critical points adjoining unimodal, and for bimodal critical points of corank 2 adjoining each other, see [131] and the table of spectra on page 389.

(3) The symmetry of the spectrum about the point $n/2 - 1$ proves the conjecture for the case in which the critical point P' is non-degenerate.

(4) From the symmetry of the spectrum about to the point $n/2 - 1$ and the conjecture, there follow the two-sided inequalities

$$\alpha_k \leqslant \alpha_k' \leqslant \alpha_{k + (\mu - \mu')}.$$

For example if we split off from a compound critical point one non-degenerate point ($\mu = \mu' + 1$) then the spectrum of the point P' alternates with the spectrum of the point P.

The relation between the spectra of the points P, P' are the same as between the semiaxes of ellipsoids in $\mathbb{R}^\mu$ and the semiaxes of its sections by subspaces $\mathbb{R}^{\mu'}$.

(5) The conjecture involves the semicontinuity of the dimensions of the spaces of the Hodge filtration, namely the semicontinuity of the numbers

$$h^r = \sum_{r \leqslant k} \sum_m h^{k,m},$$

$$h_r = \sum_{r > k} \sum_m h^{k,m}.$$

(6) In particular, for critical points of functions of two variables, these semicontinuities lead to the semicontinuity of the genus g of a fibre of the Milnor fibration and the semicontinuity of the "cogenus" $\mu - g$ (in this case the fibre is a Riemann surface of Euler characteristic $1 - \mu$ with $\mu + 1 - 2g$ "holes"). The semicontinuity of both numbers is obvious (the semicontinuity of the "cogenus" follows from the fact that the inclusion of the homologies, vanishing at the simpler point, into the homologies, vanishing at the more complicated critical point, is a monomorphism).

(7) The conjecture as formulated here is an amendment and generalisation of a conjecture about the semicontinuity of the oscillation index of a critical point of a real analytic function (see §§ 6.6, 9.2, 13.1.4, 13.3.3, [12, 13, 14]).

Let us consider a deformation of the initial critical point P. Let us suppose that in the process of deformation the critical point does not decompose, that is for each value of the parameter of the deformation there is exactly one critical point of multiplicity μ.

Theorem 14.5 (see [361, 365]). For such deformations the spectrum is constant.

Remarks.

(1) The statement of the theorem is a variant of the statement of Arnold's conjecture for the case $\mu = \mu'$.

(2) In [365] it was proved that the subspaces of the weighted and Hodge filtrations change holomorphically as the parameter of the deformation changes.

(3) From Theorem 14.6 it follows that the smallest possible order for the integral of a holomorphic form over the classes of a covariantly constant family of homologies, vanishing at the critical point (that is the first spectral number) does not change under the indicated deformations. It means that if for some deformation of the critical point the smallest possible order changes, then the multiplicity of the critical point is not preserved under the deformation (that is the critical point decomposes). From this reasoning we extract Theorem 14.7, formulated below.

Let us consider the germ of a holomorphic function $f : (\mathbb{C}^n, 0) \to (\mathbb{C}, 0)$ at a critical point of multiplicity μ and a deformation

$$F : (\mathbb{C}^n \times \mathbb{C}^k, 0 \times 0) \to (\mathbb{C}, 0)$$

of it. The $\mu = \text{const}$ *stratum* of the deformation is the germ of the set

$$(\Lambda, 0) \subset (\mathbb{C}^k, 0)$$

consisting of all values of the parameter λ for which the function $F(\cdot, \lambda)$ has a critical point of multiplicity μ with critical value zero.

Theorem 14.7 (see [368], see also [381]). The codimension of the $\mu = \text{const}$ stratum in the base of a versal deformation of the germ of a holomorphic function at a critical point of finite multiplicity is not less than the number of those spectral numbers of the mixed Hodge structure of the critical point of the germ which are less than $\alpha_1 + 1$ (where α_1 is the first spectral number).

Remark. In the case of a critical point of a semiquasihomogeneous function the codimension of the $\mu = \text{const}$ stratum in the base of a versal deformation is equal to the number of spectral numbers, indicated in Theorem 14.7. An estimate from above follows from [196], an estimate from below is given in Theorem 14.7, and the numbers, estimating the codimensions are equal. Furthermore, in the case of a critical point of a quasihomogeneous function, the reasoning mentioned in connection with Theorem 14.7 allows us explicitly to indicate the $\mu = \text{const}$ stratum.

Theorem 14.8 (see [368]). Let $f : (\mathbb{C}^n, 0) \to (\mathbb{C}, 0)$ be a quasihomogeneous germ of type $(\alpha_1, \ldots, \alpha_n)$ and weight 1. Let $\{x^m | m \in I\}$ be a set of monomials, projecting into a basis over $\mathbb{C}$ of the local algebra $\mathbb{C}\{x\}/(\partial f/\partial x)$. Let us consider a representative

$$F(x, \lambda) = f(x) + \sum_{m \in I} \lambda_m x^m$$

of a versal deformation of the germ f. Then the $\mu = \text{const}$ stratum is given by the equations

$$\{\lambda_m = 0 | m \in I, \ (\alpha, m) < 1\}.$$

For homogeneous germs Theorem 14.8 was proved in [120].

Recently Arnold's conjecture has been proved.

Let $f : (\mathbb{C}^n, 0) \to (\mathbb{C}, 0)$ be the germ of a holomorphic function at an isolated critical point, and

$$F : (\mathbb{C}^n \times \mathbb{C}^l, 0 \times 0) \to (\mathbb{C}, 0)$$

be a deformation of it.

Definition [371]. A subset $U \subset \mathbb{R}$ is called a *set of semicontinuity* for the given deformation F if it possesses the following property: for any sufficiently small $\lambda \in \mathbb{C}^l$ let $x^1, \ldots, x^s \in \mathbb{C}^n$ be critical points of the function $F(\cdot, \lambda)$, with a common critical value, then the number of spectral numbers of the initial critical point of the germ f falling in U is not less than the sum of the numbers of spectral numbers, falling in U, of the critical points $x^1, \ldots, x^s$.

Conjecture on the semicontinuity of the density of spectra (see [372]). For any deformation any interval $(\alpha, \alpha + 1)$, where $\alpha \in \mathbb{R}$ is a set of semicontinuity.

Remark. It is possible that in the formulation of the conjecture we need to change the interval $(\alpha, \alpha + 1)$ to the half-open interval $(\alpha, \alpha + 1]$.

It is clear that Arnold's conjecture follows from this conjecture.

Results about semicontinuity.

(I) The conjecture on the semicontinuity of the density of spectra is true for deformations of germs of functions of one or two variables (see [374, 375]).

(II) Any interval $(\alpha, \alpha + 1)$ for $\alpha \in (-2, -1)$ and the half-open interval $(-1, 0]$ are sets of semicontinuity for deformations of germs of functions of any number of variables (see [374, 375]).

(III) For any irrational $\alpha \in \mathbb{R}$ the set

$$\bigcup_{k \in \mathbb{Z}} (\alpha + 2k, \alpha + 2k + 1)$$

is a set of semicontinuity for deformations of germs of functions of any number of variables (see [371]).

(IV) Let $f(x_1, \ldots, x_n)$ be a quasihomogeneous polynomial of type $(\alpha_1, \ldots, \alpha_n)$ and weight 1, with an isolated critical point at the origin. Let us call a *lower deformation* of it any polynomial

$$F(x, \lambda) = f(x) + \Sigma \lambda_j \phi_j,$$

where $\{\phi_j\}$ are monomials of quasihomogeneous weight less than 1. Then for lower deformations of a polynomial the conjecture on the semicontinuity of the density of spectra is true (see [372]).

Corollaries.

(1) The complex oscillation index of a critical point of a germ of a function of one, two or three variables is upper semicontinuous for deformations (see [373]).
(2) Let us call the critical point of the germ of a function of any number of variables *sufficiently degenerate* if its complex oscillation index belongs to $(-1, 0]$. Then the complex oscillation index of a sufficiently degenerate critical point is upper semicontinuous for deformations of the germ (see [373]).

For criteria for sufficient degeneracy see [373] and in § 13.1.7.
Finally Arnold's conjecture was proved in [345].
(V) Any interval $(\alpha, \alpha + 1]$ is a set of semicontinuity (see [345]).

Section (IV) of the results gives a new result in the following question. Let $Y \subset \mathbb{C}P^n$ be an algebraic hypersurface of degree d, with only non-degenerate (simple, double) singular points.

What is the maximum number $N_n(d)$ of non-degenerate critical points which a hypersurface of degree d can have?

The complete answer to this question is known only for $n = 1, 2$: for $n = 1$ $N_1(d) = [d/2]$, for $n = 2$ $N_2(d) = d(d-1)/2$. The maximum is attained on a curve wich is a union of lines in general position. The first non-trivial case is $n = 3$.

Estimates from above.

The first result is the result of *A*. Basset (1906, [42]):

$$N_3(d) \leqslant (d(d-1)^2 - 5 - \sqrt{d(d-1)(3d-14)+25})/2,$$

with the right hand side asymptotic to $d^3/2$, as $d \to \infty$. In subsequent works (see [43, 59, 127, 340]) the estimate was improved and generalised to the case $n > 3$, but in all these works the estimating number had asymptotic $d^n/2$ as $d \to \infty$. An estimate with a new asymptotic is given in section (IV) of the previous result.

Let us call the *Arnold number* $A_n(d)$ the number of integer points strictly inside the cube $(0, d)^n$, for which

$$(n-2)d^n/2 + 1 < \Sigma k_i \leqslant nd/2.$$

For example, for $n = 3$,

$$A_3(d) = 23d^3/48 + (\text{terms of lesser degree in } d).$$

Result about estimate from above (see [372]).

Let $Y \subset \mathbb{C}P^n$ be an algebraic hypersurface of degree d, with only isolated singular points. Then the number of its nondegenerate singular points is not greater than the number $A_n(d)$. If $n = 3$, then the number of all the singular points is not greater than $A_3(d)$.

For fixed n the number $A_n(d)$ has the form

$$a_n d^n + \text{(terms of lower degree in } d).$$

It is not hard to prove that $a_n \sim \sqrt{(6/\pi n)}$ as $n \to \infty$.

In the case of a surface in $\mathbb{C}P^3$, Miyaoka has recently proved the best known upper bound on the number of non-degenerate singular point. The estimate is asymptotically $4d^3/9$.

Theorem (see [257]). Let $Y \subset \mathbb{C}P^3$ be a surface of degree d having only ordinary double points (that is, only points of type $A_\mu, D_\mu, E_6, E_7, E_8$), then the number of singularities is not greater than $\dfrac{4}{9} d(d-1)^2$.

We give a table of bounds for small d and $n = 3$; the list is basically from [257].

d	$N_3(d)$	$A_3(d)$	Miyaoka	Basset	Stagnaro	Bruce
4	16	16	16	16	16	17
6	31	31	36	34	32	32
6	$\geqslant 64$	68	66	66	64	73
7	$\geqslant 90$	104	112	114	111	108
8	$\geqslant 160$	180	174	224	178	193
9	$\geqslant 192$	246	256	270	267	256
10	$\geqslant 325$	375	360	384	380	401
11	$\geqslant 375$	480	488	535	521	500
12	$\geqslant 576$	676	645	696	693	721

In Stagnaro's bound there is a hypothesis of general position of the singular points on the surface. Thus the equality $N_3(6) = 64$ is not proved. Recently an announcement of Stagnaro has appeared stating that $N_3(6) = 66$. The equality $N_3(4) = 16$ is due to Kummer (1864), the equality $N_3(5) = 31$ to Beauville (1980,

[43]). The inequality $N_3(6) \geqslant 64$ is due to Catanese and Ceresa [68] and Stagnaro [340]; $N_3(7) \geqslant 90$ to Stagnaro [340]; $N_3(8) \geqslant 160$ to Kreiss [197] and Gallarati [121]; $N_3(9) \geqslant 192$, $N_3(11) \geqslant 375$ to Chmutov; $N_3(10) \geqslant 325$, $N_3(12) \geqslant 576$ to Kreiss [197].

Estimates from below.

S. V. Chmutov suggested a method which gives, apparently, the best possible estimate from below of the number $N_n(d)$ for large d. Chmutov suggested that as a hypersurface with a large number of singular points we should consider the hypersurface with affine equation

$$\sum_{j=1}^{n} T_d(x_j) = 0,$$

if n is even, and

$$\sum_{j=1}^{n} T_d(x_j) = 1,$$

if n is odd, where T_d is the Chebyshev polynomial of degree d, with two critical values ± 1. The number $C_n(d)$ of singular points of Chmutov's hypersurface has the form

$$c_n d^n + (\text{terms of lower degree in } d).$$

For example $c_3 = 3/8$. As $n \to \infty$, $c_n \sim \sqrt{(2/\pi n)}$.

The case $d = 3$. As $n \to \infty$, $A_n(3) \sim 2^n \sqrt{8/n\pi}$ (A. B. Givental). Using the idea of Chmutov, Givental constructed examples of cubic hypersurfaces having a large number of singularities. Let $G(x, y)$ be a polynomial of degree 3 having two critical values ± 1, such that for one critical value there are three critical points and at the other there is only one. Then the number of singular points of the cubic hypersurface in n variables (n even) with affine equation

$$\sum_{j=1}^{n/2} (-1)^j G(x_j, y_j) = 0$$

has asymptotically $g_n \sim 2^n \sqrt{16/3\pi n}$ as $n \to \infty$.

For $d=3$ and small values of n, the upper bounds and Givental's examples are given in the following table.

n	2	3	4	5	6	7	8	9
$A_n(3)$	3	4	10	15	35	56	126	196
Givental	3	4	10	15	33	54	118	189

The equality $N_5(3)=15$ is due to Togliatti (1936, [354]).

In conclusion we note another consequence of the result on the upper bound.

Asseertion (see [375]). Let $Y \subset \mathbb{CP}^2$ be an algebraic curve of degree d, with only isolated singularities. Let n_1 be the number of non-degenerate singularities, n_2 the number of cusps, and $B(d)$ the number of integral points (k_1, k_2) strictly inside the square $(0, d)^2$ for which

$$[d/6]+1 < k_1 + k_2 \leqslant 7d/6.$$

Then $n_1 + 2n_2 \leqslant B(d)$.

Note that $B(d) \sim 23d^2/36$ as $d \to \infty$.

In conclusion we note the work of Chmutov [76], which gives an upper bound on the number of singularities on a pair of level sets of a function.

14.3.3 Real singularities

Let us discuss the application of the mixed Hodge structure in vanishing cohomologies to the estimation of real characteristics of real functions. Such an application is connected with the study in algebraic geometry of the topology of real algebraic varieties.

Let us consider a non-singular real algebraic curve of degree m, lying in the real projective plane. The connected components of the curve (one-dimensional manifolds, diffeomorphic to circles) are called *ovals*. The question about the mutual arrangement of the ovals of a real algebraic curve is one of the classical questions of geometry (see Hilbert's 16th problem). Plane curves of degree two were studied already in ancient Greece, curves of degree three and four by Descartes and Newton. The study of the topology of curves of higher degree has proved to be a considerably more difficult problem: the topology of non-singular curves of degree 6 was fully studied only in 1969 and all possible arrangements of ovals of curves of degree 8 are not known even today (see [20, 143, 395]).

Side by side with the description of the arrangements of ovals of curves of small degree, results are known about the ranges within which various numerical characteristics of algebraic curves of given degree can vary (see [20]). Among results of this type is the inequality of I. G. Petrovskii, formulated below. We shall survey its generalisations.

Each oval of a curve of even degree divides the projective plane into two parts, one of which is diffeomorphic to a disc and called the *interior* of the oval, and the other is diffeomorphic to a Möbius band. An oval is called *positive* (or *even*) if it lies inside an even number of others, and *negative* (or *odd*) if it lies inside an odd number of ovals. Notation: p is the number of positive, k is the number of negative ovals.

In 1938 I. G. Petrovskii proved [281] for curves of even degree $d=2l$ the inequality

$$|2(p-k)-1|\leqslant 3l^2-3l+1;$$

in the same place is given a generalisation of this inequality for curves of odd degree. In 1949 I. G. Petrovskii and O. A. Oleinik proved [282] analogous inequalities for smooth real algebraic hypersurfaces in a space of any number of dimensions.

Namely, let us consider a real non-singular projective hypersurface $A\subset \mathbb{R}P^{n-1}$ of degree d, given by a homogeneous polynomial f in n variables. If d is even, let us denote by B_+ and B_- the parts of $\mathbb{R}P^{n-1}$ given by the conditions $f\geqslant 0$, $f\leqslant 0$, respectively.

The *Petrovskii number* is the number of integer points, strictly inside the cube $(0,d)^n$ lying in a hyperplane passing through the centre of the cube and perpendicular to a body diagonal of the cube. Notation:

$$\Pi_n(d)=\{\#k=(k_1,\ldots,k_n)|0<k_s<d,\ \Sigma k_s=dn/2\}.$$

The inequalities of Petrovskii-Oleinik consist of the following:

$$|\chi(A)-1|\leqslant \Pi_n(d),$$

if n is even; and

$$|\chi(B_+)-\chi(B_-)|\leqslant \Pi_n(d),$$

if n is odd, but d is even, where χ is the Euler characteristic. In particular, the first inequality for $n=4$ estimates the Euler characteristic of algebraic surfaces in three-dimensional projective space, the second inequality for $n=3$ is just Petrovskii's inequality.

V. I. Arnold offered the following unified form of the Petrovskii-Oleinik inequalities.

Theorem 14.9a (see [17]). The number on the left-hand side of the Petrovskii-Oleinik inequalities, in the cases of both even and odd numbers of variables n, is equal to $|\mathrm{ind}|$, where ind is the index of the singular point $0 \in \mathbb{R}^n$ of the gradient in $\mathbb{R}^n$ of the polynomial f giving the hypersurface under consideration.

Theorem 14.9b (see [17]). The number on the right-hand side of the Petrovskii-Oleinik inequality for a hypersurface given by a homogeneous polynomial f is equal to the Hodge number

$$h_{\lambda=1}^{n/2,\,n/2}$$

of the mixed Hodge structure in cohomologies vanishing at the critical point $0 \in \mathbb{C}^n$ of the polynomial f (considered as a function on $\mathbb{C}^n$), if the number of variables n is even, and equal to the Hodge number

$$h_{\lambda=1}^{(n+1)/2,\,(n+1)/2}$$

of the mixed Hodge structure in cohomologies vanishing at the critical point $0 \in \mathbb{C}^{n+1}$ of the polynomial $f(x)+y^2$, if the number of variables n of the polynomial f is odd, and the degree d of the polynomial f is even.

In this way the Petrovskii-Oleinik inequality acquires a unified form: the modulus of the index of a singular point in $\mathbb{R}^n$ of the gradient of a polynomial with real coefficients is estimated from above by the corresponding Hodge number of the mixed Hodge structure in cohomologies vanishing at the critical point of the polynomial, considered as a function on complex space.

In this form the Petrovskii-Oleinik inequality was generalised to the case of a critical point of finite multiplicity of a smooth function.

Let $f:(\mathbb{C}^n,0) \to (\mathbb{C},0)$ be the germ of a holomorphic function at a critical point of finite multiplicity. Let us suppose that the germ f, restricted to the real subspace $\mathbb{R}^n \subset \mathbb{C}^n$, takes only real values. Let us consider the vector field

$$\mathrm{grad}\ f|_{\mathbb{R}^n}$$

on $\mathbb{R}^n$ in a neighbourhood of the origin. Let us denote by ind the index of its singular point 0.

Theorem 14.10 (see [17]cf. [377]). If $n=2k$ is even, then

$$|\text{ind}| \leqslant h^{k,k}_{\lambda=1},$$

where $h^{k,k}_{\lambda=1}$ is the Hodge number of the mixed Hodge structure in cohomologies vanishing at the critical point of the germ f. If $n=2k-1$ is odd, then

$$|\text{ind}| \leqslant h^{k,k}_{\lambda=1},$$

where $h^{k,k}_{\lambda=1}$ is the Hodge number of the mixed Hodge structure in cohomologies vanishing at the critical point $0 \in \mathbb{C}^{n+1}$ of the germ $f(x)+z^2$ of the function of $n+1$ variables.

Remarks.

(1) For $n=2$ it follows from the theorem that the modulus of the index of a finite-multiplicity singular point 0 of the gradient of a real function of two variables with fixed Newton polygon does not exceed the number of interior integral points on the Newton diagram (see [17], and also § 13.3.4).

(2) As in the case of the Petrovskii-Oleinik inequality the index can be expressed in terms of the Euler characteristics of local level manifolds of the germ

$$f|_{\mathbb{R}^n}.$$

Namely, let X_t be the fibre of the Milnor fibration of the critical point of the germ f. Let us denote its real part $X_t \cap \mathbb{R}^n$ by $\mathbb{R}X_t$. Let us denote by $\tilde{\chi}_+$ (respectively $\tilde{\chi}_-$) the reduced (decreased by 1) Euler characteristic of the manifold $\mathbb{R}X_t$ for positive t (respectively, for negative t). The following lemma is easy to prove (see, for example, [17]).

Lemma 14.1. The index of the singular point 0 of the vector field

$$\text{grad } f|_{\mathbb{R}^n}$$

is related to the reduced Euler characteristics of the real local level manifolds of the germ $f|_{\mathbb{R}^n}$ by the relations

$$\text{ind} = -\tilde{\chi}_- = \begin{cases} -\tilde{\chi}_+, & \text{if } n \text{ is even,} \\ \tilde{\chi}_+, & \text{if } n \text{ is odd.} \end{cases}$$

(3) There is a unified expression for the numbers $h^{k,k}_{\lambda=1}$, appearing in the theorem, which does not depend on the parity of the number n, in terms of the spectrum of the critical point of the germ f: this number is equal to the number of spectral pairs equal to $(n/2-1, n-1)$. For $n=2k$ this result is obvious, for $n=2k-1$ we need to use the corollary of Theorem 13.7. Note that $(n/2-1, n-1)$ is the centre of symmetry of the set of spectral pairs.

(4) A prototype of the inequalities in the theorem is, side by side with the Petrovskii-Oleinik inequality, the following inequality of V. M. Kharlamov [187]

$$|\chi(A)-1| \leqslant h^{k,k} - 1,$$

where A is an arbitrary non-singular real projective manifold of dimension $2k$, and $h^{k,k}$ is the Hodge number of the pure Hodge structure in cohomologies of the complexification of the manifold A.

For other restrictions on the arrangement of a real algebraic variety see the works of V. I. Arnold, O. Ya. Viro, D. A. Gudkov, V. I. Zvonilov, V. V. Nikulin, O. A. Oleinik, I. G. Petrovskii, G. M. Polotovskii, V. A. Rokhlin, R. Thom, V. M. Kharlamov, cited at the end of the book.

(5) The estimates, indicated in the theorem serve as examples of the following general scheme of reasoning in real geometry (see V. I. Arnold [17]). For an estimate of any invariant of a real topological type there is sought a suitable invariant of a complex object, majorising the first. Invariants of complex objects are constant for almost all fibers of a complex irreducible family (since the degenerate cases correspond to a complex hypersurface in the space of parameters of the family, and the complement of such a hypersurface is connected). Therefore invariants of a complex object can be calculated in terms of discrete data of the problem (degree, Newton polyhedron, etc.). In this way the estimation of invariants of a real topological type is broken down into two problems: finding a majorising invariant of a complex object and its calculation in terms of discrete data.

(6) Let us formulate an unsolved problem [17]: give the best possible estimates (in terms of Hodge numbers (?)) for the individual Betti numbers of the local level manifold of a real smooth function in a neighbourhood of a degenerate critical point, in particular for the number b_0. Possibly it is easier to estimate the

numbers

$$b_0, \ b_0 - b_1, \ b_0 - b_1 + b_2, \ldots,$$

and also combinations of local Morse numbers

$$M_0, \ M_0 - M_1, \ M_0 - M_1 + M_2, \ldots,$$

where M_i is the number of non-degenerate critical points of index i, merging into the initial critical point for a morsification of the initial critical point.

In the concluding section let us indicate the structure of the level surface of a function of three variables in a neighbourhood of a simple or unimodal critical point.

It is well known that in a neighbourhood of a non-degenerate critical point the function, expressed in terms of suitable variables, is a quadratic form. Therefore, depending on the signature of the quadratic form, there are two possibilities. In the first case the level surfaces are spheres and empty sets (depending on the value of the level), and in the second case are one-sheeted and two-sheeted hyperboloids. It turns out that in a neighbourhood of a simple or unimodal critical point of a function of three variables the structure of the level surfaces can change in the ten ways indicated below.

The change of structure of the level surface in a neighbourhood of the critical point with zero as the critical value is the pair of surfaces of small positive and small negative level, lying in a ball of small radius with centre at the critical point. Such a pair is given by using the zero level surface to break the sphere of small radius with centre at the critical point into two parts: the set on which the function takes positive values is diffeomorphic to the small-positive-level surface, and the set on which the function takes negative values is diffeomorphic to the small-negative-level surface. The set on which the value of the function is zero is a smooth one-dimensional manifold (a union of "ovals", see [256]).

It is useful to describe such a decomposition of the sphere by a graph, the vertices of which are assigned the sign $+$ or $-$. If we are given a decomposition then its connected regions are the vertices of the graph (taken with the sign of the function in that region), two vertices being joined by an edge if the regions are adjacent. Since the sphere is simply connected it follows that this graph is a tree. For example, for the function $x^2 + y^2 - z^2$ this graph has three vertices, joined by two arcs, the outer ends of which have the sign $-$.

If the function depends on two variables (respectively, one variable) then its structure in a neighbourhood of the critical point is given by a decomposition of the circle (resp. pair of points) of small radius into two regions – the positive and negative values of the function. Such a decomposition is determined by the number of connected components of these regions – the pair of numbers

$$(b_0^+, b_0^-),$$

called then the *type of the structure*.

In the theorem formulated below we shall use the letter notation for the critical points of functions introduced in Chapter 17 of Volume 1. The letters have indices $+, -, \pm$. There are as many of these $\pm$ signs as in the corresponding formulae in Chapter 17 of Volume 1, or 1 more. In this case the extra sign is the sign of the parameter a in the corresponding formula. Moreover the order of the indices in the notation is the same as the order of the signs in the formulae. We assume also that they satisfy the restrictions indicated in Chapter 17 and in addition we require for $T_{p,q,r}$ that $p, q, r > 2$.

Theorem (S. Yu. Orevkov, see [273]).

1. Critical points of functions of one variable have structures of three types:

A_{2k+1}^+ have type $(2,0)$ – a minimum,

A_{2k+1}^- have type $(0,2)$ – a maximum,

A_{2k} have type $(1,1)$.

2. Simple and unimodal critical points with zero 2-jet of functions of two variables have level line structure of the following types:

Type $(1, 0)$	Type $(0, 1)$
X_9^{++} for $a > -2$,	X_9^{--} for $a < 2$,
X_{9+2k}^{+++},	X_{9+2k}^{---},
$Y_{2r,2s}^{+++}$,	$Y_{2r,2s}^{---}$,
$\tilde{Y}_r^+$ have type $(1, 0)$;	$\tilde{Y}_r^-$ have type $(0, 1)$;

Type $(1, 1)$	Type $(2, 2)$
D_{2k}^+,	$D_{2k+1}^\pm$,
E_6,	E_7,
E_8,	X_9^{+-},
J_{10}^+ for $a^2 < 4$,	X_9^{-+},

J_{10+2k}^{++},

J_{10+2k}^{--},

$X_{9+2k+1}^{++\pm}$,

$X_{9+2k+1}^{--\pm}$,

$Y_{2r,2s+1}^{++\pm}$,

$Y_{2r,2s+1}^{--\pm}$ have type $(1, 1)$;

$J_{10+2k+1}^{\pm}$,

$X_{9+2k}^{+\pm-}$,

$X_{9+2k}^{-\pm+}$,

$Y_{2r,2s}^{\pm+-}$,

$Y_{2r,2s}^{\pm-+}$,

$Y_{2r+1,2s+1}^{\pm\pm\pm}$ have type $(2, 2)$;

Type $(3, 3)$

D_{2k}^{-},

J_{10}^{-},

J_{10}^{+} for $a^2 > 4$,

J_{10+2k}^{+-},

J_{10+2k}^{-+},

$X_{9+2k+1}^{+-\pm}$,

$X_{9+2k+1}^{-+\pm}$,

$Y_{2r,2s+1}^{+-\pm}$,

$Y_{2r,2s+1}^{-+\pm}$ have type $(3, 3)$;

Type $(4, 4)$

X_9^{++} for $a < -2$,

X_9^{--} for $a > 2$,

X_{9+2k}^{-+-},

X_{9+2k}^{+-+},

$Y_{2r,2s}^{+--}$,

---

3. Unimodal critical points of functions of three variables have level surface structures, the types and graphs of which are indicated in figure 84.

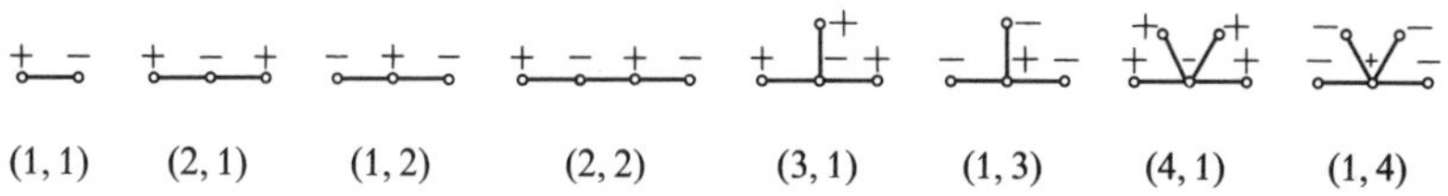

(1, 1)   (2, 1)   (1, 2)   (2, 2)   (3, 1)   (1, 3)   (4, 1)   (1, 4)

Fig. 84.

Type $(1, 1)$

P_8^{+} for $a^2 < 4$,

P_{8+2k}^{++},

P_{8+2k}^{--},

$R_{2l,2m}^{+-}$,

$R_{2l,2m}^{-+}$,

Type $(2, 1)$

P_{8+2k+1}^{++},

P_{8+2k+1}^{--},

$R_{2l,2m+1}^{-\pm}$,

$T_{2p,2q,2r}^{+--}$,

$T_{2p,2q+1,2r+1}^{-\pm\pm}$,

$R_{2l+1,2m+1}^{++}$,

$R_{2l+1,2m+1}^{--}$,

$\tilde{R}_m^{++}$,

$\tilde{R}_{2m+1}^{+-}$,

$T_{2p,2q,2r+1}^{+-\pm}$,

$T_{2p+1,2q+1,2r+1}^{\pm\pm\pm\pm}$ (even number
                             of plusses),

$\tilde{T}_{2p+1,m}^+$,

E_{12},

E_{14},

W_{12},

Q_{10},

Q_{12},

S_{12},

$U_{12}^{+\pm}$ have type $(1,1)$;

$\tilde{T}_{2p,m}^+$,

E_{13}^-,

$Z_{13}^{\pm-}$,

$W_{13}^{\pm-}$,

Q_{11}^+,

S_{11}^-,

Z_{11}^- have type $(2,1)$;

Type $(1,2)$

P_{8+2k+1}^{+-},

P_{8+2k+1}^{-+},

$R_{2l,2m+1}^{+\pm}$,

$T_{2p,2q,2r}^{++-}$,

$T_{2p,2q+1,2r+1}^{+\pm\pm}$,

$\tilde{T}_{2p,m}^-$,

E_{13}^+,

$Z_{13}^{\pm+}$,

$W_{13}^{\pm+}$,

Q_{11}^-,

S_{11}^+,

Z_{11}^+ have type $(1,2)$;

Type $(2,2)$

P_8^+ for $a^2 > 4$,

P_{8+2k}^{+-},

P_{8+2k}^{-+},

$R_{2l+1,2m+1}^{+-}$,

$R_{2l+1,2m+1}^{-+}$,

$\tilde{R}_{2m+1}^{-+}$,

$\tilde{R}_m^{--}$,

$T_{2p+1,2q+1,2r+1}^{\pm\pm\pm\pm}$ (odd number
                           of plusses),

$\tilde{T}_{2p+1,m}^-$ have type $(2,2)$;

Type $(3,1)$

$R_{2l,2m}^{--}$,

$\tilde{R}_{2m}^{+--}$,

$\tilde{R}_{2m}^{-+-}$,

Type $(1,3)$

$R_{2l,2m}^{++}$,

$\tilde{R}_{2m}^{+-+}$,

$\tilde{R}_{2m}^{-++}$,

$T^{--\pm}_{2p,2q,2r+1}$,	$T^{++\pm}_{2p,2q,2r+1}$,
Z^-_{12},	Z^-_{12},
U^{--}_{12} have type (3, 1);	U^{-+}_{12} have type (1, 3);

Type (4, 1)	Type (1, 4)
$T^{---}_{2p,2q,2r}$ have type (4,1);	$T^{+++}_{2p,2q,2r}$ have type (1,4).

Corollary. For the critical points of Section 3 of the theorem the graph of the structure is determined by the type of the structure.

Remarks.

1. (S. Yu. Oryevkov, see [274]) The theorem allows one to describe the structure of the level hypersurfaces in the neighbourhoods of critical points which are direct sums of the critical points enumerated in the theorem. Indeed if the direct summands depend the one on m and the other on k variables, then the region of positive (negative) values of the sum on the sphere S^{m+k-1} will correspond to the set

$$(M^{m-1}_\pm \times D^k) \cup (D^m \times M^{k-1}_\pm)$$

under the homomorphism

$$(S^{m-1} \times D^k) \cup (D^m \times S^{k-1}) \to S^{m+k-1},$$

where $M_\pm$ is the region of positive (negative) values of the direct summand on the corresponding sphere. In particular, under the addition of a square of a new variable (that is under the change to a stably equivalent critical point) the new region of negative values on the sphere is obtained from the old by multiplication by an interval and the new region of positive values on the sphere is obtained by gluing two copies of a ball along the old region of positive values on the sphere, taken as the boundary of the ball.

2. There are ten possible reconstructions of a level surface for a function of three variables in a neighbourhood of a simple or unimodal critical point: eight of these are shown in figure 84, the other two are (1, 0) – the minimum and (0, 1) – the maximum. This result follows easily from the theorem and the above remark.

14.3.4 Bernstein polynomials

Let $Q(x) = x_1^2 + \ldots + x_n^2$ be a quadratic form. There is an identity

$$\left(\frac{1}{4} \sum_{i=1}^{n} (\partial/\partial x_i)^2 \right) Q(x)^\lambda = \lambda \left(\lambda + \frac{n}{2} - 1 \right) Q(x)^{\lambda - 1}.$$

This identity was used by I. M. Gelfand and G. E. Shilov in [123] to determine the complex degree of a quadratic form as a generalised function. This identity served as motivation for the theorem of I. N. Bernstein, which is formulated below.

Let $f : (\mathbb{C}^n, 0) \to (\mathbb{C}, 0)$ be the germ of a holomorphic function. Let λ be an independent variable. Let us consider the set of finite sums of the form

$$\sum_{k, l \geqslant 0} a_{k,l}(x) \lambda^l f(x)^{\lambda - k},$$

where $a_{k,l} : (\mathbb{C}^n, 0) \to (\mathbb{C}, 0)$ are germs of holomorphic functions and $f^{\lambda - k}$ is a formal symbol. Let us furnish this set with the obvious relation

$$f(x) \cdot f(x)^{\lambda - k - 1} = f(x)^{\lambda - k}.$$

Let us consider the differential operator $P(x, \lambda, \partial/\partial x)$ with coefficients which are holomorphic in x and polynomial in λ:

$$P(x, \lambda, \partial/\partial x) = \sum_{k, \alpha \geqslant 0} b_{k, \alpha}(x) \lambda^k (\partial/\partial x)^\alpha.$$

These operators will act on the previous set if we put

$$\partial/\partial x_i \, f^{\lambda - k} = (\lambda - k) \partial f/\partial x_i \, f^{\lambda - k - 1}.$$

Theorem 14.11. There exist polynomials $B(\lambda)$ and a differential operator $P(x, \lambda, \partial/\partial x)$, for which

$$P(x, \lambda, \partial/\partial x) f^\lambda = B(\lambda) f^{\lambda - 1}. \tag{7}$$

This theorem was proved by I. N. Bernstein [46] for the case in which f is a polynomial, and extended by Björk [52] to the general case.

It is easy to see that the set of polynomials $B(\lambda)$, for which there exists an identity (7), generates an ideal in $\mathbb{C}[\lambda]$. The monic generator of this ideal is called the *Bernstein polynomial* of the germ f. It is clear that $b(0)=0$ (for this it is sufficient to put $\lambda=0$ in the identity (7)). Let us write

$$b(\lambda)=\lambda\tilde{b}(\lambda).$$

The polynomial $\tilde{b}(\lambda)$ is called the *reduced Bernstein polynomial*.

One of the motives for proving Theorem 14.11 is the following application of it.

Let us suppose that the germ f takes only real values on the real subspace $\mathbb{R}^n\subset\mathbb{C}^n$. Let us fix a representative f of the germ f and let us define two functions $f_\pm$ on a neighbourhood of the origin in $\mathbb{R}^n$:

$$f_+(x)=\begin{cases} f(x) & \text{for} \quad f(x)\geqslant 0 \\ 0 & \text{for} \quad f(x)<0 \end{cases}$$

$$f_-(x)=\begin{cases} 0 & \text{for} \quad f(x)\geqslant 0 \\ -f(x) & \text{for} \quad f(x)<0. \end{cases}$$

Let $\phi:\mathbb{R}^n\to\mathbb{R}$ be a smooth function with support concentrated in a sufficiently small neighbourhood of the origin. Put

$$I_\pm(\lambda,\phi)=\int_{\mathbb{R}^n} (f_\pm(x))^\lambda\phi(x)dx,$$

where $\lambda\in\mathbb{C}$ is a complex parameter, Re $\lambda>0$. The integrals $I_\pm$ can be considered as generalised functions, depending on the parameter λ on the space of such functions $\{\phi\}$. The integrals $I_\pm$ are well-defined for Re $\lambda>0$ and depend holomorphically on λ.

Theorem 14.12. The integrals $I_\pm$ can be analytically continued to $\mathbb{C}$ as meromorphic functions of the parameter λ, and their poles lie on the arithmetic progressions

$$\lambda_i,\ \lambda_i-1,\ \lambda_i-2,\ldots,$$

where $\lambda_1,\ \lambda_2,\ldots$ are the roots of the Bernstein polynomial of the germ f.

Proof. We have

$$b(\lambda)\int_{\mathbb{R}^n} f_\pm^{\lambda-1}\varphi dx=\int_{\mathbb{R}^n} [Pf_\pm^\lambda]\varphi dx=\int_{\mathbb{R}^n} f_\pm^\lambda[P^*\varphi]x;$$

where b is the Bernstein polynomial, P is a differential operator satisfying with b the identity (7), and

$$P^* = \sum (-1)^{|\alpha|} \cdot \lambda^k (\partial/\partial x)^\alpha b_{k,\alpha}(x)$$

is the conjugate operator. These equations allow us to continue analytically the integrals $I_\pm$ from the half-plane Re $\lambda > 0$ to the half-plane Re $\lambda > -1$, etc.

In [237, 238] B. Malgrange related the roots of the Bernstein polynomial with the eigenvalues of the monodromy operator in the cohomologies vanishing at the critical point of the germ f.

Let us suppose that the germ $f : (\mathbb{C}^n, 0) \rightarrow (\mathbb{C}, 0)$ has a critical point of finite multiplicity μ. We place each eigenvalue λ of the monodromy operator in correspondence with the arithmetic progression $L(\lambda)$ of all numbers α for which

$$\exp(2\pi i\alpha) = \lambda.$$

Theorem 14.13 (see [238]). The roots of the Bernstein polynomial of the germ f belong to the union of all the arithmetic progressions constructed above. Each of its roots is less than 1.

Corollary. The roots are rational numbers.

The rationality of the roots of the Bernstein polynomial of a germ with a critical point whose multiplicity is not necessarily finite was proved by Kashiwara in [180].

In [360] there is defined a filtration in the fibres of the cohomological Milnor fibration of a critical point of finite multiplicity, and the roots of the Bernstein polynomial are expressed in terms of the action of the monodromy operator on the spaces of this filtration. Obvious correspondences between this filtration and the Hodge filtration point to new inequalities, relating the roots of the Bernstein polynomial.

Let us define the above-mentioned filtration, which we shall call the *third* (after the weight and Hodge filtrations). The third filtration in the cohomologies $H^{n-1}(X(t))$ of the fibre of the Milnor fibration we shall denote by $\{G_t^k\}$.

For an arbitrary holomorphic differential n-form ω on X let us expand its geometric section in a series of covariantly constant sections:

$$s[\omega] = \sum_{p,\alpha} t^\alpha (\ln t)^p A_{p,\alpha}^\omega / p \,!;$$

see § 13.1. We put the subspace G_t^k equal to the linear span of all the values at the point t of the sections $A_{p,\alpha}^\omega$ with

$$\alpha \leqslant n-1-k, \quad p=0,1,\ldots,n-1,$$

of all the forms ω.

From the definition it easily follows that the third filtration is decreasing:

$$\ldots \subset G_t^{k+1} \subset G_t^k \subset \ldots ;$$

the terms of the third filtration are invariant relative to the monodromy operator, and for any k the subspace G_t^k contains the subspace F_t^k of the Hodge filtration.

Theorem 14.14 (see [360]). For any k let us denote by Q_k the minimum polynomial of the action of the monodromy operator on G_t^k/G_t^{k+1}. Each root λ of the polynomial Q_k we place in correspondence with the number

$$l_k(\lambda)+1-n,$$

where $l_k(\lambda)$ is defined by the conditions:

$$\exp(2\pi i l_k(\lambda))=\lambda, \quad k \leqslant l_k(\lambda) < k+1.$$

Let us consider the union of all the numbers constructed for all k. Let us denote them by $\alpha_1, \alpha_2, \ldots$. Then

$$[(s-\alpha_1)(s-\alpha_2)\ldots] \in \mathbb{C}[s]$$

is the reduced Bernstein polynomial.

Corollary. If f is the germ of a quasihomogeneous polynomial, then the roots of the reduced Bernstein polynomial of the germ are the spectrum of the critical point of the germ multiplied by -1 (proof: in this case the third filtration is the same as the Hodge filtration).

The following theorem follows from the inclusion of the terms of the Hodge filtration in the terms of the third filtration:

Theorem 14.15 (see [361, 364]).

1. Any k-fold root of the reduced Bernstein polynomial is greater than $k - n$.
2. If $\alpha \in [p, p+1)$ (where p is an integer) is a k-fold root of the reduced Bernstein polynomial, then in the interval $(-n+k-p, 1)$ among the terms of the arithmetic progressions

$$L(\exp(-2\pi i \alpha))$$

there are not less than k roots (counted with multiplicities) of the reduced Bernstein polynomial.

3. The reduced Bernstein polynomial $\tilde{b}(s)$ is divisible by $(s - \alpha)^n$ if and only if $\alpha > 0$ and the monodromy operator has a Jordan block of dimension n associated with the eigenvalue $\exp(2\pi i \alpha)$. The polynomial $\tilde{b}(s)$ is divisible by $(s - \alpha)^{n-1}$ for integral α if and only if $\alpha = 0$ and the monodromy operator has a Jordan block of dimension $n - 1$ associated with the eigenvalue 1.

The theorem asserts that a large part of the roots of the reduced Bernstein polynomial lie to the right of the point $s = 1 - n/2$ (compare with the symmetry of the spectrum in § 13.3).

The roots of the Bernstein polynomial can change under a deformation of the critical point of the germ in the $\mu = \mathrm{const}$ stratum.

Example [316]. Let

$$f(x, y) = ax^5 + y^6 + x^4 y,$$

where $a \in \mathbb{C}$ is a parameter. For $a = 0$ the roots of the reduced Bernstein polynomial are equal to $\{l/24\}$, where

$$l = -15, \ -11, \ -10, \ -7, \ -6, \ -5, \ -2, \ -1, \ 0,$$

$$1, \ 2, \ 3, \ 5, \ 6, \ 7, \ 10, \ 11, \ 15.$$

For $a \neq 0$ the roots of the reduced Bernstein polynomial are equal to $\{l/24\}$, where

$$l = -7, \ -6, \ -5, \ -3, \ -2, \ -1, \ 0, \ 1, \ 2, \ 3, \ 5,$$

$$6, \ 7, \ 9, \ 10, \ 11, \ 13, \ 14, \ 15.$$

In the example the roots jump down as $a \to 0$. In [211] a theorem is proved, which asserts that this phenomenon is typical, and explains in terms of the third filtration how the roots of the Bernstein polynomial can change under deformations of the critical point along the $\mu = $ const stratum.

We indicate the works which are related to Bernstein polynomials: [38–40, 65–67, 203, 221–228, 286, 287, 382, 406–408].

14.3.5   The mixed Hodge structure and the local algebra of a critical point

Let $f : (\mathbb{C}^n, 0) \to (\mathbb{C}, 0)$ be the germ of a holomorphic function at a critical point of multiplicity μ and let $f : X \to S$ be a specialisation of the germ. In this section we shall consider holomorphic differential n-forms on X modulo forms divisible by df, that is equivalence classes in

$$\Omega^n(X)/df \wedge \Omega^{n-1}(X),$$

where $\Omega^p(X)$ is the space of holomorphic differential p-forms on X. Each equivalence class can be put in correspondence with a section of a suitable fibration, constructed from the weight and Hodge filtration of the cohomological Milnor fibration of the critical point of the germ. This section is called the *original coefficient of the equivalence class*. The correspondence relating the equivalence class and the original coefficient establishes a relationship between the space

$$\Omega^n(X)/df \wedge \Omega^{n-1}(X)$$

and the cohomology, vanishing at the critical point of the germ. We shall give an example of the use of this relationship.

Remark. Forms in $\Omega^n(X)$ have the form

$$h dx_1 \wedge \ldots \wedge dx_n,$$

forms in $df \wedge \Omega^{n-1}(X)$ have the form

$$\sum (-1)^k h_k \partial f/\partial x_k \cdot dx_1 \wedge \ldots \wedge dx_n.$$

Therefore if X is sufficiently small (and this is always assumed), then

$$\Omega^n(X)/df \wedge \Omega^{n-1}(X)$$

is a μ-dimensional vector space over $\mathbb{C}$ (as is the local algebra $\mathbb{C}\{x\}/(\partial f/\partial x)$).

Let us indicate the construction of the original coefficient. Let us fix an equivalence class and consider the upper bound of the orders of the forms belonging to the class. We shall call the upper bound the *Hodge number* of the equivalence class.

Theorem 14.16 (see [364, § 9]). The Hodge number of a class is equal to $+\infty$ if and only if the class is the class of the zero form (that is the class is just $df \wedge \Omega^{n-1}(X)$).

Let us suppose that the chosen class does not contain the zero form. Among the forms of the class with the largest order let us consider only those forms for which the principal part is a section of the subfibration of the weight filtration with the smallest number. Let us call this smallest number the *weight number* of the class.

Let α, l be the Hodge and weight numbers of the class, respectively. The principal part of each differential form of the class satisfying the above two conditions projects to a section of the fibration

$$gr^k F gr_l W(f^*),$$

where $k = n + 1 + [-\alpha]$ (for the definition of $gr^k F gr_l W(f^*)$ see page 380).

Theorem 14.17 (see [364, § 9]). This section does not depend on the form of the class satisfying the two above conditions, and is a non-zero section.

The indicated section of the fibration

$$gr^k F gr_l W(f^*)$$

is called the *original coefficient* of the equivalence class. Forms of the class with order equal to the Hodge number of the class and for which the principal part is a section of the subfibration of the weight filtration with number equal to the weight number of the class are called *original forms* (compare with [363, 364]).

Let $\omega \in \Omega^n(X)$ be a form of order α. Let us suppose that the principal part of the form is a section of a weight subfibration with number l and is not a section of a weight subfibration with number $l-1$. Then the principal part of the form ω projects into a section of the fibration

$$gr^k F gr_l W(f*),$$

where $k = n+1+[-\alpha]$.

Theorem 14.18 (see [364, § 9]). If a section of the subfibration

$$gr^k F gr_l W(f*),$$

induced by the principal part of the form ω, is not the zero section, then the form ω is an original form in its equivalence class.

The Hodge and weight numbers of the equivalence classes define on

$$\Omega^n(X)/df \wedge \Omega^{n-1}(X)$$

additional structures.

Definition. The *spectral vector* of the class

$$[\omega] \in \Omega^n(X)/df \wedge \Omega^{n-1}(X)$$

is the ordered pair

$$V[\omega] = (\alpha[\omega], l[\omega]),$$

where $\alpha[\omega]$, $l[\omega]$ are the Hodge and weight numbers of the class $[\omega]$. If $[\omega]$ is the class of the zero form we put

$$V[\omega] = (+\infty, -\infty).$$

Let us order the spectral vectors lexicographically. Namely, put $V > V'$ if $\alpha > \alpha'$ or if $\alpha = \alpha'$ and $l < l'$.

For example,

$$(1/3, 0) < (1/2, 1) < (1/2, 0).$$

It is clear that multiplying an equivalence class by a non-zero number does not change its spectral vector; the spectral vector of a sum of classes is not less than the minimum of the spectral vectors of the summands.

For any vector $V \in \mathbb{R}^2$ let us denote by FW_V (respectively by $FW_{>V}$) the set of all classes from

$$\Omega^n(X)/df \wedge \Omega^{n-1}(X),$$

the spectral vectors of which are not less than V (respectively, are greater than V). It is clear that $FW_V \supset FW_{>V}$ and each of these subsets is a complex vector space.

Let us call the filtration

$$\{FW_V\}_{V \in \mathbb{R}^2}$$

the *Hodge-weight filtration* of the space

$$\Omega^n(X)/df \wedge \Omega^{n-1}(X).$$

Put

$$gr_V FW = FW_V/FW_{>V}.$$

Let us call the μ-dimensional complex vector space

$$gr FW = \oplus_{V \in \mathbb{R}^2} gr_V FW$$

the *graded space* of the Hodge-weight filtration.

Theorems 14.16–14.18 establish an isomorphism between the space $gr FW$ and a distinguished μ-dimensional space of sections (the space of original coefficients) of the μ-dimensional fibration

$$gr F gr W(f^*) = \oplus_{k,l} gr^k F gr_l W(f^*).$$

This isomorphism maps an element of $gr_V FW$ into its original coefficient.

To sum up the above construction informally: after factoring by the Hodge and weight filtrations, the space of cohomologies vanishing at the critical point of the germ f, and $\Omega^n(X)/df \wedge \Omega^{n-1}(X)$ are canonically isomorphic. The isomorphism is established by a mapping from an equivalence class of forms to the original coefficient of the class.

Remarks.

1. Using the indicated isomorphism, we can determine the spectral pairs of the mixed Hodge structure in the vanishing cohomologies in terms of the Hodge-weight filtration on $\Omega^n(X)/df \wedge \Omega^{n-1}(X)$. Namely we choose the pair $(\alpha, l) \in \mathbb{R}^2$ exactly the same number of times as the dimension of the space $gr_{(\alpha, l)} FW$, if $\alpha \notin \mathbb{Z}$, and of the space $gr_{(\alpha, l+1)} FW$, if $\alpha \in \mathbb{Z}$. The union of all the chosen pairs is exactly the set of all spectral pairs.

2. In each fibration

$$gr^k F gr_l W(f^*)$$

there is a Gauss-Manin connection. The monodromy operator of the connection does not have Jordan blocks. An arbitrary original coefficient, generally speaking, is not a covariantly constant section of this connection; however, the directions, determined by its values, are invariants relative to the connection.

3. On the space $gr FW$ we can introduce a mixed Hodge structure with the help of the original coefficients.

Now let us cite an example of a result, the proof of which is based on the isomorphism indicated above.

Theorem 14.19 (see [363]. Let

$$N : H^{n-1}(X, \mathbb{C}) \to H^{n-1}(X, \mathbb{C})$$

be the logarithm of the unipotent part of the monodromy operator. Let

$$\{f\} : \mathbb{C}\{x\}/(\partial f/\partial x) \to \mathbb{C}\{x\}/(\partial f/\partial x)$$

be the operation of multiplication by f. Then for any $j \geq 0$,

$$\dim(\ker(\{f\}^j)) \leq \dim(\ker(N^j)),$$

where $\dim(\ker(\cdot))$ is the dimension of the kernel of the operator.

Corollary (see [317]). If the operator $\{f\}$ does not have Jordan blocks of dimension $\geq j$, then the monodromy operator does not have such blocks.

For example, in the quasihomogeneous case $\{f\}$ is the zero operator, therefore the monodromy operator is diagonalisable.

Sketch of the proof. The Jordan structure of the operator $\{f\}$ is the same as the Jordan structure of the operation of multiplication by f in

$$\Omega^n(X)/df \wedge \Omega^{n-1}(X).$$

It can be proved that the operation of multiplication by f maps $FW_{(\alpha,l)}$ into $FW_{(\alpha+1,l-2)}$ for any (α, l) (see the proof of Lemma 13.12). Therefore $\{f\}$ induces an operator

$$gr\{f\} : grFW \to grFW,$$

mapping $gr_{(\alpha,l)}FW$ into $gr_{(\alpha+1,l-2)}FW$ for any (α, l). It is clear that the Jordan structures of the operator $\{f\}$ and the operator $gr\{f\}$ are connected by the relation

$$\dim(\ker(\{f\}^j)) \leqslant \dim(\ker((gr\{f\})^j))$$

for any $j \geqslant 0$.

By using Theorem 13.3 on mixed Hodge structures it can be further proved that under the isomorphism between the space $grFW$ and the space of original coefficients the operator $gr\{f\}$ maps into the operator N, multiplied in each term of the space $grFW$ by a corresponding non-zero number. The theorem is proved.

Remark. In [364] the following result was proved. Let us call the *length of the spectrum* of the critical point of a germ f the difference between the largest and smallest spectral numbers. If j is greater than the length of the spectrum, then $\{f\}^j = 0$, in other words $f^j \in (\partial f/\partial x)$. Since the spectrum belongs to the interval $(-1, n-1)$, we always have $f^n \in (\partial f/\partial x)$ (see [54, 219]).

Chapter 15

The period map and the intersection form

Let us be given a smooth fibration and a differential form on the space of fibration which is closed on the fibres. In such a situation there arises the period map of the form – a many-valued map from the base of the fibration to the cohomology of the fibre. A point of the base is mapped to the cohomology class of the form in the fibre over the point translated to the cohomology of a distinguished fibre. The fact that it is many-valued arises from the fact that there is not a unique choice of path for the translation.

Let us denote the form by ω. Let us choose a basis $\delta_1^0, \ldots, \delta_\mu^0$ of the integral homology of the distinguished fibre in the dimension equal to the dimension of the form. Let us extend the basis by continuity to neighbouring fibres and construct a many-valued family $\delta_1, \ldots, \delta_\mu$, continuously depending on the point of the base, of bases of the homologies of the fibres of the fibration. The basis $\delta_1^0, \ldots, \delta_\mu^0$ determines coordinates in the cohomologies of the distinguished fibre. With respect to these coordinates the period map has the form

$$\lambda \mapsto \left(\int_{\delta_1(\lambda)} \omega, \ldots, \int_{\delta_\mu(\lambda)} \omega \right),$$

where λ is a point of the base of the fibration.

The period map allows us to transfer to the base of the fibration structures existing in the space of cohomologies. For example, the intersection number of cohomology classes of the middle dimension in the cohomology of the fibre maps to a bilinear form on the tangent bundle of the base (if the differential form itself has a middle dimension).

In this chapter we shall consider the period map in the Milnor fibration, associated with a versal deformation of the critical point of the function. In this case the dimension of the base is equal to the dimension of the middle cohomology of the fibre. It can be shown that for almost all differential forms the period map is non-degenerate and in a natural sense does not depend on the differential form used to define it. This means that constructions connected with the period map are determined by the fibration and, in the end, by the critical point.

In this chapter we shall consider a bilinear form arising on the base of the fibration (that is on the complement of the discriminant) from the intersection form. It can be proved that under several conditions the bilinear form can be analytically continued to the whole base of the versal deformation.

In a series of cases this bilinear form is a symplectic structure. It can be shown that the strata of the base of the versal deformation have in this symplectic structure special Lagrange properties, reflecting the types of decomposition of the critical point into simpler ones. Several of the strata provide us with important examples of Lagrangean manifolds with singularities.

15.1   The construction

15.1.1

Definition. Let $f:(\mathbb{C}^n,0)\to(\mathbb{C},0)$ be the germ of a holomorphic function at a critical point of multiplicity μ. Let us define the fibration in which we shall study the period map. This fibration is a fibration of zero level hypersurfaces of functions constituting a minimal versal deformation of the germ.

Namely let us fix a representation of a versal deformation of the germ f in the form

$$F(x,\lambda)=f(x)+\lambda_1+\lambda_2\phi_2(x)+\ldots+\lambda_\mu\phi_\mu(x),$$

where the functions $\phi_1\equiv 1,\phi_2,\ldots,\phi_\mu$ generate a basis over $\mathbb{C}$ of the local algebra $\mathbb{C}\{x\}/(\partial f/\partial x)$. Let us choose a sufficiently small ball

$$B=\{x\in\mathbb{C}^n|\,|x|<\varrho\}.$$

Depending on ϱ let us choose a sufficiently small ball

$$\Lambda=\{\lambda\in\mathbb{C}^\mu|\,|\lambda|<\delta\}.$$

Let us denote by Σ the hypersurface of all such $\lambda\in\Lambda$, for which the local zero level set

$$X_\lambda=\{x\in B|F(x,\lambda)=0\}$$

is singular. The hypersurface Σ is called the *discriminant*.

Over the complement $\Lambda\setminus\Sigma$ of the discriminant the manifolds $\{X_\lambda\}$ form a locally trivial fibration.

Remark. This fibration is different from the Milnor fibration of the deformation F (see § 10.3). In order to obtain this fibration from the Milnor fibration of the deformation F it is necessary to restrict the Milnor fibration to the set of zero values of the deformation F.

The fibration over $\Lambda \setminus \Sigma$ we shall call the *central* Milnor fibration.

Let us denote by Ω^p the space of holomorphic p-forms on $B \times \Lambda$. Let us consider an arbitrary $(n-1)$-form $\omega \in \Omega^{n-1}$. Its restriction to an arbitrary fibre of the central Milnor fibration is a closed form. The *period map* of the form ω is the section

$$P_\omega : \lambda \mapsto [\omega | X_\lambda] \in H^{n-1}(X_\lambda, \mathbb{C})$$

of the fibration of $(n-1)$st cohomologies, associated with the central Milnor fibration (more briefly: of the central cohomological Milnor fibration). For each integer $k \geqslant 0$ the kth *associated period map* of the form ω is the section

$$P_\omega^k = (\nabla_{\partial/\partial\lambda_1})^k P_\omega$$

of the same fibration (here $\nabla_{\partial/\partial\lambda_1}$ is differentiation in the Gauss-Manin connection along the vector field $\partial/\partial\lambda_1$; remember that λ_1 is the constant term of the versal deformation).

Remark. Let $\delta_1(\lambda), \ldots, \delta_\mu(\lambda)$ be a basis in $H_{n-1}(X_\lambda, \mathbb{Z})$, depending continuously on λ. With respect to this basis

$$P_\omega^k(\lambda) = (\partial/\partial\lambda_1)^k \left(\int\limits_{\delta_1(\lambda)} \omega, \ldots, \int\limits_{\delta_\mu(\lambda)} \omega \right).$$

We say that the map P_ω^k is *non-degenerate* if the vectors

$$v_i(\lambda) = (\nabla_{\partial/\partial\lambda_i} P_\omega^k)|_\lambda, \quad i = 1, \ldots, \mu,$$

are linearly independent for all $\lambda \in \Lambda \setminus \Sigma$, sufficiently near the origin in Λ (that is if the map P_ω^k, written down in coordinates with respect to a covariantly constant basis, gives a many-valued map in $\mathbb{C}^\mu$ with Jacobian different from zero for all $\lambda \in \Lambda \setminus \Sigma$, sufficiently near to the origin); and is *infinitesimally non-degenerate* if on the λ_1-axis, passing through the origin in Λ, the determinant of the matrix, consisting of the coordinates of the vectors $\{v_i\}$ with respect to a covariantly constant basis, has as $\lambda_1 \to 0$ a zero of order $\mu(n-2k-2)/2$.

Remark. The coordinates of the vectors $\{v_i\}$ with respect to a covariantly constant basis are many-valued, however the square of the determinant of the matrix, consisting of the coordinates, is a single-valued holomorphic function in $\Lambda \setminus \Sigma$, meromorphic in Λ (see Theorem 12.2).

We can show that the property of the map P_ω^k being infinitesimally non-degenerate is determined by a finite jet of the form ω at the point $0 \times 0 \in B \times \Lambda$ (see formulae (3) and (4) on page 284–285 and Lemma 12.3).

We shall say that the property of infinitesimal non-degeneracy of the kth associated period maps is *generic* for given k, if the jets determining the infinitesimally non-degenerate maps constitute in the space of jets of sufficiently high order the complement of a proper analytic subset.

15.1.2 Non-degeneracy and stability

Theorem 15.1 (see [220], [215, § 10]). For any form $\omega \in \Omega^{n-1}$ and any $k \geqslant 0$, if the kth associated period map of the form ω is infinitesimally non-degenerate, then it is non-degenerate.

Proof. As in the proof of Theorem 12.2 we can prove that the square of the determinant of the matrix consisting of the coordinates of the vectors $\{v_i\}$ with respect to a covariantly constant basis at an arbitrary non-singular point of the discriminant has a zero of order not less than $(n - 2k - 2)$. As in Corollary 1 of Theorem 12.2 we can conclude that the square of the determinant does not map to zero on $\Lambda \setminus \Sigma$ in a sufficiently small neighbourhood of the origin in Λ. This means that in a neighbourhood of the origin in Λ the Jacobian matrix of the map P_ω^k is non-degenerate.

Theorem 15.2 (see [369], [364, § 10]). For $k = 0$ the property of infinitesimal non-degeneracy is generic. If the intersection form in

$$H_{n-1}(X_\lambda, \mathbb{C}), \quad \lambda \in \Lambda \setminus \Sigma,$$

is non-degenerate then the property of infinitesimal non-degeneracy is generic for any $k \geqslant 0$.

Remarks.

(1) In [369] there is proved a more general result: if among the spectral numbers of the critical point of a germ f there is not an integer less than k, then the property of infinitesimal non-degeneracy is generic for this k. Theorem 15.2 follows from this result and Theorem 14.3.

(2) Note the following corollary of Lemma 12.3: if for a given $k \geqslant 0$ there exists an infinitesimally non-degenerate period map P_ω^k, then the property of infinitesimal non-degeneracy for this k is generic.

Let us define the concept of equivalent period maps. An informal definition is: two period maps are said to be equivalent if there exists a diffeomorphism of the pair Λ, Σ, possessing the property: the first map is equal to the composite of the diffeomorphism and the second map. This definition requires to be stated more precisely in view of the fact that the period map is not single-valued. In addition we shall consider diffeomorphism not of all of Λ but only of a neighbourhood of the origin.

Definition. Two maps P_ω^k, P_η^k are said to be *equivalent* if there exists a neighbourhood U of the origin in Λ and a continuous map

$$H: U \times [0, 1] \to U,$$

possessing the properties:
(i) $H(\cdot, 0)$ is the identity map;
(ii) $H(\cdot, s)$ for any $s \in [0, 1]$ is a holomorphic map with non-zero Jacobian;
(iii) for any $s \in [0, 1]$ the point $H(\lambda, s)$ belongs to Σ if and only if $\lambda \in \Sigma$;
(iv) $U \cap H(U, 1)$ contains the origin;
(v) for any $\lambda \in U \setminus (U \cap \Sigma)$ the vector $P_\omega^k(\lambda)$, parallel translated in the Gauss-Manin connection along the curve $H(\lambda, \cdot)$ to the point $H(\lambda, 1)$, is equal to the value at this point of the section P_η^k.

Theorem 15.3 (see [369]).
1. Any infinitesimally non-degenerate kth associated period map P_ω^k is stable, that is a kth associated period map P_η^k, for any form η near to ω, is equivalent to P_ω^k.

2. If f is a quasihomogeneous germ, then all infinitesimally non-degenerate kth associated period maps are equivalent.

Proof. Let us suppose that the form ω depends on a parameter and for the zero value of the parameter the corresponding kth associated period map is infinitesimally non-degenerate. Let us prove that for all small values of the parameter the corresponding kth associated period maps are equivalent. To do this, let us construct a vector field, depending on the parameter, in a neighbourhood of the origin in Λ, which for each value of the parameter is tangent to Σ and the flow of which establishes simultaneously the required equivalence of all the maps with small value of the parameter. The vector field is constructed initially on $\Lambda \backslash \Sigma$ and then it is proved that the field can be holomorphically extended to Σ and that this holomorphic extension is tangent to Σ.

It is easy to see that the required field on $\Lambda \backslash \Sigma$ exists and is unique. Indeed, let us consider the kth associated period map as a many-valued map in the cohomology of the distinguished fibre. Each orbit of the required field must connect points with equal image relative to the kth associated period maps of the one-parameter family under consideration. Since for all small values of the parameter these maps in a neighbourhood of each point are diffeomorphisms into the cohomology of the distinguished fibre, it is possible to draw through each point of $\Lambda \backslash \Sigma$ – in one and only one way – a parametrised curve of points with the same image.

It is sufficient to verify the assertions that the field we have constructed can be holomorphically continued to Σ and that the holomorphic continuation is tangent to Σ near non-singular points of the discriminant. Then at an arbitrary point of the discriminant the result will follow from standard theorems about the removal of singularities in codimension 2. The verification of the result near non-singular points of the discriminant can be carried out with the help of explicit formulae, analogous to the formulae of Lemma 12.2 (for $k = 0$ with the help of the formulae of Lemma 12.2) (see [369]).

The second part of the theorem follows from the first part and a theorem of V. M. Zakalyukin [412], asserting that for a quasihomogeneous germ f a vector field on Λ, tangent to the discriminant, is necessarily equal to zero at the origin.

15.1.3   The intersection form in the cotangent bundle

To each non-degenerate period map P_ω^k there corresponds a natural isomorphism of fibrations

$$T_*(\Lambda \backslash \Sigma) \to H^{n-1}$$

and dual to it the isomorphism

$$H_{n-1} \to T^*(\Lambda \backslash \Sigma).$$

Here T_* and T^* are, respectively, the tangent and cotangent bundles, H_{n-1} and H^{n-1}, respectively, the homological and cohomological central Milnor fibrations.

Remark. Here and later it is implied that isomorphisms are defined only on a neighbourhood of the origin in Λ (see the definition of non-degeneracy).

In the fibres of the cohomological fibration there is a bilinear pairing – the intersection number of cycles of the middle dimension in X_λ. In this way a non-degenerate period map P_ω^k determines an *intersection form* Φ_ω^k on the cotangent bundle $T^*(\Lambda \setminus \Sigma)$.

Theorem 15.4 (see [369]). The form Φ_ω^k is holomorphic in $\Lambda \setminus \Sigma$. If P_ω^k is infinitesimally non-degenerate and $k \geqslant [(n-1)/2]$, then the form Φ_ω^k can be holomorphically continued to $T^*\Lambda$.

Proof. The first assertion of the theorem is obvious, since the form Φ_ω^k is induced from a constant form by a holomorphic map. It is sufficient to verify the second assertion near a non-singular point of the discriminant. Near such a point the Jacobian matrix of the kth associated period map can be written out explicitly with the help of the formulae of Lemma 12.2. The form Φ_ω^k is induced from a constant form with the help of a matrix, inverse to the conjugate of the Jacobian matrix. Therefore the theorem can be verified if the expansions in series (analogous to the series indicated in Lemma 12.2) of the coordinates of the inverse matrix contain only non-negative powers of the parameters. It can be verified directly that this is so for the indicated values of k (see [369]).

Theorem 15.5 (see [369]). The intersection form Φ_ω^k, corresponding to an infinitesimally non-degenerate P_ω^k, is stable, that is the intersection forms Φ_η^k for all forms η near to ω map to Φ_ω^k under suitable holomorphic diffeomorphisms of the pair Λ, Σ into itself. If f is a quasihomogeneous germ, then the form Φ_ω^k, corresponding to an infinitesimally non-degenerate P_ω^k, is defined invariantly up to diffeomorphism of the pair Λ, Σ to itself.

Theorem 15.5 is a direct corollary of Theorem 15.3.

15.1.4   The kernel map

Let us make explicit which object in the tangent bundle $T_*(\Lambda\setminus\Sigma)$ is induced by a non-degenerate associated period map from the intersection form in vanishing homologies.

Let $\lambda \in \Lambda\setminus\Sigma$. The intersection form S in $H_{n-1}(X_\lambda, \mathbb{C})$ determines a linear map

$$\pi: H_{n-1}(X_\lambda, \mathbb{C}) \to H^{n-1}(X_\lambda, \mathbb{C}).$$

The kernel of the map π is the same as the kernel of the form S. On the image Im of this map there is a well-defined non-degenerate bilinear form S^*:

$$S^*(\alpha, \beta) = S(\pi^{-1}(\alpha), \pi^{-1}(\beta)).$$

Let P_ω^k be a non-degenerate associated period map. The isomorphism

$$dP_\omega^k: T_*(\Lambda\setminus\Sigma) \to H^{n-1}$$

induces in the tangent bundle of $\Lambda\setminus\Sigma$ a distribution

$$\mathrm{Im}_\omega^k: \lambda \mapsto \mathrm{Im}_\omega^k(\lambda)$$

where $\mathrm{Im}_\omega^k(\lambda) \subset T_{*,\lambda}(\Lambda\setminus\Sigma)$ is a subspace isomorphic to the subspace

$$\mathrm{Im}(\lambda) \subset H^{n-1}(X_\lambda, \mathbb{C}).$$

The codimension of this distribution is equal to the dimension of the kernel of the intersection form in $H_{n-1}(X_\lambda, \mathbb{C})$.

On the planes of the distribution Im_ω^k there is a well-defined non-degenerate bilinear *intersection form* Ψ_ω^k, induced from the form S^*.

It can be shown that the distribution Im_ω^k is integrable, and, furthermore, its integral manifolds are fibres of a holomorphic map from $\Lambda\setminus\Sigma$ into a complex vector space. This map can be given by the following geometrical construction.

Let us consider the finite-dimensional complex vector space Ker of all single-valued covariantly constant sections of the central homological Milnor fibration. It is not hard to convince oneself that an arbitrary section from Ker can be obtained in the following way. We need to choose in the fibre of the central homological Milnor fibration a suitable homology class, belonging to the kernel of the intersection form, and extend it to a covariantly constant section. In particular, the dimension of the space Ker is equal to the dimension of the kernel of the intersection form in the fibres of the homological fibration.

With the space Ker we associate a complex vector space of functions $\{h_\gamma\}_{\gamma\in\mathrm{Ker}^*}$ on $\Lambda\setminus\Sigma$, where the function h_γ is defined by the formula

$$h_\gamma(\lambda) = \langle P_\omega^k(\lambda), \gamma(\lambda)\rangle = (\partial/\partial\lambda_1)^k \int\limits_{\gamma(\lambda)} \omega.$$

Let us define a holomorphic map

$$K_\omega^k \colon \Lambda\setminus\Sigma \to \mathrm{Ker}^*$$

where Ker^* is the space dual to Ker. For $\lambda\in\Lambda\setminus\Sigma$ put the value of $K_\omega^k(\lambda)$ equal to the linear function on Ker which on the vector $\gamma\in\mathrm{Ker}$ is equal to $h_\gamma(\lambda)$. We shall call the map K_ω^k the *kernel map* associated with the form ω.

Remark. Let us offer an equivalent construction of the kernel map. In the central cohomological fibration H^{n-1} there is a subfibration Im, the fibres of which are the subspaces $\{\mathrm{Im}(\lambda)\}$. This subfibration is invariant relative to the Gauss-Manin connection. Therefore the Gauss-Manin connection is defined on the quotient fibration H^{n-1}/Im. It is not hard to convince oneself that the monodromy of the connection on the quotient fibration is trivial. The period map P_ω^k is a section of the fibration H^{n-1}. The section P_ω^k induces a section of the quotient fibration. Mapping the values of the section of the quotient fibration into the distinguished fibre of the quotient fibration, we obtain a single-valued map of the base $\Lambda\setminus\Sigma$ into the space of covariantly constant sections of the fibration H^{n-1}/Im. This is also the kernel map.

Theorem 15.6. 1. For any $k\geqslant 0$, $\omega\in\Omega^{n-1}$ the kernel map K_ω^k is holomorphic on $\Lambda\setminus\Sigma$ and meromorphic on Λ. If $k=0$, then the kernel map can be holomorphically continued to Λ.

2. If the period map P_ω^k is non-degenerate then the kernel map has maximal rank on $\Lambda\setminus\Sigma$ in a neighbourhood of the origin. Furthermore the tangent planes to the fibres of the kernel map are the same as the planes of the distribution Im_ω^k.

Proof. Section 1 is a direct corollary of Theorems 10.4, 10.7. The first part of Section 2 is a direct corollary of the non-degeneracy of the period map. For a proof of the second part it is sufficient to remark that a vector $\xi\in T_{*,\lambda}(\Lambda\setminus\Sigma)$ belongs to $\mathrm{Im}_\omega^k(\lambda)$ if and only if

$$\langle dP_\omega^k(\xi), \alpha\rangle = 0$$

for every homology class $\alpha \in H_{n-1}(X_\lambda, \mathbb{C})$, belonging to the kernel of the intersection form, that is if and only if $\langle dh_\gamma, \xi \rangle = 0$ for any $\gamma \in \mathrm{Ker}$. The theorem is proved.

Corollary of Theorem 15.3. The kernel map K_ω^k and the intersection form Ψ_ω^k on its fibres corresponding to an infinitesimally non-degenerate P_ω^k, are stable, that is the pairs K_η^k, Ψ_η^k for all forms η near to ω map to the pair K_ω^k, Ψ_ω^k under a suitable holomorphic diffeomorphism of the pair Λ, Σ to itself. If f is a quasihomogeneous germ, then the pair K_ω^k, Ψ_ω^k, corresponding to an infinitesimally non-degenerate P_ω^k, is defined invariantly modulo diffeomorphisms of the pair Λ, Σ to itself.

If the number n of arguments of the germ f is equal to 2, Theorem 13.3 on mixed Hodge structures allows us to prove the non-degeneracy of the continuation to Λ of the kernel map $K_\omega^{k=0}$.

Theorem 15.7. If $n = 2$, $k = 0$ and P_ω is an infinitesimally non-degenerate period map, then the kernel map K_ω^0 can be holomorphically continued on Σ to a map with maximal rank.

Proof. It is necessary to verify that the differentials of the functions $\{h_\gamma\}_{\gamma \in \mathrm{Ker}}$ at the origin in Λ form a space of dimension equal to the dimension of the kernel of the intersection form in homologies. For this it is sufficient to verify the same fact on the λ_1-axis, passing through the origin in Λ. Let $\gamma \in \mathrm{Ker}$, then

$$dh_\gamma = \Sigma \langle V_{\partial/\partial\lambda_j} P_\omega, \gamma \rangle \, d\lambda_j.$$

Over the λ_1-axis the section $V_{\partial/\partial\lambda_j} P_\omega$ is a geometric section of a suitable 2-form (see formulae (3), (4) on page 284–285). Now the result of the theorem follows easily from the result (iii) in § 14.2.1 and the infinitesimal non-degeneracy of the map P_ω (see also Lemma 2.4 in [364]).

Let us consider outside the discriminant an arbitrary fibre of the kernel map K_ω^k, corresponding to a non-degenerate period map P_ω^k. In the tangent bundle to the fibre there is defined a non-degenerate intersection form Ψ_ω^k. If the number n of arguments of the germ f is even, then the intersection form Ψ_ω^k is a holomorphic symplectic structure on the fibre. Indeed this form is non-degenerate, antisymmetric and induced by a holomorphic map from a constant form, so in particular, is closed. For odd n the intersection form Ψ_ω^k is a (complex) holomorphic metric on the fibre with zero curvature (with respect to a connection, preserving the metric and induced by the Gauss-Manin connection).

15.1.5 The non-degenerate intersection form

Let us consider at greater length the case when the intersection form in $H_{n-1}(X_\lambda, \mathbb{C})$ is non-degenerate. In this case the intersection form for the non-degenerate map P_ω^k is defined on the whole tangent bundle $T_*(\Lambda \setminus \Sigma)$.

Theorem 15.8 (see [369]).
1. If n is even, $k = n/2 - 1$ and the intersection form Ψ_ω^k corresponds to an infinitesimally non-degenerate period map P_ω^k, then the isomorphism

$$T^*(\Lambda \setminus \Sigma) \to T_*(\Lambda \setminus \Sigma)$$

given by it defines an isomorphism of $\mathbb{C}\{\lambda\}$-modules of germs at the origin in Λ of differential 1-forms and vector fields, tangent to Σ.

For the proof of the theorem it is sufficient to verify the corresponding results near non-singular points of the discriminant; see [369].

15.2 Examples

15.2.1

Let $f : (\mathbb{C}^n, 0) \to (\mathbb{C}, 0)$ be a quasihomogeneous germ;

$$\phi_1 \equiv 1, \quad \phi_2, \ldots, \phi_\mu$$

be a set of monomials projecting into a basis over $\mathbb{C}$ of the local algebra $\mathbb{C}\{x\}/(\partial f/\partial x)$;

$$F(x, \lambda) = f(x) + \lambda_1 + \lambda_2 \phi_2(x) + \ldots + \lambda_\mu \phi_\mu(x);$$

$$\omega = x_1 dx_2 \wedge \ldots \wedge dx_\mu.$$

Then P_ω is an infinitesimally non-degenerate period map. Indeed, according to formulae (3), (4) on page 284–285, over the λ_1-axis passing through the origin, the sections

$$\{\nabla_{\partial/\partial\lambda_j} P_\omega\}, \quad j=1,\ldots,\mu,$$

are represented by the forms

$$\{-\phi_j dx_1 \wedge \ldots \wedge dx_n/df\}.$$

Now the result follows from Theorem 13.6.

15.2.2 (see [16, 369]).

Let $f = x^{\mu+1}$, $\mu \geq 1$, and

$$F(x, \lambda) = x^{\mu+1} + \lambda_2 x^{\mu-1} + \ldots + \lambda_{\mu+1}.$$

We have deliberately changed the above-mentioned numbering of the parameters of the deformation: here the suffix i of the parameter λ_i is proportional to its quasihomogeneous degree. Let us choose $\omega = x$. We cite below a formula for the components $g_{k,l}$, $k, l = 2, \ldots, \mu+1$, of the intersection form Φ_ω^0 in the cotangent bundle. Put $\lambda_0 = 1$, $\lambda_i = 0$ for $i=1$, $i < 0$ or $i > \mu+1$. Then

$$g_{k,l} = \sum_{\substack{i \geq \max(k,l) \\ i+j = k+l-2}} (i-j)\lambda_i \lambda_j + \left[1 - \min(k, l) + \frac{(k-1)(l-1)}{\mu+1} \right] \lambda_{k-1}\lambda_{l-1}.$$

15.2.3

In the case of simple germs of functions of an odd number of variables, belonging to the classes D_μ and E_6, the formulae for the metrics $\Phi_\omega^{(n-1)/2}$ on the cotangent bundle, corresponding to the quasihomogeneous form

$$\text{const} \cdot x_1 dx_2 \wedge \ldots \wedge dx_n,$$

are cited in [125] on page 14.

15.2.4

Let us give some examples of symplectic structures $\Psi_\omega^{n/2-1}$ on bases of versal deformations. Such examples are known for germs of functions of two variables with simple critical points.

Among simple critical points of functions of an even number of variables there are non-degenerate intersection forms in the vanishing homologies of critical points of types $A_{2k}, k \geqslant 1, E_6, E_8$; the intersection forms of critical points of types $D_{2k+1}, k \geqslant 2, E_7$ have one-dimensional kernel; the intersection forms of critical points of types $D_{2k}, k \geqslant 2$, have two-dimensional kernel.

We shall explain how to calculate the symplectic structure Ψ_ω^0 for germs of functions of two variables with critical points of types

$$A_{2k}, k \geqslant 1, E_6, E_8,$$

and then we shall cite the results of the calculations for critical points of small multiplicities.

Let

$$F = f(x, y) + \lambda_1 + \phi_2(x, y)\lambda_2 + \ldots + \phi_\mu(x, y)\lambda_\mu$$

be a quasihomogeneous versal deformation of the germ $f: (\mathbb{C}^2, 0) \to (\mathbb{C}, 0)$ with critical point of one of the types A_{2k}, E_6, E_8. Let $\omega = ydx$. According to example 15.2.1, the form ω generates an infinitesimally non-degenerate period map P_ω. According to Theorem 15.8 the intersection form Ψ_ω, corresponding to the period map, is a symplectic structure on Λ. The form Ψ_ω has the form

$$\sum_{k<l} g_{k,l} d\lambda_k \wedge d\lambda_l,$$

where

$$g_{k,l}(\lambda) = \langle \nabla_{\partial/\partial\lambda_k} \omega|_\lambda, \nabla_{\partial/\partial\lambda_l} \omega|_\lambda \rangle,$$

and $\langle \, , \rangle$ is the intersection form in $H^1(X_\lambda, \mathbb{C})$. We shall represent each coefficient $g_{k,l}(\lambda)$ as the residue of a suitable expression on the algebraic curve

$$Y_\lambda = \{(x, y) \in \mathbb{C}^2 | F(x, y, \lambda) = 0\}.$$

Lemma 15.1. For any $\lambda \in \Lambda \setminus \Sigma$ the natural inclusion of the fibre X_λ of the Milnor fibration in the algebraic curve Y_λ induces an isomorphism of (co)homologies.

The **proof** follows from the fact that for a quasihomogeneous simple germ f the quasihomogeneous degrees of the functions $\{\phi_j\}$ are less than the quasi-homogeneous degree of the germ f.

According to Lemma 15.1 we can calculate the coefficient $g_{k,l}(\lambda)$ as the intersection number of the cohomology classes on Y_λ which are isomorphic to the classes

$$\nabla_{\partial/\partial\lambda_k}\omega|_\lambda, \ \nabla_{\partial/\partial\lambda_l}\omega|_\lambda.$$

According to the formulae (3), (4) on page 284–285, the indicated cohomo-logy classes on Y_λ can be represented by the forms

$$\omega_k = \phi_k dx \wedge dy/d_{x,y}F,$$

$$\omega_l = \phi_l dx \wedge dy/d_{x,y}F,$$

respectively. In order to calculate the intersection number in $H^1(Y_\lambda, \mathbb{C})$ of the cohomology classes of the forms ω_k, ω_l we need to change one of these by addition of the differential of a function so that it becomes a form with compact support, and then we need to multiply the forms and integrate over Y_λ:

$$g_{k,l}(\lambda) = \int_{Y_\lambda} (\omega_k - d\alpha) \wedge \omega_l = -\int_{Y_\lambda} d\alpha \wedge \omega_l$$

(the second equality is true since the forms ω_k, ω_l are holomorphic).

Using Stokes' formula we obtain a rule for calculating the coefficients $g_{k,l}(\lambda)$: on the curve Y_λ in a neighbourhood of its unique point at infinity we represent the form ω_k in the form of the differential of a holomorphic function, namely

$$\omega_k = d\alpha,$$

then

$$g_{k,l}(\lambda) = 2\pi i \operatorname{Res}_\infty [\alpha\omega_l],$$

where $\operatorname{Res}_\infty$ is the residue at infinity.

Now let us give the answers.

(i) Let

$$f = -y^2 + x^3$$

be a germ of type A_2,

$$F = f + \lambda_1 x + \lambda_2,$$

then

$$\Psi_\omega = \text{const } d\lambda_1 \wedge d\lambda_2.$$

(ii) Let

$$f = -y^2 + x^5$$

be a germ of type A_4,

$$F = f + \lambda_1 x^3 + \lambda_2 x^2 + \lambda_3 x + \lambda_4,$$

then

$$\Psi_\omega = \text{const } [d\lambda_1 \wedge d\lambda_4 + 3 d\lambda_2 \wedge d\lambda_3 + \lambda_1 d\lambda_1 \wedge d\lambda_2].$$

(iii) Let

$$f = -y^2 + x^7$$

be a germ of type A_6,

$$F = f + \lambda_1 x^5 + \lambda_2 x^4 + \lambda_3 x^3 + \lambda_4 x^2 + \lambda_5 x + \lambda_6,$$

then

$$\Psi_\omega = \text{const } [3 \, d\lambda_1 \wedge d\lambda_6 - 9 \lambda_1 d\lambda_1 \wedge d\lambda_4 + 6 \lambda_2 d\lambda_1 \wedge d\lambda_3 +$$

$$+ (\lambda_3 + 3 \lambda_1^2) d\lambda_1 \wedge d\lambda_2 + 5 \, d\lambda_2 \wedge d\lambda_5 + 5 \lambda_1 d\lambda_2 \wedge d\lambda_3 + 15 \, d\lambda_3 \wedge d\lambda_4].$$

(iv) Let

$$f = -y^3 + x^4$$

be a germ of type E_6,

$$F = -y^3 + x^4 + \lambda_1 x^2 y + \lambda_2 xy + \lambda_3 y + \lambda_4 x^2 + \lambda_5 x + \lambda_6,$$

then

$$\Psi_\omega = \text{const} \left[-3\,d\lambda_1 \wedge d\lambda_6 + [(5/9)\lambda_1^3 - 27\lambda_4]d\lambda_1 \wedge d\lambda_4 + 2\lambda_1^2 d\lambda_1 \wedge d\lambda_3 - \right.$$

$$\left. -\lambda_1\lambda_2 d\lambda_1 \wedge d\lambda_2 - (15/2)d\lambda_2 \wedge d\lambda_5 + 15\,d\lambda_3 \wedge d\lambda_4 \right].$$

15.2.5

Let $f : (\mathbb{C}^n, 0) \to (\mathbb{C}, 0)$ be the germ of a holomorphic function at a critical point of multiplicity μ, let F be a representative of a versal deformation, and let $\omega \in \Omega^{n-1}$ be a form defining a non-degenerate period map P_ω^k for some k. Let us suppose that the intersection form in $H_{n-1}(X_\lambda, \mathbb{C})$ is non-degenerate and, consequently, there is defined on $T_*(\Lambda \setminus \Sigma)$ a non-degenerate form Ψ_ω^k. Let us suppose that f, F, ω are real on the real parts of their domains. Let us discuss to what degree the form Ψ_ω^k is real on the real part of its domain, that is on $T_*(\mathbb{R}^n \cap (\Lambda \setminus \Sigma))$.

Theorem 15.9. Under the above assumptions the intersection form Ψ_ω^k takes only real values on $T_*(\mathbb{R}^\mu \cap (\Lambda \setminus \Sigma))$ if n is odd, and takes only pure imaginary values on that set if n is even.

Proof. Let us suppose that $f|_{\mathbb{R}^n}$, $F|_{(\mathbb{R}^n \times \mathbb{R}^\mu) \cap (B \times \Lambda)}$, $\omega|_{T_*((\mathbb{R}^n \times \mathbb{R}^\mu) \cap (B \times \Lambda))}$ are real. Let λ belong to $\mathbb{R}^\mu \cap (\Lambda \setminus \Sigma)$. The map dP_ω^k maps $T_\lambda \mathbb{R}^\mu$ into a μ-dimensional real subspace $H_\mathbb{R}$ in $H^{n-1}(X_\lambda, \mathbb{C})$. We need to make explicit what value the intersection form takes on pairs of vectors from $H_\mathbb{R}$. Let us describe $H_\mathbb{R}$.

Let us consider $H^{n-1}(X_\lambda, \mathbb{R})$ as a subspace in $H^{n-1}(X_\lambda, \mathbb{C})$. On each of these spaces there acts an involution, induced by complex conjugation: $(x, \lambda) \mapsto (\bar{x}, \bar{\lambda})$, where $(x, \lambda) \in X_\lambda$. Let us decompose the space $H^{n-1}(X_\lambda, \mathbb{R})$ into a direct sum of subspaces, consisting, respectively, of invariant and antiinvariant cohomology classes:

$$H^{n-1}(X_\lambda, \mathbb{R}) = I \oplus A.$$

Lemma 15.2. We have

$$H_\mathbb{R} = I \oplus iA,$$

where $i^2 = -1$.

The **proof** follows easily from the fact that the cohomology classes

$$\{V_{\partial/\partial\lambda_j} P_\omega^k\}$$

can be represented on X_λ by holomorphic forms, real on the real part of the manifold X_λ.

Let us conclude the proof of the theorem. The involution σ of complex conjugation changes the orientation on X_λ if n is even and does not change the orientation if n is odd. Therefore

$$\langle\cdot,\cdot\rangle = -\langle\sigma\cdot,\sigma\cdot\rangle, \quad \text{if} \quad n \quad \text{is even,}$$

$$\langle\cdot,\cdot\rangle = \langle\sigma\cdot,\sigma\cdot\rangle, \quad \text{if} \quad n \quad \text{is odd.}$$

From these formulae it easily follows that for even n the restriction of the intersection form to each of the subspaces I and A is equal to zero and for odd n the cohomology classes from different subspaces I, A do not intersect. The theorem is proved.

15.2.6

Let us give some examples of kernel maps K_ω^k.
(i) Let $f = -y^2 + x^4$ be a germ of type A_3,

$$F = -y^2 + x^4 + \lambda_1 x^2 + \lambda_2 x + \lambda_3,$$

$$\omega = y\,dx.$$

In this case the kernel of the intersection form in $H_1(X_\lambda, \mathbb{C})$ is one-dimensional:

$$K_\omega^0 = \text{const} \cdot \lambda_2.$$

(ii) Let $f = -y^2 + x^6$ be a germ of type A_5,

$$F = -y^2 + x^6 + \lambda_1 x^4 + \lambda_2 x^3 + \lambda_3 x^2 + \lambda_4 x + \lambda_5,$$

$$\omega = y\,dx.$$

The kernel of the intersection form is one-dimensional:

$$K^0_\omega = \text{const} \cdot (\lambda_3 - \lambda_1^2/4).$$

(iii) Let $f = -y^2 + x^8$ be a germ of type A_7, and let F and ω be analogous to those above. Then

$$K^0_\omega = \text{const} \cdot (\lambda_4 - \lambda_2\lambda_1/2).$$

(iv) Let $f = -y^3 + x^3$ be a germ of type D_4,

$$F = -y^3 + x^3 + \lambda_1 xy + \lambda_2 y + \lambda_3 x + \lambda_4,$$

$$\omega = ydx.$$

In this case the kernel of the intersection form is two-dimensional. Modulo a linear transformation of the image the kernel map K^0_ω has the form

$$(\lambda_1, \lambda_2, \lambda_3, \lambda_4) \longmapsto (\lambda_2, \lambda_3).$$

15.2.7.

We give some examples of images of period maps.

Example. Consider the family of non-singular complex algebraic curves given by the equations

$$y^2 = x^{\mu+1} + \lambda_\mu x^{\mu-1} + \ldots + \lambda_1,$$

depending on the parameters λ. Consider on each curve a choice of base cycles $\gamma_1(\lambda), \ldots, \gamma_\mu(\lambda)$ in the one-dimensional homology, depending continuously on the parameters. The period map of the form $\omega = ydx$ takes a point λ into the μ-tuple of "areas" of the base cycles:

$$\lambda \longmapsto \left(\int_{\gamma_1(\lambda)} \omega, \ldots, \int_{\gamma_\mu(\lambda)} \omega \right).$$

Theorem (see [384]). The image of the period map is $\mathbb{C}^{\mu}\setminus\{0\}$ (so, for example, it contains vectors with all coordinates real). Furthermore, if $\mu > 1$ each non-zero vector is realized as the vector of integrals for infinitely many values of the parameters.

In [384] it is shown that the image of the period map of a form in general position is a punctured neighbourhood of 0 for the versal deformation of any simple function germ in an even number of variables. In the case of any simple function germ in an odd number of variables, the closure of the image of the period map is a neighbourhood of 0. It seems likely that the assertion holds for arbitrary germs.

Example. Consider the family of non-singular curves

$$y^2 = x^3 + \lambda_2 x + \lambda_1$$

and the period map of the form $\eta = y^{-1} dx$:

$$\lambda \mapsto \left(\int_{\gamma_1(\lambda)} \eta, \int_{\gamma_2(\lambda)} \eta \right).$$

This is the first adjoint map of the period map for the form $y\, dx$. The image of this map is $\{(z_1, z_2) \in \mathbb{C}^2 | \mathrm{Im}(z_1/z_2) > 0\}$.

Problem: Describe the image of adjoint period maps.

15.3 The restriction of the symplectic structure on the base of a versal deformation to the stratum of the discriminant contains information on the degeneracy over the stratum

Let us suppose that the number n of variables of the germ f is even and that the intersection form in $H_{n-1}(X_\lambda, \mathbb{C})$ is non-degenerate (in particular μ is even). In this case the intersection form $\Psi_\omega^{n/2-1}$ (for an infinitesimally non-degenerate period map $P_\omega^{n/2-1}$) gives a symplectic structure on Λ. Let us break up the discriminant Σ into strata according to the types of degeneracy of the zero level X_λ.

Principle. The types of degeneracy of the zero level are reflected in the Lagrangean properties of the strata of the discriminant relative to the symplectic structure $\Psi_\omega^{n/2-1}$ (see [369]).

Let us illustrate the principle with examples.

Let the point $\lambda \in \Sigma$ correspond to a variety X_λ with exactly $\mu/2$ non-degenerate critical points. Cycles which vanish at these points do not intersect, therefore the subspace generated by them in the space of homologies of the non-singular fibre is Lagrangean. It can be shown that such λ form in Λ a Lagrangean submanifold relative to the symplectic structure $\Psi_\omega^{n/2-1}$. More precisely, let

$$\Sigma_0 = \{\lambda \in \Sigma | X \text{ has } \mu/2 \text{ singular points},$$

$$\text{all of them being non-degenerate}\}.$$

Theorem 15.10 (see [369]). The space Σ_0 is a Lagrangean submanifold of the symplectic space $(\Lambda, \Psi_\omega^{n/2-1})$.

Examples.

1. Among the critical points of germs of functions of two variables the critical points which have non-degenerate intersection form are those for which the germs of the critical level curves are irreducible. In this case, by deforming the critical level curve it is not hard to convince oneself that the variety Σ_0 is not empty.

2. For germs of type A_{2k} the variety Σ_0 is isomorphic to the subvariety in the space B^{2k+1} of polynomials of the form

$$x^{2k+1} + \lambda_1 x^{2k-1} + \ldots + \lambda_{2k-1},$$

consisting of polynomials with k roots of multiplicity 2. We can show that subvarieties of the same space, consisting of polynomials with roots of multiplicity $k+1$, are Lagrangean with respect to some other symplectic structure (in the linear space of binary forms of odd degree there is exactly one (modulo multiples) non-zero SL_2-invariant exterior 2-form; it is the other symplectic structure, see [124]).

Theorem 15.11 (see [369]). Two affine algebraic varieties in B^{2k+1}

$$\Sigma_1 = \{P_{2k+1} \in B^{2k+1} | P_{2k+1} = (x-a)^{k+1} P_k, \ a \in \mathbb{C}, \ P_k \in B^k\}$$

and

$$\Sigma_2 = \{P_{2k+1} \in B^{2k+1} | P_{2k+1} = (x-a) P_k^2, \ a \in \mathbb{C}, \ P_k \in B^k\}$$

are isomorphic.

Theorem 15.11 can be proved by producing explicit formulae for the automorphism of the space B^{2k+1}, mapping one variety into the other.

Proof of Theorem 15.10. According to Tessier's theorem [348] Σ_0 is a transversal intersection of $\mu/2$ non-singular leaves of the discriminant Σ and, consequently, is a non-singular manifold of dimension $\mu/2$.

Let us denote by W a neighbourhood of the point $\lambda^0 \in \Sigma_0$ in the manifold Σ_0. Put $\mathbb{R}_+ = \{t \in \mathbb{R} | t \geqslant 0\}$,

$$U = \mathbb{R}_+ \times W = \{\lambda = \lambda^1 + te_1 | \lambda^1 \in W, t \in \mathbb{R}_+, e_1$$

is a basic vector of the λ_1-axis in $\Lambda\}$

(we fix a linear structure in Λ).

Let $\gamma_1, \ldots, \gamma_{\mu/2}$ be cycles, vanishing as $t \to 0$. Then the plane

$$\langle \gamma_1, \ldots, \gamma_{\mu/2} \rangle$$

in

$$H_{n-1}(X_\lambda, \mathbb{C}), \ \lambda \in U \setminus W,$$

is Lagrangean. We supplement $\{\gamma_i\}$ with cycles $\{\delta_i\}$ to make a symplectic basis in $H_{n-1}(X_\lambda, \mathbb{C})$. According to Lemma 12.2

$$\int\limits_{\gamma_i} \omega = t^{n/2} \varphi_i, \quad \int\limits_{\delta_i} \omega = t^{n/2} \ln t \varphi_i' + \varphi_i'',$$

where ϕ_i, ϕ_i', ϕ_i'' are analytic functions of the coordinates (t, λ^1) on U. Therefore the period map $P_\omega^{n/2-1}$ gives a family p_t of holomorphic maps of the submanifold W in

$$H^{n-1}(X_\lambda, \mathbb{C}),$$

continuously depending on the parameter $t \in \mathbb{R}_+$; moreover

$$p_0(W)|_{\langle \gamma_1, \ldots, \gamma_{\mu/2} \rangle} \equiv 0.$$

This last statement means that the restriction of the intersection form $\langle \cdot, \cdot \rangle$ in $H^{n-1}(X_\lambda, \mathbb{C})$ to $p_0(W)$ is zero. Since the form $\Psi_\omega^{n/2-1}$ is analytic in Λ we have

$$\Psi_\omega^{n/2-1}|_W = \lim_{t \to 0} \Psi_\omega^{n/2-1}|_{W+te_1} = \lim_{t \to 0} p_t^* \langle \cdot, \cdot \rangle = p_0^* \langle \cdot, \cdot \rangle = 0.$$

The theorem is proved.

Let us formulate a theorem, generalising Theorem 15.10 and proved analogously.

Let $f_1, \ldots, f_k : (\mathbb{C}^n, 0) \to (\mathbb{C}, 0)$ be germs of holomorphic functions with critical points of finite multiplicity. In the base Λ of the versal deformation of the germ $f : (\mathbb{C}^n, 0) \to (\mathbb{C}, 0)$ we choose a non-singular submanifold $\Sigma_{f_1, \ldots, f_k}$ consisting of all points $\lambda \in \Sigma$ for which $F(\cdot, \lambda)$ has k critical points

$$x_1(\lambda), \ldots, x_k(\lambda)$$

equivalent to the critical points of the germs $f_1, \ldots, f_k$, respectively.

Let us suppose that the point $\lambda \in \Lambda \setminus \Sigma$ lies sufficiently close to $\lambda_0 \in \Sigma_{f_1, \ldots, f_k}$. Then the cycles in $H_{n-1}(X_\lambda, \mathbb{C})$, vanishing at the point $x_i(\lambda_0)$, generate a subspace L_i in $H_{n-1}(X_\lambda, \mathbb{C})$. The subspace $\{L_i\}$ possesses the properties:
(i) The restriction of the intersection form $\langle \cdot, \cdot \rangle$ in $H_{n-1}(X_\lambda, \mathbb{C})$ to L_i is isomorphic to the intersection form in homologies, vanishing at the critical point of the germ f_i;
(ii) The sum L of the subspaces

$$\{L_i\}, \quad i = 1, \ldots, k,$$

is a direct sum;
(iii) $\langle L_i, L_j \rangle = 0$ for $i \neq j$;
(iv) L_i is invariant relative to the local monodromy of the point λ_0.

Let us suppose that n is even and that the spectral numbers of the germs $f_1, \ldots, f_k$ lie in the interval $(n/2 - 2, n/2)$ (this is so, for example, for germs, stably equivalent to germs of functions of two variables).

Theorem 15.12. Let us suppose that n is even and that the form $\Psi_\omega^{n/2-1}$ corresponds to an infinitesimally non-degenerate period map $P_\omega^{n/2-1}$. Then
1. (see [369]). The restriction of the symplectic structure $\Psi_\omega^{n/2-1}$ to the subspace

$$T_{\lambda_0} \Sigma_{f_1, \ldots, f_k}$$

can be induced from the intersection form in $H^{n-1}(X_\lambda, \mathbb{C})$ by a suitable linear map in the annihilator of the subspace

$$L \subset H_{n-1}(X_\lambda, \mathbb{C}).$$

2. If additionally it is known that the critical points of all the germs $f_1, \ldots, f_k$ are simple, then the restriction to

$$T_{\lambda_0} \Sigma_{f_1, \ldots, f_k}$$

of the symplectic structure $\Psi_\omega^{n/2-1}$ is isomorphic to the restriction of the intersection form in $H^{n-1}(X_\lambda, \mathbb{C})$ to the annihilator of the subspace

$$L \subset H_{n-1}(X_\lambda, \mathbb{C}).$$

Corollary of part 1.

$$\dim_{\mathbb{C}} \mathrm{Ker}\,(\Psi_\omega^{n/2-1} | T_{\lambda_0}(\Sigma_{f_1, \ldots, f_k})) \geq \sum_{i=1}^k \dim_{\mathbb{C}} \mathrm{Ker}\,(\langle\,,\,\rangle_{f_i}),$$

where $\langle\cdot,\cdot\rangle_{f_i}$ is the intersection form in homologies, vanishing at the critical point of the germ f_i.

Postscript. During the time that this book was being prepared for publication, many new works have appeared. We note here only the proof of the conjecture on semicontinuous spectra ([345, 371, 372, 375]), the theory of open swallowtails, the discovery of a connection between Lagrangean and Legendrean manifolds and singularities with simplectic and contact structures of manifolds of binary forms and polynomials ([22, 128]), the inclusion in the theory of singularities of the groups of reflections H_3 and H_4 ([235, 334, 378]), the classification of the projections of two-dimensional surfaces in general position from three-dimensional space to a plane ([25, 291, 292]) and the calculation of the ring of Legendrean cobordisms ([34–36]). You can learn about these new achievements from the survey articles [2, 3, 23, 25, 26].

References

Abbreviations

FAP = Funktsional'nyi Analiz i ego Prilozheniya
FAA = Functional Analysis and its Applications (Eng. Transl. of FAP)
UMN = Uspekhi Matematicheskykh Nauk
RMS = Russian Mathematics Survays (Eng. Transl. of most of UMN)
DAN = Doklady Akademii Nauk SSSR
SMD = Soviet Mathematics Doklady (Eng. Transl. of most of DAN)
IAN = Izvestiya Akademii Nauk SSSR, seriya Matematicheskaya
MUI = Mathematics of the USSR Izvestiya (Eng. Transl. of most of IAN)

ALGEBRAIC SURFACES

[1]   Proc Mat Inst V. A. Steklov, LXXV, Moscow, Nauka, 1965, AMS Providence, RI, 1967.

[2]   Science and technology reviews: current problems in mathematics 22, M Viniti 1983. Current problems in mathematics 22. Itogi Nauki i Tekhniki. Viniti AN SSSR, Moscow 1983.

[3]   Singularities. Proc Symp Pure Math 1983, 40:1–2.

A'CAMPO, N.

[4]   Le nombre de Lefschetz d'une monodromie. Indag Math 1973, 76, 113–118.

A'CAMPO, N.

[5]   La fonction zeta d'une monodromie. Commentarii Mathematici Helvetici 1975, 50, 233–248.

A'CAMPO, N.

[6]   Tresses, monodromie et le groupe symplectique. Commentarii Mathematici Helvetici 1979, 54, 318–327.

A'CAMPO, N.

[7]   Le groupe de monodromie du deploiement des singularités isolées de courbes planes I, Math Ann 1975, 213, 1–32.

A'CAMPO, N.

[8]   Le groupe de monodromie du deploiement des singularités isolées de courbes planes II, Proc. of ICM (Vancouver 1974) 1975, 1, 395–404.

ANDREOTTI, A.

[9]   On Torelli's theorem. Am J Math 1958, 80, 101–121.

ARNOLD, V.I.

[10]   On the arrangement of ovals of real plane algebraic curves, involutions of four-dimensional smooth manifolds and arithmetic of integral quadratic forms. FAP 1971, 5:3, 1–9. FAA 5, 169–176.

ARNOLD, V.I.

[11]   Integrals of rapidly oscillating functions and singularities of projection of Lagrangean manifolds. FAP 1972, 6:3, 61–62. FAA 6, 222–225.

ARNOLD, V.I.

[12]   Remarks on the method of stationary phase and Coxeter numbers. UMN 1973, 28:5, 17–44. RMS 28:5, 19–48.

ARNOLD, V.I.
   [13]   Critical points of smooth functions. Proc ICM (Vancouver 1974) 1975, *1*, 19–41.
ARNOLD, V.I.
   [14]   Critical points of smooth functions and their normal forms. UMN 1975, *30*:5, 3–65.
           RMS *30*:5, 1–75.
ARNOLD, V.I.
   [15]   Some unsolved problems in the theory of singularities. Proc sem S.L. Sobolev 1976, *1*,
           5–15. Eng. translation is [27].
ARNOLD, V.I.
   [16]   Wave front evolution and equivariant Morse lemma. Comm Pure Appl Math 1976,
           29, 557–582.
ARNOLD, V.I.
   [17]   The index of a singular point of a vector field, the inequalities of Petrovskii-Oleinik
           and the mixed Hodge structures. FAP 1978, *12*:1, 1–14. FAA *12*, 1–11.
ARNOLD, V.I.
   [18]   Critical points of functions on manifolds with boundary, the simple Lie groups
           B_k, C_k, F_4 and the singularities of evolutes. UMN 1978, *33*:5 (203), 91–105. RMS
           33:5, 99–116
ARNOLD, V.I.
   [19]   On some problems in singularity theory. K. Patodi memorial volume, Bombay 1979.
ARNOLD, V.I. – OLEINIK, O.A.
   [20]   The topology of real algebraic varieties. Vestnik MGU ser math 1979, *6*, 7–17.
ARNOLD, V.I.
   [21]   Normal forms of functions near degenerate critical points, the Weyl groups A_k, D_k, E_k
           and Lagrangean singularities. FAP 1972, *6*:4, 3–25. FAA *6*, 254–272.
ARNOLD, V.I.
   [22]   Singularities of Legendre varieties, of evolvents and of fronts at an obstacle. Ergodic
           theory and Dynamical Systems 1982, *2*, 301–309.
ARNOLD, V.I.
   [23]   Catastrophe theory. Moscow Izd MGU 1983. Eng. Transl. Springer 1983.
ARNOLD, V.I.
   [24]   Singularities in the calculus of variations. Science reviews: current problems in
           mathematics 1983, *22*, 3–55.
ARNOLD, V.I.
   [25]   Singularities of systems of rays. UMN 1983, *38*:2, 77–147. RMS *38*:2, 87–176.
ARNOLD, V.I.
   [26]   Singularities of ray systems. ICM, Warsaw 1983.
ARNOLD, V.I.
   [27]   Some open problems in the theory of singularities. Proc Symp Pure Math 1983, *40*:1,
           57–69.
ARNOLD, V.I.
   [28]   Vanishing inflections, FAP 1984, *18*:3, 58–59. FAA *18*, 128–130.
ARNOLD, V.I. – SHANDARIN, S.F. – ZELDOVICH, Y.B.
   [29]   The large scale structure of the universe I. Geophys Astrophys Fluid dynamics 1982,
           20, 111–130.
ARNOLD, V.I.
   [30]   Remarks on Poisson structures on the plane and other powers of the volume form.
           Trudi seminara Petrovskii, 1985, *12*, 1–14.

ARNOLD, V.I. – VARCHENKO, A.N. – GIVENTAL, A.B. – HOVANSKY, A.G.

[31]   Singularities of functions, wave fronts, caustics and multidimensional integrals. Math Phys Rev, 1984, *4*, 1–92.

ATIYAH, M.F.

[32]   Resolution of singularities and division of distributions. Comm Pure Appl Math 1970, *23*, 145–150.

ATIYAH, M.F. – BOTT, R. – GARDING, L.

[33]   Lacunae for hyperbolic differential operators with constant coefficients. Acta Math 1970, *124*, 109–189.

AUDIN, M.

[34]   Quelques calculs en corbordisme Lagrangien. Ann Inst Fourier 1985, *35*:3, 159–194.

AUDIN, M.

[35]   Classes caractéristiques d'immersion lagrangiennes définies par des varietés de caustiques (d'après Vasiliev). Seminaire Sud-Rhodanien de géométrie, I. Travaux en Cours, Hermann, Paris 1984.

AUDIN, M.

[36]   Remarques sur les nombres caractéristiques entiers de certaines immersions lagrangiennes. C.R. Acad Sci 1983, *297*, 561–563.

BABENKO, K.I.

[37]   On the asymptotics of the Dirichlet kernel of spherical means of multiple Fourier series. DAN, 1978, *243*:5, 1097–1100. SMD, *19*, 1457–1461.

BARLET, D.

[38]   Forme hermitienne canonique sur la cohomologie de la fibre de Milnor d'une hypersurface à singularité isolée. Invent Math 1985, *81*, 151–154.

BARLET, D.

[39]   Contribution effective de la monodromie aux developments asymptotiques. Ann Sci de l'ENS 1984, *17*, 293–315.

BARLET, D.

[40]   Contribution du cup-produit de la fibre de Milnor aux pôles de $|f|^s$. Ann Inst Fourier, Grenoble 1984, *34*, 75–107.

BARLET, D. – VARCHENKO, A.N.

[41]   Around the intersection form of an isolated singularity of a hypersurface. Université de Nancy I, 1986.

BASSET, A.B.

[42]   The maximum number of double points on a surface. Nature 1906, *73*, 246.

BEAUVILLE, A.

[43]   Sur les nombre maximum de points doubles d'une surface dans P^3 ($\mu(5)=31$). Journées de geométrie algébrique d'Angier, Algebraic geometry Angers 1979, Sythoff and Noordhoff 1980, 207–215.

BERNSTEIN, D.N.

[44]   The number of roots of a system of equations. FAP 1975, *9*:3, 1–4. FAA *9*, 183–185.

BERNSTEIN, D.N. – KUSHNIRENKO, A.G. – HOVANSKY, A.G.

[45]   Newton polyhedra. UMN 1976, *31*:3, 201–202.

BERNSTEIN, I.N.

[46]   Analytic continuation of generalised functions with respect to a parameter. FAP 1972, *6*:4, 26–40. FAA *6*, 273–285.

BERNSTEIN, I.N. – GELFAND, S.I.

[47]   The function P^λ is meromorphic. FAP 1969, *3*:1, 84–86. FAA *3*, 68–69.

468 References

BERRY, M.V.
[48]   Waves and Thom's theorem. Adv Phys 1976, 25, 1–26.
BERRY, M.V.
[49]   Singularities in waves and rays. Les Houches Summer School 1980, Amsterdam, North Holland 1981.
BERRY, M.V. – NYE, J.F.
[50]   Fine structure in caustic junctions. Nature, London 1977, 267, 34–36.
BERRY, M.V. – UPSTILL, C.
[51]   Catastrophe optics: morphologies of caustics and their diffraction patterns. Progress in optics XVIII, North Holland, 1980.
BJORK, J.E.
[52]   Rings of differential operators. North Holland, Amsterdam, 1979.
BOURBAKI, N.
[53]   Lie groups and Lie algebras. Hermann, Paris 1971.
BRIANÇON, J. – SKODA, H.
[54]   Sur la cloture integrale d'un ideal de germes de fonctions holomorphes en un point de $\mathbb{C}^n$. C.R. Acad Sci Paris ser A 1974, 278, 949–951.
BRIESKORN, E.
[55]   Die Monodromie der isolierten Singularitäten von Hyperflächen. Manuscripta Math 1970, 2, 103–161.
BRIESKORN, E.
[56]   Special singularities – resolution, deformation, monodromy. AMS lecture notes prepared in connection with the summer institute on algebraic geometry held at Humboldt State University, Arcata, California 1974.
BRIESKORN, E.
[57]   Sur les groupes de tresses (d'après Arnold). Sém. Bourbaki (1971/72) 22–24. Also published in: Lecture Notes in Math 317, Springer, Berlin 1973.
BRIESKORN, E.
[58]   Die Milnorgitter der exzeptionellen unimodularen Singularitäten. Bonner Math Schriften 1983, 150, 1–225.
BRUCE, J.W.
[59]   An upper bound for the number of singularities on a projective hypersurface. Bull Lond Math Soc 1981, 13:1, 47–50.
BRUCE, J.W.
[60]   Counting singularities. Proc Roy Soc Edinburgh 1982, A93:1–2, 137–159.
BRUCE, J.W. – GAFFNEY, T.
[61]   Simple singularities of mappings $(\mathbb{C}, 0) \to (\mathbb{C}^2, 0)$. J Lond Math Soc 1982, 26:3, 465–474.
BRUCE, J.W. – GIBLIN, P.J.
[62]   Smooth stable maps of discriminant varieties. Proc Lond Math Soc 1985, 50, 535–551.
BRUCE, J.W. – GIBLIN, P.J.
[63]   Outlines and their duals. Proc Lond Math Soc 1985, 50, 552–570.
BRUCE, J.W. – GIBLIN, P.J.
[64]   Two-parameter families of plane caustics by reflection. Proc Lond Math Soc 1985, 50, 571–576.
CASSOU-NOGUES, P.
[65]   Prolongement des séries de Dirichlet associées à un polynôme à 2 indeterminées. J Number Theory, 1986, 23, 1–54.

References 469

CASSOU-NOGUES, P.
[66]   Racines des polynômes de Bernstein. Ann Inst Fourier, Grenoble, 1986, *36*:4, 1–30.
CASSOU-NOGUES, P.
[67]   Séries de Dirichlet et intégrales associées à un polynôme à deux indeterminées. J Number Theory, 1986, *23*, 1–54.
CATANESE, F. – CERESA, G.
[68]   Constructing sextic surfaces with a given number of nodes. J Pure Appl Algebra, 1982, *23*, 1–12.
CATTANI, E. – KAPLAN, A.
[69]   On the local monodromy of a variation of Hodge structure. Bull AMS 1981, *4*, 116–118.
CATTANI, E. – KAPLAN, A.
[70]   Polarised mixed Hodge structures and the monodromy of a variation of Hodge structure. Preprint 1981, 1–25.
CHERN, S.S.
[71]   Complex Manifolds Without Potential Theory. Van Nostrand, Princeton, 1967. 2nd ed. Springer, New York 1979.
CHMUTOV, S.V.
[72]   Monodromy groups of singularities of functions of two variables. FAP 1981, *15*:1, 61–66. FAA *15*, 48–52.
CHMUTOV, S.V.
[73]   On the monodromy of isolated singularities. Proc 16th all-union algebra conf Leningrad 1981, *1*, 172–173.
CHMUTOV, S.V.
[74]   Monodromy groups of critical points of functions. Inven Math 1982, *67*, 123–131.
CHMUTOV, S.V.
[75]   The monodromy groups of critical points of functions II. Invent Math 1983, *73*, 491–510.
CHMUTOV, S.V.
[76]   The spectrum and equivariant deformations of critical points. UMN 1984, *39*:4, 113–114.
CLEMENS, C.H.
[77]   Picard-Lefschetz theorem for families of non-singular algebraic varieties acquiring ordinary singularities. Trans AMS 1969, *136*, 93–108.
CLEMENS, C.H.
[78]   Degeneration of Kähler manifolds. Duke Math J 1977, *44*, 215–290.
CODDINGTON, E.A. – LEVINSON, N.
[79]   Theory of ordinary differential equations. McGraw-Hill, New York 1955.
COLIN DE VERDIER, Y.
[80]   Nombre de points entiers dans une famille homothetique de domains de $\mathbb{R}^n$. Ann Sci Ecole Norm Super ser 4 1977, *10*:4, 559–575.
CORNALBA, M. – GRIFFITHS, P.
[81]   Some transcendental aspects of algebraic geometry. Proc Summer Institute AMS, Arcata, 1974.
DAMON, J.
[82]   Topological properties of discrete algebra types I. The Hilbert-Samuel function. Adv in Math supp ser 1979, *5*:, 83–118; ibid II. Real and complex algebras. Amer J Math 1979, *101*, 1219–1248.

DAMON, J.
  [83]   Topological triviality in versal unfoldings. Proc Symp Pure Math, 1983, *40*:1,
          255–266.
DAMON, J.
  [84]   Finite determinacy and topological triviality I. Invent Math 1980, *62*, 229–324; ibid II.
          Sufficient conditions and topological stability. Compositio Math 1982, *47*, 101–132.
DAMON, J.
  [85]   Topological stability in the nice dimensions. Topology 1979, *18*, 129–142.
DAMON, J.
  [86]   Finite determinacy and topological triviality II: sufficient conditions and topological
          stability. Comp Math 1982, *47*:2, 101–132.
DAMON, J. – GAFFNEY, T.
  [87]   Topological triviality of deformations of functions and Newton filtrations. Invent
          Math 1983, *72*, 335–358.
DANILOV, V.I.
  [88]   Geometry of toric manifolds. UMN 1978, *33*:2, 85–135. RMS *33*:2, 97–154.
DANILOV, V.I.
  [89]   Newton polyhedra and vanishing cohomologies. FAP 1979, *13*:2, 32–47. FAA *13*,
          103–112.
DAVYDOV, A.A.
  [90]   The boundary of attainability of many-dimensional guided systems. Proc Tbilisskogo
          univ. 1982, *232–233*:13–14, 78–96.
DELIGNE, P.
  [91]   Les immeubles des groupes de tresses generalises. Invent Math 1972, *17*, 273–302.
DELIGNE, P.
  [92]   Theorie de Hodge I. Proc ICM (Nice 1970), *1*, 425–430; ibid II. Publ Math IHES 1971,
          40, 5–58; ibid III. Publ Math IHES 1972, *44*, 5–77.
DELIGNE, P.
  [93]   Equations differentielles a points singuliers reguliers. Lecture notes in Math 163,
          Springer-Verlag 1970.
DEMAZURE, M.
  [94]   Classification des germes a point critique isolé et a nombres de modules 0 ou 1 (d'après
          V.I. Arnold). Séminaire Bourbaki, 26e année, 1973/74, 443, Fevrier 1974.
DIMCA, A.
  [95]   Germes de fonction definier sur les singularités isolées d'hypersurfaces. C.R. Acad Sci
          1983, *297*:2, 217–219.
DIMCA, A. – GIBSON, C.
  [96]   Contact unimodular germs from the plane to the plane. Quart J Math 1983, *34*:135,
          281–295.
DOLGACHEV, I.V. – NIKULIN, V.V.
  [97]   The exceptional singularities of V.I. Arnold and K-3 surfaces – The all-union
          topological conference in Minsk (thesis), Minsk 1977.
DRUCKER, D. – FROHARDT, D.
  [98]   Irreducible root systems and finite linear groups of degree two. Bull Lond Math Soc
          1982, *14*:2, 142–148.
DUFOUR, J.
  [99]   Familles des courbes planes differentiables. Topology 1983, *22*:4, 449–479.
DUISTERMAAT, J.
  [100]   Oscillatory integrals, Lagrange immersions and unfoldings of singularities. Comm
          Pure Appl Math 1974, *27*, 207–281.

References   471

DURFEE, A.H.
[101]   Fibered knots and algebraic singularities. Topology 1974, *13*, 47–59.
DURFEE, A.
[102]   Fifteen characteristics of rational double points and simple critical points.
         L'Enseignement Math 1979, *25*:1–2, 131–163.
DURFEE, A. – HAIN, R.
[103]   Mixed Hodge structure on the homotopy of links. Math Ann 1988 (to appear).
DURFEE, A.
[104]   Mixed Hodge structures on punctured neighbourhoods. Duke J Math 1983, *50*,
         1017–1040.
EBELING, W.
[105]   Quadratische Formen und Monodromee-Gruppen von Singularitäten. Math Ann
         1981, *255*, 463–498.
EHLERS, F. – LO K.-C.
[106]   Minimal characteristic exponent of the Gauss-Manin connection of isolated singular
         point and Newton polyhedron. Math Ann 1982, *259*, 431–441.
EL AMRANI, M.
[107]   Singularités des fonctions obtenues par intégration sur la fibre $x^2 - y^3 = s$ et identités
         modulaires. Bull Sci Math 2e serie, 1984, *108*, 409–421.
EL ZEIN, F.
[108]   Degenerescence diagonale I. C.R. Acad Sci 1983, *296*, 51–54; ibid II. C.R. Acad Sci
         1983, *276*, 199–202.
FEDORYUK, M.V.
[109]   The method of stationary phase for multiple integrals. J Comp Math and Math Phys
         1962, *2*, 145–150.
FEDORYUK, M.V.
[110]   The saddle-point method. Moscow, Nauka 1977.
FORSTER, O.
[111]   Riemannsche Flächen. Springer-Verlag, Berlin 1977.
FRESNEL, A.
[112]   Mémoire sur la diffraction de la lumière. Mem de l'Acad des sciences 1818, *5*,
         339–353.
FUKS, B.L.
[113]   Special chapters of the theory of analytic functions of several complex variables.
         Moscow, Physmatgiz 1963. Eng. Trans.: Amer Math Soc, Providence 1965.
FUKUDA, T.
[114]   Types topologiques des polynomes. Publ IHES 1976, *46*, 87–106.
FUKUDA, T.
[115]   Local topological properties of differentiable mappings I. Invent math 1981, *65*,
         227–250.
GABRIELOV, A.M.
[116]   Intersection matrices for certain singularities. FAP 1973, *7*:3, 18–32. FAA *7*,
         182–193.
GABRIELOV, A.M.
[117]   Bifurcation, Dynkin diagrams and the modality of isolated singularities. FAP 1974,
         8:2, 7–12. FAA *8*, 94–98.
GABRIELOV, A.M.
[118]   Dynkin diagrams of unimodal singularities. FAP 1974, *8*:3, 1–6. FAA *8*, 192–196.

GABRIELOV, A.M.

[119]   Polar curves and intersection matrices of singularities. Invent math 1979. *54*: 1, 15–22.

GABRIELOV, A.M. – KUSHNIRENKO, A.G.

[120]   Description of deformations with constant Milnor number for homogeneous functions. FAP 1975, *9*: 4, 67–68. FAA *9*, 329–331.

GALLARATI, D.

[121]   Una superficie dell'ottavo ordine con 160 nodi. Acc Sci Lett 1957, *14*, 1–7.

GARDING, L.

[122]   Sharp fronts of paired oscillatory integrals. Publ Res Inst Math Sci ser A 1977, *12*, 53–68.

GELFAND, I.M. – SHILOV, G.E.

[123]   Generalised functions and actions on them, Vol 1. Moscow, Physmatgiz 1959. Eng. Transl.: Generalized Functions. Academic Press, New York 1964.

GIVENTAL, A.B.

[124]   Manifolds of polynomials with roots of fixed codimension and generalised Newton equation. FAP 1982, *6*: 1, 13–18. FAA *16*, 10–14.

GIVENTAL, A.B.

[125]   Convolution of invariants of groups, generated by reflections and connections with simple singularities of functions. FAP 1980, *14*: 2, 4–14. FAA *14*, 81–89.

GIVENTAL, A.B.

[126]   Asymptotic of the intersection forms for quasihomogeneous singularities. FAP 1982, *16*: 4, 63–65. FAA *16*, 294–297.

GIVENTAL, A.B.

[127]   On the maximum number of singular points on a projective hypersurface. FAP 1983, *17*: 3, 73–74. FAA *17*, 223–225.

GIVENTAL, A.B.

[128]   Lagrangean varieties with singularities and irreducible sl_2-modules. UMN 1983, *38*: 6, 123–124. RMS *38*: 6, 121–122.

GODBILLON, C.

[129]   Géométrie Différentielle et Méchanique Analytique, Hermann, Paris 1969.

GORIN, E.A.

[130]   On asymptotic properties of polynomials and algebraic functions. UMN 1961, *16*: 1, 91–118. RMS *16*: 1, 93–119.

GORYUNOV, V.V.

[131]   Adjacencies of spectra of certain singularities. Vestnik MGU ser math 1981, *4*, 19–22.

GORYUNOV, V.V.

[132]   The geometry of bifurcation sets of simple projections onto a line. FAP 1981, *15*: 2, 1–8. FAA *15*, 77–82.

GORYUNOV, V.V.

[133]   The projections of zero-dimensional complete intersections onto a line and the $K(\pi, 1)$ conjecture. UMN 1982, *37*: 3, 179–180. RMS *37*: 3, 206–208.

GORYUNOV, V.V.

[134]   Bifurcation sets of some simple and quasihomogeneous singularities. FAP 1983, *17*: 2, 23–37. FAA *17*, 97–108.

GORYUNOV, V.V.

[135]   Singularities of projections of complete intersections. Current problems in mathematics, *22*, 167–206. Itogi Nauki i Tekhniki, Viniti AN SSSR, Moscow 1983.

GREUEL, G.-M. – STEENBRINK, J.

[136]   On the topology of smoothable singularities. Proc Symp Pure Math 1983, *40*: 1, 535–543.

GREUEL, G.-M.

[137]   Constant Milnor number implies constant multiplicity for quasihomogeneous singularities. Manuscripta Math 1986, 56, 159–166.

GRIFFITHS, P.

[138]   Monodromy of homology and periods of integrals on algebraic manifolds. Mimeographed notes, Princeton Univ, 1968.

GRIFFITHS, P.

[139]   On the periods of certain rational integrals I. Ann Math 1969, ser 2, 90, 460–495; ibid II. Ann Math 1969, ser 2, 90, 496–541.

GRIFFITHS, P.

[140]   Periods of integrals on algebraic manifolds: summary of main results and discussion of open problems. Bull AMS 1970, 76, 228–296.

GRIFFITHS, P. – HARRIS, J.

[141]   Principles of algebraic geometry. John Wiley and sons, New York 1978.

GRIFFITHS, P. – SCHMID, W.

[142]   Recent developments in Hodge theory, a discussion of techniques and results. Proc internat colloq on discrete subgroups of Lie groups, Bombay 1973, 31–127.

GUDKOV, D.A.

[143]   The topology of real projective algebraic varieties. UMN 1974, 29:4, 3–79. RMS 29:4, 1–79.

GUDKOV, D.A. – KRAKHNOV, A.D.

[144]   On the periodicity of the Euler characteristics of real algebraic $(M-1)$-manifolds. FAP 1973, 7:2, 15–19. FAA 7, 98–102.

GUILLEMIN, V. – STERNBERG, S.

[145]   Geometric asymptotics. Amer. Math. Soc., Providence 1975.

GUNNING, R. – ROSSI, H.

[146]   Analytic functions of several complex variables. Prentice-Hall, New York 1965.

GUSEIN-ZADE, S.M.

[147]   Intersection matrices for certain singularities of functions of two variables. FAP 1974, 8:1, 11–15. FAA 8, 10–13.

GUSEIN-ZADE, S.M.

[148]   Dynkin diagrams of singularities of functions of two variables. FAP 1974, 8:4, 23–30. FAA 8, 295–300.

GUSEIN-ZADE, S.M.

[149]   On the characteristic polynominal of the classical monodromy for a series of singularities. FAP 1976, 10:3, 78–79. FAA 10, 229–230.

GUSEIN-ZADE, S.M.

[150]   Monodromy groups of isolated singularities of hypersurfaces. UMN 1977, 32:2 (194), 23–65. RMS 32:2, 23–69.

GUSEIN-ZADE, S.M.

[151]   On distinguished bases of simple singularities. FAP 1980, 14:4, 73–74. FAA 14, 307–308.

GUSEIN-ZADE, S.M.

[152]   The index of a singular point of a gradient vector field. FAP 1984, 18:1, 7–12. FAA 18, 6–10.

GUSEIN-ZADE, S.M. – NEKHOROSHEV, N.N.

[153]   On the adjoining of a singularity A_k to points of the stratum $\mu=$ const of a singularity. FAP 1983, 17:4, 82–83. FAA 17, 312–313.

GUSEIN-ZADE, S.M.
  [154]   An equivariant analogue of the index of a gradient vector field. Lecture Notes in Math
            1214, 196–210, Springer 1986.

GUSEIN-ZADE, S.M.
  [155]   On singularities admitting perturbations with a small number of critical values. FAP
            1987, *21*:1, 76–77. FAA *21*, 66–68.

HAMM, H.
  [156]   Locale topologische Eigenschaften komplexen Räume. Math Ann 1971, *191*,
            235–252.

HAMM, H. – LÊ DŨNG TRÁNG.
  [157]   Un theorème de Zariski du type de Lefschetz. Annales Scientifiques de l'Ecole
            Normale Supérieure 1973, 317–356.

HIRONAKA, H.
  [158]   Resolution of singularities of an algebraic variety over a field of characteristic zero.
            I, II. Math Ann 1964, *79*, 109–203, 205–326.

HOVANSKY, A.G. (=KHOVANSKII, A.G.)
  [159]   Newton polyhedra and toral manifolds. FAP 1977, *11*:4, 56–67. FAA *11*, 289–295.

HOVANSKY, A.G.
  [160]   Newton polyhedra and the genus of complete intersections. FAP 1978, *12*:1, 51–61.
            FAA *12*, 38–46.

HOVANSKY, A.G.
  [161]   Newton polyhedra and the Euler-Jacobi formula. UMN 1978, *33*:6, 245–246. RMS
            33:6, 237–238.

HOVANSKY, A.G.
  [162]   The index of a polynomial vector field. FAP 1979, *13*:1, 49–58. FAA *13*, 38–45.

HOVANSKY, A.G.
  [163]   On one class of systems of transcendental equations. DAN 1980, *255*:4, 804–807.
            SMD *22*:3, 762–765.

HOVANSKY, A.G.
  [164]   Sur les racines complexes des systèmes d'equations algébriques ayant un petit nombre
            de monomes. C.R. Acad Sci Paris 1981, *292*:June, 937–940.

HOVANSKY, A.G.
  [165]   Theorème de Bezout pour les fonctions de Liouville. Preprint IHES 1981, Sept, 1–31.

HOVANSKY, A.G.
  [166]   Real analytic manifolds with properties of finiteness and complex abelian integrals.
            FAP 1984, *18*:2, 45–58. FAA *18*, 119–127.

HOVANSKY, A.G.
  [167]   Fewnomials and Pfaff manifolds. Proc Int Cong of Math, Warsaw 1984, vol. 1,
            549–564.

HOVANSKY, A.G.
  [168]   Newton polyhedra and resolutions of singularities. Current problems in mathematics
            22, 207–239. Itogi Nauki i Tekhniki, Viniti AN SSSR, Moscow 1983.

HUA-LOO-KENG
  [169]   Abschätzung von Exponentialsummen und ihre Anwendung in der Zahlentheorie.
            Teubner, Leipzig 1959.

HUMPHRIES, S.P.
  [170]   On weakly distinguished bases and free generating sets of free groups. Quart J Math
            Oxford 1985, *36*, 215–219.

Iʟ'ʏuᴛᴀ, G.G.

[171]   Monodromy and vanishing cycles of boundary singularities. FAP 1985, *19*:3, 11–21.
         FAA *19*, 173–182.

Jᴀɴssᴇɴ, W.A.M.

[172]   Skew-symmetric vanishing lattices and their monodromy groups, I. Math Ann 1983,
         266, 115–133.

Jᴀɴssᴇɴ, W.A.M.

[173]   Skew-symmetric vanishing lattices and their monodromy groups, II. Math Ann 1985,
         272, 11–22.

Jᴇᴀɴǫuᴀʀᴛɪᴇʀ, P.

[174]   Développement asymptotique de la distribution de Dirac. C.R. Acad Sci Paris 1970,
         271, 1159–1161.

Kᴀʀᴀᴛsuʙᴀ, A.A.

[175]   Ivan Matveevich Vinogradov (on his ninetieth birthday). UMN 1981, *36*:6, 3–15.
         RMS *30*:6, 1–17.

Kᴀʀᴘusʜᴋɪɴ, V.N.

[176]   Uniform estimates of oscillatory integrals in $\mathbb{R}^2$. DAN 1980, *254*:1, 28–31.

Kᴀʀᴘusʜᴋɪɴ, V.N.

[177]   Uniform estimates of oscillatory integrals. UMN 1981, *36*:4, 213.

Kᴀʀᴘusʜᴋɪɴ, V.N.

[178]   Uniform estimates of oscillatory integrals with parabolic or hyperbolic phase. Proc
         sem. I.G. Petrovskii 1983, *9*, 3–39.

Kᴀʀᴘusʜᴋɪɴ, V.N.

[179]   Theorems on uniform estimates of oscillatory integrals with phase depending on two
         variables. Proc sem I.G. Petrovskii 1984, *10*, 150–169.

Kᴀsʜɪᴡᴀʀᴀ, M.

[180]   *B*-functions and holonomic systems. Rationality of roots of *b*-functions. Invent Math
         1976. *38*, 33–53.

Kᴀᴛᴢ, N.M.

[181]   Nilpotent connections and the monodromy theorem. Application of a result of
         Turrittin. Publ Math IHES 1970, *39*, 175–232.

Kᴀᴛᴢ, N.M.

[182]   The regularity theorem in algebraic geometry. Actes CIM 1970, *1*, 437–443.

Kᴀᴢᴀʀɴᴏᴠsᴋɪɪ, B.Y.

[183]   On the zeros of exponential sums. DAN 1981, *257*:4, 804–808. SMD *23*:2, 347–351.

Kᴇᴍᴘғ, G. – Kɴuᴅsᴇɴ, F. – Muᴍғᴏʀᴅ, D. – Sᴀɪɴᴛ-Dᴏɴᴀᴛ, B.

[184]   Toroidal embeddings. Lecture notes in Math, 339, Springer-Verlag 1973.

Kʜᴀʀʟᴀᴍᴏᴠ, V.M.

[185]   The maximal number of components of a surface of degree 4 in $\mathbb{R}\,P^3$. FAP 1972, *6*:4,
         101. FAA *6*, 345–346.

Kʜᴀʀʟᴀᴍᴏᴠ, V.M.

[186]   New relations for the Euler characteristic of real algebraic varieties. FAP 1973, *7*:2,
         74–78. FAA *7*, 147–150.

Kʜᴀʀʟᴀᴍᴏᴠ, V.M.

[187]   A generalised Petrovskii inequality. FAP 1974, *8*:2, 50–56. FAA *8*, 132–137

Kʜᴀʀʟᴀᴍᴏᴠ, V.M.

[188]   A generalised Petrovskii inequality II. FAP 1975, *9*:3, 93–94. FAA *9*, 266–268.

Kʜᴀʀʟᴀᴍᴏᴠ, V.M.

[189]   Additional comparisons for the Euler characteristic of even-dimensional real
         algebraic varieties. FAP 1975, *9*:4, 51–60. FAA *9*, 134–141.

KHARLAMOV, V.M.

[190]   Topological types of non-singular surfaces of degree 4 in $\mathbb{R} P^3$. FAP 1976, *10*:4,
          55–68. FAA *10*, 295–305.

KHARLAMOV, V.M.

[191]   Isotopy types of non-singular surfaces of degree 4 in $\mathbb{R} P^3$. FAP 1978, *12*:1, 86–87.
          FAA *12*, 68–69.

KMETY, F.

[192]   Résolution des singularités des courbes. Lecture notes in Math 409, Springer-Verlag
          1974.

KNORRER, H.

[193]   Zum $K(\pi, 1)$-Problem für isolierte Singularitäten von voligen Durchschnitten. Comp
          Math 1982, *45*:3, 333–340.

KOCHMAN, F.

[194]   Bernstein polynomials and Milnor algebras. Proc NASc USA 1976, *73*:8.

KUSHNIRENKO, A.G.

[195]   Newton polyhedra and Milnor numbers. FAP 1975, *9*:1, 74–75. FAA *9*, 71–72.

KOUCHNIRENKO, A.G.

[196]   Polyedres de Newton et nombres de Milnor. Invent math 1976, *32*, 1–31.

KREISS, H.-O.

[197]   Über syzygetische Flächen. Ann Math Pura ed Appl 1956, *41*, 105–111.

KULIKOV, V.S.

[198]   The calculus of the singularities of inclusion of a general algebraic surface in the
          projective space P^3. FAP 1983, *17*:3, 15–27. FAA *17*, 176–186.

LAMOTKE, K.

[199]   Die Homologie isolierter Singularitäten. Math Zeit 1975, *143*, 27–44.

LANDIS, E.E.

[200]   Estimates of the variation of sets of small value of a smooth function in a
          neighbourhood of a minimum point. Vestnik MGU ser math 1980, *3*, 24–28.

LANDIS, E.E.

[201]   Tangential singularities in contact geometry. UMN 1982, *37*:4, 96.

LANDMAN, A.

[202]   On the Picard-Lefschetz transformation for algebraic manifolds aquiring general
          singularities. Trans AMS 1973, *181*, 89–126.

LANDO, S.K.

[203]   Normal forms for the degrees of a volume form. FAP 1985, *19*:2, 78–79. FAA *19*,
          146–148.

LAUFER, H.

[204]   Ambient deformations for exceptional sets in two-manifolds. Invent Math 1979, *55*,
          1–36.

LAZZERI, F.

[205]   Some remarks on the Picard-Lefschetz monodromy. "Quelques journées singulières",
          Centre de Math de l'Ecole Polytechnique, Paris 1974.

LÊ DŨNG TRÁNG.

[206]   Sur les noeuds algébriques. Compositio Mathematica 1972, *25*:3, 281–321.

LÊ DŨNG TRÁNG.

[207]   Le theorème de la monodromie singulière. C.R. Acad Sci Paris 1979, AB*288*:21,
          A985–A988.

LÊ DŨNG TRÁNG.

[208]   La monodromie n'a pas de points fixes. J of the faculty of science, the University of
          Tokyo sec 1A, *22*:3, 409–427.

Lê Dũng Tráng – Ramanujam, C.P.

[209]   The invariance of Milnor's number implies the invariance of the topological type. Amer J Math 1976, *98*, 67–68.

Lê Dũng Tráng – Saito, K.

[210]   The local π_1 of the complement of a hypersurface with normal crossings in codimension 1 is abelian. RIMS 1981, 1–31.

Lê Dũng Tráng – Teissier, B.

[211]   Varietes polaires locales et classes de Chern des varietes singulieres. Ann of Math 1981, *114*, 457–491.

Lê Dũng Tráng – Teissier, B.

[212]   Cycles evanescents, sections planes et conditions de Whitney II. Proc Symp Pure Math 1983, *40*:2, 65–103.

Leray. J.

[213]   Le calcul différentiel et intégral sur une variété analitique complexe. (Problème de Cauchy III.) Bull Soc Math France 1959, *87*, 81–180.

Levantovskii, A.V.

[214]   Singularities of the boundary of a region of stability. FAP 1982, *16*:1, 44–48. FAA *16*, 34–37.

Levantovskii, A.V.

[215]   On the boundary of a set of stable matrices. UMN 1982, *35*:2, 213–214.

Levantovskii, A.V.

[216]   On the singularities of the boundary of stability. Vestnic MGU ser math 1980, *6*, 20–22.

Levine, J.

[217]   Polynomial invariants of knots of codimension two. Ann of Math ser II 1966, *84*, 537–554.

Linnik, Yu.V.

[218]   Ergodic properties of algebraic fields. Leningrad Univ. press, Leningrad 1967.

Lipman, J. – Tessier, B.

[219]   Pseudo-rational local rings and a theorem of Briancon-Skoda about integral closures of ideals. Michigan Math J 1981, *28*, 97–116.

Livshits, I.S.

[220]   Automorphisms of the complement of the bifurcation set of functions for simple singularities. FAP 1981, *15*:1, 38–42. FAA *15*, 29–32.

Loeser, F.

[221]   Quelques conséquences locales de la theorie de Hodge. Ann Inst Fourier, Grenoble 1985, *35*, 75–92.

Loeser, F.

[222]   Fonctions $|f|^s$, theorie de Hodge et polynômes de Bernstein-Sato. Géometrie Algebrique et applications III. Travaux en cours, *24*, 21–33, Hermann, Paris 1987.

Loeser, F.

[223]   Exposant d'Arnold et sections planes. CR Acad Sci Paris, 1984, *298*, 485–488.

Loeser, F.

[224]   A propos de la forme Hermitienne canonique d'une singularité isolée d'hypersurface. Bull Soc Math France 1986, *114*, 385–392.

Loeser, F.

[225]   Evolutions d'intégrales et théorie de Hodge. Lecture notes in Math, *1246*, 125–142, Springer 1987

LOESER, F.
 [226]   Fonction d'Igusa p-adiques et polynômes de Bernstein. Centre de Math, Ecole
          Polytechnique, 1986.
LOESER, F.
 [227]   Une estimation asymptotique du nombre de solutions approchées d'une equation
          p-adique. Invent Math, 1986, 85, 31–38.
LOESER, F.
 [228]   Volumes de tubes autour de singularités. Duke Math J 1986, 53, 443–455.
LOESER, F.
 [229]   Un analogue local de l'inegalité de Petrowsky-Oleinik. 1987, A.M.S. preprint.
LOOIJENGA, E.
 [230]   A period mapping of certain semiuniversal deformations. Comp Math 1975, 30 : 3,
          299–316.
LOOIJENGA, E.
 [231]   The complement of the bifurcation variety of a simple singularity. Invent Math 1974,
          23 : 2, 105–116.
LOOIJENGA, E.
 [232]   The discriminant of a real simple singularity. Comp Math 1978, 37 : 1, 51–62.
LOOIJENGA, E. – STEENBRINK, J.
 [233]   Milnor number and Tjurina number of complete intersections. Math Ann 1985, 271,
          121–124.
LYASHKO, O.V.
 [234]   Decomposition of simple singularities of functions. FAP 1976, 10 : 2, 49–56. FAA 10,
          122–127.
LYASHKO, O.V.
 [235]   Classification of critical points on a manifold with singular boundary. FAP 1983,
          17 : 3, 28–36. FAA 17, 187–193.
LYASHKO, O.V.
 [236]   The geometry of bifurcation diagrams. Current problems in mathematics 22, 94–129.
          Itogi Nauki i Tekhniki, Viniti AN SSSR, Moscow 1983.
MALGRANGE, B.
 [237]   On the polynomials of I.N. Bernstein. UMN 1974, 29 : 4, 81–88.
MALGRANGE, B.
 [238]   Le polynôme de Bernstein d'une singularité isolée. Lecture Notes in Math 459,
          98–119, Springer, Berlin 1974.
MALGRANGE, B.
 [239]   Integrales asymptotiques et monodromie. Ann Sci Ecole Norm Super ser 4 1974, 7,
          405–430.
MALYSHEV, A.V.
 [240]   On the representation of integers by positive quadratic forms. Proc Steklov Inst 1962,
          65.
MANIN, Y.I.
 [241]   Algebraic curves over fields with differentiation. IAN 1958, 22, 737–756.
MANIN, Y.I.
 [242]   Rational points of algebraic curves over functional fields. IAN 1963, 27, 1395–1440.
MARKUSHEVICH, A.I.
 [243]   Introduction to the classical theory of abelian functions. Moscow, Nauka 1979.
MASLOV, V.P.
 [244]   Théorie des perturbations et méthodes asymptotiques. Izd MGU 1965. Fr. Transl.
          Dunod, Paris 1972.

MASLOV, V.P.
[245]   Operator methods. Moscow, Nauka 1973. Eng. Transl.: Mir Publishers, Moscow 1976.

MASLOV, V.P. – FODORYUK, M.V.
[246]   Semi-classical approximation in quantum mechanics. Moscow, Nauka 1976. Eng. Transl. D. Riedel, Dordrecht 1981.

MATHER, J.
[247]   Stratifications and mappings. In: Dynamical Systems, ed. M.M. PEIXOTO. Academic Press, New York 1973.

MATHER, J. – YAU, ST.S.-T.
[248]   Classification of isolated hypersurface singularities by their moduli algebras. Invent Math 1982, 69, 243–252.

MATOV, V.I.
[249]   On unimodal germs of functions on manifolds with boundary. FAP 1980, 14:1, 69–70. FAA 14, 55–57.

MATOV, V.I.
[250]   Unimodular and bimodular germs of functions on manifolds with boundary. Proc Seminar I.G. Petrovskii 1981, 7, 174–189.

MATOV, V.I.
[251]   Singularities of functions of maxima on manifolds with boundary. Proc Seminar I.G. Petrovskii 1981, 6, 195–222.

MATOV, V.I.
[252]   Topological classification of germs of functions of maxima and minima of families of functions in general position. UMN 1982, 37:4, 129–130. RMS 37:4, 127–128.

McKAY, J.
[253]   Cartan matrices, finite groups of quaternions, and Kleinian singularities. Proc AMS 1981, 81:1, 153–154.

MILNOR, J.
[254]   On the Betti numbers of real varieties. Proc AMS 1964, 15, 275–280.

MILNOR, J.
[255]   Morse theory. Princeton U.P. 1963.

MILNOR, J.
[256]   Singular points of complex hypersurfaces. Princeton U.P. 1968.

MIYAOKA, Y.
[257]   The maximal number of quotient singularities on surfaces with given numerical invariants. Math Ann 1984, 268, 159–171.

MOND, D.
[258]   On the classification of germs of maps from $\mathbb{R}^2$ to $\mathbb{R}^3$. Proc Lond Math Soc 1985, 50, 333–369.

MUMFORD, L.
[259]   Algebraic Geometry I. Complex projective varieties. Springer-Verlag, Berlin 1976.

NGUYENHUU DUC – NGUYENTIEN DAI
[260]   Stabilité de l'interaction géometrique entre deux composantes holonomes simple. C.R. Acad Sci 1980, 291, 113–116.

NIKULIN, V.V.
[261]   Integral symmetrical bilinear forms and some of their geometrical applications. IAN 1979, 43:1, 111–177.

NILSSON, N.
[262]   Some growth and ramification properties of certain integrals. Arkiv for Mat 1965, 5, 463–476.

NILSSON, N.

[263]   Monodromy and asymptotic properties of certain multiple integrals. Arkiv for Mat
          1980, *18*:2, 181–198.

NYE, J.F.

[264]   Optical caustics in the near field from liquid drops. Proc Royal Soc London ser A
          1978, *361*, 21–41.

NYE, J.F.

[265]   Optical caustics from liquid drops under gravity: observations of the parabolic and
          symbolic umbilics. Philos Trans Royal Soc London 1979, *292*, 25–44.

NYE, J.F.

[266]   The motion and structure of dislocations in wave fronts. Proc Royal Soc London ser A
          1981, *378*, 219–239.

NYE, J.F. – COOLEY, C.R. – THORNDIKE, A.S.

[267]   The structure and evolution of flow fields and other vector fields. J Phys A Math Gen
          1978, *11*:8, 1455–1490.

NYE, J.F. – POTTER, J.R.

[268]   The use of catastrophe theory to analyse the stability and toppling of icebergs. Ann of
          Glaciology 1980, *1*, 49–54.

NYE, J.F. – THORNDIKE, A.S.

[269]   Events in evolving three dimensional vector fields. J Phys A Math Gen 1980, *13*,
          1–14.

ODA, T.

[270]   K. Saito's period map for holomorphic functions with isolated critical points. Adv
          Studies in Pure Math 1987, *10*, 591–648.

OLEINIK, O.A.

[271]   Estimates of Betti numbers of real algebraic hypersurfaces. Math sbornik 1951, *28*,
          635–640.

OLEINIK, O.A.

[272]   On the topology of real algebraic curves on an algebraic surface. Math sbornik 1951,
          29, 133–156.

OREVKOV, S.Yu.

[273]   Reconstruction of level surfaces of a function in a neighbourhood of a simple or
          unimodal critical point. UMN 1984, *39*:2, 181–182, RMS, *39*:2, 187–188.

OREVKOV, S.Yu.

[274]   Reconstruction of the level hypersurfaces of the direct sum of two functions. Vestnik
          MGU, Seriya Mat 1984, *6*, 10–13.

PALAMODOV, V.P.

[275]   The asymptotics of oscillatory integrals with an isolated stationary phase point. DAN,
          1980, *250*:6, 1321–1324. SMD *21*, 323–327.

PERRON, P.

[276]   "μ constant" implique "type topologique constant" en dimension complexe trois.
          C.R. Acad Sci 1982, *295*, 735–738.

PETROV, G.S.

[277]   On the number of zeros of complete elliptic integrals. FAP 1984, *18*:2, 87–88. FAA
          18, 148–149.

PETROV, G.S.

[278]   The number of zeros of complete elliptic integrals. FAP 1984, *18*:2, 73–74. FAA *18*,
          148–149.

PETROV, G.S.
[279]   Elliptic integrals and their non-oscillation. FAP 1986, 20:1, 46–49. FAA 20, 37–40.

PETROVSKY, I.G.
[280]   Sur la topologie des courbes reelle et algebriques. C.R. Acad Sci Paris 1933, 197, 1270–1273.

PETROVSKY, I.G.
[281]   On the topology of real plane algebraic curves. Ann Math 1938, 39, 189–209.

PETROVSKII, I.G. – OLEINIK, O.A.
[282]   On the topology of real algebraic surfaces. IAN 1949, 13, 389–402.

PETROVSKII, I.G.
[283]   On the diffusion of waves and lacunae for systems of hyperbolic equations. IAN 1944, 8:3, 101–106.

PHAM, F.
[284]   Formules de Picard-Lefschetz generalise et Ramification des intégrales. Bull Soc Math France 1965, 93, 333–367.

PHAM, F.
[285]   Remarque sur l'equisingularité universelle. Faculté des Sciences, Nice 1970, 1–24.

PHAM, F.
[286]   Caustiques, phase stationnaire et microfonctions. Acta Scientarum Vietnamicarum 1977, 2:2.

PHAM, F.
[287]   Singularités des systèmes differentiels de Gauss-Manin. Birkhäuser, Boston 1979.

PHAM, F.
[288]   Vanishing homologies and the n-variable saddle-point method. invited talk at the AMS summer institute on singularities, Arcata, Calif 1981, 1–25.

PHAM, F.
[289]   Structures de Hodge mixtes associées a un germe de fonction a point isolé. Asterisque 1983, 101–102, 268–285.

PINKHAM, H.
[290]   Singularités exceptionnelles, la dualité étrange d'Arnold et les surfaces $K-3$. C.R. Acad Sci Paris 1977, 284, 615–618.

PLATONOVA, O.A.
[291]   Singularities of projection of smooth surfaces. UMN 1984, 39:2, 149–150. RMS 39:1, 177–178.

PLATONOVA, O.A.
[292]   Projections of smooth surfaces. Proc seminar I.G. Petrovskii 1984, 10, 135–149.

DU PLESSIS, A.
[293]   On the determinacy of smooth map germs. Invent Math 1980, 58, 107–160.

DU PLESSIS, A.
[294]   On the genericity of topologically finitely determined map germs. Topology 1982, 21, 131–156.

POLOTOVSKII, G.M.
[295]   Catalogue of M-decomposing curves of the sixth order. DAN 1977, 236:3, 548–551. SMD 18:5, 1241–1245.

RANDALL, J.D.
[296]   Topological sufficiency of smooth map-germs. Invent Math 1982, 67, 117–122.

RANDOL, B.
[297]   A lattice point problem I. Trans AMS 1966, 121, 257–268.

RANDOL, B.
[298]   A lattice point problem II. Trans AMS 1966, *125*, 101–113.
RANDOL, B.
[299]   On the Fourier transform of the indicator function of a planar set. Trans AMS 1969, *139*, 271–278.
RANDOL, B.
[300]   On the asymptotic behaviour of the Fourier transform of the indicator function of a convex set. Trans AMS 1969, *139*, 279–285.
ROKHLIN, V.A.
[301]   Proof of the Gudkov conjecture. FAP 1972, *6* : 2, 62–64. FAA *6*, 136–138.
ROKHLIN, V.A.
[302]   Comparisons modulo 16 in Hilbert's 16th problem. FAP 1972, *6* : 4, 58–64. FAA *6*, 301–306.
ROKHLIN, V.A.
[303]   Comparisons modulo 16 in Hibert's 16th problem II. FAP 1973, *7* : 2, 91–92. FAA *7*, 163–164.
ROKHLIN, V.A.
[304]   Complex orientation of real algebraic curves. FAP 1974, *8* : 4, 71–75. FAA *8*, 331–334.
ROKHLIN, V.A.
[305]   Complex topological characteristics of real algebraic curves. UMN 1978, *33* : 5, 77–89. RMS *33* : 5, 85–98.
SAITO, K.
[306]   Quasihomogene isolierte Singularitäten von Hyperflächen. Invent Math 1971, *14*, 123–142.
SAITO, K.
[307]   On a linear structure of a quotient variety by a finite reflection group. Preprint, Kyoto Univ, Japan, Kyoto 1979.
SAITO, K.
[308]   A characterization of the intersection form of a Milnor's fiber for a function with an isolated critical point. Proc Japan Acad ser A 1982, *58* : 2, 79–81.
SAITO, K.
[309]   On the period of primitive integrals I. RIMS 1982, 1–235.
SAITO, K.
[310]   The zeros of characteristic function χ_f for the exponents of a hypersurface isolated singular point. Advanced studies in Pure Math 1982, *1*, 193–215.
SAITO, K.
[311]   Period mapping associated to a primitive form. Preprint 1983.
SAITO, M.
[312]   Exponents of a reduced and irreducible plane curve singularity. Preprint 1982.
SAITO, M.
[313]   Exponents and the geometric genus of an isolated hypersurface singularity. Proc Symp Pure Math 1983, *40* : 2, 465–472.
SAITO, M.
[314]   Exponents and Newton polyhedra of isolated hypersurface singularities. Inst Fourier, Grenoble 1983, 1–10.
SAITO, M.
[315]   On the structure of Brieskorn lattices. Inst Fourier, Grenoble 1983, 1–28.
SCHERK, J.
[316]   On the Gauss-Manin connection of an isolated hypersurface singularity. Math Ann 1978, *238* : 1, 23–32.

SCHERK, J.
[317]   On the monodromy theorem for isolated hypersurface singularities. Invent Math 1980, *58*:3, 289–301.

SCHERK, J.
[318]   A note on two local Hodge filtrations. Preprint, University of Alberta, Edmonton 1981, 1–7.

SCHERK, J.
[319]   A propos d'une theoreme de Mather et Yau. C.R. Acad Sci 1983, *296*:12, 513–515.

SCHMID, W.
[320]   Variation of Hodge structures: the singularities of the period mappings. Invent Math 1973, *22*, 211–319.

SEBASTIANI, M.
[321]   Monodromie et polynôme de Bernstein, d'après Malgrange. Lecture Notes in Math *670*, 370–381, Springer-Verlag, Berlin 1978.

SEBASTIANI, M. – THOM, R.
[322]   Un resultat sur la monodromie. Invent Math 1971, *13*:1–2, 90–96.

SEDYKH, V.D.
[323]   Singularities of convex shells. Sib math J 1983, *24*:3, 158–175.

SEDYKH, V.D.
[324]   Functional modules of singularities of convex shells of manifolds of codimension 1 and 2. Math sb 1982, *119*:2, 233–247.

SEIDENBERG, A.
[325]   A new decision method for elementary algebra. Ann Math 1959, *60*, 356–374.

SERRE, J.-P.
[326]   Quelque problèmes globeux relatifs aux varietés de Stein. Colloque sur les fonctions de plusieur variables tenu a Bruxelles, Paris 1953, 57–68. (Russian translation: Stratifications of space and their applications MIL 1958, 363–371).

SERRE, J.-P.
[327]   Lie algebras and Lie groups. Benjamin, New York-Amsterdam 1965. (Russian Transl. of 1964 notes of course given at Harvard University.)

SHAFAREVICH, I.R.
[328]   Basic algebraic geometry. Moscow, Nauka 1972. Eng. Transl. Springer-Verlag, Berlin 1977.

SHCHERBAK, I.G.
[329]   On isolated singularities of complete intersections. Proc Seminar I.G. Petrovskii 1983, *9*, 230–239.

SHCHERBAK, I.G.
[330]   The focal set of a surface with boundary and caustics of groups, generated by reflections, B_k, C_k, F_4. FAP 1984, *18*:1, 94–95. FAA *18*, 84–85.

SHCHERBAK, I.G.
[331]   Duality of boundary singularities. UMN 1984, *39*:2, 220–221. RMS *39*:2, 195–196.

SHCHERBAK, O.P.
[332]   Singularities of frontal maps of submanifolds of projective space. UMN 1982, *37*:4, 95–96.

SHCHERBAK, O.P.
[333]   Projectively dual space curves and Legendrean singularities. Proc Tbilisskii univ. 1982, *232–233*:13–14, 280–336.

SHCHERBAK, O.P.

[334]   Singularities of families of evolvents in the neighbourhood of a point of inflection of a
          curve and the group H_3, generated by reflections. FAP 1983, 17:4, 70–72. FAA 17,
          301–302.

SHOSHITAISHVILI, A.N.

[335]   On analytic sets, biholomorphically equivalent to quasihomogeneous sets. FAP 1983,
          17:2, 92–93. FAA 17, 159–161.

SIERSMA, D.

[336]   Singularities of functions on boundaries, corners, etc. Quart J Math 1981, 32:125,
          119–127.

SIERSMA, D.

[337]   Isolated line singularities. Proc Symp Pure Math 1983, 40:2, 485–496.

SLODOWY, P.

[338]   Simple singularities and simple algebraic groups. Lecture notes in Math 815, 1–175,
          Springer-Verlag, Berlin 1980.

SPRINGER, G.

[339]   Introduction to Riemann surfaces. Addison-Wesley, Reading, Mass. 1957.

STAGNARO, E.

[340]   Sul massimo numero di punti doppi isolati di una superficie algebrica di P^3. Rend Sem
          Mat Univ Padua 1978, 59, 197–198.

STEENBRINK, J.H.M.

[341]   Intersection forms for quasihomogeneous singularities. Compositio Math 1977, 34:2,
          211–223.

STEENBRINK, J.H.M.

[342]   Limits of Hodge structures. Invent Math 1975, 31:3, 229–257.

STEENBRINK, J.H.M.

[343]   Mixed Hodge structure on the vanishing cohomology. Real and complex singularities.
          Nordic Summer School, Oslo 1976.

STEENBRINK, J.H.M.

[344]   Mixed Hodge structures associated with isolated singularities. Proc Symp Pure Math
          1983, 40:2, 513–536.

STEENBRINK, J.

[345]   Semicontinuity of the singularity spectrum. Invent Math 1985, 79, 557–565.

STEENBRINK, J.H.M. – ZUCKER, S.

[346]   Variation of mixed Hodge structure I. Invent Math 1985, 80, 489–542.

TARNOPOLSKA-WEISS, M.

[347]   On the number of lattice points in a compact n-dimensional polyhedron. Proc AMS
          1979, 74:1, 124–127.

TEISSIER, B.

[348]   Cycles evanescents, sections planes et conditions de Whitney. Asterisque 1973, 7–8,
          285–362.

TEISSIER, B.

[349]   Deformations a type topologique constante, I et II. Sem Donady-Verdier, Secretariate
          Mathematique, 45 rue d'Ulm, Paris V 1972.

TEISSIER, B.

[350]   Varietés polaires I. Invent Math 1977, 40:3, 267–292.

TERAO, H.

[351]   Generalized exponents of a free arrangement of hypersurfaces and Shepherd-Todd-
          Brieskorn formula. Invent Math 1981, 69, 159–170.

THOM, R.

[352] Sur l'homologie des varietés algébriques reélles. Differential and combinatorial topology: a symposium in honour of Marston Morse. Princeton Univ Press 1965, 255–265.

TIMOURIAN, I.G.

[353] The invariance of Milnor's number implies topological triviality. Amer J Math 1977, 99 : 2, 437–446.

TOGLIATTI, E.

[354] Sulle forme cubiche dello spazio a cinque dimensioni aventi il massimo numero finito di punti doppi. Scritti oferti a L. Bezzolari, Pavia 1936, 577–593.

TROTMAN, D.

[355] Transverse transversals and homeomorphic transversals. Topology 1985, 24, 25–39.

TYURINA, G.N.

[356] Locally semiuniversal flat deformations of isolated singularities of complex spaces. IAN 1969, 33, 1026–1058. Math USSR Izvestija 3 : 5, 967–999.

VARCHENKO, A.N.

[357] Theorems of topological equisingularity of families of algebraic varieties and families of polynominal maps. IAN 1972, 36 : 5, 957–1019.

VARCHENKO, A.N.

[358] Newton polyhedra and estimates of oscillatory integrals. FAP 1976, 10 : 3, 13–38. FAA 10,175–196.

VARCHENKO, A.N.

[359] Zeta function of monodromy and Newton's diagram. Invent Math 1976, 37, 253–262.

VARCHENKO, A.N.

[360] The Gauss-Manin connection of isolated singular point and Bernstein polynomial. Bull des sciences math 2e ser 1980, 104, 205–223.

VARCHENKO, A.N.

[361] Hodge properties of the Gauss-Manin connection. FAP 1980, 14 : 1, 46–47. FAA 14, 34–35.

VARCHENKO, A.N.

[362] The asymptotics of holomorphic forms define a mixed Hodge structure. DAN 1980, 255 : 5, 1035–1038. SMD 22 : 5, 772–775.

. VARCHENKO, A.N.

[363] On the monodromy operator in vanishing cohomologies and on the operation of multiplication by f in the local ring. DAN 1981, 260 : 2, 272–276.

VARCHENKO, A.N.

[364] Asymptotic mixed Hodge structure in vanishing cohomologies. IAN 1981, 45, 540–591. Math USSR Izvestija, 1982, 18 : 3, 469–512.

VARCHENKO, A.N.

[365] The complex singular index does not change along the stratum μ=const. FAP 1982, 16 : 1, 1–12. FAA 16, 1–9.

VARCHENKO, A.N.

[366] On the number of integer points in a region. UMN 1982, 37 : 3, 177–178. RMS 37 : 3, 223–224.

VARCHENKO, A.N.

[367] Number of lattice points in families of homothetic domains in $\mathbb{R}^r$. FAP 1983, 17 : 2, 1–6. FAA 17, 79–83.

VARCHENKO, A.N.

[368] An estimate from below of the codimension of the stratum μ =const by mixed Hodge structures. Vestnik MGU ser math 1982, 6, 28–31.

VARCHENKO, A.N. – GIVENTAL, A.B.
   [369]   Period maps and intersection forms. FAP 1982, *16*:1, 1–14. FAA *16*, 83–93.
VARCHENKO, A.N. – MOLCHANOV, S.A.
   [370]   Application of the method of stationary phase in limiting theorems for Markov
            chains. DAN 1977, *233*:1, 11–14. SMD *18*, 10–13.
VARCHENKO, A.N.
   [371]   The spectrum and decomposition of a critical point of a function. DAN 1983, *270*:2,
            267–270.
VARCHENKO, A.N.
   [372]   On the semicontinuity of spectra and estimates from above of the number of singular
            points of a projective hypersurface. DAN 1983, *270*:6, 1294–1297.
VARCHENKO, A.N.
   [373]   On the semicontinuity of the complex singular index. FAP 1983, *17*:4, 77–78. FAA
            17, 307–309.
VARCHENKO, A.N.
   [374]   On the change in discrete characteristics of critical points of functions under
            deformations. UMN 1983, *38*:5, 126–127.
VARCHENKO, A.N.
   [375]   Asymptotics of integrals and Hodge structures. Science reviews: current problems in
            mathematics 1983, *22*, 130–166. J Sov Math 1984, Vol 27.
VARCHENKO, A.N.
   [376]   Estimate of the number of zeros of a real abelian integral, depending on a parameter
            and limiting cycles. FAP 1984, *18*:2, 1–13. FAA *18*, 98–108.
VARCHENKO, A.N.
   [377]   On local residue and intersection form in vanishing cohomologies. IAN 1984, *48*:5,
            1010–1035.
VARCHENKO, A.N. – CHMUTOV, S.V.
   [378]   Finite irreducible groups, generated by reflections may be monodromy groups of
            suitable singularities. FAP 1984, *18*:3, 1–14. FAA *18*, 171–183.
VARCHENKO, A.N. – GUSEIN-ZADE, S.M.
   [379]   The topology of caustics, wave fronts and degenerations of critical points. UMN 1984,
            39:2, 190–191. RMS *39*:2, 209–210.
VARCHENKO, A.N. – HOVANSKY, A.G.
   [380]   Asymptotics of integrals over vanishing cycles and the Newton polyhedron. DAN
            1985, *283*:3, 521–525. SMD *32*, 122–127.
VARCHENKO, A.N. – CHMUTOV, S.V.
   [381]   On the tangent cone to a stratum μ=const. Vestnik MGU, Ser Mat 1985, *1*, 6–9.
VARCHENKO, A.N.
   [382]   Local classification of volume forms in the presence of a hypersurface. FAP 1985,
            19:4, 23–31. FAA *19*, 269–276.
VARCHENKO, A.N.
   [383]   On normal forms of non-smoothness of solutions of hyperbolic equations. IAN 1987,
            51:3, 652–665.
VARCHENKO, A.N.
   [384]   The image of the period map for simple singularities. Methods of contemporary
            Functional Analysis in Geometry and Topology. Voronezh State Univ. 1987, 47–60.
VARCHENKO, A.N.
   [385]   The period map and the discriminant. Mat Sbornik, 1987, *134*:1, 66–81.

VASILEV, V.A.
[386]   The asymptotics of exponential integrals, Newton diagrams and classification of minima. FAP 1977, *11*:3, 1–11. FAA *11*, 163–172.

VASILEV, V.A.
[387]   The asymptotics of exponential integrals in complex domain. FAP 1979, *13*:4, 1–12. FAA *13*, 239–247.

VASILEV, V.A.
[388]   Self-intersections of wave fronts and Legendrean (Lagrangean) characteristic numbers. FAP 1982, *16*:2, 68–69. FAA *16*, 131–133.

VASILEV, V.A.
[389]   Sharpness and the local Petrovskii condition for strictly hyperbolic operators with constant coefficients. IAN 1986, *50*:2, 251–283. MUI 1987, *28*, 233–273.

VERDIER, J.-L.
[390]   Stratification de Whitney et theorems de Bertine-Sard. Invent Math 1976, *36*, 295–312.

VINOGRADOV, I.M.
[391]   The method of trigonometrical sums in number theory. Moscow, Nauka 1971.

VINOGRADOV, I.M.
[392]   Special variants of the method of trigonometrical sums. Moscow, Nauka 1976.

VIRO, O.Y.
[393]   Generalisations of inequalities of Petrovskii and Arnold on curves with singularities. UMN 1978, *33*:3, 145–146.

VIRO, O.Y.
[394]   The construction of *M*-manifolds. UMN 1979, *34*:4, 160.

VIRO, O.Y.
[395]   Curves of degree 7, curves of degree 8, and the conjecture of Regsdeil. DAN 1980, *254*:6, 1306–1310.

VORONIN, S.M.
[396]   Analytic classification of pairs of involutions and its application. FAP 1982, *16*:2, 21–29. FAA *16*, 94–100.

WAJNRYB, B.
[397]   On the monodromy group of plane curve singularities. Math Ann 1980, *246*:2, 141–154.

WALL, C.T.C.
[398]   Are maps finitely determined in general? Bull LMS 1979, *11*, 151–154.

WALL, C.T.C.
[399]   A note on symmetry of singularities. Bull LMS 1980, *12*, 169–175.

WALL, C.T.C.
[400]   A second note on symmetry of singularities. Bull Lond Math Soc 1980, *12*, 347–354.

WALL, C.T.C.
[401]   Topological invariance of the Milnor number modulo 2. Topology 1983, *22*,345–350.

WALL, C.T.C.
[402]   Determination of the semi-nice dimensions. Math Proc Camb Phil Soc 1985, *97*, 79–88.

WALL, C.T.C.
[403]   Equivariant jets. Math Ann 1985, *272*, 41–65.

WALL, C.T.C.
[404]   Infinite-determinancy of equivariant map-germs. Math Ann 1985, *272*, 67–82.

WELLS, R.
   [405]   Differential Analysis on Complex Manifolds. 2nd ed. Springer-Verlag, Berlin 1980.
YANO, T.
   [406]   On the holonomic system of f^s and b-function. Publ RIMS 1976, *12*, 469–480.
YANO, T.
   [407]   On the structure of differential equations associated with some isolated singularities. Soc reports of Saitama Univ ser A 1977, *9*:1, 9–20.
YANO, T.
   [408]   On the theory of b-functions. Publ RIMS 1978, *14*, 111–202.
YAU, ST.S.-T.
   [409]   Solvable Lie algebras and Kac-Moody algebras arising from isolated singularities. Univ Illinois 1983, 1–94.
YAU, ST.S.-T.
   [410]   Milnor algebras and equivalence relations among holomorphic functions. Bull AMS 1983, *9*:2, 235–239.
YOSHINAGA, E. – SUZUKI, M.
   [411]   Normal forms of non-degenerated quasihomogeneous functions with inner modality ⩽4. Invent Math 1979, *55*, 185–206.
ZAKALYUKIN, V.M.
   [412]   Bifurcation of wave fronts, depending on one parameter. FAP 1976, *10*:2, 69–70. FAA *10*, 139–140.
ZARISKI, O.
   [413]   Algebraic surfaces. 2nd edition. Springer-Verlag 1971.
ZUCKER, S.
   [414]   Variation of mixed Hodge structure II. Invent Math 1985, *80*, 543–565.
ZVONILOV, V.I.
   [415]   The inequality of Kharlamov and the inequality of Petrovskii-Oleinik. FAP 1975, *9*:2, 69–70. FAA *9*, 152–153.

Subject Index